Differential and Difference Equations

Differential and Difference Equations: A Themed Issue Dedicated to Prof. Hari M. Srivastava on the Occasion of his 80th Birthday

Editor

Sotiris K. Ntouyas

MDPI • Basel • Beijing • Wuhan • Barcelona • Belgrade • Manchester • Tokyo • Cluj • Tianjin

Editor
Sotiris K. Ntouyas
University of Ioannina
Greece

Editorial Office
MDPI
St. Alban-Anlage 66
4052 Basel, Switzerland

This is a reprint of articles from the Special Issue published online in the open access journal *Axioms* (ISSN 2075-1680) (available at: https://www.mdpi.com/journal/axioms/special_issues/differ_equ).

For citation purposes, cite each article independently as indicated on the article page online and as indicated below:

LastName, A.A.; LastName, B.B.; LastName, C.C. Article Title. *Journal Name* **Year**, *Article Number*, Page Range.

ISBN 978-3-03943-068-0 (Hbk)
ISBN 978-3-03943-069-7 (PDF)

© 2020 by the authors. Articles in this book are Open Access and distributed under the Creative Commons Attribution (CC BY) license, which allows users to download, copy and build upon published articles, as long as the author and publisher are properly credited, which ensures maximum dissemination and a wider impact of our publications.

The book as a whole is distributed by MDPI under the terms and conditions of the Creative Commons license CC BY-NC-ND.

Contents

About the Editor vii

Preface to "Differential and Difference Equations: A Themed Issue Dedicated to Prof. Hari M. Srivastava on the Occasion of his 80th Birthday" ix

Irem Kucukoglu, Burcin Simsek and Yilmaz Simsek
Generating Functions for New Families of Combinatorial Numbers and Polynomials: Approach to Poisson–Charlier Polynomials and Probability Distribution Function
Reprinted from: *Axioms* **2019**, *8*, 112, doi:10.3390/axioms8040112 1

Temirkhan Aleroev
On One Problems of Spectral Theory for Ordinary Differential Equations of Fractional Order
Reprinted from: *Axioms* **2019**, *8*, 117, doi:10.3390/axioms8040117 17

Rasool Shah, Hassan Khan, Dumitru Baleanu
Fractional Whitham–Broer–Kaup Equations within Modified Analytical Approaches
Reprinted from: *Axioms* **2019**, *8*, 125, doi:10.3390/axioms8040125 25

Paolo Emilio Ricci and Pierpaolo Natalini
General Linear Recurrence Sequences and Their Convolution Formulas
Reprinted from: *Axioms* **2019**, *8*, 132, doi:10.3390/axioms8040132 47

Chenkuan Li and Hunter Plowman
Solutions of the Generalized Abel's Integral Equationsof the Second Kind with Variable Coefficients
Reprinted from: *Axioms* **2019**, *8*, 137, doi:10.3390/axioms8040137 59

Nak Eun Cho and Jacek Dziok
Harmonic Starlike Functions with Respect to Symmetric Points
Reprinted from: *Axioms* **2020**, *9*, 3, doi:10.3390/axioms9010003 69

Choonkil Park, Osama Moaaz and Omar Bazighifan
Oscillation Results for Higher Order Differential Equations
Reprinted from: *Axioms* **2020**, *9*, 14, doi:10.3390/axioms9010014 81

Leonid Shaikhet
Stability of Equilibria of Rumor Spreading Model under Stochastic Perturbations
Reprinted from: *Axioms* **2020**, *9*, 24, doi:10.3390/axioms9010024 93

Nita H Shah, Nisha Sheoran and Yash Shah
Dynamics of HIV-TB Co-Infection Model
Reprinted from: *Axioms* **2020**, *9*, 29, doi:10.3390/axioms9010029 109

Abdeljabbar Talal Yousef and Zabidin Salleh
On a Harmonic Univalent Subclass of Functions Involving a Generalized Linear Operator
Reprinted from: *Axioms* **2020**, *9*, 32, doi:10.3390/axioms9010032 123

Donal O'Regan
Coincidence Continuation Theory for Multivalued Maps with Selections in a Given Class
Reprinted from: *Axioms* **2020**, *9*, 37, doi:10.3390/axioms9020037 133

Rabha W. Ibrahim, Rafida M. Elobaid and Suzan J. Obaiys
Generalized Briot-Bouquet Differential Equation Based on New Differential Operator with Complex Connections
Reprinted from: *Axioms* **2020**, *9*, 42, doi:10.3390/axioms9020042 141

Tursun K. Yuldashev
Nonlocal Inverse Problem for a Pseudohyperbolic- PseudoellipticType Integro-Differential Equations
Reprinted from: *Axioms* **2020**, *9*, 45, doi:10.3390/axioms9020045 155

Davor Dragičević and Ciprian Preda
Lyapunov Type Theorems for Exponential Stability of Linear Skew-Product Three-Parameter Semiflows with Discrete Time
Reprinted from: *Axioms* **2020**, *9*, 47, doi:10.3390/axioms9020047 177

Ahmed Alsaedi, Abrar Broom, Sotiris K. Ntouyas and Bashir Ahmad
Nonlocal Fractional Boundary Value Problems Involving Mixed Right and Left Fractional Derivatives and Integrals
Reprinted from: *Axioms* **2020**, *9*, 50, doi:10.3390/axioms9020050 189

R. Leelavathi, G. Suresh Kumar, Ravi P. Agarwal, Chao Wang, and M.S.N. Murty
Generalized Nabla Differentiability and Integrability for Fuzzy Functions on Time Scales
Reprinted from: *Axioms* **2020**, *9*, 65, doi:10.3390/axioms9020065 205

Bashir Ahmad, Najla Alghamdi, Ahmed Alsaedi and Sotiris K. Ntouyas
Existence Results for Nonlocal Multi-Point and Multi-Term Fractional Order Boundary Value Problems
Reprinted from: *Axioms* **2020**, *9*, 70, doi:10.3390/axioms9020070 229

Ilya Boykov, Vladimir Roudnev and Alla Boykova
Approximate Methods for Solving Linear and Nonlinear Hypersingular Integral Equations
Reprinted from: *Axioms* **2020**, *9*, 74, doi:10.3390/axioms9030074 251

Osama Moaaz, Hamida Mahjoub and Ali Muhib
On the Periodicity of General Class of Difference Equations
Reprinted from: *Axioms* **2020**, *9*, 75, doi:10.3390/axioms9030075 269

Osama Moaaz , Hamida Mahjoub and Ali Muhib
Eigenfunction Families and Solution Bounds forMultiplicatively Advanced Differential Equations
Reprinted from: *Axioms* **2020**, *9*, 83, doi:10.3390/axioms9030083 281

Alexander Yeliseev
On the Regularized Asymptotics of a Solution to the Cauchy Problem in the Presence of a Weak Turning Point of the Limit Operator
Reprinted from: *Axioms* **2020**, *9*, 86, doi:10.3390/axioms9030086 309

Konstantinos Kalimeris and Athanassios S. Fokas
The Modified Helmholtz Equation on a Regular Hexagon—The Symmetric Dirichlet Problem
Reprinted from: *Axioms* **2020**, *9*, 89, doi:10.3390/axioms9030089 327

Thomas Ernst
On the Triple Lauricella–Horn–Karlsson q-Hypergeometric Functions
Reprinted from: *Axioms* **2020**, *9*, 93, doi:10.3390/axioms9030093 345

About the Editor

Sotiris K. Ntouyas is Professor Emeritus in the Department of Mathematics of the University of Ioannina, Greece. He received his BS degree and PhD from the University of Ioannina, in 1972 and 1980, respectively. His research interests include initial and boundary value problems for differential equations (ordinary, functional, with deviating arguments, neutral, partial, integrodifferential, inclusions, impulsive, fuzzy, stochastic, fractional), inequalities, asymptotic behavior and controllability. More than 640 of his papers have appeared in print or have been accepted for publication in refereed journals. He is the co-author of the books *Impulsive differential equations and inclusions, Controllability for semilinear functional differential equations and inclusions, Quantum Calculus: New Concepts, Impulsive IVPs and BVPs, Inequalities and Hadamard-Type Fractional Differential Equations, Inclusions and Inequalities*. He is a member of 21 international journals' Editorial Boards, and a reviewer for many international journals. He appears in the 2018 list, published by Clarivate Analytics, of Highly Cited Researchers.

Preface to "Differential and Difference Equations: A Themed Issue Dedicated to Prof. Hari M. Srivastava on the Occasion of his 80th Birthday"

It is our great pleasure to publish this book. All contents were peer-reviewed by multiple referees and published as papers in the Special Issue "Differential and Difference Equations: A Themed Issue Dedicated to Prof. Hari M. Srivastava on the Occasion of his 80th Birthday" in the journal *Axioms*. These studies provide new and interesting results in different branches of differential equations, so that the readers will be able to obtain the latest developments in the fields of differential equations.

Sotiris K. Ntouyas
Editor

Article

Generating Functions for New Families of Combinatorial Numbers and Polynomials: Approach to Poisson–Charlier Polynomials and Probability Distribution Function

Irem Kucukoglu [1], Burcin Simsek [2] and Yilmaz Simsek [3],*

[1] Department of Engineering Fundamental Sciences, Faculty of Engineering, Alanya Alaaddin Keykubat University, TR-07425 Antalya, Turkey; irem.kucukoglu@alanya.edu.tr
[2] Department of Statistics, University of Pittsburgh, Pittsburgh, PA 15260, USA; bus5@pitt.edu
[3] Department of Mathematics, Faculty of Science University of Akdeniz, TR-07058 Antalya, Turkey
* Correspondence: ysimsek@akdeniz.edu.tr

Received: 14 September 2019; Accepted: 9 October 2019; Published: 11 October 2019

Abstract: The aim of this paper is to construct generating functions for new families of combinatorial numbers and polynomials. By using these generating functions with their functional and differential equations, we not only investigate properties of these new families, but also derive many new identities, relations, derivative formulas, and combinatorial sums with the inclusion of binomials coefficients, falling factorial, the Stirling numbers, the Bell polynomials (i.e., exponential polynomials), the Poisson–Charlier polynomials, combinatorial numbers and polynomials, the Bersntein basis functions, and the probability distribution functions. Furthermore, by applying the p-adic integrals and Riemann integral, we obtain some combinatorial sums including the binomial coefficients, falling factorial, the Bernoulli numbers, the Euler numbers, the Stirling numbers, the Bell polynomials (i.e., exponential polynomials), and the Cauchy numbers (or the Bernoulli numbers of the second kind). Finally, we give some remarks and observations on our results related to some probability distributions such as the binomial distribution and the Poisson distribution.

Keywords: generating functions; functional equations; partial differential equations; special numbers and polynomials; Bernoulli numbers; Euler numbers; Stirling numbers; Bell polynomials; Cauchy numbers; Poisson-Charlier polynomials; Bernstein basis functions; Daehee numbers and polynomials; combinatorial sums; binomial coefficients; p-adic integral; probability distribution

MSC: Primary 05A10; 05A15; 11B73; 11B68; 11B83; Secondary 05A19; 11B37; 11S23; 26C05; 34A99; 35A99; 40C10

1. Introduction

In recent years, generating functions and their applications on functional equations and differential equations has gained high attention in various areas. These techniques allow researchers to derive various identities and combinatorial sums that yield important special numbers and polynomials. In fact, the current trend is to combine the p-adic integrals with these techniques. In most of fields of mathematics and physics, different applications of generating functions are used as an important tool. For instance, a common research topic in quantum physics is to identify a generating function that could be a solution to a differential equation.

The motivation of this paper is to outline the advantages of techniques associated with generating functions. First, generating functions are presented for new families of combinatorial numbers and polynomials. Second, we derive new identities, relations, and formulas including the Bersntein basis functions, the Stirling numbers, the Bell polynomials (i.e., exponential polynomials), the Poisson–Charlier polynomials, the Daehee numbers and polynomials, the probability distribution functions, as well as combinatorial sums including the Bernoulli numbers, the Euler numbers, the Cauchy numbers (or the Bernoulli numbers of the second kind), and combinatorial numbers.

With the followings, we briefly introduce the notations, definitions, relations, and formulas are used throughout this paper:

As usual, let $\mathbb{N}$, $\mathbb{Z}$, $\mathbb{N}_0$, $\mathbb{Q}$, $\mathbb{R}$, and $\mathbb{C}$ denote the set of natural numbers, set of integers, set of nonnegative integers, set of rational numbers, set of real numbers, and set of complex numbers, respectively. Let $\log z$ denote the principal branch of the multi-valued function $\log z$ with the imaginary part $\text{Im}(\log z)$ constrained by the interval $(-\pi, \pi]$. We also assume that:

$$0^n = \begin{cases} 1, & \text{if } n = 0 \\ 0, & \text{if } n \in \mathbb{N}. \end{cases}$$

Moreover,

$$\binom{z}{v} = \frac{z(z-1)\cdots(z-v+1)}{v!} = \frac{(z)_v}{v!} \quad (v \in \mathbb{N}, z \in \mathbb{C})$$

so that,

$$\binom{z}{0} = (z)_0 = 1$$

(cf. [1–31]).

The Poisson–Charlier polynomials $C_n(x;a)$, which are members of the family of Sheffer-type sequences, are defined as below:

$$F_{pc}(t,x;a) = e^{-t}\left(\frac{t}{a}+1\right)^x = \sum_{n=0}^{\infty} C_n(x;a) \frac{t^n}{n!}, \tag{1}$$

where,

$$C_n(x;a) = \sum_{j=0}^{n} (-1)^{n-j} \binom{n}{j} \frac{(x)_j}{a^j}, \tag{2}$$

(cf. [16], (p. 120, [18]), [24]).

Let $x \in [0,1]$ and let n and k be nonnegative integers. The Bernstein basis functions, $B_k^n(x)$, are defined by:

$$B_k^n(x) = \binom{n}{k} x^k (1-x)^{n-k}, \quad (k = 0, 1, \ldots, n) \tag{3}$$

so that,

$$\binom{n}{k} = \frac{n!}{k!(n-k)!}$$

and its generating function is given by:

$$F_B(t,x;k) = \frac{(xt)^k e^{(1-x)t}}{k!} = \sum_{n=0}^{\infty} B_k^n(x) \frac{t^n}{n!}, \tag{4}$$

where $t \in \mathbb{C}$ (cf. [1,15,20,26]).

The Stirling numbers of the first kind, $S_1(n,k)$, are defined by the following generating function:

$$F_{S_1}(t;k) = \frac{(\log(1+t))^k}{k!} = \sum_{n=k}^{\infty} S_1(n,k) \frac{t^n}{n!}, \quad (k \in \mathbb{N}_0) \tag{5}$$

so that,
$$(x)_n = \sum_{k=0}^{n} S_1(n,k) x^k \qquad (6)$$

(cf. [2–4,29,30]; see also the references cited therein).

The λ-Stirling numbers of the second kind, $S_2(n,k;\lambda)$, are defined with generating function given below (cf. [21,30]):

$$F_{S_2}(t;v;\lambda) = \frac{(\lambda e^t - 1)^v}{v!} = \sum_{n=0}^{\infty} S_2(n,v;\lambda) \frac{t^n}{n!}, \qquad (v \in \mathbb{N}_0). \qquad (7)$$

Notice here that, when $\lambda = 1$, this reduces to the Stirling numbers of the second kind, $S_2(n,v)$, whose generating function is given below:

$$F_{S_2}(t;v) = \frac{(e^t - 1)^v}{v!} = \sum_{n=0}^{\infty} S_2(n,v) \frac{t^n}{n!}, \qquad (v \in \mathbb{N}_0), \qquad (8)$$

namely, $S_2(n,v) = S_2(n,v;1)$ (cf. [2,5,21,30]).

The Bell polynomials (i.e., exponential polynomials), $Bl_n(x)$, is defined by:

$$Bl_n(x) = \sum_{v=1}^{n} S_2(n,v) x^v \qquad (9)$$

so that the generating function for the Bell polynomials is given by:

$$F_{Bell}(t,x) = e^{(e^t - 1)x} = \sum_{n=0}^{\infty} Bl_n(x) \frac{t^n}{n!} \qquad (10)$$

(cf. [4,18]).

The numbers $Y_n^{(k)}(\lambda)$ and the polynomials $Y_n^{(k)}(x;\lambda)$ are defined by the following generating functions, respectively:

$$\mathcal{F}(t,k;\lambda) = \left(\frac{2}{\lambda(1+\lambda t) - 1}\right)^k = \sum_{n=0}^{\infty} Y_n^{(k)}(\lambda) \frac{t^n}{n!}, \qquad (11)$$

and,

$$\mathcal{F}(t,x,k;\lambda) = \mathcal{F}(t,k;\lambda)(1+\lambda t)^x = \sum_{n=0}^{\infty} Y_n^{(k)}(x;\lambda) \frac{t^n}{n!}, \qquad (12)$$

where $k \in \mathbb{N}_0$ and λ is real or complex number (cf. [14]).

Substituting $k = 1$ into Equation (11), we have:

$$Y_n(\lambda) = Y_n^{(1)}(\lambda)$$

(cf. [23]).

Substituting $k = 1$ and $\lambda = -1$ into Equation (11), we get the following well-known relation between the numbers $Y_n(\lambda)$ and the Changhee numbers of the first kind, Ch_n:

$$Ch_n = (-1)^{n+1} Y_n(-1).$$

Thus we have,

$$Ch_n = \frac{(-1)^n n!}{n+1} = \sum_{k=0}^{n} S_1(n,k) E_k \qquad (13)$$

where the Changhee numbers of the first kind, Ch_n are defined means of the following generating function:

$$\frac{2}{t+1} = \sum_{n=0}^{\infty} Ch_n \frac{t^n}{n!} \quad (14)$$

(*cf.* [9], see also [7]).

The Daehee polynomials, $D_n(x)$, is defined by the following generating functions (*cf.* [8]):

$$F_D(x,t) = \frac{\log(1+t)}{t}(1+t)^x = \sum_{n=0}^{\infty} D_n(x) \frac{t^n}{n!} \quad (15)$$

which, for $x = 0$, corresponds the generating functions of the Daehee number, $D_n = D_n(0)$, given by the following explicit formula:

$$D_n = \frac{(-1)^n n!}{n+1}. \quad (16)$$

The combinatorial numbers, $y_1(n,k;\lambda)$, are defined by the following generating function:

$$F_{y_1}(t,k;\lambda) = \frac{1}{k!}\left(\lambda e^t + 1\right)^k = \sum_{n=0}^{\infty} y_1(n,k;\lambda) \frac{t^n}{n!} \quad (17)$$

where $k \in \mathbb{N}_0$ and $\lambda \in \mathbb{C}$ (*cf.* [22]).

Use the preceding generating function for the combinatorial numbers, $y_1(n,k;\lambda)$ to compute the following explicit formula:

$$y_1(n,k;\lambda) = \frac{1}{k!}\sum_{j=0}^{k}\binom{k}{j}\lambda^j j^n \quad (18)$$

(*cf.* [22] (Theorem 1, Equation (9))).

Note that the following equality holds true:

$$y_1(n,k;\lambda) = \frac{1}{k!}\frac{d^n}{dt^n}\left(\lambda e^t + 1\right)^k\bigg|_{t=0} \quad (19)$$

(*cf.* [31] (p. 64)).

When $\lambda = 1$, if we multiply the numbers $y_1(n,k;\lambda)$ by $k!$, then Equation (18) is reduced to the following combinatorial numbers (*cf.* [6,19,22]):

$$B(k,n) = \sum_{j=0}^{k}\binom{k}{j}j^n$$

which satisfies the following differential equation:

$$B(k,n) = \frac{d^n}{dt^n}\left(e^t + 1\right)^k\bigg|_{t=0} \quad (20)$$

(*cf.* [6], (Equation (2), p. 2 [22])).

The combinatorial numbers $B(n,k)$ have various kinds of combinatorial applications. For instance, Ross [19] (pp. 18–20, Exercises 10–12) gave the following applications for solutions of exercises 10–12:

From a group of n people, suppose that we want to choose a committee of k, $k \leq n$, one of whom is to be designated as chairperson.

How many different selections are there in which the chairperson and the secretary are the same? Ross [19] (p. 18, Exercise 12) gave the following answer: $B(1,n) = n2^{n-1}$.

By using the preceding idea summarized above, the following combinatorial identities are obtained:

$$\sum_{k=0}^{n}\binom{n}{k}k = n2^{n-1}$$

$$\sum_{k=0}^{n}\binom{n}{k}k^2 = 2^{n-2}n(n+1)$$

and,

$$\sum_{k=0}^{n}\binom{n}{k}k^3 = 2^{n-3}n^2(n+3)$$

(cf. (pp. 18–20, Exercises 10–12 [19]), [22,25]). Observe that these numbers are also arised from Equation (20).

Next, we present the outline of the present paper: In Section 2, we construct generating functions for new families of combinatorial numbers and polynomials. By using these generating functions, we not only investigate properties of these new families, but also provide some new identities and relations with the inclusion of the Bersntein basis functions, combinatorial numbers, and the Stirling numbers. In Section 3, we obtain some derivative formulas and recurrence relations for these new families of combinatorial numbers and polynomials by using differential equations that are a result of these generating functions and their partial derivatives. In Section 4, by using functional equations of the generating functions, we derive some formulas and combinatorial sums including binomials coefficients, falling factorial, the Stirling numbers, the Bell polynomials (i.e., exponential polynomials), the Poisson–Charlier polynomials, combinatorial numbers and polynomials, and the Bersntein basis functions. In Section 5, by applying the p-adic integrals and Riemann integral to some new formulas derived by the authors of this paper, some combinatorial sums comprising the binomial coefficients, falling factorial, the Bernoulli numbers, the Euler numbers, the Stirling numbers, the Bell polynomials (i.e. exponential polynomials), and the Cauchy numbers (or the Bernoulli numbers of the second kind) are presented. In Section 6, we give some remarks and observations on our results related to some probability distributions such as the binomial distribution and the Poisson distribution. In Section 7, we conclude our findings.

2. New Families of the Combinatorial Numbers and Polynomials

In this section, we define new families of the combinatorial numbers and polynomials by the following generating functions, respectively:

$$\mathcal{G}(t,k;\lambda) = 2^{-k}(\lambda(1+\lambda t)-1)^k = \sum_{n=0}^{\infty} Y_n^{(-k)}(\lambda)\frac{t^n}{n!} \qquad (21)$$

and,

$$\mathcal{G}(t,x,k;\lambda) = \mathcal{G}(t,k;\lambda)(1+\lambda t)^x = \sum_{n=0}^{\infty} Q_n(x;\lambda,k)\frac{t^n}{n!} \qquad (22)$$

where $k \in \mathbb{N}$ and λ is a real or complex number.

Combining Equations (21) and (22), we get:

$$\sum_{n=0}^{\infty} Q_n(x;\lambda,k)\frac{t^n}{n!} = \sum_{n=0}^{\infty}\sum_{j=0}^{n}\binom{n}{j}\lambda^{n-j}Y_j^{(-k)}(\lambda)(x)_{n-j}\frac{t^n}{n!}. \qquad (23)$$

Comparing coefficient of $\frac{t^n}{n!}$ on both sides of the above equation, we arrive at the following theorem:

Theorem 1.

$$Q_n(x; \lambda, k) = \sum_{j=0}^{n} \binom{n}{j} \lambda^{n-j} Y_j^{(-k)}(\lambda) (x)_{n-j}. \tag{24}$$

By the binomial theorem, we have:

$$\sum_{n=0}^{\infty} Y_n^{(-k)}(\lambda) \frac{t^n}{n!} = 2^{-k} \sum_{n=0}^{\infty} \binom{k}{n} \lambda^{2n} (\lambda - 1)^{k-n} t^n.$$

Comparing the coefficient of t^n on both sides of the above equation, we arrive at the following theorem:

Theorem 2. *Let k and n be nonnegative integers. Then:*

$$Y_n^{(-k)}(\lambda) = \begin{cases} 2^{-k} n! \binom{k}{n} \lambda^{2n} (\lambda-1)^{k-n} & \text{if } n \le k \\ 0 & \text{if } n > k. \end{cases} \tag{25}$$

By Equation (25), a few values of the numbers $Y_n^{(-k)}(\lambda)$ are computed as follows:

$$\begin{aligned}
Y_0^{(-k)}(\lambda) &= 2^{-k}(\lambda-1)^k, \\
Y_1^{(-k)}(\lambda) &= 2^{-k}\binom{k}{1}\lambda^2(\lambda-1)^{k-1}, \\
Y_2^{(-k)}(\lambda) &= 2^{-k}2!\binom{k}{2}\lambda^4(\lambda-1)^{k-2},
\end{aligned}$$

$$\vdots$$

$$Y_j^{(-k)}(\lambda) = 2^{-k}j!\binom{k}{j}\lambda^{2j}(\lambda-1)^{k-j} \quad \text{for } j \le k,$$

$$\vdots$$

$$\begin{aligned}
Y_k^{(-k)}(\lambda) &= 2^{-k}k!\lambda^{2k}, \\
Y_j^{(-k)}(\lambda) &= 0 \quad \text{for } j > k.
\end{aligned}$$

By Equations (24) and (25), we also compute a few values of the polynomials $Q_n(x; \lambda, k)$ as follows:

$$\begin{aligned}
Q_0(x; \lambda, k) &= 2^{-k}(\lambda-1)^k, \\
Q_1(x; \lambda, k) &= 2^{-k}(\lambda-1)^k \lambda x + 2^{-k}k\lambda^2(\lambda-1)^{k-1}, \\
Q_2(x; \lambda, k) &= 2^{-k}(\lambda-1)^k \lambda^2 x^2 + \left(-2^{-k}(\lambda-1)^k \lambda^2 + 2^{-k+1}k\lambda^3(\lambda-1)^{k-1}\right) x \\
&\quad + 2^{-k}k(k-1)\lambda^4(\lambda-1)^{k-1}.
\end{aligned}$$

By Equation (3), we arrive at a computation formula, for the numbers $Y_n^{(-k)}(\lambda)$, in terms of the Bernstein basis functions by the following corollary:

Corollary 1. *Let n and k be nonnegative integers and $\lambda \in [0,1]$. Then,*

$$Y_n^{(-k)}(\lambda) = \begin{cases} 2^{-k} n! (-1)^{k-n} \lambda^n B_n^k(\lambda) & \text{if } n \le k \\ 0 & \text{if } n > k. \end{cases} \tag{26}$$

Replacing $1 + \lambda t$ by $e^{\log(1+\lambda t)}$ leads Equation (21) to be:

$$\sum_{n=0}^{\infty} Y_n^{(-k)}(\lambda) \frac{t^n}{n!} = \frac{(-1)^k}{2^k} \left(-\lambda e^{\log(1+\lambda t)} + 1\right)^k. \tag{27}$$

By combining Equation (17) with the above quation, we get:

$$\sum_{n=0}^{\infty} Y_n^{(-k)}(\lambda) \frac{t^n}{n!} = \frac{(-1)^k k!}{2^k} \sum_{m=0}^{\infty} y_1(m,k;-\lambda) \frac{(\log(1+\lambda t))^m}{m!} \tag{28}$$

which follows from Equation (5) that:

$$\sum_{n=0}^{\infty} Y_n^{(-k)}(\lambda) \frac{t^n}{n!} = \frac{(-1)^k k!}{2^k} \sum_{n=0}^{\infty} \sum_{m=0}^{n} \lambda^n y_1(m,k;-\lambda) S_1(n,m) \frac{t^n}{n!}. \tag{29}$$

Therefore, by comparing coefficient of $\frac{t^n}{n!}$ on both sides of the above equation, we arrive at the following theorem:

Theorem 3.

$$Y_n^{(-k)}(\lambda) = \frac{(-1)^k k!}{2^k} \sum_{m=0}^{n} \lambda^n y_1(m,k;-\lambda) S_1(n,m). \tag{30}$$

Combining Equations (25) with (30) yields the following corollary:

Corollary 2.

$$\sum_{m=0}^{n} y_1(m,k;-\lambda) S_1(n,m) = \begin{cases} \frac{(-1)^k \lambda^n (\lambda-1)^{k-n}}{(k-n)!} & \text{if } n \leq k \\ 0 & \text{if } n > k \end{cases} \tag{31}$$

If we also combine Equations (26) with (30), then we have the following result:

Corollary 3. *Let n and k nonnegative integer with $n \leq k$. Then,*

$$\sum_{m=0}^{n} y_1(m,k;-\lambda) S_1(n,m) = (-1)^n \frac{n!}{k!} B_n^k(\lambda). \tag{32}$$

On the other hand, since the following equality holds true (*cf.* [13]):

$$S_2(n,k;\lambda) = (-1)^k y_1(n,k;-\lambda), \tag{33}$$

Equation (31) leads the following corollary:

Corollary 4.

$$\sum_{m=0}^{n} S_2(m,k;\lambda) S_1(n,m) = \begin{cases} \frac{\lambda^n (\lambda-1)^{k-n}}{(k-n)!} & \text{if } n \leq k \\ 0 & \text{if } n > k \end{cases} \tag{34}$$

3. Derivative Formulas and Recurrence Relations Arising from Differential Equations of Generating Functions

In this section, by using differential equations involving the generating functions $\mathcal{G}(t,k;\lambda)$ and $\mathcal{G}(t,x,k;\lambda)$ and their partial derivatives with respect to the parameters t, λ, and x, we obtain some derivative formulas and recurrence relations for the numbers $Y_n^{(-k)}(\lambda)$ and the polynomials $Q_n(x;\lambda,k)$.

Differentiating both sides of Equation (21) with respect to λ, we get the following partial derivative equation:

$$\frac{\partial}{\partial \lambda}\{\mathcal{G}(t,k;\lambda)\} = \frac{k}{2}(2\lambda t + 1)\mathcal{G}(t, k-1; \lambda). \tag{35}$$

Also, if we differentiate both sides of Equation (21) with respect to t, then we get the following partial derivative equation:

$$\frac{\partial}{\partial t}\{\mathcal{G}(t,k;\lambda)\} = \frac{k\lambda^2}{2}\mathcal{G}(t, k-1; \lambda). \tag{36}$$

By combining Equation (35) with the RHS of Equation (21), we obtain:

$$\sum_{n=0}^{\infty} \frac{d}{d\lambda}\{Y_n^{(-k)}(\lambda)\}\frac{t^n}{n!} = \frac{k}{2}\sum_{n=0}^{\infty}\left(2n\lambda Y_{n-1}^{(-k+1)}(\lambda) + Y_n^{(-k+1)}(\lambda)\right)\frac{t^n}{n!}. \tag{37}$$

Comparing the coefficients of $\frac{t^n}{n!}$ on both sides of the above equation, we arrive at the following theorem:

Theorem 4. *Let* $n \in \mathbb{N}$. *Then, we have:*

$$\frac{d}{d\lambda}\{Y_n^{(-k)}(\lambda)\} = \frac{k}{2}\left(2n\lambda Y_{n-1}^{(-k+1)}(\lambda) + Y_n^{(-k+1)}(\lambda)\right). \tag{38}$$

By combining Equation (36) with the RHS of Equation (21), we get:

$$\frac{\partial}{\partial t}\sum_{n=0}^{\infty} Y_n^{(-k)}(\lambda)\frac{t^n}{n!} = \frac{k\lambda^2}{2}\sum_{n=0}^{\infty} Y_n^{(-k+1)}(\lambda)\frac{t^n}{n!}. \tag{39}$$

which, by comparing the coefficients of $\frac{t^n}{n!}$ on both sides of the above equation, yields the following theorem:

Theorem 5. *Let* $n \in \mathbb{N}_0$. *Then, we have:*

$$Y_{n+1}^{(-k)}(\lambda) = \frac{k\lambda^2}{2}Y_n^{(-k+1)}(\lambda). \tag{40}$$

Differentiating both sides of Equation (22) with respect to λ, we get the following partial derivative equation:

$$\frac{\partial}{\partial \lambda}\{\mathcal{G}(t,x,k;\lambda)\} = \frac{k}{2}(2\lambda t + 1)\mathcal{G}(t, x, k-1; \lambda) + xt\mathcal{G}(t, x-1, k; \lambda). \tag{41}$$

Furthermore, if we differentiate both sides of the Equation (22) with respect to t, then we also get the following partial derivative equation:

$$\frac{\partial}{\partial t}\{\mathcal{G}(t,x,k;\lambda)\} = \frac{k\lambda^2}{2}\mathcal{G}(t, x, k-1; \lambda) + x\lambda\mathcal{G}(t, x-1, k; \lambda). \tag{42}$$

Additionally, when we differentiate both sides of Equation (22) with respect to x, we also get the following partial derivative equation:

$$\frac{\partial}{\partial x}\{\mathcal{G}(t,x,k;\lambda)\} = \log(1 + \lambda t)\mathcal{G}(t, x, k; \lambda). \tag{43}$$

By combining Equation (41) with the RHS of Equation (22), we get:

$$\sum_{n=0}^{\infty} \frac{\partial}{\partial \lambda}\{Q_n(x;\lambda,k)\}\frac{t^n}{n!} = \frac{k}{2}(2\lambda t + 1)\sum_{n=0}^{\infty} Q_n(x;\lambda,k-1)\frac{t^n}{n!} + xt\sum_{n=0}^{\infty} Q_n(x-1;\lambda,k)\frac{t^n}{n!}$$

which yields:

$$\sum_{n=0}^{\infty} \frac{\partial}{\partial \lambda}\{Q_n(x;\lambda,k)\}\frac{t^n}{n!} = \frac{k}{2}\sum_{n=0}^{\infty}(2n\lambda Q_{n-1}(x;\lambda,k-1) + Q_n(x;\lambda,k-1))\frac{t^n}{n!}$$
$$+ x\sum_{n=0}^{\infty} n Q_{n-1}(x-1;\lambda,k)\frac{t^n}{n!}.$$

Comparing the coefficients of $\frac{t^n}{n!}$ on both sides of the above equation, we arrive at the following theorem:

Theorem 6. *Let $n \in \mathbb{N}$. Then, we have:*

$$\frac{\partial}{\partial \lambda}\{Q_n(x;\lambda,k)\} = kn\lambda Q_{n-1}(x;\lambda,k-1) + \frac{k}{2}Q_n(x;\lambda,k-1) + xn Q_{n-1}(x-1;\lambda,k). \tag{44}$$

By combining Equation (42) with the RHS of Equation (22), we get:

$$\frac{\partial}{\partial t}\sum_{n=0}^{\infty} Q_n(x;\lambda,k)\frac{t^n}{n!} = \frac{k\lambda^2}{2}\sum_{n=0}^{\infty} Q_n(x;\lambda,k-1)\frac{t^n}{n!} + x\lambda \sum_{n=0}^{\infty} Q_n(x-1;\lambda,k)\frac{t^n}{n!}$$

which, by comparing the coefficients of $\frac{t^n}{n!}$ on both sides of the above equation, yields the following theorem:

Theorem 7. *Let $n \in \mathbb{N}_0$. Then, we have:*

$$Q_{n+1}(x;\lambda,k) = \frac{k\lambda^2}{2}Q_n(x;\lambda,k-1) + x\lambda Q_n(x-1;\lambda,k). \tag{45}$$

By combining Equation (43) with the RHS of Equation (22) and the Taylor series of the function $\log(1 + \lambda t)$, we get:

$$\sum_{n=0}^{\infty} \frac{\partial}{\partial x}\{Q_n(x;\lambda,k)\}\frac{t^n}{n!} = \sum_{n=1}^{\infty}(-1)^{n-1}\frac{\lambda^n t^n}{n}\sum_{n=0}^{\infty} Q_n(x;\lambda,k)\frac{t^n}{n!}.$$

Applying the Cauchy product rule to the above equation yields:

$$\sum_{n=0}^{\infty} \frac{\partial}{\partial x}\{Q_n(x;\lambda,k)\}\frac{t^n}{n!} = t\sum_{n=0}^{\infty}\left(\sum_{j=0}^{n}(-1)^j \binom{n}{j}\frac{j!\lambda^{j+1}}{j+1}Q_{n-j}(x;\lambda,k)\right)\frac{t^n}{n!}.$$

Comparing the coefficients of $\frac{t^n}{n!}$ on both sides of the above equation, we arrive at the following theorem:

Theorem 8. *Let $n \in \mathbb{N}$. Then, we have:*

$$\frac{\partial}{\partial x}\{Q_n(x;\lambda,k)\} = n\sum_{j=0}^{n-1}(-1)^j \binom{n-1}{j}\frac{j!\lambda^{j+1}}{j+1}Q_{n-1-j}(x;\lambda,k). \tag{46}$$

Remark 1. *Substituting Equation (16) into Equation (46) yields the following formula including Daehee numbers:*

$$\frac{\partial}{\partial x}\{Q_n(x;\lambda,k)\} = n\sum_{j=0}^{n-1}\binom{n-1}{j}\lambda^{j+1}D_j Q_{n-1-j}(x;\lambda,k). \tag{47}$$

By Equation (15), another form of the partial differential Equation (43) is given by:

$$\frac{\partial}{\partial x}\{\mathcal{G}(t,x,k;\lambda)\} = \lambda t \mathcal{G}(t,k;\lambda) F_D(x,\lambda t). \tag{48}$$

By combining Equation (48) with the RHS of the Equations (15) and (22), we get:

$$\sum_{n=0}^{\infty} \frac{\partial}{\partial x}\{Q_n(x;\lambda,k)\}\frac{t^n}{n!} = \lambda t \sum_{n=0}^{\infty} Y_n^{(-k)}(\lambda) \frac{t^n}{n!} \sum_{n=0}^{\infty} \lambda^n D_n(x) \frac{t^n}{n!}.$$

Applying the Cauchy product rule to the above equation yields:

$$\sum_{n=0}^{\infty} \frac{\partial}{\partial x}\{Q_n(x;\lambda,k)\}\frac{t^n}{n!} = \lambda t \sum_{n=0}^{\infty} \left(\sum_{j=0}^{n} \lambda^j \binom{n}{j} Y_{n-j}^{(-k)}(\lambda) D_j(x) \right) \frac{t^n}{n!}.$$

Comparing the coefficients of $\frac{t^n}{n!}$ on both sides of the above equation, we arrive at the following theorem:

Theorem 9.

$$\frac{\partial}{\partial x}\{Q_n(x;\lambda,k)\} = n \sum_{j=0}^{n-1} \lambda^{j+1} \binom{n-1}{j} Y_{n-j}^{(-k)}(\lambda) D_j(x).$$

4. Some Identities and Relations Derived from Functional Equations of Generating Functions

In this section, by using functional equations of the aforementioned generating functions, we derive some formulas and combinatorial sums including binomials coefficients, falling factorial, the Stirling numbers, the Bell polynomials (i.e., exponential polynomials), the Poisson–Charlier polynomials, combinatorial numbers and polynomials, and the Bersntein basis functions.

Now, we set the following functional equation:

$$F_{pc}(t,x;a) = \mathcal{F}\left(t,x,k;\frac{1}{a}\right) \mathcal{F}\left(t,-k;\frac{1}{a}\right) e^{-t}. \tag{49}$$

Combining the above equation with the Equations (1), (11), and (12), we get:

$$\sum_{n=0}^{\infty} C_n(x;a) \frac{t^n}{n!} = \sum_{n=0}^{\infty} Y_n^{(k)}\left(x;\frac{1}{a}\right) \frac{t^n}{n!} \sum_{n=0}^{\infty} Y_n^{(-k)}\left(\frac{1}{a}\right) \frac{t^n}{n!} \sum_{n=0}^{\infty} (-1)^n \frac{t^n}{n!}. \tag{50}$$

Applying the Cauchy product rule to the above equation yields:

$$\sum_{n=0}^{\infty} C_n(x;a) \frac{t^n}{n!} = \sum_{n=0}^{\infty} \left(\sum_{l=0}^{n} \sum_{j=0}^{l} (-1)^{n-l} \binom{n}{l}\binom{l}{j} Y_j^{(k)}\left(x;\frac{1}{a}\right) Y_{l-j}^{(-k)}\left(\frac{1}{a}\right) \right) \frac{t^n}{n!}. \tag{51}$$

Therefore, by comparing the coefficient of $\frac{t^n}{n!}$ on both sides of the above equation, we arrive at the following theorem:

Theorem 10.

$$C_n(x;a) = \sum_{l=0}^{n} \sum_{j=0}^{l} (-1)^{n-l} \binom{n}{l}\binom{l}{j} Y_j^{(k)}\left(x;\frac{1}{a}\right) Y_{l-j}^{(-k)}\left(\frac{1}{a}\right). \tag{52}$$

Moreover, we also set the following functional equation:

$$F_B(t,-x;k) F_{pc}(t,x;a) = \frac{(-1)^k (xt)^k}{k!} e^{xt} \left(\frac{t}{a}+1\right)^x.$$

Combining the above equation with the Equations (1) and (4) yields:

$$\sum_{n=0}^{\infty} B_k^n(-x)\frac{t^n}{n!} \sum_{n=0}^{\infty} C_n(x;a)\frac{t^n}{n!} = \frac{(-1)^k (xt)^k}{k!} \sum_{n=0}^{\infty} \frac{(xt)^n}{n!} \sum_{n=0}^{\infty} \binom{x}{n}\frac{t^n}{a^n}.$$

Applying the Cauchy product rule to the above equation yields:

$$\sum_{n=0}^{\infty} \left(\sum_{j=0}^{n} \binom{n}{j} B_k^j(-x) C_{n-j}(x;a) \right) \frac{t^n}{n!} = \frac{(-1)^k x^k}{k!} \sum_{n=0}^{\infty} \left((n)_k \sum_{j=0}^{n-k} \binom{n-k}{j} \frac{x^{n-k-j}(x)_j}{a^j} \right) \frac{t^n}{n!}.$$

Therefore, by comparing the coefficient of $\frac{t^n}{n!}$ on both sides of the above equation, we arrive at the following theorem:

Theorem 11.

$$\sum_{j=0}^{n} \binom{n}{j} B_k^j(-x) C_{n-j}(x;a) = \frac{(-1)^k (n)_k}{k!} \sum_{j=0}^{n-k} \binom{n-k}{j} \frac{x^{n-j}(x)_j}{a^j}. \tag{53}$$

Additionaly, we also have the following functional equation:

$$\frac{(xt)^k}{k!} F_{pc}(-t,x;a) e^{-xt} = F_B(t,x;k) \left(-\frac{t}{a} + 1 \right)^x.$$

Combining the above equation with Equations (1) and (4) yields:

$$\frac{(xt)^k}{k!} \sum_{n=0}^{\infty} (-1)^n C_n(x;a) \frac{t^n}{n!} \sum_{n=0}^{\infty} (-x)^n \frac{t^n}{n!} = \sum_{n=0}^{\infty} B_k^n(x) \frac{t^n}{n!} \sum_{n=0}^{\infty} (-1)^n \binom{x}{n} \frac{t^n}{a^n}.$$

Applying the Cauchy product rule to the above equation yields:

$$\sum_{n=0}^{\infty} \left(\frac{(-1)^{n-k}(n)_k}{k!} \sum_{j=0}^{n-k} \binom{n-k}{j} x^{n-j} C_j(x;a) \right) \frac{t^n}{n!} = \sum_{n=0}^{\infty} \left(\sum_{j=0}^{n} (-1)^j \binom{n}{j} \frac{(x)_j B_k^{n-j}(x)}{a^j} \right) \frac{t^n}{n!}.$$

Therefore, by comparing coefficient of $\frac{t^n}{n!}$, we arrive at the following theorem:

Theorem 12.

$$\frac{(-1)^{n-k}(n)_k}{k!} \sum_{j=0}^{n-k} \binom{n-k}{j} x^{n-j} C_j(x;a) = \sum_{j=0}^{n} (-1)^j \binom{n}{j} \frac{(x)_j B_k^{n-j}(x)}{a^j}. \tag{54}$$

By substituting $t \to a(e^t - 1)$ into Equation (1), we also get the following functional equation:

$$F_{pc}\left(a(e^t - 1), x; a \right) = e^{tx} F_{Bell}(t, -a). \tag{55}$$

By Equations (1) and (10), we thus get:

$$\sum_{n=0}^{\infty} a^n C_n(x;a) \frac{(e^t - 1)^n}{n!} = \sum_{n=0}^{\infty} x^n \frac{t^n}{n!} \sum_{n=0}^{\infty} Bl_n(-a) \frac{t^n}{n!}. \tag{56}$$

Applying the Cauchy product rule to the above equation and combining Equation (8) with the final equation yields:

$$\sum_{m=0}^{\infty}\sum_{n=0}^{m} a^n C_n(x;a) S_2(m,n) \frac{t^m}{m!} = \sum_{m=0}^{\infty}\left(\sum_{j=0}^{m}\binom{m}{j}x^{m-j} Bl_j(-a)\right)\frac{t^m}{m!}. \tag{57}$$

Therefore, by comparing the coefficient of $\frac{t^m}{m!}$, we arrive at the following theorem:

Theorem 13.
$$\sum_{n=0}^{m} a^n C_n(x;a) S_2(m,n) = \sum_{j=0}^{m}\binom{m}{j}x^{m-j} Bl_j(-a). \tag{58}$$

5. Some Identities and Relations Arising from the p-adic Integrals and Riemann Integral

In this section, by applying the p-adic integrals and Riemann integral to some of our results, we derive some combinatorial sums including the binomial coefficients, falling factorial, the Bernoulli numbers, the Euler numbers, the Stirling numbers, the Bell polynomials (i.e., exponential polynomials), and the Cauchy numbers (or the Bernoulli numbers of the second kind).

Let $\mathbb{Z}_p$ denote a set of p-adic integers. Let $f(x)$ be a uniformly differentiable function on $\mathbb{Z}_p$. The Volkenborn integral (or p-adic bosonic integral) of the function $f(x)$ is given by:

$$\int_{\mathbb{Z}_p} f(x) d\mu_1(x) = \lim_{N\to\infty} \frac{1}{p^N} \sum_{x=0}^{p^N-1} f(x), \tag{59}$$

where,
$$\mu_1(x) = \mu_1(x+p^N\mathbb{Z}_p) = \frac{1}{p^N}$$

(cf. [17]; see also [11,12]).

It is known that the bosonic p-adic integral of the function $f(x) = x^n$ gives the Bernoulli numbers as follows (cf. [11,17]):

$$B_n = \int_{\mathbb{Z}_p} x^n d\mu_1(x) \tag{60}$$

where B_n denotes the Bernoulli numbers of the first kind defined by means of the following generating function:

$$\frac{t}{e^t-1} = \sum_{n=0}^{\infty} B_n \frac{t^n}{n!}, \quad (t<|2\pi|) \tag{61}$$

which arise in not only analytic number theory, but also other related areas (cf. [5–31]).

The fermionic p-adic integral of the function $f(x)$ is given by (cf. [12]):

$$\int_{\mathbb{Z}_p} f(x) d\mu_{-1}(x) = \lim_{N\to\infty} \sum_{x=0}^{p^N-1} (-1)^x f(x) \tag{62}$$

where $p \neq 2$ and,

$$\mu_{-1}(x) = \mu_{-1}\left(x+p^N\mathbb{Z}_p\right) = \frac{(-1)^x}{p^N}$$

(cf. [10,12]).

The fermionic p-adic integral of the function $f(x) = x^n$ gives the Euler numbers as follows (cf. [11]):
$$E_n = \int_{\mathbb{Z}_p} x^n d\mu_{-1}(x), \qquad (63)$$

where E_n denotes the Euler numbers of the first kind defined by means of the following generating function:
$$\frac{2}{e^t + 1} = \sum_{n=0}^{\infty} E_n \frac{t^n}{n!}, \quad (t < |\pi|) \qquad (64)$$

(cf. [5–30]).

It is known that the following p-adic bosonic and fermionic integral representations for the Poisson–Charlier polynomials hold true (see [24] (Equations (33) and (35), pp. 944–945)):
$$\int_{\mathbb{Z}_p} C_n(x;a) d\mu_1(x) = \sum_{k=0}^{n} (-1)^n \binom{n}{k} \frac{k!}{(k+1)a^k} \qquad (65)$$

and,
$$\int_{\mathbb{Z}_p} C_n(x;a) d\mu_{-1}(x) = \sum_{k=0}^{n} (-1)^n \binom{n}{k} \frac{k!}{(2a)^k}. \qquad (66)$$

By applying the bosonic p-adic integral to Equation (58) and combining the final equation with Equations (60) and (65), we arrive at the following theorem:

Theorem 14.
$$\sum_{n=0}^{m}\sum_{k=0}^{n} (-1)^n \frac{(n)_k a^{n-k} S_2(m,n)}{k+1} = \sum_{j=0}^{m} \binom{m}{j} B_m {}_jBl_j(-a). \qquad (67)$$

By applying the fermionic p-adic integral to Equation (58) and combining the final equation with Equations (63) and (66), we also arrive at the following theorem:

Theorem 15.
$$\sum_{n=0}^{m}\sum_{k=0}^{n} (-1)^n \frac{(n)_k a^{n-k} S_2(m,n)}{2^k} = \sum_{j=0}^{m} \binom{m}{j} E_{m-j} Bl_j(-a). \qquad (68)$$

Moreover, by integrating Equation (58) with respect to x from 0 to 1, we have:
$$\sum_{n=0}^{m} a^n S_2(m,n) \int_0^1 C_n(x;a) dx = \sum_{j=0}^{m} \binom{m}{j} Bl_j(-a) \int_0^1 x^{m-j} dx. \qquad (69)$$

On the other hand, by integrating Equation (2) with respect to x from 0 to 1, we also have:
$$\int_0^1 C_n(x;a) dx = \sum_{j=0}^{n} (-1)^{n-j} \binom{n}{j} \frac{1}{a^j} \int_0^1 (x)_j dx, \qquad (70)$$

By making use of the following definition of the well-known Cauchy numbers (or the Bernoulli numbers of the second kind) $b_n(0)$ (cf. [4]):
$$b_n(0) = \int_0^1 (x)_n dx, \qquad (71)$$

Equation (70) yields:

$$\int_0^1 C_n(x;a)\,dx = \sum_{j=0}^n (-1)^{n-j} \binom{n}{j} \frac{b_j(0)}{a^j}. \tag{72}$$

Combining the above equation with Equation (69), we arrive at the following theorem:

Theorem 16.

$$\sum_{n=0}^m \sum_{j=0}^n (-1)^{n-j} \binom{n}{j} a^{n-j} S_2(m,n) b_j(0) = \sum_{j=0}^m \binom{m}{j} \frac{Bl_j(-a)}{m-j+1}. \tag{73}$$

6. Applications in the Probability Distribution Function

In this section, we investigate some applications of the numbers $Y_n^{(-k)}(\lambda)$. Assume that $0 < p \leq 1$ and $n = 0, 1, 2, \ldots, k$. We set the following discrete probability distribution:

$$f(p;k,n) = \frac{(-1)^{k-n} 2^k}{n! p^n} Y_n^{(-k)}(p) \tag{74}$$

where p is a probability of success, k is number of trials, n is number of successes in k trials, and $n = 0, 1, 2, \ldots, k$. Therefore, $f(p;k,n)$ is binomially distributed with parameters (k, p).

Properties of Discrete Probability Distribution $f(p;k,n)$

Here, we give some properties of discrete probability distribution $f(p;k,n)$. We examine the properties of the probability distribution $f(p;k,n)$ with a random variable with parameters k, n, and p as follows:

For all k, n, p with $0 \leq n \leq k$ and $0 < p \leq 1$, $0 \leq f(p;k,n) \leq 1$. That is $f(p;k,n) \geq 0$.
The probability distribution function $f(p;k,n)$ satisfies that:

$$\sum_{n=0}^\infty f(p;k,n) = 1.$$

Computing the distribution function $f(p;k,n)$. Suppose that X is a binomial with parameters (k, p). To computing its distribution function:

$$P(X \leq j) = \sum_{n=0}^j f(p;k,n),$$

where $j = 0, 1, \ldots, k$.

In order to compute its expected value and variance for random variable with parameters k and p:

$$E[X^v] = \sum_{n=0}^k n^v f(p;k,n) \tag{75}$$

Observe that the probability distribution function $f(p;k,n)$ is a modification of the binomial probability distribution function with parameters (k, p). Substituting $v = 1$ into Equation (75), $E[X] = kp$. Substituting $v = 2$ into Equation (75), variance $E[X^2] - (E[X])^2 = kp(1-p)$.

If we take $k \to \infty$, then the distribution $f(p;k,n)$ goes to the Poisson distribution. On the other hand the Poisson–Charlier polynomials are orthogonal with respect to the Poisson distribution (*cf.* [18,24]).

7. Conclusions

Applications of generating functions are used in many areas, and we used them to study new families of combinatorial numbers and polynomials. We then studied properties of these new families, which yielded a handful of new identities and relations. Namely, these identities were related to numerous special numbers, special polynomials, and special functions such as the Bersntein basis functions, the Stirling numbers, the Bell polynomials (or exponential polynomials), the Poisson–Charlier polynomials, and the probability distribution functions. Furthermore, we should note that newly defined combinatorial numbers in this paper gave a different approach to the binomial (or Newton) distribution and the Poisson distribution, as well as combinatorial sums including the Bernoulli numbers, the Euler numbers, the Cauchy numbers (or the Bernoulli numbers of the second kind), and combinatorial numbers. This is why the results of this paper have the potential to be used in numerous areas such as mathematics, probability, physics, and in other associated areas.

Author Contributions: Investigation, I.K., B.S., Y.S.; wirting-original draft, I.K., B.S., Y.S.; writing-review and editing, I.K., B.S., Y.S.

Funding: This research received no external funding.

Acknowledgments: This paper is dedicated to Hari Mohan Srivastava on the occasion of his 80th Birthday. Yilmaz Simsek was supported by the Scientific Research Project Administration of Akdeniz University.

Conflicts of Interest: The authors declare no conflict of interest.

References

1. Acikgoz, M.; Araci, S. On generating function of the Bernstein polynomials. *Proc. Int. Conf. Numer. Anal. Appl. Math. Am. Inst. Phys. Conf. Proc.* **2010**, *CP1281*, 1141–1143.
2. Bona, M. *Introduction to Enumerative Combinatorics*; The McGraw-Hill Companies Inc.: New York, NY, USA, 2007.
3. Charalambides, C.A. *Enumerative Combinatorics*; Chapman and Hall (CRC Press Company): London, UK; New York, NY, USA, 2002.
4. Comtet, L. *Advanced Combinatorics: The Art of Finite and Infinite Expansions*; D. Reidel Publishing Company: Dordrecht, The Netherlands; Boston, MA, USA, 1974.
5. Djordjevic, G.B.; Milovanović, G.V. *Special Classes of Polynomials*; Faculty of Technology, University of Nis: Leskovac, Serbia, 2014.
6. Golombek, R. Aufgabe 1088. *El. Math.* **1994**, *49*, 126–127.
7. Jordan, C. *Calculus of Finite Differences*, 2nd ed.; Chelsea Publishing Company: New York, NY, USA, 1950.
8. Kim, D.S.; Kim, T. Daehee numbers and polynomials. *Appl. Math. Sci. (Ruse)* **2013**, *7*, 5969–5976. [CrossRef]
9. Kim, D.S.; Kim, T.; Seo, J. A note on Changhee numbers and polynomials. *Adv. Stud. Theor. Phys.* **2013**, *7*, 993–1003. [CrossRef]
10. Kim, M.-S. On Euler numbers, polynomials and related p-adic integrals. *J. Number Theory* **2009**, *129*, 2166–2179. [CrossRef]
11. Kim, T. q-Volkenborn integration. *Russ. J. Math. Phys.* **2002**, *19*, 288–299.
12. Kim, T. q-Euler numbers and polynomials associated with p-adic q-integral and basic q-zeta function. *Trend Math. Inf. Cent. Math. Sci.* **2006**, *9*, 7–12.
13. Kucukoglu, I.; Simsek, Y. Observations on identities and relations for interpolation functions and special numbers. *Adv. Stud. Contemp. Math.* **2018**, *28*, 41–56.
14. Kucukoglu, I.; Simsek, B.; Simsek, Y. An approach to negative hypergeometric distribution by generating function for special numbers and polynomials. *Turk. J. Math.* **2019**, *43*, 2337–2353.
15. Lorentz, G.G. *Bernstein Polynomials*; Chelsea Publishing Company: New York, NY, USA, 1986.
16. Ozmen, N.; Erkus-Duman, E. On the Poisson–Charlier polynomials. *Serdica Math. J.* **2015**, *41*, 457–470.
17. Schikhof, W.H. *Ultrametric Calculus: An Introduction to p-Adic Analysis*; Cambridge Studies in Advanced Mathematics 4; Cambridge University Press: Cambridge, UK, 1984.
18. Roman, S. *The Umbral Calculus*; Dover Publishing Inc.: New York, NY, USA, 2005.
19. Ross, S.M. *A First Course in Probability*, 8th ed.; Pearson Education, Inc.: London, UK, 2010.

20. Simsek, Y. Functional equations from generating functions: A novel approach to deriving identities for the Bernstein basis functions. *Fixed Point Theory Appl.* **2013**, *2013*, 1–13. [CrossRef]
21. Simsek, Y. Generating functions for generalized Stirling type numbers, array type polynomials, Eulerian type polynomials and their applications. *Fixed Point Theory Appl.* **2013**, *2013*, 343–355. [CrossRef]
22. Simsek, Y. New families of special numbers for computing negative order Euler numbers and related numbers and polynomials. *Appl. Anal. Discr. Math.* **2018**, *12*, 1–35. [CrossRef]
23. Simsek, Y. Construction of Some New Families of Apostol-type Numbers and Polynomials via Dirichlet Character and *p*-adic *q*-integrals. *Turk. J. Math.* **2018**, *42*, 557–577. [CrossRef]
24. Simsek, Y. Formulas for Poisson–Charlier, Hermite, Milne-Thomson and other type polynomials by their generating functions and *p*-adic integral approach. *RACSAM Rev. R. Acad. A* **2019**, *113*, 931–948. [CrossRef]
25. Simsek, Y. Generating functions for finite sums involving higher powers of binomial coefficients: Analysis of hypergeometric functions including new families of polynomials and numbers. *J. Math. Anal. Appl.* **2019**, *477*, 1328–1352. [CrossRef]
26. Simsek, Y.; Acikgoz, M. A new generating function of (*q*-) Bernstein-type polynomials and their interpolation function. *Abstr. Appl. Anal.* **2010**, 769095. [CrossRef]
27. Srivastava, H.M. Some formulas for the Bernoulli and Euler polynomials at rational arguments. *Math. Proc. Camb. Philos. Soc.* **2000**, *129*, 77–84. [CrossRef]
28. Srivastava, H.M. Some generalizations and basic (or *q*-) extensions of the Bernoulli, Euler and Genocchi polynomials. *Appl. Math. Inf. Sci.* **2011**, *5*, 390–444.
29. Srivastava, H.M.; Choi, J. *Series Associated with the Zeta and Related Functions*; Kluwer Acedemic Publishers: Dordrecht, The Netherlands; Boston, MA, USA; London, UK, 2001.
30. Srivastava, H.M.; Choi, J. *Zeta and q-Zeta Functions and Associated Series and Integrals*; Elsevier Science Publishers: Amsterdam, The Netherlands; London, UK; New York, NY, USA, 2012.
31. Xu, A. On an open problem of Simsek concerning the computation of a family special numbers. *Appl. Anal. Discret. Math.* **2019**, *13*, 61–72. [CrossRef]

© 2019 by the authors. Licensee MDPI, Basel, Switzerland. This article is an open access article distributed under the terms and conditions of the Creative Commons Attribution (CC BY) license (http://creativecommons.org/licenses/by/4.0/).

Article

On One Problems of Spectral Theory for Ordinary Differential Equations of Fractional Order

Temirkhan Aleroev

NRU MGSU, Moscow 129337, Russia; aleroev@mail.ru or t.s.aleroev@gmail.com

Received: 8 September 2019; Accepted: 15 October 2019; Published: 18 October 2019

Abstract: The present paper is devoted to the spectral analysis of operators induced by fractional differential equations and boundary conditions of Sturm-Liouville type. It should be noted that these operators are non-self-adjoint. The spectral structure of such operators has been insufficiently explored. In particular, a study of the completeness of systems of eigenfunctions and associated functions has begun relatively recently. In this paper, the completeness of the system of eigenfunctions and associated functions of one class of non-self-adjoint integral operators corresponding boundary value problems for fractional differential equations is established. The proof is based on the well-known Theorem of M.S. Livshits on the spectral decomposition of linear non-self-adjoint operators, as well as on the sectoriality of the fractional differentiation operator. The results of Dzhrbashian-Nersesian on the asymptotics of the zeros of the Mittag-Leffler function are used.

Keywords: Mittag-Leffler function; spectrum; eigenvalue; fractional derivative

1. Introduction

The present paper is devoted to the spectral analysis of operators induced by the fractional differential equations and boundary conditions of Sturm-Liouville type. It should be noted that these operators are non-self-adjoint. The spectral structure of such operators has been insufficiently explored. In particular, a study of the completeness of systems of eigenfunctions and associated functions has begun relatively recently. In this paper, the completeness of the system of eigenfunctions and associated functions of one class of non-self-adjoint integral operators corresponding boundary value problems for fractional differential equations is established. The proof is based on the well-known Theorem of M.S. Livshits on the spectral decomposition of linear non-self-adjoint operators, as well as on the sectoriality of the fractional differentiation operator. The results of Dzhrbashian-Nersesian on the asymptotics of the zeros of the Mittag-Leffler function are used.

2. Results

Reference [1] studied the operator in the space $L_2(0,1)$

$$-A_\rho u = \int_0^1 G(x,t)u(t)\,dt = \frac{1}{\Gamma(\rho^{-1})}\left[\int_0^x (x-t)^{\frac{1}{\rho}-1}u(t)\,dt - \int_0^1 x^{\frac{1}{\rho}-1}(1-t)^{\frac{1}{\rho}-1}u(t)\,dt\right],$$

which was first considered in References [2,3], where $0 < \rho < 2$ and

$$G(x,t) = \begin{cases} \dfrac{(1-t)^{\frac{1}{\rho}-1}x^{\frac{1}{\rho}-1} - (x-t)^{\frac{1}{\rho}-1}}{\Gamma(\rho^{-1})}, & 0 \leq t \leq x \leq 1 \\ \dfrac{(1-t)^{\frac{1}{\rho}-1}x^{\frac{1}{\rho}-1}}{\Gamma(\rho^{-1})}, & 0 \leq x \leq t \leq 1 \end{cases}$$

is the Green function of the following problem S (with $\lambda = 0$):

$$\frac{1}{\Gamma(n-\rho^{-1})} \frac{d^n}{dx^n} \int_0^x (x-s)^{n-\rho^{-1}-1} u(s) ds + \lambda u = 0,$$

$(n - 1 \leq \rho^{-1} < n,\ n = [\rho^{-1}] + 1$, where $[\rho^{-1}]$ is the integer part of ρ^{-1})

$$u(0) = 0, u'(0) = 0, \cdots, u^{(n-2)}(0) = 0, u(1) = 0.$$

In particular, in this paper, we provide very important proof of the completeness of the system of eigenfunctions and associated functions in $L_2(0,1)$ of the operator A_ρ for $1 < \rho < 2$ based on fact that the operator of fractional differentiation is sectorial and for $0 < \rho < 1$ (this fact plays a main role in solving boundary value problems for advection-diffusion equation of fractional order by the method of separation of variables [4] since we can write out both the exact solution in the form of an infinite series by eigenfunctions and the approximate solution replacing the infinite sums by sums of the first n terms), a proof based on the well-known Livshits theorem [5] (researching of case for $1 < \rho < 2$ published in this paper firstly):

Theorem 1 (Livshits). *If $K(x,y)$ ($a \leq x,y \leq b$) – is a limited kernel and "real part" $\frac{1}{2}(K + K^*)$ of it is non-negative kernel, then the inequality is hold*

$$\sum_{j=1}^\infty Re(\frac{1}{\lambda_j}) \leq \int_a^b ReK(t,t)dt,$$

where λ_j – is the characteristic numbers of kernel K. The system of main functions of the kernel K is complete in domain of values of the integral operator Kf if and only if, when there is an equal sign in inequality above.

In his paper [6] M. M. Dzhrbashian wrote, that "the question about the completeness of the systems of eigenfunctions of the operator A_ρ or a finer question about whether these systems compose a basis in $L_2(0,1)$, has a certain interest but its solving is apparently associated with significant analytic difficulties". The questions of the completeness of the systems of eigenfunctions and associated functions for similar problems were studied by A. V. Agibalova in [7,8]. Undoubtedly, we shall note the fundamental results of M. M. Malamud and L. L. Oridoroga [9–12] obtained in this direction. In [13,14] (see also [2,15]), using the theorem of Matsaev and Palant, it was established that the system of eigenfunctions of the operator A_ρ is complete in $L_2(0,1)$. And this fact used by M. Ali, S. Aziz and S.A. Malik in their paper [16].

As noted above, in this paper, a similar result was obtained using the well-known Livshchits theorem [5]. The following proof of the completeness of the system of eigenfunctions is simpler than the previously presented proofs, which makes the results of this paper very significant.

Next, we need one definition.

Definition 1. *If a series of s-numbers [17] of the completely continuous operator is convergent, that is, $\sum_{k=1}^\infty s_k(A) < \infty$ then such operator called as trace-class operator.*

Lemma 1. *Let $0 < \rho < 2$, then the operator A_ρ is trace-class and*

$$sp(A_\rho) = \frac{\Gamma(\rho^{-1})}{\Gamma(2\rho^{-1})}.$$

Proof of Lemma 1. To find the trace spA_ρ of the operator A_ρ, let's rewrite A_ρ as $A_\rho u = A_1 u - A_0 u$ where

$$A_0 u = \frac{1}{\Gamma(\rho^{-1})} \int_0^x (x-t)^{\frac{1}{\rho}-1} u(t) dt,$$

$$A_1 u = \frac{1}{\Gamma(\rho^{-1})} \int_0^1 x^{\frac{1}{\rho}-1}(1-t)^{\frac{1}{\rho}-1} u(t) dt.$$

Clearly, for $0 < \rho < 2$, the operators A_0 and A_1 are trace class. Hence

$$sp A_\rho = sp(A_1 - A_0) = sp(A_1) - sp(A_0).$$

Moreover, it is clear that $sp(A_0) = 0$. Thus

$$sp A_\rho = sp(A_1).$$

Since operator A_1 is one-dimensional, it is easy to find a trace. Consider the equation

$$u(x) - \frac{\lambda}{\Gamma(\rho^{-1})} \int_0^1 x^{\frac{1}{\rho}-1}(1-t)^{\frac{1}{\rho}-1} u(t) dt = 0$$

The Fredhold determinant

$$d(\lambda) = |1 - \lambda K_{11}|,$$

where

$$K_{11} = \frac{1}{\Gamma(\rho^{-1})} \int_0^1 t^{\frac{1}{\rho}-1}(1-t)^{\frac{1}{\rho}-1} dt = \frac{\Gamma(2-\nu)}{\Gamma(4-2\nu)} \ (\nu = 2 - \rho^{-1}).$$

From above follow that

$$sp(A_1) = \frac{\Gamma(2-\nu)}{\Gamma(4-2\nu)}$$

which proves the Lemma 1.
□

Remark 1. *Of course, for $\rho > 1/2$, nuclearity of the operator A_ρ follows from well-known Dzhrbaschian-Nersisian lemma ([18], p. 142).*

Lemma 2 (Dzhrbaschian-Nersisian). *1. All zeros of functions $E_\rho(z;\mu) = \sum_{n=0}^\infty \frac{z^n}{\Gamma(\mu+n/\rho)}$ (where $\rho > \frac{1}{2}, \rho \neq 1; Im\mu = 0$) with largest absolute values, are prime.*
2. The following asymptotic formulas are valid

$$\gamma_k^\pm = e^{\pm i\frac{\pi}{2\rho}}(2\pi k)^{1/\rho}\left(1 + O\left(\frac{\log k}{k}\right)\right), k \to \infty,$$

and the fact that the value λ_j is an eigenvalue of the operator A_ρ if and only if $E_\rho(\lambda_j; \frac{1}{\rho}) = 0$.

Now we give the main result of paper.

Theorem 2. *The system of eigenfunctions and associated functions of the operator A_ρ, where $0 < \rho < 1$, is complete in $L_2(0,1)$.*

Proof of Theorem 2. We denote the kernel of A_ρ as $K(x,y)$. In [13] the authors have proved that this kernel is non-negative by the following way: Let us rewrite A_ρ as

$$A_\rho u = \frac{1}{\Gamma(\rho^{-1})}\left[\int_0^1 (x-xt)^{\frac{1}{\rho}-1}u(t)dt - \int_0^x (x-t)^{\frac{1}{\rho}-1}u(t)\,dt\right].$$

Clearly, for $\rho < 1$, the kernel of A_ρ is non-negative.

By the same way, we may show that the kernel $K^*(x,y)$ for adjoint operator

$$A_\rho^* u = \frac{1}{\Gamma(\rho^{-1})}\left[\int_0^1 (t-xt)^{\frac{1}{\rho}-1}u(x)dx - \int_x^1 (t-x)^{\frac{1}{\rho}-1}u(x)\,dx\right]$$

is non-negative too. Thus $\frac{1}{2}(K+K^*)$ is non-negative. Let us show that the following expression holds

$$\sum_{j=1}^\infty Re(\frac{1}{\lambda_j}) = \int_0^1 ReK(t,t)dt.$$

If $\lambda_j = \alpha_j + i\beta_j$ is eigenvalue of the operator A_ρ, then complex conjugate $\overline{\lambda}_j = \alpha_j - i\beta_j$ is eigenvalue of the operator A_ρ too. Thus

$$spA_\rho = \sum_{j=1}^\infty \frac{1}{\lambda_j} = \sum_{j=1}^\infty Re(\frac{1}{\lambda_j}).$$

So, taking to account lemma 1, we obtain that the system of eigenfunctions and associated functions of the operator A_ρ for $0 < \rho < 1$, is complete in $L_2(0,1)$. □

Remark 2. *For $(\frac{1}{\rho}-1) > 0$ the kernel of the operator A_ρ is continuous. Therefore, as it was showed by Lalesko [5], the Fredholm determinant of this kernel is whole function of zero kind. In this case [5],*

$$\sum_{i=1}^\infty \frac{1}{\lambda_j} = \int_0^1 K(t,t)dt,$$

that is, the equation

$$\sum_{j=1}^\infty Re(\frac{1}{\lambda_j}) = \int_0^1 ReK(t,t)dt$$

we can get by the obvious way.

Theorem 3. *The system of eigenfunctions and associated functions of the operator A_ρ, where $1 < \rho < 2$, is complete in $L_2(0,1)$.*

Proof of Theorem 3. For $1 < \rho < 2$ the kernel of the operator A_ρ is not fixed-sign, thus we cannot use the Livshits theorem, used above. To prove the formulated theorem, let us consider the value of the the form $(A_\rho u, \overline{u})$ [19]. Let us introduce the following designation

$$A_\rho u = \frac{1}{\Gamma(\rho^{-1})}\left[\int_0^1 (x-xt)^{\frac{1}{\rho}-1}u(t)dt - \int_0^x (x-t)^{\frac{1}{\rho}-1}u(t)\,dt\right] = v(x).$$

So,
$$(A_\rho u, \overline{u}) = (v, D_{0x}^{1/\rho} v) = \int_0^1 v(x)[D_{0x}^{1/\rho} v]dx = \int_0^\varepsilon v(x)[D_{0x}^{1/\rho} v]dx + \int_\varepsilon^1 v(x)[D_{0x}^{1/\rho} v]dx$$

where
$$D_{0x}^\alpha f(x) = \frac{1}{\Gamma(n-\alpha)} \left(\frac{d}{dx}\right)^n \int_0^x \frac{f(t)}{(x-t)^{\alpha-n+1}},$$

$n = [\alpha] + 1$, $[\alpha]$ is the integer part of α, called the operator of fractional differentiation in the Sturm-Liouville sense of order α. As was mentioned in Reference [19] (see also the references therein), the study of forms
$$\int_\varepsilon^1 v(x)[D_{0x}^{1/\rho} v]dx$$

was provided in the paper and there, in particular, were established the values of those forms lying in $|arg\lambda| < \frac{\pi\rho}{2}$. Clearly, for small values ε, the operator A_ρ is sectorial. Since the operator A_ρ is sectorial and a trace-class operator, by Lidskii's Theorem [20] the system of eigenfunctions and associated functions of A_ρ are complete in $L_2(0,1)$. □

Corollary 1. *Since the operator A_ρ does not generate any associated functions [21], we prove the completeness of system*
$$\chi_n(x) = x^{\frac{1}{\rho}-1} E_\rho(\lambda_n x^{\frac{1}{\rho}}; \frac{1}{\rho})$$

in $L_2(0,1)$ (but the system of these eigenfunctions, unfortunately, is not orthogonal, therefore, for solving inverse problems, and in Reference [16] the corresponding biorthogonal system was used).

By the same method, we can provide spectral analysis of the operator
$$A_\rho^{[\alpha^{-1},\rho]} u = \frac{1}{\Gamma(\rho^{-1})} \int_0^1 x^{\frac{1}{\rho}-1}(1-t)^{\alpha-1} u(t) dt - \frac{1}{\Gamma(\rho^{-1})} \int_0^x (x-t)^{\frac{1}{\rho}-1} u(t) dt,$$

considered in Reference [13] (and see the references therein).

Theorem 4. *Let $0 < \rho < 2$, $\alpha < \frac{1}{\rho}$. Then, the system of eigenfunctions and associated functions of the operator $A_\rho^{[\alpha^{-1},\rho]}$ is complete in $L_2(0,1)$.*

Proof. We carry out the proof of Theorem 4 in the same way as the proof of Theorem 3. It can easily be shown that the kernel $M(x,t)$ of the operator $A_\rho^{[\alpha^{-1},\rho]}$ is non-negative. Elementary calculations show that the kernel $M^*(x,t)$ of the operator adjoint to the operator $A_\rho^{[\alpha^{-1},\rho]}$ will be non-negative too. Thus $\frac{1}{2}(M + M^*)$ will be non-negative too. The fact that
$$\sum_{j=1}^\infty Re(\frac{1}{\mu_j}) = \int_0^1 Re M(t,t) dt$$

where μ_j are eigenvalues of the operator $A_\rho^{[\alpha^{-1},\rho]}$, shown in the same way as in Theorem 2. □

3. Conclusions

In the present paper, by way of the Livshits Theorem we provide proof of the completeness of the eigenfunctions and associated functions of the operators, generated by the ordinary differential expressions of the fractional order and boundary conditions of Sturm-Liouville type.

Funding: This research received no external funding.

Conflicts of Interest: The author declares no conflict of interest.

References

1. Aleroev, T.S.; Aleroeva, H.T. On a class of non-selfadjoint operators, corresponding to differential equations of fractional order. *Russ. Math.* **2014**, *58*, 3–12. [CrossRef]
2. Aleroev, T.S. Completeness of the system of eigenfunctions of a fractional-order differential operator. *Differ. Equ.* **2000**, *36*, 918—919. [CrossRef]
3. Aleroev, T.S. Boundary-Value Problems for Differential Equations with Fractional Derivatives. Ph.D. Thesis, Moscow State University, Moscow, Russia, 2000.
4. Aleroev, T.S.; Aleroeva, H.T.; Huang, J.; Tamm, M.V.; Tang, Y.; Zhao, Y. Boundary value problems of fractional Fokker-Planck equations. *Comput. Math. Appl.* **2017**, *73*, 959–969. [CrossRef]
5. Livshits, M.S. On spectral decomposition of linear non-selfadjoint operators. *Am. Math. Soc.* **1954**, *34*, 145–199.
6. Dzhrbashian, M.M. The boundary-value problem for a differential fractional-order operator of the Sturm–Liouville type. *Izv. Akad. Nauk ArmSSR Ser. Mat.* **1970**, *5*, 71–96.
7. Agibalova, A.V. On the completeness of a system of eigenfunctions and associated functions of differential operators of the orders $(2 - \varphi)$ and $(1 - \varphi)$. *J. Math. Sci.* **2011**, *174* 425–436. [CrossRef]
8. Agibalova, A.V. On the completeness of the systems of root functions of a fractional-order differential operator with matrix coefficients. *Matematicheskie Zametki*, **2010**, *88*, 317–320. [CrossRef]
9. Malamud, M.M. Similarity of Volterra operators and related problems in the theory of differential equations of fractional orders. (Russian). *Trans. Moscow Math. Soc.* **1994**, *55*, 57–122.
10. Malamud, M.M; Oridoroga, L.L. Analog of the Birkhoff theorem and the completeness results for fractional order differential equations. *Russ. J. Math. Phys.* **2001**, *8*, 287–308.
11. Malamud, M.M; Oridoroga, L.L.On some questions of the spectral theory of ordinary differential fractional-order equation. *Dopov. Nats. Akad. Nauk Ukr. Mat. Prirodozn. Tekh. Nauki* **1998**, *9*, 39–47.
12. Malamud, M.M. Spectral theory of fractional order integration operators, their direct sums, and similarity problem to these operators of their weak perturbations. In *Handbook of Fractional Calculus with Applications. Volume 1: Basic Theory*; Kochubei, A., Luchko, Y., Eds.; Walter de Gruyter GmbH: Berlin, Germany; Boston, MA, USA, 2019.
13. Aleroev, T.S. On one class of operators associated with differential equations of fractional order. *Sib. Math. J.* **2005**, *46*, 963–968. [CrossRef]
14. Aleroev, T.S.; Aleroeva, H.T. Problems of Sturm-Liouville type for differential equations with fractional derivatives. In *Handbook of Fractional Calculus with Applications. Volume 2: Fractional Differential Equations*; Kochubei, A., Luchko, Y., Eds.; Walter de Gruyter GmbH: Berlin, Germany; Boston, MA, USA, 2019.
15. Aleroev, T.S.; Aleroeva, H.T.; Nie, N.-M.; Tang, Y.-F. Boundary value problems for differential equations of fractional order. *Mem. Differ. Equ. Math. Phys.* **2010**, *49*, 19–82. [CrossRef]
16. Ali, M.; Aziz, S.; Malik, S.A. Inverse source problems for a space–time fractional differential equation. *Inverse Probl. Sci. Eng.* **2019**. [CrossRef]
17. Gohberg, I.C.; Krein, M.G. *Introduction to the Theory of Linear Non-selfadjoint Operators*; American Mathematical Society: Providence, RI, USA, 1969.
18. Dzhrbashian, M.M. *Integral Transforms and Representation of Functions in The Comples Domain*; Nauka: Moscow, Russia, 1966.
19. Aleroev, T.S.; Kekharsaeva E.R.Boundary value problems for differential equations with fractional derivatives. *Integral Transform. Spec. Funct.* **2017**, *28*, 900–908. [CrossRef]

20. Lidskii, V.B. Conditions for completeness of a system of root subspaces for non-selfadjoint operators with discrete spectrum. *Tr. Mosk. Mat. Obs.* **1959**, *8*, 83–120.
21. Aleroev, T.; Kirane, M.; Tang, Y.F. Boundary value problems for differential equations of fractional order. *J. Math. Sci.* **2013**, *194*, 499–512. [CrossRef]

© 2019 by the author. Licensee MDPI, Basel, Switzerland. This article is an open access article distributed under the terms and conditions of the Creative Commons Attribution (CC BY) license (http://creativecommons.org/licenses/by/4.0/).

Article

Fractional Whitham–Broer–Kaup Equations within Modified Analytical Approaches

Rasool Shah[1], Hassan Khan[1] and Dumitru Baleanu[2,3,*]

[1] Department of Mathematics, Abdul Wali Khan University, Mardan 23200, Pakistan; rasoolshah@awkum.edu.pk (R.S.); hassanmath@awkum.edu.pk (H.K.)
[2] Department of Mathematics, Faculty of Arts and Sciences, Cankaya University, Ankara 06530, Turkey
[3] Institute of Space Sciences, 077125 Magurele, Romania
* Correspondence: dumitru@cankaya.edu.tr

Received: 12 October 2019; Accepted: 5 November 2019; Published: 7 November 2019

Abstract: The fractional traveling wave solution of important Whitham–Broer–Kaup equations was investigated by using the q-homotopy analysis transform method and natural decomposition method. The Caputo definition of fractional derivatives is used to describe the fractional operator. The obtained results, using the suggested methods are compared with each other as well as with the exact results of the problems. The comparison shows the best agreement of solutions with each other and with the exact solution as well. Moreover, the proposed methods are found to be accurate, effective, and straightforward while dealing with the fractional-order system of partial differential equations and therefore can be generalized to other fractional order complex problems from engineering and science.

Keywords: q-Homotopy analysis transform method; Natural decomposition method; Whitham–Broer–Kaup equations; Caputo derivative

1. Introduction

The modern, broadly considered concept of fractional calculus was developed from a question raised by L'Hospital to Gottfried Wilhelm Leibniz in 1695. L'Hospital insisted on knowing about the outcome of the derivative of order $\alpha = \frac{1}{2}$, which laid down the foundation of a powerful fractional calculus [1,2]. Since then, the new theory of fractional calculus has gained the full attention of mathematicians, physicists, biologists, engineers, and economists in many areas of applied science. In modern decades, researchers have recognized that fractional-order differential equations contributed, in a natural way, to the study of different physical problems, such as diffusion processes, signal processing, viscoelastic systems, control processing, fractional stochastic systems, biology and ecology, quantum mechanics, wave theory, biophysics, and other research fields [3,4].

Partial differential equations (PDEs) involving non-linearities explain different phenomena in applied sciences, technology, and engineering, ranging from gravity to mechanics. In general, non-linear PDEs are important tools that can be used in various fields such as plasma physics, mathematical biology, solid state physics, and fluid dynamics for modeling nonlinear dynamic phenomena [5]. The majority of dynamic schemes can be denoted by an acceptable array of PDEs. It is also well-appreciated that PDEs, such as Poincare and Calabi conjecture models, are utilized to solve mathematical difficulties.

It has been found that the non-linear development of shallow water waves in the fluid dynamics is described by utilizing the coupled scheme Whitham–Broer–Kaup equations (WBKEs) [6]. The coupled scheme of the above equations was developed by Whitham, Broer, and Kaup [7–9]. The above equation defines the propagation of shallow water waves with specific diffusion families.

In the classical order, the major equations of the said phenomena are given as:

$$D_\eta^\delta \mu(\alpha,\eta) + \mu(\alpha,\eta)\frac{\partial \mu(\alpha,\eta)}{\partial \alpha} + \frac{\partial \mu(\alpha,\eta)}{\partial \alpha} + g\frac{\partial v(\alpha,\eta)}{\partial \alpha} = 0,$$

$$D_\eta^\delta v(\alpha,\eta) + \mu(\alpha,\eta)\frac{\partial v(\alpha,\eta)}{\partial \alpha} + v(\alpha,\eta)\frac{\partial \mu(\alpha,\eta)}{\partial \alpha} + p\frac{\partial^3 \mu(\alpha,\eta)}{\partial \alpha^3} - g\frac{\partial^2 v(\alpha,\eta)}{\partial \alpha^2} = 0, \quad 0 < \delta \leq 1.$$

Here $\mu(\alpha,\eta)$ and $v(\alpha,\eta)$ describe the straight velocity and height, which deviate from the equilibrium situation of the fluid, respectively, and p and q are constants expressed in various diffusion forces. Investigating solutions to such nonlinear PDEs over the last several decades it is an important research area [10]. Several scientists have developed numerous mathematical techniques to explore the approximate solutions to nonlinear PDEs. Aminikhah and Biazar [11] used the HPM (homotopy perturbation method) to solve the coupled model of Brusselator and Burger equations. Noor and Mohyud-Din [12] utilized HPM to examine the solutions of different classical orders of PDEs. Ahmad et al. [5] studied a coupled scheme result of WBKEs by the Adomian decomposition method (ADM). Whitham–Broer–Kaup equations are solved by other researchers using different analytical and numerical methods, such as the hyperbolic function method [13], residual power series method (RPSM) [14], Adomian decomposition method [15], reduced differential transformation method [16], homotopy perturbation method [17,18], exp-function method [19], Lie Symmetry analysis [20,21], G'/G^2-Expansion method [22], and homotopy analysis method [23]. Recently, Amjad et al. [10] used the result of a standard order coupled of fractional-order Whitham–Broer–Kaup equation by the Laplace decomposition method.

Singh et al. [24] suggested the q-HATM, which is a well-designed mixture of Laplace transform and q-HAM. The future system monitors and manipulates the sequences result, which converges quickly to the exact solutions for the problem. The strength of the proposed technique is its ability to combine two powerful algorithms to solve both numerically and analytically linear and non-linear fractional-order differential equations. A future procedure has several study properties that include a non-local effect, straightforward result system, promising broad convergence area, and free of any perturbation, discretization and assumption. It is worth disclosing that, using semi-analytical techniques, the Laplace transform takes less C.P.U. time to determine the solutions of complex nonlinear models and phenomena that occur in technology and science. The solution q-HATM includes two auxiliary parameters h and n, which aims to help us modify and control the solution's convergence [25]. Recently, with the help of q-HATM, several researchers studied different phenomena in different fields for example, Singh et al. studied to find the advection-dispersion equation solution [26] and Srivastava et al. used an arbitrary order vibration equation model [27].

The natural decomposition method (NDM) is a mixture of the Adomian decomposition method and the natural transform method (NTM). In 2014, S. Maitama and M. Rawashdeh first implemented the NDM [28,29] to solve linear and non-linear ordinary differential equations (ODEs) and PDEs that occur in several fields of science. A huge quantity of physical models have been studied using NDM, such as the study of fractional order diffusion equations [30], fractional order delay PDEs [31] nonlinear PDEs [32,33], the fractional uncertain flow of a system of polytropic gas [34], fractional-order physical schemes [35], fractional wave and heat problems [36], and fractional telegraph equation [37].

In the current research article, two analytical methods, namely the natural decomposition method and q-homotopy analysis transform method are used to solve the fractional-order Whitham–Broer–Kaup equation. The solutions obtained by the proposed techniques are very simple and straightforward. Moreover, the accuracy of the present methods is sufficient to obtain the analytical solution of the targeted problems. The obtained solutions are compared and to found to be in a good agreement with the exact solution for the problem. This article introduces an approximate analytical solution of a multi dimensional, time fractional model of the Whitham–Broer–Kaup equation by implementing NDM and q-HATM.

2. Preliminaries Concepts

Definition 1. *The Laplace transformation of a Caputo fractional derivative $D^\delta g(\eta)$ is described as:*

$$\pounds[D^\delta g(\eta)] = s^\delta \mathcal{R}(s) - \sum_{j=0}^{m-1} s^{\delta-(j+1)}[g^{(j)}(0^+)] \quad m-1 \leq \delta < m$$

Definition 2. *The natural transformation of the $g(\eta)$ function is represented by $N^+[g(\eta)]$ for $\eta \in R$ and identified by:*

$$N^+[g(\eta)] = \mathcal{R}(s,u) = \int_{-\infty}^{\infty} e^{-s\eta} g(\eta) d\eta; \quad s,u \in (-\infty,\infty),$$

where the natural transformation variables are s and u. If $g(\eta)Q(\eta)$ is described on the real positive axis, the natural transformation is described as:

$$N^+[g(\eta)Q(\eta)] = N^+[g(\eta)] = \mathcal{R}^+(s,u) = \int_0^\infty e^{-s\eta} g(\eta) d\eta; \quad s,u \in (0,\infty), \quad \text{and} \quad \eta \in R$$

where $Q(\eta)$ represents the function of Heaviside. Simply, for $u = 1$, the equation is reduced to the Laplace transformation, and for $s = 1$, the equation is the Sumudu transformation.

Theorem 1. *Let $\mathcal{R}(s,u)$ be the natural transformation of the function $g(\eta)$, then the natural transform $\mathcal{R}_\delta(s,u)$ of the Riemann–Liouville fractional derivative of $g(\eta)$ is symbolized by $D^\delta g(\eta)$ and is presented as:*

$$N^+[D^\delta g(\eta)] = \mathcal{R}_\delta(s,u) = \frac{s^\delta}{u^\delta} \mathcal{R}(s,u) - \sum_{j=0}^{m-1} \frac{s^j}{u^{\delta-j}} [D^{\delta-j-1} g(\eta)]_{\eta=0},$$

where δ is the order and m be any positive integer. Furthermore, $m - 1 \leq \delta < m$.

Theorem 2. *Let $\mathcal{R}(s,u)$ be the natural transformation of the $g(\eta)$, then the natural transformation $\mathcal{R}_\delta(s,u)$ of the Caputo fractional derivative of $g(\eta)$ is symbolized by $D^\delta g(\eta)$ and is represented as:*

$$N^+[^c D^\delta g(\eta)] = \mathcal{R}_\delta^c(s,u) = \frac{s^\delta}{u^\delta} \mathcal{R}(s,u) - \sum_{j=0}^{m-1} \frac{s^{\delta-(j+1)}}{u^{\delta-j}} [D^j g(\eta)]_{\eta=0} \quad m-1 \leq \delta < m$$

Definition 3. *The fractional derivative of $g \in C_{-1}^m$ in the Caputo sense is represented as:*

$$D_\eta^\delta g(\eta) = \begin{cases} \frac{\partial^m g(\eta)}{\partial \eta^m}, & \delta = m \in N, \\ \frac{1}{\Gamma(m-\delta)} \int_0^\eta (\eta-\phi)^{m-\delta-1} g^m(\phi) \partial\phi, & m-1 < \delta < m, \quad m \in N. \end{cases}$$

Definition 4. *Function of Mittag–Leffler, $E_\delta(b)$ for $\delta > 0$ is defined as:*

$$E_\delta(b) = \sum_{m=0}^{\infty} \frac{b^m}{\Gamma(\delta m + 1)} \quad \delta > 0 \quad b \in \mathbb{C},$$

3. The Procedure of NDM

In this section, we describe the NDM solution scheme for fractional partial differential equations.

$$\begin{aligned} D_\eta^\delta \mu(\alpha,\eta) + \mathcal{R}_1(\mu,\nu) + \mathcal{N}_1(\mu,\nu) - \mathcal{P}_1(\alpha,\eta) &= 0, \\ D_\eta^\delta \nu(\alpha,\eta) + \mathcal{R}_2(\mu,\nu) + \mathcal{N}_2(\mu,\nu) - \mathcal{P}_2(\alpha,\eta) &= 0, \quad 0 < \delta \leq 1, \end{aligned} \quad (1)$$

with the initial condition:
$$\mu(\alpha,0) = g_1(\alpha), \quad \nu(\alpha,0) = g_2(\alpha). \tag{2}$$

where is $D_\eta^\delta = \frac{\partial^\delta}{\partial \eta^\delta}$ the Caputo fractional derivative of order δ, $\mathcal{R}_1$, $\mathcal{R}_2$ and $\mathcal{N}_1$, $\mathcal{N}_2$ are linear and non-linear functions or operators, respectively, and $\mathcal{P}_1, \mathcal{P}_2$ are source functions.

Applying the natural transform to Equation (1),
$$\begin{aligned} N^+[D_\eta^\delta \mu(\alpha,\eta)] + N^+[\mathcal{R}_1(\mu,\nu) + \mathcal{N}_1(\mu,\nu) - \mathcal{P}_1(\alpha,\eta)] &= 0, \\ N^+[D_\eta^\delta \nu(\alpha,\eta)] + N^+[\mathcal{R}_2(\mu,\nu) + \mathcal{N}_2(\mu,\nu) - \mathcal{P}_2(\alpha,\eta)] &= 0. \end{aligned} \tag{3}$$

Using the differentiation property of natural transform, we get:
$$\begin{aligned} N^+[\mu(\alpha,\eta)] &= \frac{u^\delta}{s^\delta} \sum_{k=0}^{m-1} \frac{s^{\delta-k-1}}{u^{\delta-k}} \frac{\partial^k \mu(\alpha,\eta)}{\partial^k \eta}\Big|_{\eta=0} + \frac{u^\delta}{s^\delta} N^+[\mathcal{P}_1(\alpha,\eta)] - \frac{u^\delta}{s^\delta} N^+\{\mathcal{R}_1(\mu,\nu) + \mathcal{N}_1(\mu,\nu)\}], \\ N^+[\nu(\alpha,\eta)] &= \frac{u^\delta}{s^\delta} \sum_{k=0}^{m-1} \frac{s^{\delta-k-1}}{u^{\delta-k}} \frac{\partial^k \nu(\alpha,\eta)}{\partial^k \eta}\Big|_{\eta=0} + \frac{u^\delta}{s^\delta} N^+[\mathcal{P}_2(\alpha,\eta)] - \frac{u^\delta}{s^\delta} N^+\{\mathcal{R}_2(\mu,\nu) + \mathcal{N}_2(\mu,\nu)\}], \end{aligned} \tag{4}$$

NDM describes the solution of infinite series $\mu(\alpha,\eta)$ and $\nu(\alpha,\eta)$,
$$\mu(\alpha,\eta) = \sum_{m=0}^{\infty} \mu_m(\alpha,\eta), \quad \nu(\alpha,\eta) = \sum_{m=0}^{\infty} \nu_m(\alpha,\eta), \tag{5}$$

Adomian polynomials of non-linear terms of $\mathcal{N}_1$ and $\mathcal{N}_2$ are represented as:
$$\mathcal{N}_1(\mu,\nu) = \sum_{m=0}^{\infty} \mathcal{A}_m, \quad \mathcal{N}_2(\mu,\nu) = \sum_{m=0}^{\infty} \mathcal{B}_m, \tag{6}$$

All forms of non-linearity of the Adomian polynomials can be defined as:
$$\begin{aligned} \mathcal{A}_m &= \frac{1}{m!} \left[\frac{\partial^m}{\partial \lambda^m} \left\{ \mathcal{N}_1 \left(\sum_{k=0}^{\infty} \lambda^k \mu_k, \sum_{k=0}^{\infty} \lambda^k \nu_k \right) \right\} \right]_{\lambda=0}, \\ \mathcal{B}_m &= \frac{1}{m!} \left[\frac{\partial^m}{\partial \lambda^m} \left\{ \mathcal{N}_2 \left(\sum_{k=0}^{\infty} \lambda^k \mu_k, \sum_{k=0}^{\infty} \lambda^k \nu_k \right) \right\} \right]_{\lambda=0}, \end{aligned} \tag{7}$$

Substituting Equations (13) and (14) into Equation (12) gives:
$$\begin{aligned} N^+[\sum_{m=0}^{\infty} \mu_m(\alpha,\eta)] &= \frac{u^\delta}{s^\delta} \sum_{k=0}^{m-1} \frac{s^{\delta-k-1}}{u^{\delta-k}} \frac{\partial^k \mu(\alpha,\eta)}{\partial^k \eta}\Big|_{\eta=0} + \frac{u^\delta}{s^\delta} N^+\{\mathcal{P}_1(\alpha,\eta)\} - \frac{u^\delta}{s^\delta} N^+\{\mathcal{R}_1(\sum_{m=0}^{\infty} \mu_m, \sum_{m=0}^{\infty} \nu_m) + \sum_{m=0}^{\infty} \mathcal{A}_m\}, \\ N^+[\sum_{m=0}^{\infty} \nu_m(\alpha,\eta)] &= \frac{u^\delta}{s^\delta} \sum_{k=0}^{m-1} \frac{s^{\delta-k-1}}{u^{\delta-k}} \frac{\partial^k \nu(\alpha,\eta)}{\partial^k \eta}\Big|_{\eta=0} + \frac{u^\delta}{s^\delta} N^+\{\mathcal{P}_2(\alpha,\eta)\} - \frac{u^\delta}{s^\delta} N^+\{\mathcal{R}_2(\sum_{m=0}^{\infty} \mu_m, \sum_{m=0}^{\infty} \nu_m) + \sum_{m=0}^{\infty} \mathcal{B}_m\}, \end{aligned} \tag{8}$$

Applying the inverse natural transformation of Equation (16),
$$\begin{aligned} \sum_{m=0}^{\infty} \mu_m(\alpha,\eta) &= N^-[\frac{u^\delta}{s^\delta} \sum_{k=0}^{m-1} \frac{s^{\delta-k-1}}{u^{\delta-k}} \frac{\partial^k \mu(\alpha,\eta)}{\partial^k \eta}\Big|_{\eta=0} + \frac{u^\delta}{s^\delta} N^+\{\mathcal{P}_1(\alpha,\eta)\}] - N^-[\frac{u^\delta}{s^\delta} N^+\{\mathcal{R}_1(\sum_{m=0}^{\infty} \mu_m, \sum_{m=0}^{\infty} \nu_m) + \sum_{m=0}^{\infty} \mathcal{A}_m\}], \\ \sum_{m=0}^{\infty} \nu_m(\alpha,\eta) &= N^-[\frac{u^\delta}{s^\delta} \sum_{k=0}^{m-1} \frac{s^{\delta-k-1}}{u^{\delta-k}} \frac{\partial^k \nu(\alpha,\eta)}{\partial^k \eta}\Big|_{\eta=0} + \frac{u^\delta}{s^\delta} N^+\{\mathcal{P}_2(\alpha,\eta)\}] - N^-[\frac{u^\delta}{s^\delta} N^+\{\mathcal{R}_2(\sum_{m=0}^{\infty} \mu_m, \sum_{m=0}^{\infty} \nu_m) + \sum_{m=0}^{\infty} \mathcal{B}_m\}], \end{aligned} \tag{9}$$

we define the following terms,

$$\mu_0(\alpha,\eta) = N^-[\frac{u^\delta}{s^\delta}\sum_{k=0}^{m-1}\frac{s^{\delta-k-1}}{u^{\delta-k}}\frac{\partial^k\mu(\alpha,\eta)}{\partial^k\eta}|_{\eta=0} + \frac{u^\delta}{s^\delta}N^+\{\mathcal{P}_1(\alpha,\eta)\}],$$
$$\nu_0(\alpha,\eta) = N^-[\frac{u^\delta}{s^\delta}\sum_{k=0}^{m-1}\frac{s^{\delta-k-1}}{u^{\delta-k}}\frac{\partial^k\nu(\alpha,\eta)}{\partial^k\eta}|_{\eta=0} + \frac{u^\delta}{s^\delta}N^+\{\mathcal{P}_2(\alpha,\eta)\}],$$
(10)

$$\mu_1(\alpha,\eta) = -N^-[\frac{u^\delta}{s^\delta}N^+\{\mathcal{R}_1(\mu_0,\nu_0) + \mathcal{A}_0\}],$$
$$\nu_1(\alpha,\eta) = -N^-[\frac{u^\delta}{s^\delta}N^+\{\mathcal{R}_2(\mu_0,\nu_0) + \mathcal{B}_0\}],$$

the general for $m \geq 1$, is given by:

$$\mu_{m+1}(\alpha,\eta) = -N^-[\frac{u^\delta}{s^\delta}N^+\{\mathcal{R}_1(\mu_m,\nu_m) + \mathcal{A}_m\}],$$
$$\nu_{m+1}(\alpha,\eta) = -N^-[\frac{u^\delta}{s^\delta}N^+\{\mathcal{R}_2(\mu_m,\nu_m) + \mathcal{B}_m\}],$$

4. Fundamental Idea of q-Homotopy Analysis Transform Method

To introduce the basic concept of the current method, we consider a fractional-order nonlinear PDEs of the form:

$$D_\eta^\delta \mu(\alpha,\beta,\eta) + R\mu(\alpha,\beta,\eta) + N\mu(\alpha,\beta,\eta) = f(\alpha,\beta,\eta), \quad 1 < \delta \leq n, \quad (11)$$

where $D_\eta^\delta \mu(\alpha,\beta,\eta)$ denote's the Caputo's fractional derivative R and N are linear and non-linear functions or operators. Using the differentiation property of the Laplace transform on Equation (12), we get:

$$s^\delta \pounds[\mu(\alpha,\beta,\eta)] - \sum_{k=0}^{m-1}s^{\delta-k-1}\frac{\partial^k\mu(\alpha,\beta,\eta)}{\partial^k\eta}|_{\eta=0} + \pounds[R\mu(\alpha,\beta,\eta) + N\mu(\alpha,\beta,\eta)] = \pounds[f(\alpha,\beta,\eta)], \quad (12)$$

$$\pounds[D_\eta^\delta \mu(\alpha,\beta,\eta)] = \pounds[\mu(\alpha,\beta,\eta)] - \frac{1}{s^\delta}\sum_{k=0}^{m-1}s^{\delta-k-1}\frac{\partial^k\mu(\alpha,\beta,\eta)}{\partial^k\eta}|_{\eta=0}. \quad (13)$$

On simplifying Equation (13), we have:

$$\pounds\mu(\alpha,\beta,\eta) - \frac{1}{s^\delta}\sum_{k=0}^{m-1}s^{\delta-k-1}\frac{\partial^k\mu(\alpha,\beta,\eta)}{\partial^k\eta}|_{\eta=0} + \frac{1}{s^\delta}\pounds[R\mu(\alpha,\beta,\eta) + N\mu(\alpha,\beta,\eta) - f(\alpha,\beta,\eta)] = 0. \quad (14)$$

We can describe the non-linear operator as:

$$N[\phi(\alpha,\beta,\eta;q)] = \pounds[\phi(\alpha,\beta,\eta;q)] - \frac{1}{s^\delta}\sum_{k=0}^{m-1}s^{\delta-k-1}\frac{\partial^k\phi(\alpha,\beta,\eta:q)}{\partial^k\eta}|_{\eta=0} + \frac{1}{s^\delta}\pounds[R\phi(\alpha,\beta,\eta;q)]$$
$$+ \frac{1}{s^\delta}\pounds[N\phi(\alpha,\beta,\eta;q)] - \frac{1}{s^\delta}\pounds[f(\alpha,\beta,\eta)], \quad (15)$$

where $q \in [0,\frac{1}{n}]$, and $\phi(\alpha,\beta,\eta;q)$ is real function of α, β, η, and q. The concept of a nonzero auxiliary function of homotopy is the following:

$$(1-nq)\pounds[\phi(\alpha,\beta,\eta;q) - \mu_0(\alpha,\beta,\eta)] = \hbar q H(\alpha,\beta,\eta)N[\phi(\alpha,\beta,\eta,q)], \quad (16)$$

where £ a sign of the Laplace transformation, $q \in [0, \frac{1}{n}](n \geq 1)$ is the embedding parameter, $\hbar \neq 0$ is an auxiliary parameter, $H(\alpha, \beta, \eta)$ signifies a nonzero auxiliary function, $\phi(\alpha, \beta, \eta; q)$ is an unidentified function, and $\mu_0(\alpha, \beta, \eta)$ is an initial guess of $\mu(\alpha, \beta, \eta)$. The subsequent outcomes hold correspondingly for $q = 0$ and $q = \frac{1}{n}$.

$$\phi(\alpha, \beta, \eta; 0) = \mu_0(\alpha, \beta, \eta), \phi(\alpha, \beta, \eta, \frac{1}{n}) = \mu(\alpha, \beta, \eta). \tag{17}$$

Thus, by intensifying q from 0 to $\frac{1}{n}$, the result $\phi(\alpha, \beta, \eta; q)$ converge from $\mu_0(\alpha, \beta, \eta)$ to the solution $\mu(\alpha, \beta, \eta)$. Expand the function $\phi(\alpha, \beta, \eta, q)$ in sequences form by using the Taylor theorem near to q, where one can get:

$$\phi(\alpha, \beta, \eta; q) = \mu_0(\alpha, \beta, \eta) + \sum_{m=1}^{\infty} \mu_m(\alpha, \beta, \eta) q^m, \tag{18}$$

where,

$$\mu_0(\alpha, \beta, \eta) = \frac{1}{m!} \frac{\partial^m \phi(\alpha, \beta, \eta; q)}{\partial q^m} \bigg|_{q=0} \tag{19}$$

On selecting the auxiliary linear operator, $\mu_0(\alpha, \beta, \eta)$, n and $\hbar$, the series (19) converge at $q = \frac{1}{n}$ and then it produces one of the results for Equation (12):

$$\mu(\alpha, \beta, \eta) = \mu_0(\alpha, \beta, \eta) + \sum_{m=1}^{\delta} \mu_m(\alpha, \beta, \eta) (\frac{1}{n})^m \tag{20}$$

Now, differentiating the zero-th order distortion Equation (17) m-times with respect to q and then dividing by $m!$ and lastly taking $q = 0$, which provides:

$$£[\mu_m(\alpha, \beta, \eta) - K_m \mu_{m-1}(\alpha, \beta, \eta)] = \hbar \Re_m(\vec{\mu}_{m-1}), \tag{21}$$

where,

$$\vec{\mu_m} = \mu_0(\alpha, \beta, \eta) + \mu_1(\alpha, \beta, \eta)......., \mu_m(\alpha, \beta, \eta). \tag{22}$$

Using the inverse Laplace transformation on Equation (22), it produces:

$$\mu_m(\alpha, \beta, \eta) = K_m \mu_{m-1}(\alpha, \beta, \eta) + \hbar £^{-1}[\Re_m(\vec{\mu}_{m-1})] \tag{23}$$

where,

$$\Re(\vec{\mu}_{m-1}) = £[\mu_{m-1}(\alpha, \beta, \eta)] - (1 - \frac{K_m}{n})(\sum_{k=0}^{m-1} s^{\delta-k-1} \frac{\partial^k \mu(\alpha, \beta, \eta)}{\partial^k \eta} \bigg|_{\eta=0} + \frac{1}{s^\delta} £[f(\alpha, \beta, \eta)])$$
$$+ \frac{1}{s^\delta} £[\Re(\mu_{m-1} + H_{m-1}], \tag{24}$$

And,

$$k_m = \begin{cases} 0, m \leq 1 \\ 1, m > 1 \end{cases} \tag{25}$$

In Equation (25), H_m denotes a homotopy polynomial and is defined as:

$$H_m = \mu_0(\alpha, \beta, \eta) = \frac{1}{m!} \frac{\partial^m \phi(\alpha, \beta, \eta; q)}{\partial q^m} \bigg|_{q=0} \quad and \quad \phi(\alpha, \beta, \eta; q) = \phi_0 + q\phi_1 + q^2 \phi_2 + ... \tag{26}$$

By Equations (24) and (25), we have:

$$\mu_m(\alpha,\beta,\eta) = (k_m + \hbar)\mu_{m-1}(\alpha,\beta,\eta) - (1 - \frac{k_m}{n})\pounds^{-1}[\sum_{k=0}^{m-1} s^{\delta-k-1}\frac{\partial^k \mu(\alpha,\beta,\eta)}{\partial^k \eta}|_{\eta=0} + \frac{1}{s^\delta}\pounds[f(\alpha,\beta,\eta)]] \qquad (27)$$
$$+ \hbar \pounds^{-1}[\frac{1}{s^\delta}\pounds[\Re(\mu_{m-1} + H_{m-1})]].$$

On solving Equation (28) for $m = 1,2,3,4,\ldots\ldots$ with the help of $\mu_0(\alpha,\beta,\eta) = \mu(x,y,0)$ and Equation (25), we get the iterative terms of $\mu_m(\alpha,\beta,\eta)$. The q-homotopy analysis transform method series solution is given by:

$$\mu(\alpha,\beta,\eta) = \sum_{m=0}^{\infty} \mu_m(\alpha,\beta,\eta). \qquad (28)$$

5. Numerical Examples

Example 1. *Consider the coupled system of the fractional-order Whitham–Broer–Kaup equations with:*

$$D_\eta^\delta \mu(\alpha,\eta) + \mu(\alpha,\eta)\frac{\partial \mu(\alpha,\eta)}{\partial \alpha} + \frac{\partial \mu(\alpha,\eta)}{\partial \alpha} + \frac{\partial v(\alpha,\eta)}{\partial \alpha} = 0,$$
$$D_\eta^\delta v(\alpha,\eta) + \mu(\alpha,\eta)\frac{\partial v(\alpha,\eta)}{\partial \alpha} + v(\alpha,\eta)\frac{\partial \mu(\alpha,\eta)}{\partial \alpha} + 3\frac{\partial^3 \mu(\alpha,\eta)}{\partial \alpha^3} - \frac{\partial^2 v(\alpha,\eta)}{\partial \alpha^2} = 0, \qquad (29)$$
$$0 < \delta \le 1, \quad -1 < \eta \le 1, \quad -10 \le \alpha \le 10,$$

with the initial condition:

$$\begin{cases} \mu(\alpha,0) = \frac{1}{2} - 8\tanh(-2\alpha), \\ v(\alpha,0) = 16 - 16\tanh^2(-2\alpha). \end{cases} \qquad (30)$$

*Firstly, we will solve this scheme by using the **NDM**.*

After the natural transformation of Equation (29), we get:

$$N^+\left\{\frac{\partial^\delta \mu(\alpha,\eta)}{\partial \eta^\delta}\right\} = -N^+\left[\mu(\alpha,\eta)\frac{\partial \mu(\alpha,\eta)}{\partial \alpha} + \frac{\partial \mu(\alpha,\eta)}{\partial \alpha} + \frac{\partial v(\alpha,\eta)}{\partial \alpha}\right],$$
$$N^+\left\{\frac{\partial^\delta v(\alpha,\eta)}{\partial \eta^\delta}\right\} = -N^+\left[\mu(\alpha,\eta)\frac{\partial v(\alpha,\eta)}{\partial \alpha} + v(\alpha,\eta)\frac{\partial \mu(\alpha,\eta)}{\partial \alpha} + 3\frac{\partial^3 \mu(\alpha,\eta)}{\partial \alpha^3} - \frac{\partial^2 v(\alpha,\eta)}{\partial \alpha^2}\right],$$
$$\frac{s^\delta}{u^\delta}N^+\{\mu(\alpha,\eta)\} - \frac{s^{\delta-1}}{u^\delta}\mu(\alpha,0) = -N^+\left[\mu(\alpha,\eta)\frac{\partial \mu(\alpha,\eta)}{\partial \alpha} + \frac{\partial \mu(\alpha,\eta)}{\partial \alpha} + \frac{\partial v(\alpha,\eta)}{\partial \alpha}\right]$$
$$\frac{s^\delta}{u^\delta}N^+\{v(\alpha,\eta)\} - \frac{s^{\delta-1}}{u^\delta}v(\alpha,0) = -N^+\left[\mu(\alpha,\eta)\frac{\partial v(\alpha,\eta)}{\partial \alpha} + v(\alpha,\eta)\frac{\partial \mu(\alpha,\eta)}{\partial \alpha} + 3\frac{\partial^3 \mu(\alpha,\eta)}{\partial \alpha^3} - \frac{\partial^2 v(\alpha,\eta)}{\partial \alpha^2}\right],$$

The above algorithm is reduced to be simplified:

$$N^+\{\mu(\alpha,\eta)\} = \frac{1}{s}\{\mu(\alpha,0)\} - \frac{u^\delta}{s^\delta}N^+\left[\mu(\alpha,\eta)\frac{\partial \mu(\alpha,\eta)}{\partial \alpha} + \frac{\partial \mu(\alpha,\eta)}{\partial \alpha} + \frac{\partial v(\alpha,\eta)}{\partial \alpha}\right],$$
$$N^+\{v(\alpha,\eta)\} = \frac{1}{s}\{v(\alpha,0)\} - \frac{u^\delta}{s^\delta}N^+\left[\mu(\alpha,\eta)\frac{\partial v(\alpha,\eta)}{\partial \alpha} + v(\alpha,\eta)\frac{\partial \mu(\alpha,\eta)}{\partial \alpha} + 3\frac{\partial^3 \mu(\alpha,\eta)}{\partial \alpha^3} - \frac{\partial^2 v(\alpha,\eta)}{\partial \alpha^2}\right], \qquad (31)$$

Applying inverse natural transformation, we get:

$$\mu(\alpha,\eta) = \mu(\alpha,0) - N^-\left[\frac{u^\delta}{s^\delta}N^+\left[\mu(\alpha,\eta)\frac{\partial \mu(\alpha,\eta)}{\partial \alpha} + \frac{\partial \mu(\alpha,\eta)}{\partial \alpha} + \frac{\partial v(\alpha,\eta)}{\partial \alpha}\right]\right],$$
$$v(\alpha,\eta) = v(\alpha,0) - N^-\left[\frac{u^\delta}{s^\delta}N^+\left[\mu(\alpha,\eta)\frac{\partial v(\alpha,\eta)}{\partial \alpha} + v(\alpha,\eta)\frac{\partial \mu(\alpha,\eta)}{\partial \alpha} + 3\frac{\partial^3 \mu(\alpha,\eta)}{\partial \alpha^3} - \frac{\partial^2 v(\alpha,\eta)}{\partial \alpha^2}\right]\right], \qquad (32)$$

Assume that the unknown functions $\mu(\alpha,\eta)$ and $\nu(\alpha,\eta)$ infinite series solution is as follows:

$$\mu(\alpha,\eta) = \sum_{m=0}^{\infty} \mu_m(\alpha,\eta), \text{ and } \nu(\alpha,\eta) = \sum_{m=0}^{\infty} \nu_m(\alpha,\eta)$$

Remember that $\mu\mu_\alpha = \sum_{m=0}^{\infty} \mathcal{A}_m$, $\mu\nu_\alpha = \sum_{m=0}^{\infty} \mathcal{B}_m$ and $\nu\mu_\alpha = \sum_{m=0}^{\infty} \mathcal{C}_m$ are the Adomian polynomials and the nonlinear terms were characterized. Using such terms, Equation (32) can be rewritten in the form:

$$\sum_{m=0}^{\infty} \mu_m(\alpha,\eta) = \mu(\alpha,0) - N^- \left[\frac{u^\delta}{s^\delta} N^+ \left[\sum_{m=0}^{\infty} \mathcal{A}_m + \frac{\partial \mu(\alpha,\eta)}{\partial \alpha} + \frac{\partial \nu(\alpha,\eta)}{\partial \alpha} \right] \right],$$

$$\sum_{m=0}^{\infty} \nu_m(\alpha,\eta) = \nu(\alpha,0) - N^- \left[\frac{u^\delta}{s^\delta} N^+ \left[\sum_{m=0}^{\infty} \mathcal{B}_m + \sum_{m=0}^{\infty} \mathcal{C}_m + 3\frac{\partial^3 \mu(\alpha,\eta)}{\partial \alpha^3} - \frac{\partial^2 \nu(\alpha,\eta)}{\partial \alpha^2} \right] \right],$$

$$\sum_{m=0}^{\infty} \mu_m(\alpha,\eta) = \frac{1}{2} - 8\tanh(-2\alpha) - N^- \left[\frac{u^\delta}{s^\delta} N^+ \left[\sum_{m=0}^{\infty} \mathcal{A}_m + \frac{\partial \mu(\alpha,\eta)}{\partial \alpha} + \frac{\partial \nu(\alpha,\eta)}{\partial \alpha} \right] \right], \quad (33)$$

$$\sum_{m=0}^{\infty} \nu_m(\alpha,\eta) = 16 - 16\tanh^2(-2\alpha) - N^- \left[\frac{u^\delta}{s^\delta} N^+ \left[\sum_{m=0}^{\infty} \mathcal{B}_m + \sum_{m=0}^{\infty} \mathcal{C}_m + 3\frac{\partial^3 \mu(\alpha,\eta)}{\partial \alpha^3} - \frac{\partial^2 \nu(\alpha,\eta)}{\partial \alpha^2} \right] \right],$$

According to Equation (7), all forms of non-linearity the Adomian polynomials can be defined as:

$$\mathcal{A}_0 = \mu_0 \frac{\partial \mu_0}{\partial \alpha}, \quad \mathcal{A}_1 = \mu_0 \frac{\partial \mu_1}{\partial \alpha} + \mu_1 \frac{\partial \mu_0}{\partial \alpha}, \quad \mathcal{B}_0 = \mu_0 \frac{\partial \nu_0}{\partial \beta}, \quad \mathcal{B}_1 = \mu_0 \frac{\partial \nu_1}{\partial \beta} + \mu_1 \frac{\partial \nu_0}{\partial \beta},$$

$$\mathcal{C}_0 = \nu_0 \frac{\partial \mu_0}{\partial \alpha}, \quad \mathcal{C}_1 = \nu_0 \frac{\partial \mu_1}{\partial \alpha} + \nu_1 \frac{\partial \mu_0}{\partial \alpha},$$

Thus, we can easily obtain the recursive relationship by comparing two sides of Equation (33):

$$\mu_0(\alpha,\eta) = \frac{1}{2} - 8\tanh(-2\alpha), \quad \nu_0(\alpha,\eta) = 16 - 16\tanh^2(-2\alpha),$$

For $m = 0$,

$$\mu_1(\alpha,\eta) = -8\operatorname{sech}^2(-2\alpha)\frac{\eta^\delta}{\Gamma(\delta+1)}, \quad \nu_1(\alpha,\eta) = -32\operatorname{sech}^2(-2\alpha)\tanh(-2\alpha)\frac{\eta^\delta}{\Gamma(\delta+1)},$$

For $m = 1$,

$$\mu_2(\alpha,\eta) = -16\operatorname{sech}^2(-2\alpha)\left(4\operatorname{sech}^2(-2\alpha) - 8\tanh^2(-2\alpha) + 3\tanh(-2\alpha)\right)\frac{\eta^{2\delta}}{\Gamma(2\delta+1)},$$

$$\nu_2(\alpha,\eta) = -32\operatorname{sech}^2(-2\alpha)\{40\operatorname{sech}^2(-2\alpha)\tanh(-2\alpha) + 96\tanh(-2\alpha) - 2\tanh^2(-2\alpha) - 32\tanh^3(-2\alpha)$$

$$- 25\operatorname{sech}^2(-2\alpha)\}\frac{\eta^{2\delta}}{\Gamma(2\delta+1)},$$

In the same procedure, the remaining μ_m and ν_m ($m \geq 2$) components of the NDM solution can be obtained smoothly. We therefore determine the sequence of alternatives as:

$$\mu(\alpha,\eta) = \sum_{m=0}^{\infty} \mu_m(\alpha,\beta) = \mu_0(\alpha,\beta) + \mu_1(\alpha,\beta) + \mu_2(\alpha,\beta) + \mu_3(\alpha,\beta) + \cdots$$

$$\nu(\alpha,\eta) = \sum_{m=0}^{\infty} \nu_m(\alpha,\beta) = \nu_0(\alpha,\beta) + \nu_1(\alpha,\beta) + \nu_2(\alpha,\beta) + \nu_3(\alpha,\beta) + \cdots$$

$$\mu(\alpha,\eta) = \frac{1}{2} - 8\tanh(-2\alpha) - 8\sec h^2(-2\alpha)\frac{\eta^\delta}{\Gamma(\delta+1)}$$

$$- 16\sec h^2(-2\alpha)\left(4\sec h^2(-2\alpha) - 8\tanh^2(-2\alpha) + 3\tanh(-2\alpha)\right)\frac{\eta^{2\delta}}{\Gamma(2\delta+1)} - \cdots$$

$$\nu(\alpha,\eta) = 16 - 16\tanh^2(-2\alpha) - 32\sec h^2(-2\alpha)\tanh(-2\alpha)\frac{\eta^\delta}{\Gamma(\delta+1)}$$

$$- 32\sec h^2(-2\alpha)\{40\sec h^2(-2\alpha)\tanh(-2\alpha) + 96\tanh(-2\alpha) - 2\tanh^2(-2\alpha) - 32\tanh^3(-2\alpha)$$

$$- 25\sec h^2(-2\alpha)\}\frac{\eta^{2\delta}}{\Gamma(2\delta+1)} - \cdots$$

In Figures 1 and 2, the exact and natural decomposition method (NDM) solutions at an integer-order $\delta = 1$ are represented for both $\mu(\alpha,\eta)$ and $\nu(\alpha,\eta)$ of Example 1. It is observed that NDM solutions are in good contact with the exact solution of the problems. In Figures 3 and 4, various fractional-order solutions of Example 1, at different fractional-orders, $\delta = 1, 0.8, 0.6, 0.4$ and $\eta = 1$ are plotted. It is investigated that for Example 1, the fractional-order solutions are convergent to an integer-order solution for both $\mu(\alpha,\eta)$ and $\nu(\alpha,\eta)$.

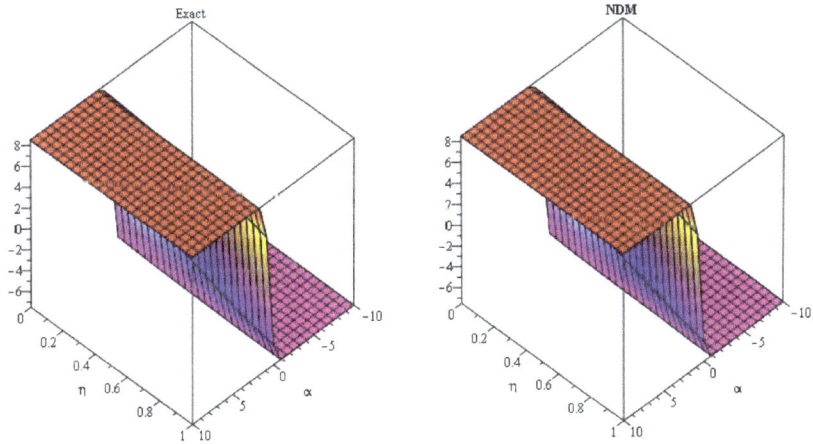

Figure 1. Exact and NDM solution of $\mu(\alpha,\eta)$ at $\delta = 1$.

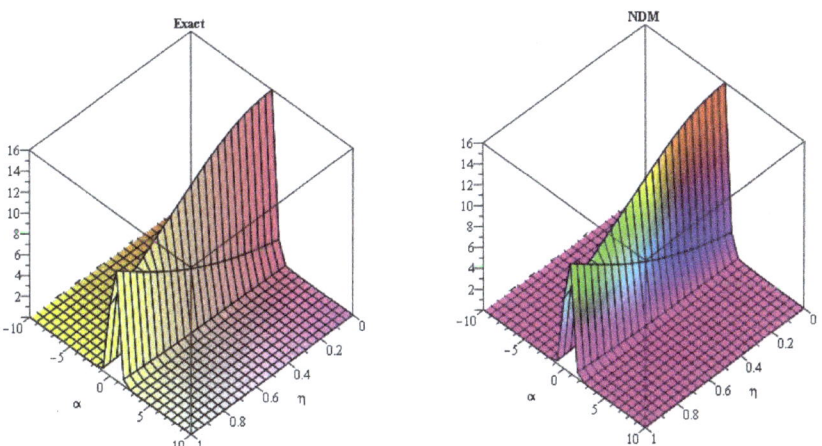

Figure 2. Exact and NDM solution of $\nu(\alpha,\eta)$ at $\delta = 1$.

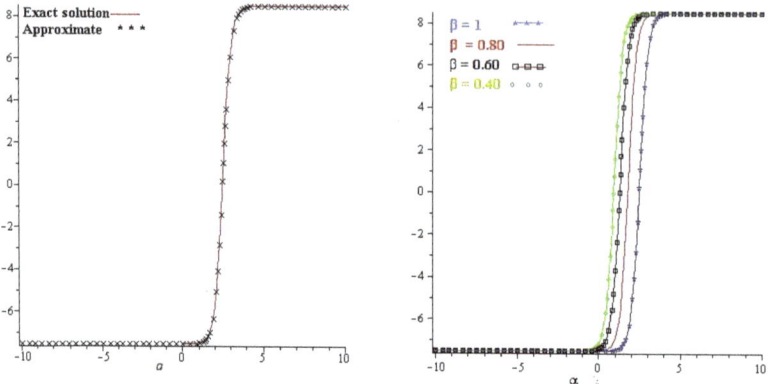

Figure 3. Exact and NDM solution of $\mu(\alpha, \eta)$ at different fractional order $\delta = 1, 0.8, 0.6, 0.4$, and $\eta = 1$.

Figure 4. Exact and NDM solution of $\nu(\alpha, \eta)$ at different fractional order $\delta = 1, 0.8, 0.6, 0.4$, and $\eta = 1$.

5.1. q-Homotopy Analysis Transform Method

The Example 1 approximate solution with the help of **q-HATM**.
After the Laplace transformation of Equation (29), we get:

$$£\{\mu(\alpha,\eta)\} = \frac{1}{s}\{\mu(\alpha,0)\} - \frac{1}{s^\delta}£\left[\mu(\alpha,\eta)\frac{\partial\mu(\alpha,\eta)}{\partial\alpha} + \frac{\partial\mu(\alpha,\eta)}{\partial\alpha} + \frac{\partial\nu(\alpha,\eta)}{\partial\alpha}\right],$$
$$£\{\nu(\alpha,\eta)\} = \frac{1}{s}\{\nu(\alpha,0)\} - \frac{1}{s^\delta}£\left[\mu(\alpha,\eta)\frac{\partial\nu(\alpha,\eta)}{\partial\alpha} + \nu(\alpha,\eta)\frac{\partial\mu(\alpha,\eta)}{\partial\alpha} + 3\frac{\partial^3\mu(\alpha,\eta)}{\partial\alpha^3} - \frac{\partial^2\nu(\alpha,\eta)}{\partial\alpha^2}\right].$$
(34)

By the help of Equation (34) we define the nonlinear operator as:

$$N^1[\phi_1(\alpha,\eta;q),\phi_2(\alpha,\eta;q)]$$
$$= £\left[\phi_1(\alpha,\eta;q) - \frac{1}{s}\{\phi_1(\alpha,0)\} + \frac{1}{s^\delta}\left\{\phi_1(\alpha,\eta)\frac{\partial\phi_1(\alpha,\eta)}{\partial\alpha} + \frac{\partial\phi_1(\alpha,\eta)}{\partial\alpha} + \frac{\partial\phi_2(\alpha,\eta)}{\partial\alpha}\right\}\right],$$
$$N^2[\phi_1(\alpha,\eta;q),\phi_2(\alpha,\eta;q)]$$
$$= £\left[\phi_2(\alpha,\eta;q) - \frac{1}{s}\{\phi_2(\alpha,0)\} + \frac{1}{s^\delta}\left\{\phi_1(\alpha,\eta)\frac{\partial\phi_2(\alpha,\eta)}{\partial\alpha} + \phi_2(\alpha,\eta)\frac{\partial\phi_1(\alpha,\eta)}{\partial\alpha} + 3\frac{\partial^3\phi_1(\alpha,\eta)}{\partial\alpha^3} - \frac{\partial^2\phi_2(\alpha,\eta)}{\partial\alpha^2}\right\}\right].$$
(35)

By applying proposed algorithm, the deformation equation of m-th order is given as:

$$\pounds[\mu_m(\alpha,\eta) - K_m\mu_{m-1}(\alpha,\eta)] = h\Re_{1,m}[\overrightarrow{\mu}_{m-1}, \overrightarrow{v}_{m-1}],$$
$$\pounds[v_m(\alpha,\eta) - K_m v_{m-1}(\alpha,\eta)] = h\Re_{2,m}[\overrightarrow{\mu}_{m-1}, \overrightarrow{v}_{m-1}],$$
(36)

$$\Re_{1,m}[\overrightarrow{\mu}_{m-1}, \overrightarrow{v}_{m-1}] = \pounds[\mu_{m-1}(\alpha,\eta)] - (1-\frac{k_m}{n})\frac{1}{s}\{\frac{1}{2} - 8\tanh(-2\alpha)\}$$
$$+ \frac{1}{s^\delta}\{\sum_{j=0}^{m-1}\mu_j(\alpha,\eta)\frac{\partial\mu_{m-1-j}(\alpha,\eta)}{\partial\alpha} + \frac{\partial\mu_{m-1}(\alpha,\eta)}{\partial\alpha} + \frac{\partial v_{m-1}(\alpha,\eta)}{\partial\alpha}\}],$$

$$\Re_{2,m}[\overrightarrow{\mu}_{m-1}, \overrightarrow{v}_{m-1}] = \pounds[v_{m-1}(\alpha,\eta)] - \frac{1}{s}\{16 - 16\tanh^2(-2\alpha)\}$$
$$+ \frac{1}{s^\delta}\{\sum_{j=0}^{m-1}\mu_j(\alpha,\eta)\frac{\partial v_{m-1-j}(\alpha,\eta)}{\partial\alpha} + \sum_{j=0}^{m-1}v_j(\alpha,\eta)\frac{\partial\mu_{m-j-1}(\alpha,\eta)}{\partial\alpha} + 3\frac{\partial^3\mu_{m-1}(\alpha,\eta)}{\partial\alpha^3} - \frac{\partial^2 v_{m-1}(\alpha,\eta)}{\partial\alpha^2}\}].$$
(37)

By applying inverse Laplace transform on Equation (36), we get:

$$\mu_m(\alpha,\eta) = K_m\mu_{m-1}(\alpha,\eta) + hL^{-1}\Re_{1,m}[\overrightarrow{\mu}_{m-1}, \overrightarrow{v}_{m-1}],$$
$$v_m(\alpha,\eta) = K_m v_{m-1}(\alpha,\eta)] + hL^{-1}\Re_{2,m}[\overrightarrow{\mu}_{m-1}, \overrightarrow{v}_{m-1}],$$
(38)

By the help of given initial condition, we have:

$$\mu_0(\alpha,\eta) = \frac{1}{2} - 8\tanh(-2\alpha),$$
$$v_0(\alpha,\eta) = 16 - 16\tanh^2(-2\alpha).$$
(39)

To find the value of $\mu_0(\alpha,\eta)$ and $v_0(\alpha,\eta)$, set $m = 1$ in Equation (38), then we get:

$$\mu_1(\alpha,\eta) = K_1\mu_0(\alpha,\eta) + h\pounds^{-1}\Re_{1,1}[\overrightarrow{\mu}_0, \overrightarrow{v}_0],$$
$$v_1(\alpha,\eta) = K_1 v_0(\alpha,\eta)] + h\pounds^{-1}\Re_{2,1}[\overrightarrow{\mu}_0, \overrightarrow{v}_0].$$
(40)

From Equation (37) for $m = 1$, we get:

$$\Re_{1,1}[\overrightarrow{\mu}_0, \overrightarrow{v}_0] = \pounds[\mu_0(\alpha,\eta)] - (1-\frac{k_1}{n})\frac{1}{s}\{\frac{1}{2} - 8\tanh(-2\alpha)\}$$
$$+ \frac{1}{s^\delta}\pounds[\{\mu_0(\alpha,\eta)\frac{\partial\mu_0(\alpha,\eta)}{\partial\alpha} + \frac{\partial\mu_0(\alpha,\eta)}{\partial\alpha} + \frac{\partial v_0(\alpha,\eta)}{\partial\alpha}\}],$$

$$\Re_{2,1}[\overrightarrow{\mu}_0, \overrightarrow{v}_0] = \pounds[v_0(\alpha,\eta)] - (1-\frac{k_1}{n})\frac{1}{s}\{16 - 16\tanh^2(-2\alpha)\}$$
$$+ \frac{1}{s^\delta}\pounds[\{\mu_0(\alpha,\eta)\frac{\partial v_0(\alpha,\eta)}{\partial\alpha} + v_0(\alpha,\eta)\frac{\partial\mu_0(\alpha,\eta)}{\partial\alpha} + 3\frac{\partial^3\mu_0(\alpha,\eta)}{\partial\alpha^3} - \frac{\partial^2 v_0(\alpha,\eta)}{\partial\alpha^2}\}].$$
(41)

Then by using Equations (25) and (41) in Equation (40), we get:

$$\mu_1(\alpha,\eta) = h\pounds^{-1}[\frac{1}{s}\{\frac{1}{2} - 8\tanh(-2\alpha)\} - (1-\frac{0}{n})\frac{1}{s}\{\frac{1}{2} - 8\tanh(-2\alpha)\}$$
$$+ \frac{1}{s^\delta}\pounds[\{\mu_0(\alpha,\eta)\frac{\partial\mu_0(\alpha,\eta)}{\partial\alpha} + \frac{\partial\mu_0(\alpha,\eta)}{\partial\alpha} + \frac{\partial v_0(\alpha,\eta)}{\partial\alpha}\}]],$$

$$v_1(\alpha,\eta) = h\pounds^{-1}[\frac{1}{s}\{16 - 16\tanh^2(-2\alpha)\} - (1-\frac{0}{n})\frac{1}{s}\{16 - 16\tanh^2(-2\alpha)\}$$
$$+ \frac{1}{s^\delta}\pounds[\{\mu_0(\alpha,\eta)\frac{\partial v_0(\alpha,\eta)}{\partial\alpha} + v_0(\alpha,\eta)\frac{\partial\mu_0(\alpha,\eta)}{\partial\alpha} + 3\frac{\partial^3\mu_0(\alpha,\eta)}{\partial\alpha^3} - \frac{\partial^2 v_0(\alpha,\eta)}{\partial\alpha^2}\}]],$$

$$\mu_1(\alpha,\eta) = -8h\sec h^2(-2\alpha)\frac{\eta^\delta}{\Gamma(\delta+1)}, \quad v_1(\alpha,\eta) = -32h\sec h^2(-2\alpha)\tanh(-2\alpha)\frac{\eta^\delta}{\Gamma(\delta+1)}.$$
(42)

Similarly from Equations (40) and (41) for m = 2, we have:

$$\mu_2(\alpha,\eta) = n\mu_1(\alpha,\eta) + h\pounds^{-1}[\pounds[\mu_1(\alpha,\eta)] - (1-\frac{n}{n})\frac{1}{s}\{\frac{1}{2} - 8\tanh(-2\alpha)\} + \frac{1}{s^\delta}\pounds[\{\mu_0(\alpha,\eta)\frac{\partial\mu_1(\alpha,\eta)}{\partial\alpha}$$
$$+ \mu_1(\alpha,\eta)\frac{\partial\mu_0(\alpha,\eta)}{\partial\alpha} + \frac{\partial\mu_1(\alpha,\eta)}{\partial\alpha} + \frac{\partial v_1(\alpha,\eta)}{\partial\alpha}\}]], \quad (43)$$

$$v_2(\alpha,\eta) = nv_1(\alpha,\eta) + h\pounds^{-1}[\pounds[\mu_1(\alpha,\eta)] - (1-\frac{n}{n})\frac{1}{s}\{16 - 16\tanh^2(-2\alpha)\} + \frac{1}{s^\delta}\pounds[\{\mu_0(\alpha,\eta)\frac{\partial v_1(\alpha,\eta)}{\partial\alpha}$$
$$+ \mu_1(\alpha,\eta)\frac{\partial v_0(\alpha,\eta)}{\partial\alpha} + v_0(\alpha,\eta)\frac{\partial\mu_1(\alpha,\eta)}{\partial\alpha} + v_1(\alpha,\eta)\frac{\partial\mu_0(\alpha,\eta)}{\partial\alpha} + 3\frac{\partial^3\mu_1(\alpha,\eta)}{\partial\alpha^3} - \frac{\partial^2 v_1(\alpha,\eta)}{\partial\alpha^2}\}]].$$

In the case of simplified, the above calculation eliminates as described:

$$\mu_2(\alpha,\eta) = -8(n+h)h\sec h^2(-2\alpha)\frac{\eta^\delta}{\Gamma(\delta+1)} - 16h^2\sec h^2(-2\alpha)(4\sec h^2(-2\alpha)$$
$$- 8\tanh^2(-2\alpha) + 3\tanh(-2\alpha))\frac{\eta^{2\delta}}{\Gamma(2\delta+1)},$$

$$v_2(\alpha,\eta) = -32(n+h)h\sec h^2(-2\alpha)\tanh(-2\alpha)\frac{\eta^\delta}{\Gamma(\delta+1)} - 32h^2\sec h^2(-2\alpha)\{40\sec h^2(-2\alpha)\tanh(-2\alpha)$$
$$+ 96\tanh(-2\alpha) - 2\tanh^2(-2\alpha) - 32\tanh^3(-2\alpha) - 25\sec h^2(-2\alpha)\}\frac{\eta^{2\delta}}{\Gamma(2\delta+1)},$$

The rest of the iterative terms can be used in the same way. Formerly, the family of q-homotopy analysis transform technique series result of Equation (29) is assumed by:

$$\mu(\alpha,\eta) = \mu_0(\alpha,\eta) + \sum_{m=1}^{\infty}\mu_m(\alpha,\eta)(\frac{1}{n})^m,$$
$$v(\alpha,\eta) = v_0(\alpha,\eta) + \sum_{m=1}^{\infty}v_m(\alpha,\eta)(\frac{1}{n})^m, \quad (44)$$

The exact solution of Equation (29) at $\delta = 1$,

$$\mu(\alpha,\eta) = \frac{1}{2} - 8\tanh\left\{-2\left(\alpha - \frac{\eta}{2}\right)\right\},$$
$$v(\alpha,\eta) = 16 - 16\tanh^2\left\{-2\left(\alpha - \frac{\eta}{2}\right)\right\}. \quad (45)$$

In Figure 5, the graph of exact and q-HATM solutions for $\mu(\alpha,\eta)$ of Example 1 are displayed. It is observed that, the solutions of q-HATM are in good agreement with the exact and NDM solutions. Similarly Figure 6, express the exact and q-HATM solutions for $v(\alpha,\eta)$. The plot representation also confirmed the higher accuracy of the proposed method with the exact solution for $v(\alpha,\eta)$. Furthermore, the graphical representations of the solutions of the proposed method have reflected its applicability and reliability. This provides the motivation to apply the current techniques for other fractional-order partial differential equations.

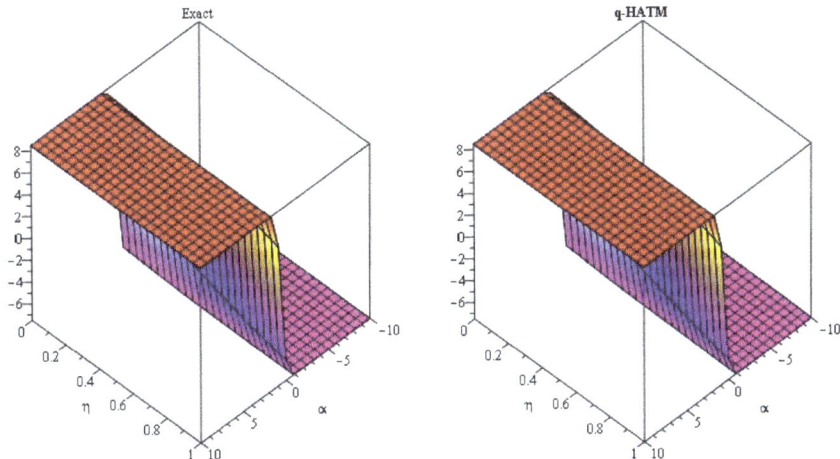

Figure 5. Exact and q-HATM solution of $\mu(\alpha, \eta)$ at $\delta = 1$.

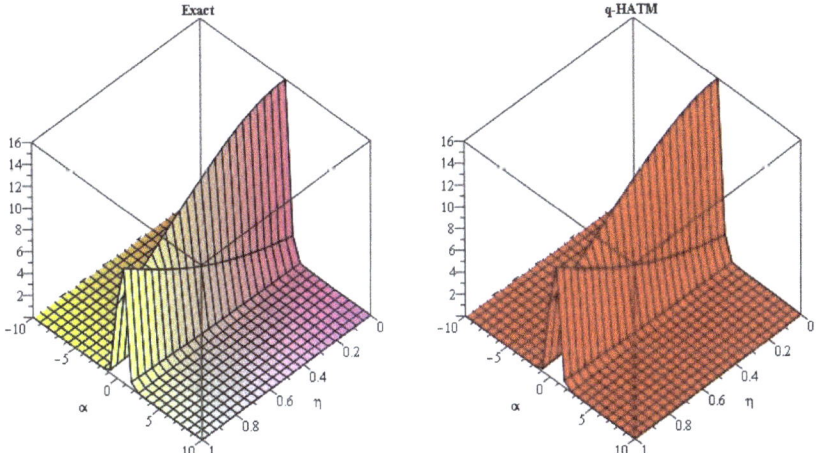

Figure 6. Exact and q-HATM solution of $\nu(\alpha, \eta)$ at $\delta = 1$.

Example 2. *Consider the coupled system of fractional-order Whitham–Broer–Kaup equations with:*

$$D^\delta_\eta \mu(\alpha, \eta) + \mu(\alpha, \eta)\frac{\partial \mu(\alpha, \eta)}{\partial \alpha} + \frac{1}{2}\frac{\partial \mu(\alpha, \eta)}{\partial \alpha} + \frac{\partial \nu(\alpha, \eta)}{\partial \alpha} = 0,$$
$$D^\delta_\eta \nu(\alpha, \eta) + \mu(\alpha, \eta)\frac{\partial \nu(\alpha, \eta)}{\partial \alpha} + \nu(\alpha, \eta)\frac{\partial \mu(\alpha, \eta)}{\partial \alpha} - \frac{1}{2}\frac{\partial^2 \nu(\alpha, \eta)}{\partial \alpha^2} = 0, \qquad (46)$$
$$0 < \delta \leq 1, \quad 0 < \eta \leq 1, \quad -100 \leq \alpha \leq 100,$$

with the initial condition:

$$\begin{cases} \mu(\alpha, 0) = \xi - \kappa \coth[\kappa(\alpha + \theta)], \\ \nu(\alpha, 0) = -\kappa^2 \operatorname{cosech}^2[\kappa(\alpha + \theta)]. \end{cases} \qquad (47)$$

*Firstly, we will solve this scheme by using the **NDM**.*

After the natural transformation of Equation (46), we get:

$$N^+\left\{\frac{\partial^\delta \mu(\alpha,\eta)}{\partial \eta^\delta}\right\} = -N^+\left[\mu(\alpha,\eta)\frac{\partial \mu(\alpha,\eta)}{\partial \alpha} + \frac{1}{2}\frac{\partial \mu(\alpha,\eta)}{\partial \alpha} + \frac{\partial v(\alpha,\eta)}{\partial \alpha}\right],$$

$$N^+\left\{\frac{\partial^\delta v(\alpha,\eta)}{\partial \eta^\delta}\right\} = -N^+\left[\mu(\alpha,\eta)\frac{\partial v(\alpha,\eta)}{\partial \alpha} + v(\alpha,\eta)\frac{\partial \mu(\alpha,\eta)}{\partial \alpha} - \frac{1}{2}\frac{\partial^2 v(\alpha,\eta)}{\partial \alpha^2}\right],$$

$$\frac{s^\delta}{u^\delta}N^+\{\mu(\alpha,\eta)\} - \frac{s^{\delta-1}}{u^\delta}\mu(\alpha,0) = -N^+\left[\mu(\alpha,\eta)\frac{\partial \mu(\alpha,\eta)}{\partial \alpha} + \frac{1}{2}\frac{\partial \mu(\alpha,\eta)}{\partial \alpha} + \frac{\partial v(\alpha,\eta)}{\partial \alpha}\right]$$

$$\frac{s^\delta}{u^\delta}N^+\{v(\alpha,\eta)\} - \frac{s^{\delta-1}}{u^\delta}v(\alpha,0) = -N^+\left[\mu(\alpha,\eta)\frac{\partial v(\alpha,\eta)}{\partial \alpha} + v(\alpha,\eta)\frac{\partial \mu(\alpha,\eta)}{\partial \alpha} - \frac{1}{2}\frac{\partial^2 v(\alpha,\eta)}{\partial \alpha^2}\right],$$

The above algorithm is reduced to the simplified form as:

$$N^+\{\mu(\alpha,\eta)\} = \frac{1}{s}\{\mu(\alpha,0)\} - \frac{u^\delta}{s^\delta}N^+\left[\mu(\alpha,\eta)\frac{\partial \mu(\alpha,\eta)}{\partial \alpha} + \frac{1}{2}\frac{\partial \mu(\alpha,\eta)}{\partial \alpha} + \frac{\partial v(\alpha,\eta)}{\partial \alpha}\right],$$

$$N^+\{v(\alpha,\eta)\} = \frac{1}{s}\{v(\alpha,0)\} - \frac{u^\delta}{s^\delta}N^+\left[\mu(\alpha,\eta)\frac{\partial v(\alpha,\eta)}{\partial \alpha} + v(\alpha,\eta)\frac{\partial \mu(\alpha,\eta)}{\partial \alpha} - \frac{1}{2}\frac{\partial^2 v(\alpha,\eta)}{\partial \alpha^2}\right],$$

(48)

Applying the inverse natural transformation, we get:

$$\mu(\alpha,\eta) = \mu(\alpha,0) - N^-\left[\frac{u^\delta}{s^\delta}N^+\left[\mu(\alpha,\eta)\frac{\partial \mu(\alpha,\eta)}{\partial \alpha} + \frac{1}{2}\frac{\partial \mu(\alpha,\eta)}{\partial \alpha} + \frac{\partial v(\alpha,\eta)}{\partial \alpha}\right]\right],$$

$$v(\alpha,\eta) = v(\alpha,0) - N^-\left[\frac{u^\delta}{s^\delta}N^+\left[\mu(\alpha,\eta)\frac{\partial v(\alpha,\eta)}{\partial \alpha} + v(\alpha,\eta)\frac{\partial \mu(\alpha,\eta)}{\partial \alpha} - \frac{1}{2}\frac{\partial^2 v(\alpha,\eta)}{\partial \alpha^2}\right]\right],$$

(49)

Assume that the unknown functions $\mu(\alpha,\eta)$ and $v(\alpha,\eta)$ infinite series solution is as follows:

$$\mu(\alpha,\eta) = \sum_{m=0}^{\infty}\mu_m(\alpha,\eta), \quad \text{and} \quad v(\alpha,\eta) = \sum_{m=0}^{\infty}v_m(\alpha,\eta),$$

Remember that $\mu\mu_\alpha = \sum_{m=0}^{\infty}\mathcal{A}_m$, $\mu v_\alpha = \sum_{m=0}^{\infty}\mathcal{B}_m$ and $v\mu_\alpha = \sum_{m=0}^{\infty}\mathcal{C}_m$ are the Adomian polynomials and the nonlinear terms were characterized. Using such terms, Equation (49) can be rewritten in the form:

$$\sum_{m=0}^{\infty}\mu_m(\alpha,\eta) = \mu(\alpha,0) - N^-\left[\frac{u^\delta}{s^\delta}N^+\left[\sum_{m=0}^{\infty}\mathcal{A}_m + \frac{1}{2}\frac{\partial \mu(\alpha,\eta)}{\partial \alpha} + \frac{\partial v(\alpha,\eta)}{\partial \alpha}\right]\right],$$

$$\sum_{m=0}^{\infty}v_m(\alpha,\eta) = v(\alpha,0) - N^-\left[\frac{u^\delta}{s^\delta}N^+\left[\sum_{m=0}^{\infty}\mathcal{B}_m + \sum_{m=0}^{\infty}\mathcal{C}_m - \frac{1}{2}\frac{\partial^2 v(\alpha,\eta)}{\partial \alpha^2}\right]\right],$$

$$\sum_{m=0}^{\infty}\mu_m(\alpha,\eta) = \xi - \kappa\coth[\kappa(\alpha+\theta)] - N^-\left[\frac{u^\delta}{s^\delta}N^+\left[\sum_{m=0}^{\infty}\mathcal{A}_m + \frac{1}{2}\frac{\partial \mu(\alpha,\eta)}{\partial \alpha} + \frac{\partial v(\alpha,\eta)}{\partial \alpha}\right]\right],$$

$$\sum_{m=0}^{\infty}v_m(\alpha,\eta) = -\kappa^2\operatorname{cosech}^2[\kappa(\alpha+\theta)] - N^-\left[\frac{u^\delta}{s^\delta}N^+\left[\sum_{m=0}^{\infty}\mathcal{B}_m + \sum_{m=0}^{\infty}\mathcal{C}_m - \frac{1}{2}\frac{\partial^2 v(\alpha,\eta)}{\partial \alpha^2}\right]\right],$$

(50)

According to Equation (7), all forms of non-linearity the Adomian polynomials can be defined as:

$$\mathcal{A}_0 = \mu_0\frac{\partial \mu_0}{\partial \alpha}, \quad \mathcal{A}_1 = \mu_0\frac{\partial \mu_1}{\partial \alpha} + \mu_1\frac{\partial \mu_0}{\partial \alpha}, \quad \mathcal{B}_0 = \mu_0\frac{\partial v_0}{\partial \beta}, \quad \mathcal{B}_1 = \mu_0\frac{\partial v_1}{\partial \beta} + \mu_1\frac{\partial v_0}{\partial \beta},$$

$$\mathcal{C}_0 = v_0\frac{\partial \mu_0}{\partial \alpha}, \quad \mathcal{C}_1 = v_0\frac{\partial \mu_1}{\partial \alpha} + v_1\frac{\partial \mu_0}{\partial \alpha},$$

Thus, we can easily obtain the recursive relationship by comparing two sides of Equation (50):

$$\mu_0(\alpha,\eta) = \xi - \kappa\coth[\kappa(\alpha+\theta)], \quad \nu_0(\alpha,\eta) = -\kappa^2 cosech^2[\kappa(\alpha+\theta)],$$

For $m = 0$,

$$\mu_1(\alpha,\eta) = -\xi\kappa^2 cosech^2[\kappa(\alpha+\theta)]\frac{\eta^\delta}{\Gamma(\delta+1)},$$

$$\nu_1(\alpha,\eta) = -\xi\kappa^2 cosech^2[\kappa(\alpha+\theta)]\coth[\kappa(\alpha+\theta)]\frac{\eta^\delta}{\Gamma(\delta+1)},$$

For $m = 1$,

$$\mu_2(\alpha,\eta) = \xi\kappa^4 cosech^2[\kappa(\alpha+\theta)]\left\{\frac{2\xi\kappa\Gamma(2\delta+1)\eta^{3\delta}}{(\Gamma(\delta+1))^2\Gamma(3\delta+1)} - \frac{(3\coth^2([\kappa(\alpha+\theta)]-1))\eta^{2\delta}}{\Gamma(2\delta+1)}\right\},$$

$$\nu_2(\alpha,\eta) = \frac{1}{\Gamma(\delta+1)}[2\xi\kappa^5 cosech^2[\kappa(\alpha+\theta)]][\frac{\xi\kappa cosech^2(3\coth^2([\kappa(\alpha+\theta)]-1))\eta^{3\delta}}{\Gamma(\delta+1)\Gamma(3\delta+1)}$$
$$+\frac{2\xi\kappa cosech^2\coth^2([\kappa(\alpha+\theta)])\eta^{3\delta}}{\Gamma(\delta+1)\Gamma(3\delta+1)} - \frac{2\xi\coth(3cosech^2([\kappa(\alpha+\theta)]-1))\eta^{2\delta}}{\Gamma(2\delta+1)}].$$

In the same procedure, the remaining μ_m and ν_m ($m \geq 2$) components of the NDM solution can be obtained smoothly. Thus, we determine the sequence of alternatives as:

$$\mu(\alpha,\eta) = \sum_{m=0}^{\infty}\mu_m(\alpha,\beta) = \mu_0(\alpha,\beta) + \mu_1(\alpha,\beta) + \mu_2(\alpha,\beta) + \mu_3(\alpha,\beta) + \cdots$$

$$\nu(\alpha,\eta) = \sum_{m=0}^{\infty}\nu_m(\alpha,\beta) = \nu_0(\alpha,\beta) + \nu_1(\alpha,\beta) + \nu_2(\alpha,\beta) + \nu_3(\alpha,\beta) + \cdots$$

$$\mu(\alpha,\eta) = \xi - \kappa\coth[\kappa(\alpha+\theta)] - \xi\kappa^2 cosech^2[\kappa(\alpha+\theta)]\frac{\eta^\delta}{\Gamma(\delta+1)}$$
$$+ \xi\kappa^4 cosech^2[\kappa(\alpha+\theta)]\left\{\frac{2\xi\kappa\Gamma(2\delta+1)\eta^{3\delta}}{(\Gamma(\delta+1))^2\Gamma(3\delta+1)} - \frac{(3\coth^2([\kappa(\alpha+\theta)]-1))\eta^{2\delta}}{\Gamma(2\delta+1)}\right\} - \cdots$$

$$\nu(\alpha,\eta) = -\kappa^2 cosech^2[\kappa(\alpha+\theta)] - \xi\kappa^2 cosech^2[\kappa(\alpha+\theta)]\coth[\kappa(\alpha+\theta)]\frac{\eta^\delta}{\Gamma(\delta+1)}$$
$$+ \frac{1}{\Gamma(\delta+1)}[2\xi\kappa^5 cosech^2[\kappa(\alpha+\theta)]][\frac{\xi\kappa cosech^2(3\coth^2([\kappa(\alpha+\theta)]-1))\eta^{3\delta}}{\Gamma(\delta+1)\Gamma(3\delta+1)}$$
$$+ \frac{2\xi\kappa cosech^2\coth^2([\kappa(\alpha+\theta)])\eta^{3\delta}}{\Gamma(\delta+1)\Gamma(3\delta+1)} - \frac{2\xi\coth(3cosech^2([\kappa(\alpha+\theta)]-1))\eta^{2\delta}}{\Gamma(2\delta+1)}] - \cdots$$

Figures 7 and 8 describe the graphical behavior of both the unknown variables $\mu(\alpha,\eta)$ and $\nu(\alpha,\eta)$ of Example 2 at an integer-order $\delta = 1$ respectively. The procedures of NDM and q-HATM are implemented to obtain the desire accuracy. The higher accuracy and rate of convergence are achieved by the proposed techniques as shown in Figure 9. The plot analysis demonstrates the validity and accuracy of the proposed techniques and considered to be the best techniques to solve other fractional-order problems.

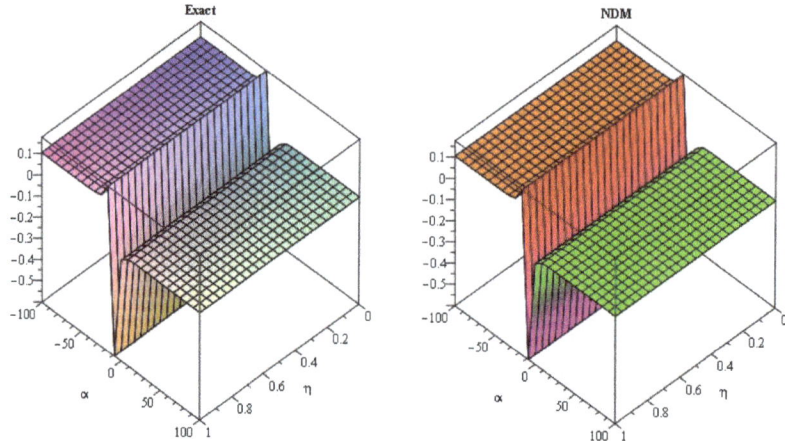

Figure 7. Exact and NDM solution of $\mu(\alpha, \eta)$ at $\delta = 1$.

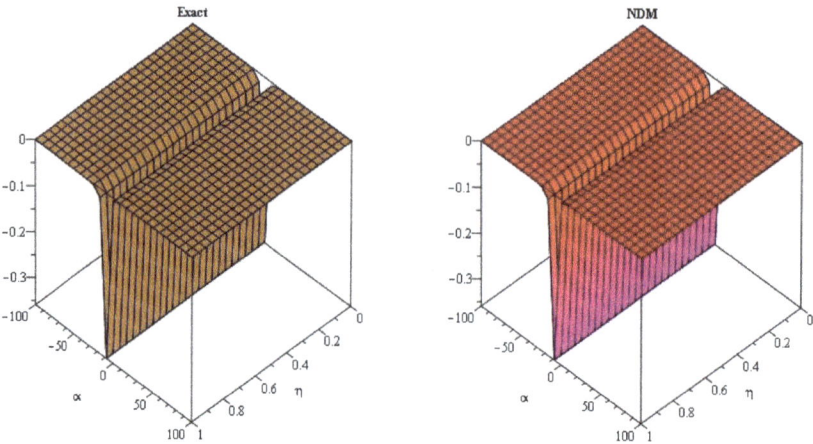

Figure 8. Exact and NDM solution of $\nu(\alpha, \eta)$ at $\delta = 1$.

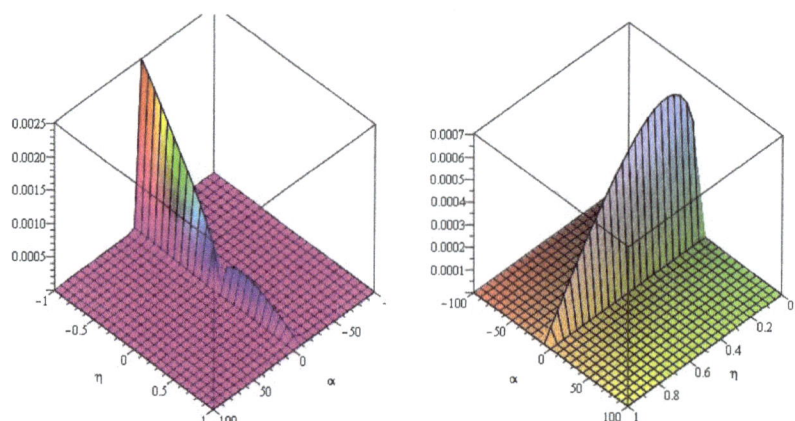

Figure 9. Error plot of $\mu(\alpha, \eta)$ and $\nu(\alpha, \eta)$.

5.2. q-Homotopy Analysis Transform Method

The Example 1 approximate solution with the help of **q-HATM**.
By taking the Laplace transformation of Equation (46), we get:

$$\pounds\{\mu(\alpha,\eta)\} = \frac{1}{s}\{\mu(\alpha,0)\} - \frac{1}{s^\delta}\pounds\left[\mu(\alpha,\eta)\frac{\partial\mu(\alpha,\eta)}{\partial\alpha} + \frac{1}{2}\frac{\partial\mu(\alpha,\eta)}{\partial\alpha} + \frac{\partial\nu(\alpha,\eta)}{\partial\alpha}\right],$$
$$\pounds\{\nu(\alpha,\eta)\} = \frac{1}{s}\{\nu(\alpha,0)\} - \frac{1}{s^\delta}\pounds\left[\mu(\alpha,\eta)\frac{\partial\nu(\alpha,\eta)}{\partial\alpha} + \nu(\alpha,\eta)\frac{\partial\mu(\alpha,\eta)}{\partial\alpha} - \frac{1}{2}\frac{\partial^2\nu(\alpha,\eta)}{\partial\alpha^2}\right], \tag{51}$$

Using Equation (51) we define the nonlinear operator as:

$$N^1[\phi_1(\alpha,\eta;q),\phi_2(\alpha,\eta;q)]$$
$$= \pounds\left[\phi_1(\alpha,\eta;q) - \frac{1}{s}\{\phi_1(\alpha,0)\} + \frac{1}{s^\delta}\left\{\phi_1(\alpha,\eta)\frac{\partial\phi_1(\alpha,\eta)}{\partial\alpha} + \frac{1}{2}\frac{\partial\phi_1(\alpha,\eta)}{\partial\alpha} + \frac{\partial\phi_2(\alpha,\eta)}{\partial\alpha}\right\}\right],$$
$$N^2[\phi_1(\alpha,\eta;q),\phi_2(\alpha,\eta;q)]$$
$$= \pounds\left[\phi_2(\alpha,\eta;q) - \frac{1}{s}\{\phi_2(\alpha,0)\} + \frac{1}{s^\delta}\left\{\phi_1(\alpha,\eta)\frac{\partial\phi_2(\alpha,\eta)}{\partial\alpha} + \phi_2(\alpha,\eta)\frac{\partial\phi_1(\alpha,\eta)}{\partial\alpha} - \frac{1}{2}\frac{\partial^2\phi_2(\alpha,\eta)}{\partial\alpha^2}\right\}\right]. \tag{52}$$

By applying the proposed algorithm, the deformation equation of m-th order is given as:

$$\pounds[\mu_m(\alpha,\eta) - K_m\mu_{m-1}(\alpha,\eta)] = h\Re_{1,m}[\vec{\mu}_{m-1},\vec{\nu}_{m-1}],$$
$$\pounds[\nu_m(\alpha,\eta) - K_m\nu_{m-1}(\alpha,\eta)] = h\Re_{2,m}[\vec{\mu}_{m-1},\vec{\nu}_{m-1}], \tag{53}$$

$$\Re_{1,m}[\vec{\mu}_{m-1},\vec{\nu}_{m-1}] = \pounds[\mu_{m-1}(\alpha,\eta) - (1-\frac{k_m}{n})\frac{1}{s}\{\xi - \kappa\coth[\kappa(\alpha+\theta)]\}$$
$$+ \frac{1}{s^\delta}\{\sum_{j=0}^{m-1}\mu_j(\alpha,\eta)\frac{\partial\mu_{m-1-j}(\alpha,\eta)}{\partial\alpha} + \frac{1}{2}\frac{\partial\mu_{m-1}(\alpha,\eta)}{\partial\alpha} + \frac{\partial\nu_{m-1}(\alpha,\eta)}{\partial\alpha}\}],$$
$$\Re_{2,m}[\vec{\mu}_{m-1},\vec{\nu}_{m-1}] = \pounds[\nu_{m-1}(\alpha,\eta) - \frac{1}{s}\{-\kappa^2\operatorname{cosech}^2[\kappa(\alpha+\theta)]\}$$
$$+ \frac{1}{s^\delta}\{\sum_{j=0}^{m-1}\mu_j(\alpha,\eta)\frac{\partial\nu_{m-1-j}(\alpha,\eta)}{\partial\alpha} + \sum_{j=0}^{m-1}\nu_j(\alpha,\eta)\frac{\partial\mu_{m-j-1}(\alpha,\eta)}{\partial\alpha} - \frac{1}{2}\frac{\partial^2\nu_{m-1}(\alpha,\eta)}{\partial\alpha^2}\}]. \tag{54}$$

By applying the inverse Laplace transform on Equation (53), we get:

$$\mu_m(\alpha,\eta) = K_m\mu_{m-1}(\alpha,\eta) + hL^{-1}\Re_{1,m}[\vec{\mu}_{m-1},\vec{\nu}_{m-1}],$$
$$\nu_m(\alpha,\eta) = K_m\nu_{m-1}(\alpha,\eta)] + hL^{-1}\Re_{2,m}[\vec{\mu}_{m-1},\vec{\nu}_{m-1}]. \tag{55}$$

By the help of the given initial condition, we have:

$$\mu_0(\alpha,\eta) = \xi - \kappa\coth[\kappa(\alpha+\theta)],$$
$$\nu_0(\alpha,\eta) = -\kappa^2\operatorname{cosech}^2[\kappa(\alpha+\theta)]. \tag{56}$$

To find the value of $\mu_0(\alpha,\eta)$ and $\nu_0(\alpha,\eta)$, set $m=1$ in Equation (38), then we get:

$$\mu_1(\alpha,\eta) = K_1\mu_0(\alpha,\eta) + h\pounds^{-1}\Re_{1,1}[\vec{\mu}_0,\vec{\nu}_0],$$
$$\nu_1(\alpha,\eta) = K_1\nu_0(\alpha,\eta)] + h\pounds^{-1}\Re_{2,1}[\vec{\mu}_0,\vec{\nu}_0], \tag{57}$$

From Equation (54) for $m = 1$, we conclude:

$$\Re_{1,1}[\vec{\mu}_0, \vec{v}_0] = \pounds[\mu_0(\alpha,\eta)] - (1 - \frac{k_1}{n})\frac{1}{s}\{\xi - \kappa\coth[\kappa(\alpha+\theta)]\}$$
$$+ \frac{1}{s^\delta}\pounds[\{\mu_0(\alpha,\eta)\frac{\partial\mu_0(\alpha,\eta)}{\partial\alpha} + \frac{1}{2}\frac{\partial\mu_0(\alpha,\eta)}{\partial\alpha} + \frac{\partial v_0(\alpha,\eta)}{\partial\alpha}\}],$$
$$\Re_{2,1}[\vec{\mu}_0, \vec{v}_0] = \pounds[v_0(\alpha,\eta)] - (1 - \frac{k_1}{n})\frac{1}{s}\{-\kappa^2 cosech^2[\kappa(\alpha+\theta)]\}$$
$$+ \frac{1}{s^\delta}\pounds[\{\mu_0(\alpha,\eta)\frac{\partial v_0(\alpha,\eta)}{\partial\alpha} + v_0(\alpha,\eta)\frac{\partial\mu_0(\alpha,\eta)}{\partial\alpha} - \frac{1}{2}\frac{\partial^2 v_0(\alpha,\eta)}{\partial\alpha^2}\},]$$
(58)

Then by using Equations (25) and (58) in Equation (57), we get

$$\mu_1(\alpha,\eta) = h\pounds^{-1}[\frac{1}{s}\{\xi - \kappa\coth[\kappa(\alpha+\theta)]\} - (1 - \frac{0}{n})\frac{1}{s}\{\xi - \kappa\coth[\kappa(\alpha+\theta)]\}$$
$$+ \frac{1}{s^\delta}\pounds[\{\mu_0(\alpha,\eta)\frac{\partial\mu_0(\alpha,\eta)}{\partial\alpha} + \frac{1}{2}\frac{\partial\mu_0(\alpha,\eta)}{\partial\alpha} + \frac{\partial v_0(\alpha,\eta)}{\partial\alpha}\}]],$$
$$v_1(\alpha,\eta) = h\pounds^{-1}[\frac{1}{s}\{-\kappa^2 cosech^2[\kappa(\alpha+\theta)]\} - (1 - \frac{0}{n})\frac{1}{s}\{-\kappa^2 cosech^2[\kappa(\alpha+\theta)]\}$$
$$+ \frac{1}{s^\delta}\pounds[\{\mu_0(\alpha,\eta)\frac{\partial v_0(\alpha,\eta)}{\partial\alpha} + v_0(\alpha,\eta)\frac{\partial\mu_0(\alpha,\eta)}{\partial\alpha} - \frac{1}{2}\frac{\partial^2 v_0(\alpha,\eta)}{\partial\alpha^2}\}]],$$
(59)

$$\mu_1(\alpha,\eta) = -\xi h\kappa^2 cosech^2[\kappa(\alpha+\theta)]\frac{\eta^\delta}{\Gamma(\delta+1)}, \quad v_1(\alpha,\eta) = -\xi h\kappa^2 cosech^2[\kappa(\alpha+\theta)]\coth[\kappa(\alpha+\theta)]\frac{\eta^\delta}{\Gamma(\delta+1)},$$

Similarly from Equations (57) and (58) for $m = 2$, we have:

$$\mu_2(\alpha,\eta) = n\mu_1(\alpha,\eta) + h\pounds^{-1}[\pounds[\mu_1(\alpha,\eta)] - (1 - \frac{n}{n})\frac{1}{s}\{\xi - \kappa\coth[\kappa(\alpha+\theta)]\} + \frac{1}{s^\delta}\pounds[\{\mu_0(\alpha,\eta)\frac{\partial\mu_1(\alpha,\eta)}{\partial\alpha}$$
$$+ \mu_1(\alpha,\eta)\frac{\partial\mu_0(\alpha,\eta)}{\partial\alpha} + \frac{1}{2}\frac{\partial\mu_1(\alpha,\eta)}{\partial\alpha} + \frac{\partial v_1(\alpha,\eta)}{\partial\alpha}\}]],$$
$$v_2(\alpha,\eta) = nv_1(\alpha,\eta) + h\pounds^{-1}[\pounds[\mu_1(\alpha,\eta)] - (1 - \frac{n}{n})\frac{1}{s}\{-\kappa^2 cosech^2[\kappa(\alpha+\theta)]\} + \frac{1}{s^\delta}\pounds[\{\mu_0(\alpha,\eta)\frac{\partial v_1(\alpha,\eta)}{\partial\alpha}$$
$$+ \mu_1(\alpha,\eta)\frac{\partial v_0(\alpha,\eta)}{\partial\alpha} + v_0(\alpha,\eta)\frac{\partial\mu_1(\alpha,\eta)}{\partial\alpha} + v_1(\alpha,\eta)\frac{\partial\mu_0(\alpha,\eta)}{\partial\alpha} - \frac{1}{2}\frac{\partial^2 v_1(\alpha,\eta)}{\partial\alpha^2}\}]].$$
(60)

In simplified, the above calculation eliminates as described:

$$\mu_2(\alpha,\eta) = -\xi(n+h)h\kappa^2 cosech^2[\kappa(\alpha+\theta)]\frac{\eta^\delta}{\Gamma(\delta+1)}$$
$$+ \xi h^2\kappa^4 cosech^2[\kappa(\alpha+\theta)]\left\{\frac{2\xi\kappa\Gamma(2\delta+1)\eta^{3\delta}}{(\Gamma(\delta+1))^2\Gamma(3\delta+1)} - \frac{(3\coth^2([\kappa(\alpha+\theta)]-1))\eta^{2\delta}}{\Gamma(2\delta+1)}\right\},$$
$$v_2(\alpha,\eta) = -\xi(n+h)h\kappa^2 cosech^2[\kappa(\alpha+\theta)]\coth[\kappa(\alpha+\theta)]\frac{\eta^\delta}{\Gamma(\delta+1)}$$
$$+ \frac{1}{\Gamma(\delta+1)}h^2[2\xi\kappa^5 cosech^2[\kappa(\alpha+\theta)]][\frac{\xi\kappa cosech^2(3\coth^2([\kappa(\alpha+\theta)]-1))\eta^{3\delta}}{\Gamma(\delta+1)\Gamma(3\delta+1)}$$
$$+ \frac{2\xi\kappa cosech^2\coth^2([\kappa(\alpha+\theta)])\eta^{3\delta}}{\Gamma(\delta+1)\Gamma(3\delta+1)} - \frac{2\xi\coth(3cosech^2([\kappa(\alpha+\theta)]-1))\eta^{2\delta}}{\Gamma(2\delta+1)}].$$

The rest of the iterative terms can be used in the same way. Formerly, the family of q-homotopy analysis transform technique series result of Equation (46) is assumed by:

$$\mu(\alpha,\eta) = \mu_0(\alpha,\eta) + \sum_{m=1}^{\infty} \mu_m(\alpha,\eta)(\frac{1}{n})^m,$$
$$v(\alpha,\eta) = v_0(\alpha,\eta) + \sum_{m=1}^{\infty} v_m(\alpha,\eta)(\frac{1}{n})^m. \tag{61}$$

The exact solution of Equation (46) at $\delta = 1$ and taking $\xi = 0.005$, $\theta = 10$ and $\kappa = 0.1$.

$$\mu(\alpha,\eta) == \xi - \kappa \coth[\kappa(\alpha + \theta - \xi\eta)],$$
$$v(\alpha,\eta) = -\kappa^2 \text{cosech}^2[\kappa(\alpha + \theta - \xi\eta)]. \tag{62}$$

The solutions $\mu(\alpha,\eta)$ and $v(\alpha,\eta)$ are also obtained by using q-HATM and found to be in good agreement with the exact solution of problems. For better understanding the results for both the variables $\mu(\alpha,\eta)$ and $v(\alpha,\eta)$ of Example 2 are plotted in Figures 10 and 11 respectively where the higher accuracy is observed.

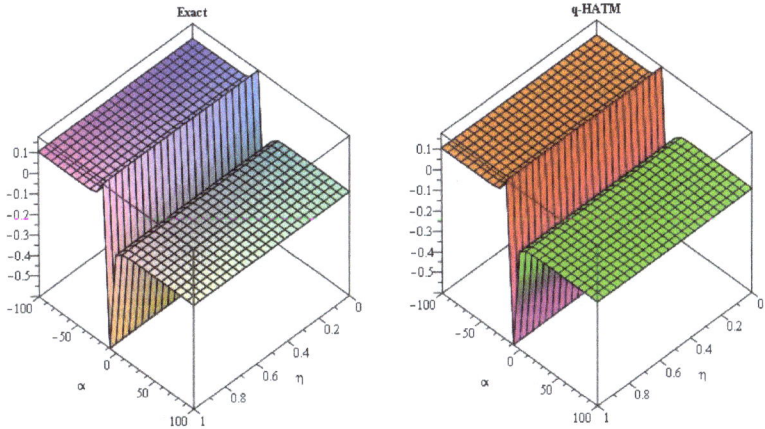

Figure 10. Exact and q-HATM solution of $\mu(\alpha,\eta)$ at $\delta = 1$.

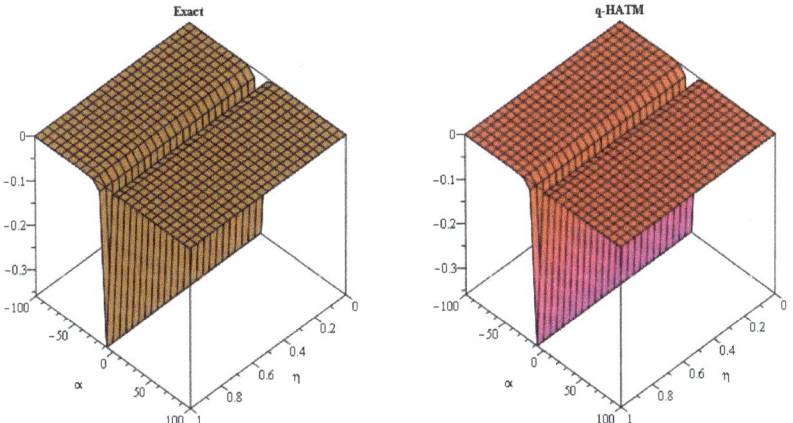

Figure 11. Exact and q-HATM solution of $v(\alpha,\eta)$ at $\delta = 1$.

6. Conclusions

In this paper, we studied the factional view of Whitham–Broer–Kaup equations by using two analytical powerful techniques. With the help of the Laplace and natural transformations, the procedure strengthened and became easy for implementation. A very close contact of the obtained solutions with the exact solution of the problem was observed. It was found that the rate of convergence of the proposed methods was sufficient for solving fractional-order partial differential equations. Therefore, the proposed techniques could be extended to solve other complicated fractional-order problems.

Author Contributions: conceptualization, R.S. and H.K.; methodology, R.S.; software, H.K.; validation, D.B., H.K. and R.S.; formal analysis, R.S.; investigation, H.K.; resources, D.B.; data curation, R.S.; writing—original draft preparation, R.S.; writing—review and editing, H.K.; visualization, D.B.; supervision, D.B.

Funding: This research received no external funding.

Conflicts of Interest: The authors declare no conflict of interest.

References

1. Miller, K.S.; Ross, B. *An Introduction to the Fractional Calculus and Fractional Differential Equations*; Wiley: Hoboken, NJ, USA, 1993.
2. Rudolf, H. (Ed.) *Applications of Fractional Calculus in Physics*; World Scientific: Singapore, 2000.
3. Katugampola, U.N. A new approach to generalized fractional derivatives. *Bull. Math. Anal. Appl.* **2014**, *6*, 1–15.
4. Srivastava, H.M.; Baleanu, D.; Li, C. *Preface: Recent Advances in Fractional Dynamics*; AIP Publishing LLC: College Park, MD, USA, 2016.
5. Ahmad, J.; Mushtaq, M.; Sajjad, N. Exact Solution of Whitham Broer-Kaup Shallow Water Wave Equations. *J. Sci. Arts* **2015**, *15*, 5.
6. Kupershmidt, B.A. Mathematics of dispersive water waves. *Commun. Math. Phys.* **1985**, *99*, 51–73. [CrossRef]
7. Whitham, G.B. Variational methods and applications to water waves. *Proc. R. Soc. Lond. Ser. Math. Phys. Sci.* **1967**, *299*, 6–25.
8. Broer, L.J.F. Approximate equations for long water waves. *Appl. Sci. Res.* **1975**, *31*, 377–395. [CrossRef]
9. Kaup, D.J. A higher-order water-wave equation and the method for solving it. *Prog. Theor. Phys.* **1975**, *54*, 396–408. [CrossRef]
10. Ali, A.; Shah, K.; Khan, R.A. Numerical treatment for traveling wave solutions of fractional Whitham–Broer–Kaup equations. *Alex. Eng. J.* **2018**, *57*, 1991–1998. [CrossRef]
11. Biazar, J.; Aminikhah, H. Study of convergence of homotopy perturbation method for systems of partial differential equations. *Comput. Math. Appl.* **2009**, *58*, 2221–2230. [CrossRef]
12. Mohyud-Din, S.T.; Noor, M.A. Homotopy perturbation method for solving partial differential equations. *Z. Naturforschung* **2009**, *64*, 157–170. [CrossRef]
13. Xie, F.; Yan, Z.; Zhang, H. Explicit and exact traveling wave solutions of Whitham–Broer–Kaup shallow water equations. *Phys. Lett. A* **2001**, *285*, 76–80. [CrossRef]
14. Wang, L.; Chen, X. Approximate analytical solutions of time fractional Whitham–Broer–Kaup equations by a residual power series method. *Entropy* **2015**, *17*, 6519–6533. [CrossRef]
15. El-Sayed, S.M.; Kaya, D. Exact and numerical traveling wave solutions of Whitham–Broer–Kaup equations. *Appl. Math. Comput.* **2005**, *167*, 1339–1349. [CrossRef]
16. Saha Ray, S. A novel method for travelling wave solutions of fractional Whitham–Broer–Kaup, fractional modified Boussinesq and fractional approximate long wave equations in shallow water. *Math. Methods Appl. Sci.* **2015**, *38*, 1352–1368. [CrossRef]
17. Mohyud-Din, S.T.; Yıldırım, A.; Demirli, G. Traveling wave solutions of Whitham–Broer–Kaup equations by homotopy perturbation method. *J. King Saud-Univ.-Sci.* **2010**, *22*, 173–176. [CrossRef]
18. Kadem, A.; Baleanu, D. *On Fractional Coupled Whitham–Broer–Kaup Equations*; Publishing House of the Romanian Academy: Wellington, New Zealand, 2011.
19. Iqbal, M. A fractional Whitham–Broer–Kaup equation and its possible application to Tsunami prevention. *Therm. Sci.* **2017**, *21*, 1847–1855.

20. Ghehsareh, H.R.; Majlesi, A.; Zaghian, A. Lie Symmetry analysis and Conservation Laws for time fractional coupled Whitham–Broer–Kaup equations. *UPB Sci. Bull. Ser. A Appl. Math. Phys.* **2018**, *80*, 153–168.
21. Zhang, Z.; Yong, X.; Chen, Y. Symmetry analysis for whitham-Broer-Kaup equations. *J. Nonlinear Math. Phys.* **2008**, *15*, 383–397. [CrossRef]
22. Arshed, S.; Sadia, M. G'/G^2-Expansion method: New traveling wave solutions for some nonlinear fractional partial differential equations. *Opt. Quantum Electron.* **2018**, *50*, 123. [CrossRef]
23. Rani, A.; Ul-Hassan, Q.M.; Ashraf, M.; Ayub, K.; Khan, M.Y. A novel technique for solving nonlinear WBK equations of fractionl-order. *J. Sci. Arts* **2018**, *18*, 301–316.
24. Singh, J.; Kumar, D.; Swroop, R. Numerical solution of time-and space-fractional coupled Burgers' equations via homotopy algorithm. *Alex. Eng. J.* **2016**, *55*, 1753–1763. [CrossRef]
25. Veeresha, P.; Prakasha, D.G. Solution for fractional Zakharov–Kuznetsov equations by using two reliable techniques. *Chin. J. Phys.* **2019**, *60*, 313–330. [CrossRef]
26. Singh, J.; Secer, A.; Swroop, R.; Kumar, D. A reliable analytical approach for a fractional model of advection-dispersion equation. *Nonlinear Eng.* **2019**, *8*, 107–116. [CrossRef]
27. Srivastava, H.M.; Kumar, D.; Singh, J. An efficient analytical technique for fractional model of vibration equation. *Appl. Math. Model.* **2017**, *45*, 192–204. [CrossRef]
28. Rawashdeh, M.; Maitama, S. Finding exact solutions of nonlinear PDEs using the natural decomposition method. *Math. Methods Appl. Sci.* **2017**, *40*, 223–236. [CrossRef]
29. Rawashdeh, M.S. The fractional natural decomposition method: Theories and applications. *Math. Methods Appl. Sci.* **2017**, *40*, 2362–2376. [CrossRef]
30. Shah, R.; Khan, H.; Mustafa, S.; Kumam, P.; Arif, M. Analytical Solutions of Fractional-Order Diffusion Equations by Natural Transform Decomposition Method. *Entropy* **2019**, *21*, 557. [CrossRef]
31. Shah, R.; Khan, H.; Kumam, P.; Arif, M.; Baleanu, D. Natural Transform Decomposition Method for Solving Fractional-Order Partial Differential Equations with Proportional Delay. *Mathematics* **2019**, *7*, 532. [CrossRef]
32. Rawashdeh, M.S.; Maitama, S. Solving nonlinear ordinary differential equations using the NDM. *J. Appl. Anal. Comput.* **2015**, *5*, 77–88.
33. Shah, R.; Khan, H.; Farooq, U.; Baleanu, D.; Kumam, P.; Arif, M. A New Analytical Technique to Solve System of Fractional-Order Partial Differential Equations. *IEEE Access* **2019**, *7*, 150037–150050. [CrossRef]
34. Cherif, M.H.; Ziane, D.; Belghaba, K. Fractional natural decomposition method for solving fractional system of nonlinear equations of unsteady flow of a polytropic gas. *Nonlinear Stud.* **2018**, *25*, 753–764.
35. Abdel-Rady, A.S.; Rida, S.Z.; Arafa, A.A.M.; Abedl-Rahim, H.R. Natural transform for solving fractional models. *J. Appl. Math. Phys.* **2015**, *3*, 1633. [CrossRef]
36. Khan, H.; Shah, R.; Kumam, P.; Arif, M. Analytical Solutions of Fractional-Order Heat and Wave Equations by the Natural Transform Decomposition Method. *Entropy* **2019**, *21*, 597. [CrossRef]
37. Eltayeb, H.; Abdalla, Y.T.; Bachar, I.; Khabir, M.H. Fractional Telegraph Equation and Its Solution by Natural Transform Decomposition Method. *Symmetry* **2019**, *11*, 334. [CrossRef]

© 2019 by the authors. Licensee MDPI, Basel, Switzerland. This article is an open access article distributed under the terms and conditions of the Creative Commons Attribution (CC BY) license (http://creativecommons.org/licenses/by/4.0/).

Article

General Linear Recurrence Sequences and Their Convolution Formulas

Paolo Emilio Ricci [1] and Pierpaolo Natalini [2],*

[1] Section of Mathematics, International Telematic University UniNettuno, Corso Vittorio Emanuele II, 39, 00186 Roma, Italy; paoloemilioricci@gmail.com
[2] Dipartimento di Matematica e Fisica, Università degli Studi Roma Tre, Largo San Leonardo Murialdo, 1, 00146 Roma, Italy
* Correspondence: natalini@mat.uniroma3.it

Received: 29 September 2019; Accepted: 15 November 2019; Published: 19 November 2019

Abstract: We extend a technique recently introduced by Chen Zhuoyu and Qi Lan in order to find convolution formulas for second order linear recurrence polynomials generated by $\left(\frac{1}{1+at+bt^2}\right)^x$. The case of generating functions containing parameters, even in the numerator is considered. Convolution formulas and general recurrence relations are derived. Many illustrative examples and a straightforward extension to the case of matrix polynomials are shown.

Keywords: liner recursions; convolution formulas; Gegenbauer polynomials; Humbert polynomials; classical polynomials in several variables; classical number sequences

AMS 2010 Mathematics Subject Classifications: 33C99; 65Q30; 11B37

1. Introduction

Generating functions [1] constitute a bridge between continuous analysis and discrete mathematics. Linear recurrence relations are satisfied by many special polynomials of classical analysis. A wide scenario including special sequences of polynomials and numbers, combinatorial analysis, and application of mathematics is related to the above mentioned topics.

It would be impossible to list in the Reference section all of even the most important articles dedicated to these subjects. As a first example, we recall the Chebyshev polynomials of the first and second kind, which are powerful tools used in both theoretical and applied mathematics. Their links with the Lucas and Fibonacci polynomials have been studied and many properties have been derived. Connections with Bernoulli polynomials have been highlighted in [2].

In particular, the important calculation of sums of several types of polynomials have been recently studied (see e.g., [3–5] and the references therein). This kind of subject has attracted many scholars. For example, W. Zhang [6] proved an identity involving Chebyshev polynomials and their derivatives.

Fibonacci and Lucas polynomias and their extensions have been studied for a long time, in particular within the Fibonacci Association, which has contributed to the study of this and similar subjects. As an applications of a results proved by Y. Zhang and Z. Chen [3], Y. Ma and W. Zhang [4] obtained some identities involving Fibonacci numbers and Lucas numbers.

Convolution techniques are connected with combinatorial identities, and many results have been obtained in this direction [2,7,8]. Convolution sums using second kind Chebyshev polynomials are contained in [7].

Recently, Taekyun Kim et al. [8] studied properties of Fibonacci numbers by introducing the so called convolved Fibonacci numbers. By using the genereting function:

$$\left(\frac{1}{1-t-bt^2}\right)^x = \sum_{n=0}^{\infty} p_n(x)\frac{t^n}{n!},$$

for $x \in \mathbf{R}$ and $r \in \mathbf{N}$, they proved the interesting relation

$$p_n(x) = \sum_{\ell=0}^{n} p_\ell(r) p_{n-\ell}(x-r) = \sum_{\ell=0}^{n} p_{n-\ell}(r) p_\ell(x-r).$$

Furthermore, they derived a link between $p_n(x)$ and a particular combination of sums of Fibonacci numbers, so that complex sums of Fibonacci numbers have been converted to the easier calculation of $p_n(x)$.

In a recent article Chen Zhuoyu and Qi Lan [9] introduced convolution formulas for second order linear recurrence sequences related to the generating function [1] of the type

$$f(t) = \frac{1}{1+at+bt^2},$$

deriving coefficient expressions for the series expansion of the function $f^x(t)$, ($x \in \mathbf{R}$). In this article, motivated by this research, we continue the study of possible applications of the considered method, by analyzing the general situation of a generating function of the type

$$G(t,x) = \left(\frac{1}{1+a_1 t + a_2 t^2 + \cdots + a_r t^r}\right)^x,$$

and we deduce the recurrence relation for the generated polynomials.

Several illustrative examples are shown in Section 6. In the last section the results are extended, in a straightforward way, to the case of matrix polynomials.

2. Generating Functions

We start from the generating function considered by Chen Zhuoyu and Qi Lan:

$$G(t,x) = \left(\frac{1}{1+at+bt^2}\right)^x = \left[\frac{1}{(1-\alpha t)(1-\beta t)}\right]^x = \tag{1}$$

$$= \exp\left\{-x \log\left[(1-\alpha t)(1-\beta t)\right]\right\},$$

with

$$a = -(\alpha+\beta), \qquad b = \alpha\beta \tag{2}$$

$$G(t,x) = \sum_{k=0}^{\infty} g_k(x;\alpha,\beta)\frac{t^k}{k!} = \tag{3}$$

$$= \exp\left\{-x \log(1-\alpha t)\right\} \cdot \exp\left\{-x \log(1-\beta t)\right\} = G_\alpha(t,x) \cdot G_\beta(t,x),$$

where

$$G_\alpha(t,x) = \exp\left[-x\log(1-\alpha t)\right] = \sum_{k=0}^{\infty} p_k(x,\alpha)\frac{t^k}{k!}, \quad (4a)$$

$$G_\beta(t,x) = \exp\left[-x\log(1-\beta t)\right] = \sum_{k=0}^{\infty} q_k(x,\beta)\frac{t^k}{k!}. \quad (4b)$$

Note that, by Equation (2) we could write, in equivalent form:

$$g_k(x;\alpha,\beta) = g_k(x;a,b), \quad p_k(x,\alpha) = p_k(x,a), \quad q_k(x,\beta) = q_k(x,b), \quad (5)$$

but, in what follows, we put for shortness:

$$g_k(x;\alpha,\beta) = g_k(x), \quad p_k(x,\alpha) = p_k(x), \quad q_k(x,\beta) = q_k(x). \quad (6)$$

By Equations (3), (4a) and (4b) we find the convolution formula:

$$g_k(x) = \sum_{h=0}^{k} \binom{k}{h} p_{k-h}(x)\, q_h(x). \quad (7)$$

3. Recurrence Relation

Note that

$$\frac{\partial G(t,x)}{\partial t} = \frac{\partial G_\alpha(t,x)}{\partial t} \cdot G_\beta(t,x) + G_\alpha(t,x) \cdot \frac{\partial G_\beta(t,x)}{\partial t} =$$

$$= \left(\frac{\alpha x}{1-\alpha t} + \frac{\beta x}{1-\beta t}\right) G(t,x) = -x\left(\frac{a+2bt}{1+at+bt^2}\right) G(t,x), \quad (8)$$

as can be derived directly from Equation (1).

Then we have

$$(1+at+bt^2)\frac{\partial G(t,x)}{\partial t} = -x\,a\,G(t,x) - 2\,b\,x\,t\,G(t,x), \quad (9)$$

$$\sum_{k=0}^{\infty} g_{k+1}(x)\frac{t^k}{k!} + a\sum_{k=0}^{\infty} g_{k+1}(x)\frac{t^{k+1}}{k!} + b\sum_{k=0}^{\infty} g_{k+1}(x)\frac{t^{k+2}}{k!} =$$

$$= -x\,a\sum_{k=0}^{\infty} g_k(x)\frac{t^k}{k!} - 2\,b\,x\sum_{k=0}^{\infty} g_k(x)\frac{t^{k+1}}{k!},$$

that is

$$\sum_{k=0}^{\infty} g_{k+1}(x)\frac{t^k}{k!} + a\sum_{k=1}^{\infty} k g_k(x)\frac{t^k}{k!} + b\sum_{k=2}^{\infty} k(k-1)g_{k-1}(x)\frac{t^k}{k!} =$$

$$= -a\,x\sum_{k=0}^{\infty} g_k(x)\frac{t^k}{k!} - 2\,b\,x\sum_{k=1}^{\infty} k g_{k-1}(x)\frac{t^k}{k!},$$

and therefore, we can conclude with the theorem:

Theorem 1. *The sequence $\{g_k(x)\}_{k\in\mathbb{N}}$ satisfies the linear recurrence relation*

$$g_k(x) + a(x+k-1)g_{k-1}(x) + b(k-1)(k+2x-2)g_{k-2}(x) = 0. \tag{10}$$

3.1. Properties of the Basic Generating Function

We consider now a few properties of the basic generating functions $G_\alpha(t,x)$. According to the definition (4a), the polynomials $p_k(x)$ are recognized as associated Sheffer polynomials [10] and quasi-monomials, according to the Dattoli [11,12] definition.

3.1.1. Differential Equation

We have:

$$G_\alpha(t,x) = \exp[-xH(t)] = \sum_{k=0}^{\infty} p_k(x,\alpha) \frac{t^k}{k!}, \tag{11}$$

where

$$H(t) = -\log(1-\alpha t), \qquad H'(t) = \frac{\alpha}{1-\alpha t}, \tag{12}$$

and its functional inverse is given by

$$H^{-1}(t) = \frac{1}{\alpha}\left(1 - e^{-t}\right), \tag{13}$$

so that, recalling the results by Y. Ben Cheikh [13], we find the derivative and multiplication operators of the quasi-monomials $p_k(x)$, in the form:

$$\hat{P} = \frac{1}{\alpha}\left(1 - e^{-D_x}\right) \qquad \hat{M} = xH'\left(H^{-1}(D_x)\right) = \alpha\, x\, e^{D_x}, \tag{14}$$

and we can conclude that

Theorem 2. *The polynomials $p_k(x)$ satisfy the differential equation:*

$$\hat{M}\hat{P}\, p_n(x) = x\left(e^{D_x} - 1\right) p_n(x) = n\, p_n(x), \tag{15}$$

that is, $\forall n \geq 1$:

$$x\left(\frac{1}{n!} p_n^{(n)} + \frac{1}{(n-1)!} p_n^{(n-1)} + \cdots + p_n'(x)\right) = n\, p_n(x). \tag{16}$$

3.1.2. Differential Identity

Differentiating Equation (11) with respect to x, we find

$$\frac{\partial G_\alpha(t,x)}{\partial x} = -G_\alpha(t,x)\log(1-\alpha t) = \sum_{k=1}^{\infty} p_k'(x,\alpha) \frac{t^k}{k!} \tag{17}$$

that is

$$\sum_{k=1}^{\infty} p_k'(x,\alpha) \frac{t^k}{k!} = -\log(1-\alpha t) \sum_{k=0}^{\infty} p_k(x,\alpha) \frac{t^k}{k!},$$

$$\sum_{k=1}^{\infty} p'_k(x,\alpha) \frac{t^k}{k!} = \sum_{k=1}^{\infty} \frac{(\alpha t)^k}{k} \sum_{k=0}^{\infty} p_k(x,\alpha) \frac{t^k}{k!} = \sum_{k=1}^{\infty} \frac{(\alpha t)^k}{k} \left[1 + \sum_{k=1}^{\infty} p_k(x,\alpha) \frac{t^k}{k!}\right],$$

$$\sum_{k=1}^{\infty} p'_k(x,\alpha) \frac{t^k}{k!} = \sum_{k=1}^{\infty} (k-1)! \, \alpha^k \frac{t^k}{k} + \sum_{k=1}^{\infty} (k-1)! \, \alpha^k \frac{t^k}{k} \sum_{k=1}^{\infty} p_k(x,\alpha) \frac{t^k}{k!},$$

$$\sum_{k=1}^{\infty} p'_k(x,\alpha) \frac{t^k}{k!} = \sum_{k=1}^{\infty} (k-1)! \, \alpha^k \frac{t^k}{k!} + \sum_{k=1}^{\infty} (k-1)! \, \alpha^k \frac{t^k}{k!} \sum_{k=1}^{\infty} p_k(x,\alpha) \frac{t^k}{k!} =$$

$$= \sum_{k=1}^{\infty} (k-1)! \, \alpha^k \frac{t^k}{k!} + \sum_{k=1}^{\infty} \sum_{h=1}^{k} (k-h-1)! \, \alpha^{k-h} \, p_h(x,\alpha) \frac{t^k}{k!},$$

so that we can conclude with the theorem:

Theorem 3. *The polynomials $p_k(x)$ satisfy the differential identity:*

$$p'_k(x,\alpha) = (k-1)! \, \alpha^k + \sum_{h=1}^{k} (k-h-1)! \, \alpha^{k-h} \, p_h(x,\alpha). \tag{18}$$

3.2. Extension by Convolution

We now consider the case of a generating function of the type:

$$G(t,x) = \left(\frac{1+ct}{1+at+bt^2}\right)^x = \sum_{k=0}^{\infty} q_k(x;c;a,b) = \frac{\sum_{k=0}^{\infty} p_k(x;c)}{\sum_{k=0}^{\infty} g_k(x;a,b)}. \tag{19}$$

A straightforward consequence is the convolution formula for the resulting polynomials:

$$p_k(x;c) = \sum_{h=0}^{k} \binom{k}{h} g_{k-h}(x;a,b) \, q_h(x;c;a,b), \tag{20}$$

so that the $q_h(x;c;a,b)$ can be found recursively by solving the infinite system

$$\begin{cases} q_0(x;c;a,b) = 1, \\ q_k(x;c;a,b) = p_k(x;a,b) - \sum_{h=0}^{k-1} \binom{k}{h} g_{k-h}(x;a,b) \, q_h(x;c;a,b). \end{cases} \tag{21}$$

Noting that $p_0(x;a,b) = g_0(x;c;a,b) = 1$, the very first polynomials are given by

$$\begin{aligned}
q_0(x;c;a,b) &= 1, \\
q_1(x;c;a,b) &= p_1(x;a,b) - g_1(x;c;a,b), \\
q_2(x;c;a,b) &= p_2(x;a,b) - 2g_1(x;c;a,b) \, p_1(x;a,b) + 2g_1^2(x;c;a,b) - g_2(x;c;a,b), \\
q_3(x;c;a,b) &= p_3(x;a,b) - 3g_1(x;c;a,b) \, p_2(x;a,b) + 6g_1^2(x;c;a,b) \, p_1(x;a,b) \\
&\quad - 6g_1^3(x;c;a,b) + 6g_1(x;c;a,b) \, g_2(x;c;a,b) - 3g_2(x;c;a,b) \, p_1(x;a,b) \\
&\quad - g_3(x;c;a,b).
\end{aligned} \tag{22}$$

Further values can be obtained by using symbolic computation.

4. The General Case

Note that the above results can be extended to the general case, considering the generating function:

$$G(t,x) = \left(\frac{1}{1+a_1t+a_2t^2+\cdots+a_rt^r}\right)^x = \left[\frac{1}{(1-\alpha_1t)(1-\alpha_2t)\cdots(1-\alpha_rt)}\right]^x = \qquad(23)$$

$$= \exp\{-x\log[(1-\alpha_1t)(1-\alpha_2t)\cdots(1-\alpha_rt)]\} = \sum_{k=0}^{\infty} g_k(x;\alpha_1,\alpha_2,\ldots,\alpha_r)\frac{t^k}{k!},$$

where

$$a_1 = \sigma_1 = -(\alpha_1+\alpha_2+\cdots+\alpha_r),$$
$$\ldots,$$
$$a_s = \sigma_r = (-1)^s \sum_{j_1,j_2,\ldots,j_s} \alpha_{j_1}\alpha_{j_2}\cdots\alpha_{j_s}, \qquad(24)$$
$$\ldots,$$
$$a_r = \sigma_r = \alpha_1\alpha_2\cdots\alpha_r,$$

are the elementary symmetric functions of the zeros.

Putting as before:

$$G_{\alpha_h}(t,x) = \exp[-x\log(1-\alpha_h t)] = \sum_{k=0}^{\infty} p_{1,k}(x,\alpha_h)\frac{t^k}{k!}, \quad (h=1,2,\ldots,r), \qquad(25)$$

since

$$G(t,x) = G_{\alpha_1}(t,x) \cdot G_{\alpha_2}(t,x) \cdots G_{\alpha_r}(t,x),$$

we find the result:

Theorem 4. *The sequence $\{g_k(x)\}_{k\in\mathbb{N}}$ satisfies the convolution formula:*

$$g_k(x) = \sum_{\substack{k_1+k_2+\cdots+k_r=k \\ 0\leq k_i\leq k}} \binom{k}{k_1,k_2,\ldots,k_r} p_{1,k_1}(x)p_{2,k_2}(x)\cdots p_{r,k_r}(x), \qquad(26)$$

where, according to our position,

$$g_k(x) = g_k(x;\alpha_1,\alpha_2,\ldots,\alpha_r), \quad p_{1,k_1}(x) = p_{1,k_1}(x,\alpha_1),\ldots, p_{r,k_r}(x) = p_{r,k_r}(x,\alpha_r).$$

5. The General Recurrence Relation

From Equation (17) we find:

$$\frac{\partial G(t,x)}{\partial t} = -x\left(\frac{a_1+2a_2t+\cdots+ra_rt^{r-1}}{1+a_1t+a_2t^2+\cdots+a_rt^r}\right)G(t,x), \qquad(27)$$

$$(1+a_1t+a_2t^2+\cdots+a_rt^r)\frac{\partial G(t,x)}{\partial t} = -x\left(a_1+2a_2t+\cdots+ra_rt^{r-1}\right)G(t,x),$$

$$\sum_{k=0}^{\infty} g_{k+1}(x) \frac{t^k}{k!} + a_1 \sum_{k=0}^{\infty} g_{k+1}(x) \frac{t^{k+1}}{k!} + a_2 \sum_{k=0}^{\infty} g_{k+1}(x) \frac{t^{k+2}}{k!} + \cdots + a_r \sum_{k=0}^{\infty} g_{k+1}(x) \frac{t^{k+r}}{k!} =$$

$$= -a_1 x \sum_{k=0}^{\infty} g_k(x) \frac{t^k}{k!} - 2 a_2 x \sum_{k=0}^{\infty} g_k(x) \frac{t^{k+1}}{k!} - \cdots - r a_r x \sum_{k=0}^{\infty} g_k(x) \frac{t^{k+r-1}}{k!},$$

that is

$$\sum_{k=0}^{\infty} g_{k+1}(x) \frac{t^k}{k!} + a_1 \sum_{k=1}^{\infty} k g_k(x) \frac{t^k}{k!} + a_2 \sum_{k=2}^{\infty} k(k-1) g_{k-1}(x) \frac{t^k}{k!} + \cdots$$

$$+ a_r \sum_{k=r}^{\infty} k(k-1) \cdots (k-r+1) g_{k-r+1}(x) \frac{t^k}{k!} =$$

$$= -a_1 x \sum_{k=0}^{\infty} g_k(x) \frac{t^k}{k!} - 2 a_2 x \sum_{k=0}^{\infty} k g_{k-1}(x) \frac{t^k}{k!} - \cdots$$

$$- r a_r x \sum_{k=0}^{\infty} k(k-1) \cdots (k-r+2) g_{k-r+1}(x) \frac{t^k}{k!}.$$

Therefore, we can conclude that

Theorem 5. *The sequence* $\{g_k(x)\}_{k \in \mathbb{N}}$ *satisfies the linear recurrence relation*

$$g_k(x) + a_1(x+k-1) g_{k-1}(x) + a_2(k-1)(2x+k-2) g_{k-2}(x) + \cdots \qquad (28)$$

$$+ a_r(k-1)(k-2) \cdots (k-r+1)(rx+k-r) g_{k-r}(x) = 0.$$

Extension to the General Case

We now generalize the convolution formula in Section 3.2, putting for shortness

$$[c]_{r-1} = c_1, c_2, \ldots, c_{r-1}, \quad [a]_r = a_1, a_2, \ldots, a_r,$$

and considering the generating function:

$$G(t,x) = \left(\frac{1 + c_1 t + c_2 t^2 + \cdots + c_{r-1} t^{r-1}}{1 + a_1 t + a_2 t^2 + \cdots + a_r t^r} \right)^x = \sum_{k=0}^{\infty} q_k(x; [c]_{r-1}; [a]_r) \frac{t^k}{k!} =$$

$$= \frac{\sum_{k=0}^{\infty} p_k(x; [c]_{r-1}) \frac{t^k}{k!}}{\sum_{k=0}^{\infty} g_k(x; [a]_r) \frac{t^k}{k!}}, \qquad (29)$$

so that we find the convolution formula:

$$p_k(x; [c]_{r-1}) = \sum_{h=0}^{k} \binom{k}{h} g_{k-h}(x; [a]_r) q_h(x; [c]_{r-1}; [a]_r), \qquad (30)$$

and the $q_h(x; [c]_{r-1}; [a]_r)$ can be found recursively by solving the infinite system

$$\begin{cases} q_0(x; [c]_{r-1}; [a]_r) = 1, \\ q_k(x; [c]_{r-1}; [a]_r) = p_k(x; [a]_r) - \sum_{h=0}^{k-1} \binom{k}{h} g_{k-h}(x; [a]_r) q_h(x; [c]_{r-1}; [a]_r). \end{cases} \qquad (31)$$

6. Illustrative Examples—Second Order Recurrences

- **Gegenbauer polynomials** [14], defined by

$$(1 - 2yt + t^2)^{-\lambda} = \sum_{k=0}^{\infty} C_k^{(\lambda)}(y)\, t^k,$$

$x = \lambda,\ a = -2y,\ b = 1,\ g_k(\lambda; -2y, 1) = k!\, C_k^{(\lambda)}(y)$.

- **Sinha polynomials** [15], defined by

$$[1 - 2yt + (2y-1)t^2]^{-\nu} = \sum_{k=0}^{\infty} S_k^{(\nu)}(y)\, t^k,$$

$x = \nu,\ a = -2y,\ b = (2y-1),\ g_k(\nu; -2y, 2y-1) = k!\, S_k^{(\nu)}(y)$.

- **Fibonacci polynomials** [16], defined by

$$\frac{t}{1 - yt - t^2} = \sum_{k=0}^{\infty} F_k(y)\, t^k, \qquad F_k(1) = F_k \quad (\text{Fibonacci numbers}).$$

We have:

$$\frac{t}{1 - yt - t^2} = t \sum_{k=0}^{\infty} g_k(1; -y, -1) \frac{t^k}{k!} = \sum_{k=0}^{\infty} k!\, F_k(y) \frac{t^k}{k!},$$

so that

$$\sum_{k=1}^{\infty} k\, g_{k-1}(1; -y, -1) \frac{t^k}{k!} = \sum_{k=0}^{\infty} k!\, F_k(y) \frac{t^k}{k!}.$$

Since $F_0(y) = 0$, we find

$$F_k(y) = \frac{1}{(k-1)!}\, g_{k-1}(1; -y, -1).$$

- **Lucas polynomials** [16], defined by

$$\frac{2 - yt}{1 - yt - t^2} = \sum_{k=0}^{\infty} L_k(y)\, t^k, \qquad L_k(1) = L_k \quad (\text{Lucas numbers}).$$

We have:

$$\frac{2 - yt}{1 - yt - t^2} = 2 \sum_{k=0}^{\infty} g_k(1; -y, -1) \frac{t^k}{k!} - y \sum_{k=1}^{\infty} k\, g_{k-1}(1; -y, -1) \frac{t^k}{k!} = \sum_{k=0}^{\infty} k!\, L_k(y) \frac{t^k}{k!}.$$

Since $L_0(y) = 0$, we find

$$L_k(y) = \left(\frac{2}{k!} - \frac{y}{(k-1)!} \right) g_{k-1}(1, -y, -1).$$

Illustrative Examples—Higher Order Recurrences

- **Humbert polynomials** [14], defined by

$$(1 - 3yt + t^3)^{-\lambda} = \sum_{k=1}^{\infty} u_k(y)\, t^k,$$

$x = \lambda$, $a_1 = -3y$, $a_2 = 0$, $a_3 = 1$, $g_k(\lambda; -3y, 0, 1) = k!\, u_k(y)$.

- **First kind Chebyshev polynomials in several variables** [17–20], defined by

$$\frac{r - (r-1)u_1 t + (r-2)u_2 t^2 + \cdots + (-1)^{r-1} u_{r-1} t^{r-1}}{1 - u_1 t + u_2 t^2 - \cdots + (-1)^{r-1} u_{r-1} t^{r-1} + (-1)^r t^r} = \sum_{k=0}^{\infty} T_k(u_1, \ldots, u_{r-1})\, t^k,$$

$x = 1$, $c_1 = -\frac{r-1}{r} u_1, \ldots, c_{r-1} = \frac{(-1)^{r-1}}{r} u_{r-1}$, $a_1 = -u_1, \ldots, a_{r-1} = (-1)^{r-1} u_{r-1}, a_r = (-1)^r$,

$$q_k(1; [c]_{r-1}; [a]_r) = \frac{1}{r}\, k!\, T_k(u_1, \ldots, u_{r-1}).$$

- **Second kind Chebyshev polynomials in several variables** [17–20], defined by

$$\frac{1}{1 - u_1 t + u_2 t^2 - \cdots + (-1)^{r-1} u_{r-1} t^{r-1} + (-1)^r t^r} = \sum_{k=0}^{\infty} U_k(u_1, \ldots, u_{r-1})\, t^k,$$

$x = 1$, $a_1 = -u_1, \ldots, a_{r-1} = (-1)^{r-1} u_{r-1}, a_r = (-1)^r$, $g_k(1; [a]_r) = k!\, U_k(u_1, \ldots, u_{r-1})$.

- **Tribonacci polynomials** [21], defined by

$$\frac{t}{1 - y^2 t - y t^2 - t^3} = \sum_{k=0}^{\infty} \tau_k(y)\, t^k.$$

We have:

$$\frac{t}{1 - y^2 t - y t^2 - t^3} = t \sum_{k=0}^{\infty} g_k(1, -y^2, -y, -1) \frac{t^k}{k!} = \sum_{k=0}^{\infty} k!\, \tau_k(y) \frac{t^k}{k!},$$

so that

$$\sum_{k=1}^{\infty} k\, g_{k-1}(1, -y^2, -y, -1) \frac{t^k}{k!} = \sum_{k=0}^{\infty} k!\, \tau_k(y) \frac{t^k}{k!}.$$

Since $\tau_0(y) = 0$, we find

$$\tau_k(y) = \frac{1}{(k-1)!}\, g_{k-1}(1, -y^2, -y, -1).$$

7. Extension to Matrix Polynomials

Extensions to Matrix polynomials have become a fashionable subject recently (see e.g., [22] and the references therein).

The above results can be easily extended to Matrix polynomials assuming, in Equations (1), (7), (10), (17), (20), and (22), instead of x, a complex $N \times N$ matrix A, satisfying the condition:

A is stable, that is, denoting by $\sigma(A)$ the spectrum of A, this results in: $\forall \lambda \in \sigma(A)$, $\Re \lambda > 0$.

Since all powers of a matrix A commute, even every matrix polynomial commute. More generally, if $\sigma(A) \subset \Omega$, where Ω is an open set of the complex plane, for any holomorphic functions f and g, this results in:

$$f(A)g(A) = g(A)f(A),$$

that is, the involved matrix functions commute.

Under these conditions, considering the generating function:

$$G(t, A) = \left(\frac{1}{1 + a_1 t + a_2 t^2 + \cdots + a_r t^r} \right)^A =$$
$$= \exp\left\{ -A \log\left[1 + a_1 t + a_2 t^2 + \cdots + a_r t^r \right] \right\} = \sum_{k=0}^{\infty} g_k(A; a_1, a_2, \ldots, a_r) \frac{t^k}{k!}, \quad (32)$$

recalling positions (18), and putting as before:

$$G_{\alpha_h}(t, A) = \exp\left[-A \log(1 - \alpha_h t) \right] = \sum_{k=0}^{\infty} p_{1,k}(A, \alpha_h) \frac{t^k}{k!}, \quad (h = 1, 2, \ldots, r), \quad (33)$$

we find the result:

Theorem 6. *The sequence $\{g_k(A)\}_{k \in \mathbb{N}}$ satisfies the convolution formula:*

$$g_k(A; \alpha_1, \alpha_2, \ldots, \alpha_r) =$$
$$= \sum_{\substack{k_1 + k_2 + \cdots + k_r = k \\ 0 \leq k_i \leq k}} \binom{k}{k_1, k_2, \ldots, k_r} p_{1,k_1}(A, \alpha_1) p_{2,k_2}(A, \alpha_2) \cdots p_{r,k_r}(A, \alpha_r). \quad (34)$$

Furthermore, denoting by I the identity matrix, we can proclaim the theorem:

Theorem 7. *The sequence $\{g_k(A) := g_k(A; a_1, a_2, \ldots, a_r)\}_{k \in \mathbb{N}}$ satisfies the linear recurrence relation*

$$g_k(A) + a_1 [A + (k-1)I] g_{k-1}(A) + a_2 (k-1)[2A + (k-2)I] g_{k-2}(A) + \ldots$$
$$+ a_r (k-1)(k-2) \cdots (k-r+1)[rA + (k-r)I] g_{k-r}(A) = 0. \quad (35)$$

8. Conclusions

Starting from the results by Chen Zhuoyu and Qi Lan [9], we have shown convolution formulas and linear recurrence relations satisfied by a generating function containing several parameters. This can be used for number sequences (assuming $x = 1$) or polynomial sequences, depending on several parameters. Illustrative examples are shown both in case of second order or high order recurrence relations.

An extension to the case of matrix polynomials is also included.

Author Contributions: The authors claim to have contributed equally and significantly in this paper. Both authors read and approved the final manuscript.

Funding: This research received no external funding.

Acknowledgments: The authors are grateful to the anonymous referee for his careful reading of the manuscript, which permitted to correct the article.

Conflicts of Interest: The authors declare that they have not received funds from any institution and that they have no conflict of interest.

References

1. Srivastava, H.M.; Manocha, H.L. *A Treatise on Generating Functions*; Halsted Press (Ellis Horwood Limited): Chichester, UK; John Wiley and Sons: New York, NY, USA; Chichester, UK; Brisbane, Australia; Toronto, ON, Canada, 1984.
2. Kuş, S.; Tuglu, N.; Kim, T. Bernoulli F-polynomials and Fibo-Bernoulli matrices. *Adv. Differ. Equ.* **2019**, *2019*, 145. [CrossRef]
3. Zhang, Y.; Chen, Z. A New Identity Involving the Chebyshev Polynomials. *Mathematics* **2018**, *6*, 244. [CrossRef]
4. Ma, Y.; Zhang, W. Some Identities Involving Fibonacci Polynomials and Fibonacci Numbers. *Mathematics* **2018**, *6*, 334. [CrossRef]
5. Shen, S.; Chen, L. Some Types of Identities Involving the Legendre Polynomials. *Mathematics* **2019**, *7*, 114. [CrossRef]
6. Zhang, W. Some identities involving the Fibonacci numbers and Lucas numbers. *Fibonacci Q.* **2004**, *42*, 149–154.
7. Wang, S.Y. Some new identities of Chebyshev polynomials and their applications. *Adv. Differ. Equ.* **2015**, *2015*, 335.
8. Kim, T.; Dolgy, D.V.; Kim, D.S.; Seo, J.J. Convolved Fibonacci numbers and their applications. *Ars Combin.* **2017**, *135*, 119–131.
9. Chen, Z.; Qi, L. Some convolution formulae related to the second-order linear recurrence sequences. *Symmetry* **2019**, *11*, 788. [CrossRef]
10. Sheffer, I.M. Some properties of polynomials sets of zero type. *Duke Math. J.* **1939**, *5*, 590–622. [CrossRef]
11. Dattoli, G. Hermite-Bessel and Laguerre-Bessel functions: A by-product of the monomiality principle. In *Advanced Special Functions and Applications, Proceedings of the Melfi School on Advanced Topics in Mathematics and Physics, Melfi, Italy, 9–12 May 1999*; Cocolicchio, D., Dattoli, G., Srivastava, H.M., Eds.; Aracne Editrice: Rome, Italy, 2000; pp. 147–164.
12. Dattoli, G.; Ricci, P.E.; Srivastava, H.M. (Eds.) Advanced Special Functions and Related Topics in Probability and in Differential Equations. In Proceedings of the Melfi School on Advanced Topics in Mathematics and Physics, Melfi, Italy, 24–29 June 2001; In *Appl. Math. Comput.* **2003**, *141*, 1–230.
13. Ben Cheikh, Y. Some results on quasi-monomiality. *Appl. Math. Comput.* **2003**, *141*, 63–76. [CrossRef]
14. Boas, R.P.; Buck, R.C. *Polynomial Expansions of Analytic Functions*; Springer: Berlin/Heidelberg, Germany; Gottingen, Germany; New York, NY, USA, 1958.
15. Sinha, S.K. On a polynomial associated with Gegenbauer polynomial. *Proc. Nat. Acad. Sci. India Sect. A* **1989**, *54*, 439–455.
16. Koshy, T. *Fibonacci and Lucas Numbers with Applications*; Wiley: New York, NY, USA, 2001.
17. Lidl, R. Tschebyscheffpolynome in mehreren variabelen. *J. Reine Angew. Math.* **1975**, *273*, 178–198.
18. Ricci, P.E. I polinomi di Tchebycheff in più variabili. *Rend. Mat. (Ser. 6)* **1978**, *11*, 295–327.
19. Dunn, K.B.; Lidl, R. Multi-dimensional generalizations of the Chebyshev polynomials. I, II. *Proc. Jpn. Acad.* **1980**, *56*, 154–165. [CrossRef]
20. Bruschi, M.; Ricci, P.E. I polinomi di Lucas e di Tchebycheff in più variabili. *Rend. Mat. (Ser. 6)* **1980**, *13*, 507–530.
21. Goh, W; He, M.X.; Ricci, P.E. On the universal zero attractor of the Tribonacci-related polynomials. *Calcolo* **2009**, *46*, 95–129. [CrossRef]
22. Srivastava, H.M.; Khan, W.A.; Hiba, H. Some expansions for a class of generalized Humbert Matrix polynomials. *Rev. R. Acad. Cienc. Exactas Fís. Nat. Ser. A. Matem.* **2019**, To appear. [CrossRef]

© 2019 by the authors. Licensee MDPI, Basel, Switzerland. This article is an open access article distributed under the terms and conditions of the Creative Commons Attribution (CC BY) license (http://creativecommons.org/licenses/by/4.0/).

Article

Solutions of the Generalized Abel's Integral Equations of the Second Kind with Variable Coefficients

Chenkuan Li * and Hunter Plowman

Department of Mathematics and Computer Science, Brandon University, Brandon, MB R7A 6A9, Canada; plowmahh10@brandonu.ca
* Correspondence: lic@brandonu.ca

Received: 16 November 2019; Accepted: 3 December 2019; Published: 5 December 2019

Abstract: Applying Babenko's approach, we construct solutions for the generalized Abel's integral equations of the second kind with variable coefficients on R and R^n, and show their convergence and stability in the spaces of Lebesgue integrable functions, with several illustrative examples.

Keywords: Riemann–Liouville fractional integral; Mittag–Leffler function; Babenko's approach; generalized Abel's integral equation

MSC: 45E10; 26A33

1. Introduction

In 1823, Abel studied a physical problem regarding the relationship between kinetic and potential energies for falling bodies and constructed the integral equation [1–4]

$$g(x) = \int_c^x (x-t)^{-1/2} u(t) dt, \ c > 0,$$

where $g(x)$ is given and $u(x)$ is unknown. Later on, he worked on a more general integral equation given as

$$g(x) = \frac{1}{\Gamma(\alpha)} \int_0^x (x-t)^{\alpha-1} u(t) dt, \ 0 < \alpha < 1, \ a \le x \le b,$$

which is called Abel's integral equation of the first kind. Abel's integral equation of the second kind is generally given as

$$u(x) - \frac{\lambda}{\Gamma(\alpha)} \int_0^x (x-t)^{\alpha-1} u(t) dt = g(x), \ \alpha > 0 \qquad (1)$$

where λ is a constant.

Abel's integral equations are related to a wide range of physical problems, such as heat transfer [5], nonlinear diffusion [6], the propagation of nonlinear waves [7], and applications in the theory of neutron transport and traffic theory. There are many studies [8–14] on Abel's integral equations, including their variants and generalizations [15,16]. In 1930, Tamarkin investigated integrable solutions of Abel's integral equations under certain conditions by several integral operators [17]. Sumner [18] studied Abel's integral equations using the convolutional transform. Minerbo and Levy [19] found a numerical solution of Abel's integral equation by orthogonal polynomials. In 1985, Hatcher [20] worked on a nonlinear Hilbert problem of power type, solved in closed form by representing a sectionally holomorphic function by means of an integral with power kernel, and transformed the problem to one of solving a generalized Abel's integral equation. Using a modification of Mikusinski operational calculus, Gorenflo and Luchko [21] obtained an explicit solution of the generalized Abel's integral equation of the second kind, in terms of the Mittag–Leffler function of several variables.

$$u(x) - \sum_{i=1}^{m} \lambda_i (I^{\alpha_i \mu} u)(x) = g(x), \quad \alpha_i > 0, m \geq 1, \mu > 0, x > 0$$

where λ_i is a constant for $i = 1, 2, \cdots, m$, and I^μ is the Riemann–Liouville fractional integral of order $\mu \in R^+$ with initial point zero [22],

$$(I^\mu u)(x) = \frac{1}{\Gamma(\mu)} \int_0^x (x - t)^{\mu - 1} u(t) dt.$$

Lubich [10] constructed the numerical solution for the following Abel's integral equation of the second kind based on fractional powers of linear multistep methods

$$u(x) = g(x) + \frac{1}{\Gamma(\alpha)} \int_0^x (x - t)^{\alpha - 1} f(t, u(t)) dt \quad \text{on } R^n$$

where $x \in [0, T]$ and $\alpha > 0$. The case $\alpha = 1/2$ is encountered in a variety of problems in physics and chemistry [23]. Pskhu [24] considered the following generalized Abel's integral equation with constant coefficients a_k for $k = 1, 2, \cdots, n$

$$\sum_{k=1}^{n} a_k I^{\alpha_k} u(x) = g(x),$$

where $\alpha_k \geq 0$ and $x \in (0, a)$, and constructed an explicit solution based on the Wright function

$$\phi(\alpha, \beta; z) = \sum_{n=0}^{\infty} \frac{z^n}{n! \Gamma(\alpha n + \beta)}, \quad \alpha > -1, \beta \in \mathbb{C}$$

and convolution. Li et al. [25–27] recently studied Abel's integral Equation (1) for any arbitrary $\alpha \in R$ in the generalized sense based on fractional calculus of distributions, inverse convolutional operators and Babenko's approach [28]. They obtained several new and interesting results that cannot be realized in the classical sense or by the Laplace transform. Many applied problems from physical science lead to integral equations which can be converted to the form of Abel's integral equations for analytic or distributional solutions in the case where classical ones do not exist [15,27].

Letting $\alpha_1 > \alpha_2 > \cdots > \alpha_n > 0$ and $a > 0$, we consider the generalized Abel's integral equation of the second kind with variable coefficients

$$u(x) - \sum_{k=1}^{n} a_k(x) I^{\alpha_k} u(x) = g(x), \quad (2)$$

where $x \in (0, a)$, $a_i(x)$ is Lebesgue integrable and bounded on $(0, a)$ for $i = 1, 2, \cdots, n$, $g(x)$ is a given function in $L(0, a)$ and $u(x)$ is the unknown function. Clearly, Equation (2) turns to be

$$u(x) - a_1 I^{\alpha_1} u(x) = g(x) \quad (3)$$

if $n = 1$ and $a_1(x) = a_1$ (constant). Equation (3) is the classical Abel's integral equation of the second kind, with the solution given by Hille and Tamarkin [29]

$$u(x) = g(x) + a_1 \int_0^x (x - t)^{\alpha_1 - 1} E_{\alpha_1, \alpha_1}(a_1 (x - t)^{\alpha_1}) g(t) dt,$$

where

$$E_{\alpha, \beta}(z) = \sum_{n=0}^{\infty} \frac{z^n}{\Gamma(\alpha n + \beta)}, \quad \alpha, \beta > 0$$

is the Mittag–Leffler function.

Following a similar approach, we also establish a convergent and stable solution for the generalized Abel's integral equation on R^n with variable coefficients

$$u(x) - a_1(x) I_1^{\alpha_1} a_2(x) I_2^{\alpha_2} \cdots a_n(x) I_n^{\alpha_n} u(x) = g(x),$$

where $x = (x_1, x_2, \cdots, x_n)$ and I_k^α is the partial Riemann–Liouville fractional integral of order $\alpha \in R^+$ with respect to x_k, with initial point 0,

$$(I_k^\alpha u)(x) = \frac{1}{\Gamma(\alpha)} \int_0^{x_k} (x_k - t)^{\alpha - 1} u(x_1, \cdots, x_{k-1}, t, x_{k+1}, \cdots, x_n) dt$$

where $k = 1, 2, \cdots, n$.

2. The Main Results

Theorem 1. *Let $x \in (0, a)$, $a_i(x)$ be Lebesgue integrable and bounded on $(0, a)$ for $i = 1, 2, \cdots, n$, and $g(x)$ be a given function in $L(0, a)$. Then the generalized Abel's integral equation of the second kind with variable coefficients*

$$u(x) - \sum_{k=1}^n a_k(x) I^{\alpha_k} u(x) = g(x)$$

has the following convergent and stable solution in $L(0, a)$

$$u(x) = \sum_{m=0}^\infty \left(\sum_{k=1}^n a_k(x) I^{\alpha_k} \right)^m g(x),$$

where $\alpha_1 > \alpha_2 > \cdots > \alpha_n > 0$.

Proof. Clearly,

$$u(x) - \sum_{k=1}^n a_k(x) I^{\alpha_k} u(x) = \left(1 - \sum_{k=1}^n a_k(x) I^{\alpha_k} \right) u(x) = g(x)$$

which implies, by Babenko's approach (treating the operator like a variable), that

$$u(x) = \frac{1}{1 - \sum_{k=1}^n a_k(x) I^{\alpha_k}} g(x) = \sum_{m=0}^\infty \left(\sum_{k=1}^n a_k(x) I^{\alpha_k} \right)^m g(x)$$

$$= \sum_{m=0}^\infty \sum_{m_1 + m_2 + \cdots + m_n = m} \left(\frac{m!}{m_1!, m_2!, \cdots, m_n!} \right) (a_1(x) I^{\alpha_1})^{m_1} \cdots (a_n(x) I^{\alpha_n})^{m_n} g(x).$$

Let $\|f\|$ be the usual norm of $f \in L(0, a)$, given by

$$\|f\| = \int_0^a |f(x)| dx < \infty.$$

Then, we have from [30]

$$\|I^{\alpha_i} g\| = \|\Phi_{\alpha_i} * g\| \leq \|\Phi_{\alpha_i}\| \|g\|$$

where

$$\Phi_{\alpha_i} = \frac{x_+^{\alpha_i - 1}}{\Gamma(\alpha_i)}.$$

This implies that

$$\|I^{\alpha_i}\| \leq \|\Phi_{\alpha_i}\| = \frac{1}{\Gamma(\alpha_i)} \int_0^a x^{\alpha_i - 1} = \frac{a^{\alpha_i}}{\Gamma(\alpha_i + 1)}.$$

Since $a_i(x)$ is bounded over $(0, a)$, there exists $M > 0$ such that
$$\sup_{x \in (0,a)} |a_i(x)| \leq M$$
for all $i = 1, 2, \cdots, n$. Therefore,
$$\|u\| \leq \sum_{m=0}^{\infty} M^m \sum_{m_1+m_2+\cdots+m_n=m} \binom{m!}{m_1!, m_2!, \cdots, m_n!} \cdot$$
$$\|I^{m_1\alpha_1}\| \|I^{m_2\alpha_2}\| \cdots \|I^{m_n\alpha_n}\| \|g\|$$
$$\leq \sum_{m=0}^{\infty} M^m \sum_{m_1+m_2+\cdots+m_n=m} \binom{m!}{m_1!, m_2!, \cdots, m_n!} \cdot$$
$$\frac{a^{m_1\alpha_1+\cdots+m_n\alpha_n}}{\Gamma(m_1\alpha_1+1)\cdots\Gamma(m_n\alpha_n+1)} \|g\|.$$

Let
$$A = \max\{a, 1\}.$$
Then,
$$a^{m_1\alpha_1+\cdots+m_n\alpha_n} \leq A^{m_1\alpha_1+\cdots+m_n\alpha_n} \leq A^{\alpha_1 m}$$
as $\alpha_1 > \alpha_2 > \cdots > \alpha_n > 0$. On the other hand,
$$\Gamma(m_1\alpha_1+1)\cdots\Gamma(m_n\alpha_n+1) \geq \Gamma(m_1\alpha_n+1)\cdots\Gamma(m_n\alpha_n+1) \geq \left(\frac{1}{2}\right)^{n-1} \Gamma\left(\alpha_n \frac{m}{n}+1\right),$$
since there exists $m_i \geq m/n$ for some i by noting that $m_1 + m_2 + \cdots + m_n = m$, and the factor $\Gamma(m_j\alpha_n+1) \geq 1/2$ for $j \neq i$. Hence,
$$\frac{1}{\Gamma(m_1\alpha_1+1)\cdots\Gamma(m_n\alpha_n+1)} \leq \frac{2^{n-1}}{\Gamma\left(\alpha_n \frac{m}{n}+1\right)},$$
and
$$\|u\| \leq 2^{n-1}\|g\| \sum_{m=0}^{\infty} \frac{M^m n^m A^{\alpha_1 m}}{\Gamma\left(\alpha_n \frac{m}{n}+1\right)} = 2^{n-1}\|g\| \sum_{m=0}^{\infty} \frac{(MnA^{\alpha_1})^m}{\Gamma\left(\alpha_n \frac{m}{n}+1\right)}$$
$$= 2^{n-1}\|g\| E_{\alpha_n/n, 1}(MnA^{\alpha_1}) < \infty$$
by using
$$\sum_{m_1+m_2+\cdots+m_n=m} \binom{m!}{m_1!, m_2!, \cdots, m_n!} = n^m.$$
Furthermore, the solution
$$u(x) = \sum_{m=0}^{\infty} \left(\sum_{k=1}^{n} a_k(x) I^{\alpha_k}\right)^m g(x)$$
is stable from the last inequality. This completes the proof of Theorem 1. □

3. Illustrative Examples

Let α and β be arbitrary real numbers. Then it follows from [31]
$$\Phi_\alpha * \Phi_\beta = \Phi_{\alpha+\beta}.$$

Example 1. *Assume $\alpha > 0$. Then Abel's integral equation with a variable coefficient*

$$u(x) - x^\alpha I^{2.5} u(x) = x, \quad x \in (0, a)$$

has the following stable solution

$$u(x) = x + \sum_{m=1}^{\infty} \frac{\Gamma(\alpha + 4.5)\Gamma(2\alpha + 7) \cdots \Gamma(m\alpha + 4.5 + (m-1)2.5)}{\Gamma(4.5)\Gamma(\alpha + 7) \cdots \Gamma((m-1)\alpha + 4.5 + (m-1)2.5)} \Phi_{m\alpha + 4.5 + (m-1)2.5}(x)$$

in $L(0,a)$.

Indeed,

$$u(x) = x + \sum_{m=1}^{\infty} (x^\alpha I^{2.5})^m \cdot x = x + \sum_{m=1}^{\infty} (x^\alpha \Phi_{2.5})^m * \Phi_2.$$

Clearly,

$$x^\alpha \Phi_{2.5} * \Phi_2 = x^\alpha \Phi_{4.5} = \frac{x^{\alpha + 3.5}}{\Gamma(4.5)} = \frac{\Gamma(\alpha + 4.5)}{\Gamma(4.5)} \Phi_{\alpha + 4.5},$$

$$(x^\alpha \Phi_{2.5}) * \frac{\Gamma(\alpha + 4.5)}{\Gamma(4.5)} \Phi_{\alpha + 4.5} = \frac{\Gamma(\alpha + 4.5)}{\Gamma(4.5)} x^\alpha \Phi_{\alpha + 7} = \frac{\Gamma(\alpha + 4.5)}{\Gamma(4.5)} \frac{x^{2\alpha + 6}}{\Gamma(\alpha + 7)}$$

$$= \frac{\Gamma(\alpha + 4.5)}{\Gamma(4.5)} \frac{\Gamma(2\alpha + 7)}{\Gamma(\alpha + 7)} \Phi_{2\alpha + 7},$$

$$\cdots,$$

$$(x^\alpha \Phi_{2.5})^m * \Phi_2 = \frac{\Gamma(\alpha + 4.5)\Gamma(2\alpha + 7) \cdots \Gamma(m\alpha + 4.5 + (m-1)2.5)}{\Gamma(4.5)\Gamma(\alpha + 7) \cdots \Gamma((m-1)\alpha + 4.5 + (m-1)2.5)}$$

$$\Phi_{m\alpha + 4.5 + (m-1)2.5}$$

where $m \geq 1$.

Example 2. *Let $a > 0$. Then Abel's integral equation*

$$u(x) - xI^{0.5} u(x) - x^{0.5} I u(x) = x^{-0.5}, \quad x \in (0, a)$$

has the following stable solution

$$u(x) = x^{-0.5} + \sqrt{\pi} \sum_{m=1}^{\infty} \sum_{k=0}^{m} C_k B_{m,k} \Phi_{2 + 1.5(m-1)}(x)$$

in $L(0,a)$, where

$$C_k = \begin{cases} 1 & \text{if } k = 0, \\ \dfrac{\Gamma(2)\Gamma(3.5) \cdots \Gamma(2 + 1.5(k-1))}{\Gamma(1.5)\Gamma(3) \cdots \Gamma(1.5 + 1.5(k-1))} & \text{if } k \geq 1 \end{cases}$$

and

$$B_{m,k} = \begin{cases} 1 & \text{if } k = m, \\ \dfrac{\Gamma(2 + 1.5k)\Gamma(2 + 1.5(k+1)) \cdots \Gamma(2 + 1.5(m-1))}{\Gamma(1 + 1.5k)\Gamma(1 + 1.5(k+1)) \cdots \Gamma(1 + 1.5(m-1))} & \text{if } k < m. \end{cases}$$

Indeed,

$$u(x) = x^{-0.5} + \sum_{m=1}^{\infty} (xI^{0.5} + x^{0.5} I)^m \cdot x^{-0.5}$$

$$= x^{-0.5} + \sqrt{\pi} \sum_{m=1}^{\infty} \sum_{k=0}^{m} \binom{m}{k} (x \Phi_{0.5})^{m-k} * (x^{0.5} \Phi_1)^k * \Phi_{0.5}.$$

Clearly,

$$(x^{0.5}\Phi_1) * \Phi_{0.5} = x^{0.5}\Phi_{1.5} = \frac{x}{\Gamma(1.5)} = \frac{\Gamma(2)}{\Gamma(1.5)}\Phi_2,$$

$$(x^{0.5}\Phi_1)^2 * \Phi_{0.5} = (x^{0.5}\Phi_1) * \frac{\Gamma(2)}{\Gamma(1.5)}\Phi_2 = \frac{\Gamma(2)}{\Gamma(1.5)}x^{0.5}\Phi_3$$

$$= \frac{\Gamma(2)}{\Gamma(1.5)}\frac{x^{2.5}}{\Gamma(3)} = \frac{\Gamma(2)\Gamma(3.5)}{\Gamma(1.5)\Gamma(3)}\Phi_{3.5},$$

$$\cdots,$$

$$(x^{0.5}\Phi_1)^k * \Phi_{0.5} = \frac{\Gamma(2)\Gamma(3.5)\cdots\Gamma(2+1.5(k-1))}{\Gamma(1.5)\Gamma(3)\cdots\Gamma(1.5+1.5(k-1))}\Phi_{0.5+1.5k} = C_k\Phi_{0.5+1.5k}$$

where C_k is defined as above. Furthermore,

$$(x\Phi_{0.5}) * \Phi_{0.5+1.5k} = x\Phi_{1+1.5k} = \frac{x^{1+1.5k}}{\Gamma(1+1.5k)} = \frac{\Gamma(2+1.5k)}{\Gamma(1+1.5k)}\Phi_{2+1.5k},$$

$$(x\Phi_{0.5})^2 * \Phi_{0.5+1.5k} = \frac{\Gamma(2+1.5k)}{\Gamma(1+1.5k)} x \Phi_{2.5+1.5k} = \frac{\Gamma(2+1.5k)}{\Gamma(1+1.5k)} x \Phi_{1+1.5(k+1)}$$

$$= \frac{\Gamma(2+1.5k)\Gamma(2+1.5(k+1))}{\Gamma(1+1.5k)\Gamma(1+1.5(k+1))}\Phi_{2+1.5(k+1)},$$

$$\cdots,$$

$$(x\Phi_{0.5})^{m-k} * \Phi_{0.5+1.5k} = \frac{\Gamma(2+1.5k)\Gamma(2+1.5(k+1))\cdots\Gamma(2+1.5(m-1))}{\Gamma(1+1.5k)\Gamma(1+1.5(k+1))\cdots\Gamma(1+1.5(m-1))} \cdot$$

$$\Phi_{2+1.5(m-1)} = B_{m,k}\Phi_{2+1.5(m-1)}$$

where $B_{m,k}$ is defined above.

Remark 1. *As far as we know, the solution for the generalized Abel's integral equation with variable coefficients over the interval $(0, a)$ is obtained for the first time. However, this approach seems unworkable if the interval is unbounded, as the Riemann–Liouville fractional integral operator is therefore unbounded. In the proof and computations of the above examples, we should point out that the convolution operations are prior to functional multiplications, according to our approach.*

Assuming that $\omega_i > 0$ for all $i = 1, 2, \cdots, n$, and $\Omega = (0, \omega_1) \times (0, \omega_2) \times \cdots \times (0, \omega_n)$, we can derive the following theorem by a similar procedure.

Theorem 2. *Let $\alpha_k \geq 0$ for $k = 1, 2, \cdots, n$ and there is at least one $\alpha_i > 0$ for some $1 \leq i \leq n$. Then the generalized Abel's integral equation of the second kind with variable coefficients on R^n for a given function $g \in L(\Omega)$*

$$u(x) - a_1(x)I_1^{\alpha_1}a_2(x)I_2^{\alpha_2}\cdots a_n(x)I_n^{\alpha_n}u(x) = g(x)$$

has the following convergent and stable solution in $L(\Omega)$

$$u(x) = \sum_{m=0}^{\infty} \left(a_1(x)I_1^{\alpha_1}a_2(x)I_2^{\alpha_2}\cdots a_n(x)I_n^{\alpha_n}\right)^m g(x), \tag{4}$$

where $a_k(x)$ is Lebesgue integrable and bounded on Ω for $k = 1, 2, \cdots, n$.

Proof. Clearly,

$$u(x) - a_1(x)I_1^{\alpha_1}\cdots a_n(x)I_n^{\alpha_n}u(x) = \left(1 - a_1(x)I_1^{\alpha_1}\cdots a_n(x)I_n^{\alpha_n}\right)u(x) = g(x),$$

and
$$u(x) = \frac{1}{1 - a_1(x)I_1^{\alpha_1} \cdots a_n(x)I_n^{\alpha_n}} g(x)$$
$$= \sum_{m=0}^{\infty} \left(a_1(x)I_1^{\alpha_1} a_2(x)I_2^{\alpha_2} \cdots a_n(x)I_n^{\alpha_n}\right)^m g(x).$$

It remains to show that the above is convergent and stable in $L(\Omega)$. Let
$$W = \left(a_1(x)I_1^{\alpha_1} \cdots a_n(x)I_n^{\alpha_n}\right)^m$$
$$= \left(a_1(x)I_1^{\alpha_1} \cdots a_n(x)I_n^{\alpha_n}\right) \cdots \left(a_1(x)I_1^{\alpha_1} \cdots a_n(x)I_n^{\alpha_n}\right).$$

Since $a_k(x)$ is bounded on Ω for $k = 1, 2, \cdots, n$, there exists $M > 0$ such that
$$\sup_{x \in \Omega} |a_k(x)| \leq M.$$

Let $\|f\|$ be the usual norm of $f \in L(\Omega)$, given by
$$\|f\| = \int_{\Omega} |f(x)|dx = \int_{\Omega} |f(x_1, x_2, \cdots, x_n)|dx_1 dx_2 \cdots dx_n < \infty.$$

Then, it follows from [30] for $k = 1, 2, \cdots, n$
$$\|I_k^{\alpha_k} g\| = \|\Phi_{k,\alpha_k} * g\| \leq \|\Phi_{k,\alpha_k}\| \|g\|$$

where
$$\Phi_{k,\alpha_k} = \frac{(x_k)_+^{\alpha_k - 1}}{\Gamma(\alpha_k)}.$$

This implies for $\alpha_k > 0$ that
$$\|I_k^{\alpha_k}\| \leq \|\Phi_{k,\alpha_k}\| = \int_{\Omega} \frac{(x_k)_+^{\alpha_k - 1}}{\Gamma(\alpha_k)} dx_1 dx_2 \cdots dx_n$$
$$= \omega_1 \cdots \omega_{k-1} \frac{\omega_k^{\alpha_k}}{\Gamma(\alpha_k + 1)} \omega_{k+1} \cdots \omega_n \leq \lambda^{n-1} \frac{\omega_k^{\alpha_k}}{\Gamma(\alpha_k + 1)}$$

where
$$\lambda = \max\{\omega_1, \omega_2, \cdots, \omega_n\} > 0.$$

In particular for $\alpha_k = 0$,
$$\|I_k^0\| \leq \lambda^{n-1}.$$

Therefore,
$$\|W\| \leq M^{nm} \|I_1^{\alpha_1 m}\| \cdots \|I_n^{\alpha_n m}\|$$
$$\leq M^{nm} \lambda^{n^2 - n} \frac{\omega_1^{\alpha_1 m}}{\Gamma(\alpha_1 m + 1)} \cdots \frac{\omega_n^{\alpha_n m}}{\Gamma(\alpha_n m + 1)}$$
$$\leq M^{nm} \lambda^{n^2 - n} S^{nm} \frac{1}{\Gamma(\alpha_1 m + 1)} \cdots \frac{1}{\Gamma(\alpha_n m + 1)},$$

where
$$S = \max\{\omega_1^{\alpha_1}, \cdots, \omega_n^{\alpha_n}\}.$$

Without loss of generality, we assume that $\alpha_1 > 0$. Then,

$$\Gamma(\alpha_1 m + 1) \cdots \Gamma(\alpha_n m + 1) \geq \frac{1}{2^{n-1}} \Gamma(\alpha_1 m + 1)$$

since

$$\Gamma(\alpha_k m + 1) \geq 1/2$$

for $k = 2, \cdots, n$. This infers that

$$\|u(x)\| \leq \lambda^{n^2-n} 2^{n-1} \|g\| \sum_{m=0}^{\infty} \frac{(M^n S^n)^m}{\Gamma(\alpha_1 m + 1)} < +\infty$$

by the Mittag–Leffler function. Furthermore, the solution

$$u(x) = \sum_{m=0}^{\infty} \left(a_1(x) I_1^{\alpha_1} a_2(x) I_2^{\alpha_2} \cdots a_n(x) I_n^{\alpha_n}\right)^m g(x)$$

is stable from the last inequality. This completes the proof of Theorem 2. □

In particular, let $g(x) = \phi_1(x_1) \cdots \phi_n(x_n) \in L(\Omega)$. Then

$$u(x) - a_1(x_1) I_1^{\alpha_1} a_2(x_2) I_2^{\alpha_2} \cdots a_n(x_n) I_n^{\alpha_n} u(x) = \phi_1(x_1) \cdots \phi_n(x_n)$$

has the following convergent and stable solution

$$u(x) = \sum_{m=0}^{\infty} \left(a_1(x_1) I_1^{\alpha_1}\right)^m \phi_1(x_1) \cdots \left(a_n(x_n) I_n^{\alpha_n}\right)^m \phi_n(x_n)$$

in $L(\Omega)$.

4. Conclusions

We establish the convergent and stable solutions for the following generalized Abel's integral equations of the second kind with variable coefficients

$$u(x) - \sum_{k=1}^{n} a_k(x) I^{\alpha_k} u(x) = g(x), \quad x \in (0, a) \subset R$$

$$u(x) - a_1(x) I_1^{\alpha_1} a_2(x) I_2^{\alpha_2} \cdots a_n(x) I_n^{\alpha_n} u(x) = g(x), \quad x \in \Omega \subset R^n$$

in the spaces of Lebesgue integrable functions, and provide applicable examples based on convolutions and gamma functions.

Author Contributions: The order of the author list reflects contributions to the paper.

Funding: This work is partially supported by NSERC (Canada 2019-03907).

Conflicts of Interest: The authors declare no conflict of interest.

References

1. Li, M.; Zhao, W. Solving Abel's type integral equation with Mikusinski's operator of fractional order. *Adv. Math. Phys.* **2013**, *2013*, 806984. [CrossRef]
2. Wazwaz, A.M. *Linear and Nonlinear Integral Equations: Methods and Applications*; Higher Education: Beijing, China; Springer: Berlin, Germany, 2011.
3. Wazwaz, A.M. *A First Course in Integral Equations*; World Scientific Publishing: Singapore, 1997.
4. Wazwaz, A.M.; Mehanna, M.S. The combined Laplace-Adomian method for handling singular integral equation of heat transfer. *Int. J. Nonlinear Sci.* **2010**, *10*, 248–252.

5. Mann, W.R.; Wolf, F. Heat transfer between solids and gases under nonlinear boundary conditions. *Q. Appl. Math.* **1951**, *9*, 163–184. [CrossRef]
6. Goncerzewicz, J.; Marcinkowska, H.; Okrasinski, W.; Tabisz, K. On percolation of water from a cylindrical reservoir into the surrounding soil. *Appl. Math.* **1978**, *16*, 249–261. [CrossRef]
7. Keller, J.J. Propagation of simple nonlinear waves in gas filled tubes with friction. *Z. Angew. Math. Phys.* **1981**, *32*, 170–181. [CrossRef]
8. Gorenflo, R.; Mainardi, F. Fractional Calculus: Integral and Differential Equations of Fractional Order. In *Fractals and Fractional Calculus in Continuum Mechanics*; Springer: New York, NY, USA, 1997; pp. 223–276.
9. Avazzadeh, Z.; Shafiee, B.; Loghmani, G.B. Fractional calculus for solving Abel's integral equations using Chebyshev polynomials. *Appl. Math. Sci.* **2011**, *5*, 2207–2216.
10. Lubich, C. Fractional linear multistep methods for Abel-Volterra integral equations of the second kind. *Math. Comput.* **1985**, *45*, 463–469. [CrossRef]
11. Huang, L.; Huang, Y.; Li, X.F. Approximate solution of Abel integral equation. *Comput. Math. Appl.* **2008**, *56*, 1748–1757. [CrossRef]
12. Brunner, H. *Collocation Methods for Volterra Integral and Related Functional Differential Equations*; Cambridge Monographs on Applied and Computational Mathematics, 15; Cambridge Univ. Press: Cambridge, UK, 2004.
13. Mandal, N.; Chakrabart, A.; Mandal, B.N. Solution of a system of generalized Abel integral equations using fractional calculus. *Appl. Math. Lett.* **1996**, *9*, 1–4. [CrossRef]
14. Srivastava, H.M.; Saxena, R.K. Operators of fractional integration and their applications. *Appl. Math. Comput.* **2001**, *118*, 1–52. [CrossRef]
15. Podlubny, I. *Fractional Differential Equations*; Academic Press: New York, NY, USA, 1999.
16. Srivastava, H.M.; Buschman, R.G. *Theory and Applications of Convolution Integral Equations*; Kluwer Academic Publishers: Dordrecht, The Netherlands; Boston, MA, USA; London, UK, 1992.
17. Tamarkin, J.D. On integrable solutions of Abel's integral equation. *Ann. Math.* **1930**, *31*, 219–229. [CrossRef]
18. Sumner, D.B. Abel's integral equation as a convolution transform. *Proc. Am. Math. Soc.* **1956**, *7*, 82–86.
19. Minerbo, G.N.; Levy, M.E. Inversion of Abel's integral equation by means of orthogonal polynomials. *SIAM J. Numer. Anal.* **1969**, *6*, 598–616. [CrossRef]
20. Hatcher, J.R. A nonlinear boundary problem. *Proc. Am. Math. Soc.* **1985**, *95*, 441–448. [CrossRef]
21. Gorenflo, R.; Luchko, Y. Operational method for solving generalized Abel integral equation of second kind. *Integr. Transform. Spec. Funct.* **1997**, *5*, 47–58. [CrossRef]
22. Kilbas, A.A.; Srivastava, H.M.; Trujillo, J.J. *Theory and Applications of Fractional Differential Equations*; Elsevier: North-Holland, The Netherlands, 2006.
23. Brunner, H. A survey of recent advances in the numerical treatment of Volterra integral and integro-differential equations. *J. Comput. Appl. Math.* **1982**, *8*, 213–219. [CrossRef]
24. Pskhu, A. On solution representation of generalized Abel integral equation. *J. Math.* **2013**, *2013*, 106251. [CrossRef]
25. Li, C.; Li, C.P.; Clarkson, K. Several results of fractional differential and integral equations in distribution. *Mathematics* **2018**, *6*, 97. [CrossRef]
26. Li, C.; Humphries, T.; Plowman, H. Solutions to Abel's integral equations in distributions. *Axioms* **2018**, *7*, 66. [CrossRef]
27. Li, C.; Clarkson, K. Babenko's Approach Abel's Integral Equations. *Mathematics* **2018**, *6*, 32.
28. Babenko, Y.I. *Heat and Mass Transfer*; Khimiya: Leningrad, Russia, 1986. (In Russian)
29. Hille, E.; Tamarkin, J.D. On the theory of linear integral equations. *Ann. Math.* **1930**, *31*, 479–528. [CrossRef]
30. Barros-Neto, J. *An Introduction to the Theory of Distributions*; Marcel Dekker, Inc.: New York, NY, USA, 1973.
31. Gel'fand, I.M.; Shilov, G.E. *Generalized Functions*; Academic Press: New York, NY, USA, 1964; Volume I.

© 2019 by the authors. Licensee MDPI, Basel, Switzerland. This article is an open access article distributed under the terms and conditions of the Creative Commons Attribution (CC BY) license (http://creativecommons.org/licenses/by/4.0/).

Article

Harmonic Starlike Functions with Respect to Symmetric Points

Nak Eun Cho [1,*] and Jacek Dziok [2]

[1] Department of Applied Mathematics, College of Natural Sciences, Pukyong National University, Busan 608-737, Korea
[2] Faculty of Mathematics and Natural Sciences, University of Rzeszow, Poland; jdziok@ur.edu.pl
* Correspondence: necho@pknu.ac.kr

Received: 2 October 2019; Accepted: 21 November 2019; Published: 22 December 2019

Abstract: In the paper we define classes of harmonic starlike functions with respect to symmetric points and obtain some analytic conditions for these classes of functions. Some results connected to subordination properties, coefficient estimates, integral representation, and distortion theorems are also obtained.

Keywords: harmonic functions; janowski functions; starlike functions; extreme points; subordination

---

1. Introduction

We denote by $\mathcal{H}$ the class of complex-valued harmonic functions in the unit disc $\mathbb{U} := \{z : |z| < r\}$. Then $f \in \mathcal{H}$ if $f = h + \overline{g}$, where h, g are functions analytic in $\mathbb{U}$. Let $\mathcal{H}_0$ be the class of function $f \in \mathcal{H}$ with the following normalization:

$$f(z) = z + \sum_{n=2}^{\infty} a_n z^n + \overline{\sum_{n=2}^{\infty} b_n z^n} \quad (z \in \mathbb{U}) \qquad (1)$$

and let $\mathcal{S}_\mathcal{H}$ denote the class of functions $f \in \mathcal{H}_0$, which are orientation preserving and univalent in $\mathbb{U}$.

For functions $f_1, f_2 \in \mathcal{H}$ of the forms:

$$f_k(z) = \sum_{n=0}^{\infty} a_{k,n} z^n + \overline{\sum_{n=1}^{\infty} b_{k,n} z^n} \quad (z \in \mathbb{U},\ k \in \{1,2\}) \qquad (2)$$

by $f_1 * f_2$ we denote the Hadamard product or convolution of f_1 and f_2, defined by:

$$(f_1 * f_2)(z) = \sum_{n=0}^{\infty} a_{1,n} a_{2,n} z^n + \overline{\sum_{n=1}^{\infty} b_{1,n} b_{2,n} z^n} \quad (z \in \mathbb{U}).$$

We say that a function $f : \mathbb{U} \to \mathbb{C}$ is subordinate to a function $F : \mathbb{U} \to \mathbb{C}$, and write $f(z) \prec F(z)$ (or simply $f \prec F$), if there exists a complex-valued function ω which maps $\mathbb{U}$ into oneself with $\omega(0) = 0$, such that $f = F \circ \omega$. In particular, if F is univalent in $\mathbb{U}$, we have the following equivalence:

$$f(z) \prec F(z) \iff [f(0) = F(0) \text{ and } f(\mathbb{U}) \subset F(\mathbb{U})].$$

In 1956 Sakaguchi [1] introduced the class $\mathcal{S}^{**}$ of analytic univalent functions in $\mathbb{U}$ which are starlike with respect to symmetrical points. An analytic function f is said to be starlike with respect to symmetric points if:

$$\operatorname{Re} \frac{z f'(z)}{f(z) - f(-z)} > 0 \quad (z \in \mathbb{U}). \qquad (3)$$

If $f \in \mathcal{S}^{**}$ then the angular velocity of $f(z)$ about the point $f(-z)$ is positive as z traverses the circle $|z| = r$ in a positive direction.

Let A and B be two distinct complex parameters and let $0 \leq \alpha < 1$. In [2] (see also [3]) it is defined the class $\mathcal{S}_\mathcal{H}^*(A,B)$ of Janowski harmonic starlike functions $f \in \mathcal{S}_\mathcal{H}$ such that:

$$\frac{D_\mathcal{H} f(z)}{f(z)} \prec \frac{1+Az}{1+Bz}, \qquad (4)$$

where,

$$D_\mathcal{H} f(z) := zh'(z) - \overline{zg'(z)} \quad (z \in \mathbb{U}).$$

The classes $\mathcal{S}_\mathcal{H}^*(\alpha) := \mathcal{S}_\mathcal{H}^*(2\alpha-1,1)$ and $\mathcal{S}_\mathcal{H}^c(\alpha) := \mathcal{S}_\mathcal{H}^c(2\alpha-1,1)$ are studied by Jahangiri [4] (see also [5]). In particular, we obtain the classes $\mathcal{S}_\mathcal{H}^c := \mathcal{S}_\mathcal{H}^c(0)$ and $\mathcal{S}_\mathcal{H}^* := \mathcal{S}_\mathcal{H}^*(0)$ of functions $f \in \mathcal{S}_\mathcal{H}$ which are convex in $\mathbb{U}(r)$ or starlike in $\mathbb{U}(r)$, respectively, for any $r \in (0,1]$.

Motivated by Sakaguchi [1], we define the class $\mathcal{S}_\mathcal{H}^{**}(A,B)$ of functions $f \in \mathcal{H}_0$ such that:

$$\frac{2D_\mathcal{H} f(z)}{f(z) - f(-z)} \prec \frac{1+Az}{1+Bz}. \qquad (5)$$

In particular, the class $SH^*(\alpha) := \mathcal{S}_\mathcal{H}^{**}(2\alpha - 1, 1)$ was introduced by Ahuja and Jahangiri [6] (see also [7,8]). The class $\mathcal{HS}_s^*(b,\alpha) := \mathcal{S}_\mathcal{H}^{**}(2b(\alpha-1)+1,1)$ was investigated by Janteng and Halim [9].

In the present paper we obtain some analytic conditions for defined classes of functions. Some results connected to subordination properties, coefficient estimates, integral representation, and distortion theorems are also obtained. These results generalize the results obtained in [6,9] (see also [7,8]).

2. Analytic Criteria

Theorem 1. *Let $Tf(z) := f(z) - f(-z)$. If $f \in \mathcal{S}_\mathcal{H}^{**}(A,B)$, then $Tf \in \mathcal{S}_\mathcal{H}^*(A,B)$.*

Proof. Let $f \in \mathcal{S}_\mathcal{H}^{**}(A,B)$ and $H(z) := \frac{1+Az}{1+Bz}$. Then:

$$\frac{2D_\mathcal{H} f(z)}{f(z) - f(-z)} \prec H(z)$$

and

$$\frac{2D_\mathcal{H}(-f)(z)}{f(z) - f(-z)} = \frac{2D_\mathcal{H} f(-z)}{f(-z) - f(z)} \prec H(-z) \prec H(z).$$

Thus, we have:

$$\frac{2D_\mathcal{H} f(z)}{Tf(z)} \in H(\mathbb{U}) \quad \text{and} \quad \frac{2D_\mathcal{H}(-f)(z)}{Tf(z)} \in H(\mathbb{U}) \quad (z \in \mathbb{U}).$$

Since H is the convex function in $\mathbb{U}$, we have:

$$\frac{1}{2}\frac{2D_\mathcal{H} f(z)}{Tf(z)} + \frac{1}{2}\frac{2D_\mathcal{H}(-f)(z)}{Tf(z)} = \frac{D_\mathcal{H}(Tf)(z)}{Tf(z)} \in H(\mathbb{U}) \quad (z \in \mathbb{U}),$$

or equivalently:

$$\frac{D_\mathcal{H}(Tf)(z)}{Tf(z)} \prec H(z),$$

which implies that:

$$Tf \in \mathcal{S}_\mathcal{H}^*(A,B).$$

□

Let $\mathcal{V} \subset \mathcal{H}$, $\mathbb{U}_0 := \mathbb{U} \setminus \{0\}$. Due to Ruscheweyh [10] we define the dual set of $\mathcal{V}$ by:

$$\mathcal{V}^* := \left\{ f \in \mathcal{H}_0 : \bigwedge_{q \in \mathcal{V}} (f * q)(z) \neq 0 \ (z \in \mathbb{U}_0) \right\}.$$

Theorem 2. *We have:*

$$\mathcal{S}_{\mathcal{H}}^{**}(A, B) = \{\psi_\xi : |\xi| = 1\}^*,$$

where,

$$\psi_\xi(z) \ : \ = z \frac{\xi(B-A) + (2 + A\xi + B\xi)z}{(1+z)(1-z)^2} \tag{6}$$

$$-\bar{z} \frac{2 + (A+B)\xi - (B\_A)\xi\bar{z}}{(1+\bar{z})(1-\bar{z})^2} \ (z \in \mathbb{U}).$$

Proof. Let $f \in \mathcal{H}_0$ be of the form (1). Then $f \in \mathcal{S}_{\mathcal{H}}^{**}(A, B)$ if and only if it satisfies Equation (5) or equivalently:

$$\frac{2D_{\mathcal{H}} f(z)}{f(z) - f(-z)} \neq \frac{1 + A\xi}{1 + B\xi} \ (z \in \mathbb{U}_0, \ |\xi| = 1). \tag{7}$$

Since,

$$D_{\mathcal{H}} h(z) = h(z) * \frac{z}{(1-z)^2}, \ \frac{h(z) - h(-z)}{2} = h(z) * \frac{z}{1 - z^2}$$

the above inequality yields:

$$(1 + B\xi) D_{\mathcal{H}} f(z) - (1 + A\xi) \frac{f(z) - f(-z)}{2}$$
$$= (1 + B\xi) D_{\mathcal{H}} h(z) - (1 + A\xi) \frac{h(z) - h(-z)}{2}$$
$$- \left\{ (1 + B\xi) \overline{D_{\mathcal{H}} g(z)} + (1 + A\xi) \overline{\frac{g(z) - g(-z)}{2}} \right\}$$
$$= h(z) * \left(\frac{(1 + B\xi)z}{(1-z)^2} - \frac{(1 + A\xi)z}{1 - z^2} \right)$$
$$- \overline{g(z)} * \left(\frac{(1 + B\xi)\bar{z}}{(1-\bar{z})^2} + \frac{(1 + A\xi)\bar{z}}{1 - \bar{z}^2} \right)$$
$$= f(z) * \psi_\xi(z) \neq 0 \ (z \in \mathbb{U}_0, \ |\xi| = 1).$$

Thus, $f \in \mathcal{S}_{\mathcal{H}}^{**}(A, B)$ if and only if $f(z) * \psi_\xi(z) \neq 0$ for $z \in \mathbb{U}_0$, $|\xi| = 1$, i.e., $\mathcal{S}_{\mathcal{H}}^{**}(A, B) = \{\psi_\xi : |\xi| = 1\}^*$. □

Theorem 3. *If a function $f \in \mathcal{H}$ of the form (1) satisfies the condition:*

$$\sum_{n=2}^{\infty} (|\alpha_n| |a_n| + |\beta_n| |b_n|) \leq B - A, \tag{8}$$

where $-B \leq A < B \leq 1$ and

$$\alpha_n = n(1+B) - (1+A)(1 - (-1)^n)/2, \ \beta_n = n(1+B) + (1+A)(1 - (-1)^n)/2, \tag{9}$$

then $f \in \mathcal{S}_{\mathcal{H}}^*(A, B)$.

Proof. The result of Lewy [11] gives that the f is orientation preserving and locally univalent if:

$$|h'(z)| > |g'(z)| \quad (z \in \mathbb{U}). \tag{10}$$

By Equation (9) we have:

$$|\alpha_n|/(B-A) \geq n, \ |\beta_n|/(B-A) \geq n \quad (n = 2, 3, \cdots). \tag{11}$$

Therefore, by Equation (8) we obtain:

$$\sum_{n=2}^{\infty} n \left(|a_n| + |b_n| \right) \leq 1 \tag{12}$$

and

$$|h'(z)| - |g'(z)| \geq 1 - \sum_{n=2}^{\infty} n |a_n| |z|^n - \sum_{n=2}^{\infty} n |b_n| |z|^n \geq 1 - |z| \sum_{n=2}^{\infty} \left(n |a_n| + n |b_n| \right)$$

$$\geq 1 - \frac{|z|}{B-A} \sum_{n=2}^{\infty} \left(|\alpha_n| |a_n| + |\beta_n| |b_n| \right) \geq 1 - |z| > 0 \quad (z \in \mathbb{U}).$$

Therefore, by Equation (10) the function f is locally univalent and sense-preserving in $\mathbb{U}$. Moreover, if $z_1, z_2 \in \mathbb{U}, z_1 \neq z_2$, then:

$$\left| \frac{z_1^n - z_2^n}{z_1 - z_2} \right| = \left| \sum_{l=1}^{n} z_1^{l-1} z_2^{n-l} \right| \leq \sum_{l=1}^{n} |z_1|^{l-1} |z_2|^{n-l} < n \quad (n = 2, 3, \cdots).$$

Let $f \in \mathcal{H}_0$ be a function of the form (1). Without loss of generality, we can assume that f is not an identity function. Then there exist $n \in \mathbb{N}_2$ such that $a_n \neq 0$ or $b_n \neq 0$. Thus, by Equation (12) we get:

$$|f(z_1) - f(z_2)| \geq |h(z_1) - h(z_2)| - |g(z_1) - g(z_2)|$$

$$= \left| z_1 - z_2 - \sum_{n=2}^{\infty} a_n (z_1^n - z_2^n) \right| - \left| \sum_{n=2}^{\infty} \overline{b_n (z_1^n - z_2^n)} \right|$$

$$\geq |z_1 - z_2| - \sum_{n=2}^{\infty} |a_n| |z_1^n - z_2^n| - \sum_{n=2}^{\infty} |b_n| |z_1^n - z_2^n|$$

$$= |z_1 - z_2| \left(1 - \sum_{n=2}^{\infty} |a_n| \left| \frac{z_1^n - z_2^n}{z_1 - z_2} \right| - \sum_{n=2}^{\infty} |b_n| \left| \frac{z_1^n - z_2^n}{z_1 - z_2} \right| \right)$$

$$> |z_1 - z_2| \left(1 - \sum_{n=2}^{\infty} n |a_n| - \sum_{n=2}^{\infty} n |b_n| \right) \geq 0.$$

This leads to the univalence of f, i.e., $f \in \mathcal{S}_{\mathcal{H}}$. Therefore, $f \in \mathcal{S}_{\mathcal{H}}^{**}(A, B)$ if and only if there exists a complex-valued function ω, $\omega(0) = 0$, $|\omega(z)| < 1$ ($z \in \mathbb{U}$) such that:

$$\frac{2 D_{\mathcal{H}} f(z)}{f(z) - f(-z)} = \frac{1 + A\omega(z)}{1 + B\omega(z)} \quad (z \in \mathbb{U}),$$

or equivalently:

$$\left| \frac{2 D_{\mathcal{H}} f(z) - f(z) + f(-z)}{2 B D_{\mathcal{H}} f(z) - A(f(z) - f(-z))} \right| < 1 \quad (z \in \mathbb{U}). \tag{13}$$

Thus for $z \in \mathbb{U} \setminus \{0\}$ it suffices to show that:

$$\left| D_{\mathcal{H}} f(z) - \frac{f(z) - f(-z)}{2} \right| - \left| BD_{\mathcal{H}} f(z) - A \frac{f(z) - f(-z)}{2} \right| < 0.$$

Indeed, letting $|z| = r$ $(0 < r < 1)$ we have:

$$\left| D_{\mathcal{H}} f(z) - \frac{f(z) - f(-z)}{2} \right| - \left| BD_{\mathcal{H}} f(z) - A \frac{f(z) - f(-z)}{2} \right|$$

$$= \left| \sum_{n=2}^{\infty} \left(n - \frac{1 - (-1)^n}{2} \right) a_n z^n - \sum_{n=2}^{\infty} \left(n + \frac{1 - (-1)^n}{2} \right) \overline{b_n} \overline{z}^n \right|$$

$$- \left| (B - A)z + \sum_{n=2}^{\infty} \left(Bn - A \frac{1 - (-1)^n}{2} \right) a_n z^n + \sum_{n=2}^{\infty} \left(Bn + A \frac{1 - (-1)^n}{2} \right) \overline{b_n} \overline{z}^n \right|$$

$$\leq \sum_{n=2}^{\infty} \left(n - \frac{1 - (-1)^n}{2} \right) |a_n| r^n + \sum_{n=2}^{\infty} \left(n + \frac{1 - (-1)^n}{2} \right) |b_n| r^n - (B - A) r$$

$$+ \sum_{n=2}^{\infty} \left(Bn - A \frac{1 - (-1)^n}{2} \right) |a_n| r^n + \sum_{n=2}^{\infty} \left(Bn + A \frac{1 - (-1)^n}{2} \right) |b_n| r^n$$

$$\leq r \left\{ \sum_{n=2}^{\infty} (|\alpha_n| |a_n| + |\beta_n| |b_n|) r^{n-1} - (B - A) \right\} < 0.$$

Hence $f \in \mathcal{S}_{\mathcal{H}}^{**}(A, B)$. □

Motivated by Silverman [12] we denote by $\mathcal{T}$ the class of functions $f \in \mathcal{H}_0$ of the form (1) such that $a_n = -|a_n|$, $b_n = |b_n|$ $(n = 2, 3, \cdots)$, i.e.,

$$f = h + \overline{g}, \quad h(z) = z - \sum_{n=2}^{\infty} |a_n| z^n, \quad g(z) = \sum_{n=2}^{\infty} |b_n| \overline{z}^n \quad (z \in \mathbb{U}). \tag{14}$$

Moreover, let us define:

$$\mathcal{S}_{\mathcal{T}}^{**}(A, B) := \mathcal{T} \cap \mathcal{S}_{\mathcal{H}}^{**}(A, B), \quad -B \leq A < B \leq 1.$$

Now, we show that the condition (8) is also the sufficient condition for a function $f \in \mathcal{T}$ to be in the class $\mathcal{S}_{\mathcal{T}}^{**}(A, B)$.

Theorem 4. *Let $f \in \mathcal{T}$ be a function of the form (14). Then $f \in \mathcal{S}_{\mathcal{T}}^{**}(A, B)$ if and only if condition (8) holds true.*

Proof. In view of Theorem 3 we need only show that each function $f \in \mathcal{S}_{\mathcal{T}}^{**}(A, B)$ satisfies the coefficient inequality of Equation (8). If $f \in \mathcal{S}_{\mathcal{T}}^{**}(A, B)$, then it satisfies Equation (13) or equivalently:

$$\left| \frac{\sum_{n=2}^{\infty} \left\{ \left(n - \frac{1 - (-1)^n}{2} \right) |a_n| z^n + \left(n + \frac{1 - (-1)^n}{2} \right) |b_n| \overline{z}^n \right\}}{(B - A)z - \sum_{n=2}^{\infty} \left\{ \left(Bn - A \frac{1 - (-1)^n}{2} \right) |a_n| z^n + \left(Bn + A \frac{1 - (-1)^n}{2} \right) |b_n| \overline{z}^n \right\}} \right| < 1 \quad (z \in \mathbb{U}).$$

Therefore, putting $z = r$ $(0 \leq r < 1)$ we obtain:

$$\frac{\sum_{n=2}^{\infty} \left\{ \left(n - \frac{1 - (-1)^n}{2} \right) |a_n| + \left(n + \frac{1 - (-1)^n}{2} \right) |b_n| r^{n-1} \right\}}{(B - A) - \sum_{n=2}^{\infty} \left\{ \left(Bn - A \frac{1 - (-1)^n}{2} \right) |a_n| + \left(Bn + A \frac{1 - (-1)^n}{2} \right) |b_n| r^{n-1} \right\}} < 1. \tag{15}$$

It is clear that the denominator of the left hand side cannot vanish for $r \in [0,1)$. Moreover, it is positive for $r = 0$, and in consequence for $r \in (0,1)$. Thus, by Equation (25) we have:

$$\sum_{n=2}^{\infty} (\alpha_n |a_n| + \beta_n |b_n|) r^{n-1} < B - A \quad (0 \leq r < 1). \tag{16}$$

The sequence of partial sums $\{S_n\}$ associated with the series $\sum_{n=2}^{\infty} (\alpha_n |a_n| + \beta_n |b_n|)$ is nondecreasing sequence. Moreover, by Equation (16) it is bounded by $B - A$. Hence, the sequence $\{S_n\}$ is convergent and

$$\sum_{n=2}^{\infty} (\alpha_n |a_n| + \beta_n |b_n|) = \lim_{n \to \infty} S_n \leq B - A,$$

which yields the assertion (8). □

Example 1. *For the function:*

$$f(z) = z - \sum_{n=2}^{\infty} \frac{B-A}{2^n \alpha_n} z^n - \sum_{n=2}^{\infty} \frac{B-A}{2^n \beta_n} \overline{z}^n \quad (z \in \mathbb{U})$$

we have,

$$\sum_{n=2}^{\infty} (\alpha_n |a_n| + \beta_n |b_n|) = \sum_{n=2}^{\infty} \frac{B-A}{2^n} + \frac{B-A}{2^n} = (B-A) \sum_{n=1}^{\infty} \frac{1}{2^n} = B - A.$$

Thus, $f \in \mathcal{S}_T^{**}(A,B)$.

3. Topological Properties

Let us consider a metric on $\mathcal{H}$ in which a sequence $\{f_n\}$ in $\mathcal{H}$ converges to f if and only if it converges to f uniformly on each compact subset of $\mathbb{U}$. The metric induces the usual topology on $\mathcal{H}$. It is easy to verify that the obtained topological space is complete. Let $\mathcal{B}$ be a subset of the space $\mathcal{H}$.

We say that a function $f \in \mathcal{B}$ is *the extreme point of $\mathcal{B}$* if it cannot be presented as nontrivial convex combination of two functions from $\mathcal{B}$. We denote by $E\mathcal{B}$ the set of extreme points of $\mathcal{B}$.

We say that $\mathcal{B}$ is *locally uniformly bounded* if for each r, $0 < r < 1$, there exists $K = K(r) > 0$ such that:

$$|f(z)| \leq K \quad (f \in \mathcal{B}, |z| \leq r).$$

We say that a set $\mathcal{B}$ is *convex* if it includes all of convex combinations of two functions from $\mathcal{B}$. Let $\overline{co}\mathcal{B}$ denote *the closed convex hull of $\mathcal{B}$* i.e., the intersection of all closed convex subsets of $\mathcal{H}$ that contain $\mathcal{B}$.

Let $\mathcal{B} \subset \mathcal{H}$ be a convex set and $\mathcal{L}$ be a real-valued functional on $\mathcal{H}$. We say that $\mathcal{L}$ is *convex functional* on $\mathcal{B}$ if:

$$\mathcal{L}(af + (1-a)g) \leq a\mathcal{L}(f) + (1-a)\mathcal{L}(g) \quad (f,g \in \mathcal{B}, 0 \leq a \leq 1).$$

By using the Krein-Milman theorem (see [13]) we get the following lemma.

Lemma 1. *Let $\mathcal{B}$ be a non-empty compact set on the space $\mathcal{H}$. Then $E\mathcal{B}$ is non-empty and $\overline{co}E\mathcal{B} = \overline{co}\mathcal{B}$.*

Motivated by Hallenbeck and MacGregor ([14], p. 45) we can formulate the following lemma.

Lemma 2. *Let $\mathcal{B}$ be a non-empty convex compact set on the space $\mathcal{H}$ and let $\mathcal{L}$ be a real-valued, convex, and continuous functional on $\mathcal{B}$. Then $\max\{\mathcal{L}(f) : f \in \mathcal{B}\} = \max\{\mathcal{L}(f) : f \in E\mathcal{B}\}$.*

Proof. We observe that there exists max $\{\mathcal{L}(f) : f \in \mathcal{B}\} =: K$, since $\mathcal{J}$ is the continuous functional on the compact set $\mathcal{B}$. Thus, the set $H := \{f \in \mathcal{B} : \mathcal{L}(f) = K\}$ is non-empty compact subset of $\mathcal{B}$ and, by Lemma 1, we get that H has an extreme point f_0. Let,

$$f_0 = af_1 + (1-a)f_2,$$

where $f_1, f_2 \in \mathcal{B}$ and $0 < a < 1$. Thus,

$$K = \mathcal{L}(f_0) \le a\mathcal{L}(f_1) + (1-a)\mathcal{L}(f_2) = aK + (1-a)K = K$$

and, in consequence, $\mathcal{L}(f_1) = \mathcal{L}(f_2) = K$, i.e., $f_1, f_2 \in H$. Since f_0 is an extreme point of H we get $f_1 = f_2 = f_0 \in E\mathcal{B}$. Thus, we obtain that there exists max $\{\mathcal{L}(f) : f \in E\mathcal{B}\} = K$, and the proof is complete. □

We observe that $\mathcal{H}$ is a complete metric space. Therefore, by Montel's theorem (see [15]) we get the following lemma.

Lemma 3. *A set $\mathcal{B}$ is compact on $\mathcal{H}$ if and only if $\mathcal{B}$ is locally uniformly bounded and closed on $\mathcal{H}$.*

Theorem 5. *The class $\mathcal{S}_\mathcal{T}^{**}(A,B)$ is compact and convex subset on $\mathcal{H}$.*

Proof. Let $f_k \in \mathcal{S}_\mathcal{T}^{**}(A,B)$ be functions of the form:

$$f_k(z) = z - \sum_{n=2}^{\infty}\left(|a_{k,n}|z^n - |b_{k,n}|\bar{z}^n\right) \quad (z \in \mathbb{U},\ k = 1,2,\ldots) \tag{17}$$

and let $0 \le \gamma \le 1$. Since,

$$\gamma f_1(z) + (1-\gamma)f_2(z) = z - \sum_{n=2}^{\infty}\left\{(\gamma|a_{1,n}| + (1-\gamma)|a_{2,n}|)z^n - (\gamma|b_{1,n}| + (1-\gamma)|b_{2,n}|)\bar{z}^n\right\},$$

and by Theorem 4 we have:

$$\sum_{n=2}^{\infty}\left\{\alpha_n(\gamma|a_{1,n}| + (1-\gamma)|a_{2,n}|) + \beta_n(\gamma|b_{1,n}| + (1-\gamma)|b_{2,n}|)\right\}$$

$$= \gamma\sum_{n=2}^{\infty}\left\{\alpha_n|a_{1,n}| + \beta_n|b_{1,n}|\right\} + (1-\gamma)\sum_{n=2}^{\infty}\left\{\alpha_n|a_{2,n}| + \beta_n|b_{2,n}|\right\}$$

$$\le \gamma(B-A) + (1-\gamma)(B-A) = B - A,$$

the function $\phi = \gamma f_1 + (1-\gamma)f_2$ belongs to the class $\mathcal{S}_\mathcal{T}^{**}(A,B)$. Hence, the class is convex. Furthermore, for $f \in \mathcal{S}_\mathcal{T}^{**}(A,B)$, $|z| \le r$, $0 < r < 1$, we have:

$$|f(z)| \le r + \sum_{n=2}^{\infty}(|a_n| + |b_n|)r^n \le r + \sum_{n=2}^{\infty}(\alpha_n|a_n| + \beta_n|b_n|) \le r + (B-A). \tag{18}$$

Thus, we conclude that the class $\mathcal{S}_\mathcal{T}^{**}(A,B)$ is locally uniformly bounded. By Lemma 3, we only need to show that it is closed, i.e., if $f_k \in \mathcal{S}_\mathcal{T}^{**}(A,B)$ $(k \in \mathbb{N})$ and $f_k \to f$, then $f \in \mathcal{S}_\mathcal{T}^{**}(A,B)$. Let f_k and f are given by Equations (17) and (14), respectively. Using Theorem 4 we have:

$$\sum_{n=2}^{\infty}(\alpha_n|a_{k,n}| + \beta_n|b_{k,n}|) \le B - A \quad (k \in \mathbb{N}). \tag{19}$$

Since $f_k \to f$, we conclude that $|a_{k,n}| \to |a_n|$ and $|b_{k,n}| \to |b_n|$ as $k \to \infty$ ($n \in \mathbb{N}$). The sequence of partial sums $\{S_n\}$ associated with the series $\sum_{n=2}^{\infty} (\alpha_n |a_n| + \beta_n |b_n|)$ is nondecreasing sequence. Moreover, by Equation (19) it is bounded by $B - A$. Therefore, the sequence $\{S_n\}$ is convergent and

$$\sum_{n=2}^{\infty} (\alpha_n |a_n| + \beta_n |b_n|) = \lim_{n \to \infty} S_n \leq B - A.$$

This gives the condition (8), and, in consequence, $f \in \mathcal{S}_\mathcal{T}^{**}(A, B)$, which completes the proof. □

Theorem 6. *We have:*

$$E\mathcal{S}_\mathcal{T}^{**}(A, B) = \{h_n : n \in \mathbb{N}\} \cup \{g_n : n \in \{2, 3 \ldots\}\},$$

where,

$$h_1(z) = z, \quad h_n(z) = z - \frac{B-A}{\alpha_n} z^n, \quad g_n(z) = z + \frac{B-A}{\beta_n} \bar{z}^n \tag{20}$$

$$(n = 2, 3, \ldots; z \in \mathbb{U}).$$

Proof. Let $0 < a < 1$ and $g_n = af_1 + (1-a)f_2$, where $f_1, f_2 \in \mathcal{S}_\mathcal{T}^{**}(A, B)$ are given by Equation (17). Thus, by Equation (8) we get $|b_{1,n}| = |b_{2,n}| = (B-A)/\beta_n$, and consequently $a_{1,k} = a_{2,k} = 0$ ($k \in \{2, 3 \ldots\}$) and $b_{1,k} = b_{2,k} = 0$ ($k \in \{2, 3 \ldots\} \setminus \{n\}$). Thus, $g_n = f_1 = f_2$, and, in consequence, $g_n \in E\mathcal{S}_\mathcal{T}^{**}(A, B)$. In the same way, we prove that the functions h_n of the form (20) are the extreme points of the class $\mathcal{S}_\mathcal{T}^{**}(A, B)$. Suppose that $f \in E\mathcal{S}_\mathcal{T}^{**}(A, B)$ and f is not of the form (20). Then there exists $k \in \{2, 3, \ldots\}$ such that:

$$0 < |a_k| < (B-A)/\alpha_n \text{ or } 0 < |b_k| < (B-A)/\beta_n.$$

If $0 < |a_k| < (B-A)/\alpha_n$ and

$$a = \frac{|a_k| \alpha_k}{B-A}, \quad \varphi = \frac{1}{1-a}(f - ah_k),$$

then we obtain $0 < a < 1$, $h_k, \varphi \in \mathcal{S}_\mathcal{T}^{**}(A, B)$, $h_k \neq \varphi$, and

$$f = ah_k + (1-a)\varphi.$$

Therefore, $f \notin E\mathcal{S}_\mathcal{T}^{**}(A, B)$. Similarly, if $0 < |b_k| < (B-A)/\beta_n$ and

$$a = \frac{|b_k| \beta_k}{B-A}, \quad \phi(z) = \frac{1}{1-a}(f - ag_k),$$

then we obtain $0 < a < 1$, $g_k, \varphi \in \mathcal{S}_\mathcal{T}^{**}(A, B)$, $g_k \neq \phi$ and

$$f = ag_k + (1-a)\phi.$$

Thus we get $f \notin E\mathcal{S}_\mathcal{T}^{**}(A, B)$, which completes the proof of Theorem 6. □

4. Applications

It is clear that if the class:

$$\mathcal{F} = \{f_n \in \mathcal{H} : n \in \mathbb{N}\}$$

is locally uniformly bounded, then:

$$\overline{co}\mathcal{F} = \left\{ \sum_{n=1}^{\infty} \gamma_n f_n : \sum_{n=1}^{\infty} \gamma_n = 1, \gamma_n \geq 0 \ (n \in \mathbb{N}) \right\}. \tag{21}$$

Corollary 1.

$$\mathcal{S}_{\mathcal{T}}^{**}(A,B) = \left\{ \sum_{n=1}^{\infty} (\gamma_n h_n + \delta_n g_n) : \sum_{n=1}^{\infty} (\gamma_n + \delta_n) = 1, \delta_1 = 0, \gamma_n, \delta_n \geq 0 \ (n \in \mathbb{N}) \right\}, \tag{22}$$

where h_n, g_n are defined by Equation (20).

Proof. By Theorem 5 and Lemma 1 we have:

$$\mathcal{S}_{\mathcal{T}}^{**}(A,B) = \overline{co}\mathcal{S}_{\mathcal{T}}^{**}(A,B) = \overline{co}E\mathcal{S}_{\mathcal{T}}^{**}(A,B).$$

Thus, by Theorem 6 and Equation (21) we have Equation (22). □

We observe, that the following real-valued functionals are convex and continuous on $\mathcal{H}$:

$$\mathcal{L}(f) = |a_n|, \ \mathcal{L}(f) = |b_n|, \ \mathcal{L}(f) = |f(z)|, \ \mathcal{L}(f) = |D_{\mathcal{H}}f(z)| \quad (f \in \mathcal{H}),$$

and

$$\mathcal{L}(f) = \left(\frac{1}{2\pi} \int_{U}^{2\pi} \left| f\left(re^{i\theta}\right) \right|^{\gamma} d\theta \right)^{1/\gamma} \quad (f \in \mathcal{H}, 0 < r < 1, \gamma \geq 1).$$

Thus, by using Theorem 6 and Lemma 2 we obtain the following two corollaries.

Corollary 2. *If $f \in \mathcal{S}_{\mathcal{T}}^{**}(A,B)$ is a function of the form (14), then:*

$$|a_n| \leq \frac{B-A}{\alpha_n}, \ |b_n| \leq \frac{B-A}{\beta_n} \quad (n = 2, 3, \ldots), \tag{23}$$

with α_n, β_n defined by Equation (9). The result is sharp. The functions h_n, g_n of the form (20) are the extremal functions.

Proof. Since For the extremal functions h_n and g_n we have $|a_n| = \frac{B-A}{\alpha_n}$ and $|b_n| = \frac{B-A}{\beta_n}$. Thus, by Lemma 2 we have Equation (23). □

Example 2. *In particular, since $\frac{B-A+1}{\alpha_3} > \frac{B-A}{\alpha_3}$ the polynomial:*

$$w(z) = z - z^2 - \frac{B-A+1}{\alpha_3}z^3 \quad (z \in \mathbb{U})$$

does not belong to the class $\mathcal{S}_{\mathcal{T}}^{**}(A,B)$.

Corollary 3. *Let $f \in \mathcal{S}_{\mathcal{T}}^{**}(A,B), |z| = r < 1$. Then,*

$$r - \frac{B-A}{2(1+B)}r^2 \leq |f(z)| \leq r + \frac{B-A}{2(1+B)}r^2 \tag{24}$$

and

$$r - \frac{B-A}{1+B}r^2 \leq |D_{\mathcal{H}}f(z)| \leq r + \frac{B-A}{1+B}r^2. \tag{25}$$

The result is sharp. The function h_2 of the form (20) is the extremal function.

Proof. For the extremal functions h_n and g_n of the form (20) we have:

$$|h_n(z)| \leq r + \frac{B-A}{\alpha_n} r^n \leq r + \frac{B-A}{2(1+B)} r^2 \quad (n = 2, 3 \ldots),$$

$$|g_n(z)| \leq r + \frac{B-A}{\beta_n} r^n \leq r + \frac{B-A}{2(1+B)} r^2 \quad (n = 2, 3 \ldots),$$

$$|h_n(z)| \geq r - \frac{B-A}{\alpha_n} r^n \geq r - \frac{B-A}{2(1+B)} r^2 \quad (n = 2, 3 \ldots),$$

$$|g_n(z)| \geq r - \frac{B-A}{\beta_n} r^n \geq r - \frac{B-A}{2(1+B)} r^2 \quad (n = 2, 3 \ldots).$$

Thus, by Lemma 2 we have Equation (24). Similarly, we prove Equation (25). □

Due to Littlewood [16] we consider the integral means inequalities for functions from the class $S_T^{**}(A, B)$.

Lemma 4. *[16] Let $f, g \in \mathcal{A}$. If $f \prec g$, then,*

$$\int_0^{2\pi} |f(re^{i\theta})|^\gamma d\theta \leq \int_0^{2\pi} |g(re^{i\theta})|^\gamma d\theta \quad (0 < r < 1, \gamma > 0).$$

Lemma 5. *Let $0 < r < 1$, $\gamma > 0$. Then,*

$$\frac{1}{2\pi} \int_0^{2\pi} |h_n(re^{i\theta})|^\gamma d\theta \leq \frac{1}{2\pi} \int_0^{2\pi} |h_2(re^{i\theta})|^\gamma d\theta \quad (n = 1, 2, \cdots) \tag{26}$$

and

$$\frac{1}{2\pi} \int_0^{2\pi} |g_n(re^{i\theta})|^\gamma d\theta \leq \frac{1}{2\pi} \int_0^{2\pi} |h_2(re^{i\theta})|^\gamma d\theta \quad (n = 2, 3, \cdots), \tag{27}$$

where h_n and g_n are defined by Equation (20).

Proof. Let h_n and g_n are defined by Equation (20) and let $\tilde{g}_n(z) = z + \frac{B-A}{\beta_n} z^n$ $(n = 2, 3, \cdots)$. Since $\frac{h_n(z)}{z} \prec \frac{h_2(z)}{z}$ and $\frac{\tilde{g}(z)}{z} \prec \frac{h_2(z)}{z}$, by Lemma 4 we have:

$$\int_0^{2\pi} |h_n(re^{i\theta})|^\gamma d\theta \leq \int_0^{2\pi} |h_2(re^{i\theta})|^\gamma d\theta,$$

$$\int_0^{2\pi} |g_n(re^{i\theta})|^\gamma d\theta = \int_0^{2\pi} |\tilde{g}_n(re^{i\theta})|^\gamma d\theta \leq \int_0^{2\pi} |h_2(re^{i\theta})|^\gamma d\theta,$$

which complete the proof. □

Corollary 4. *If $f \in S_T^{**}(A, B)$ then:*

$$\frac{1}{2\pi} \int_0^{2\pi} |f(re^{i\theta})|^\gamma d\theta \leq \frac{1}{2\pi} \int_0^{2\pi} |h_2(re^{i\theta})|^\gamma d\theta$$

and

$$\frac{1}{2\pi}\int_0^{2\pi}\left|D_{\mathcal{H}}f(re^{i\theta})\right|^\gamma d\theta \leq \frac{1}{2\pi}\int_0^{2\pi}\left|D_{\mathcal{H}}h_2(re^{i\theta})\right|^\gamma d\theta,$$

where $\gamma \geq 1$, $0 < r < 1$ and h_2 is the function defined by Equation (20).

Remark 1. *Some new and also well-known results can be obtained by choosing the parameters A, B in the defined classes of functions (see for example [6–9]). In particular, for $A = 2\alpha - 1$, $B = 1$ we have results obtained by Ahuja and Jahangiri [6] (see also [7,8]), for $A = 2b(\alpha - 1) + 1$, $B = 1$ we have results obtained by Janteng and Halim [9].*

Author Contributions: Conceptualization, N.E.C. and J.D.; methodology, N.E.C. and J.D.; formal analysis, N.E.C. and J.D.; investigation, N.E.C. and J.D.; writing—original draft preparation, J.D.; writing—review and editing, N.E.C.; funding acquisition, N.E.C. All authors have read and agreed to the published version of the manuscript.

Acknowledgments: The work was supported by the Centre for Innovation and Transfer of Natural Sciences and Engineering Knowledge, University of Rzeszów and by the Basic Science Research Program through the National Research Foundation of Korea (NRF) funded by the Ministry of Education, Science and Technology (No. 2019R1I1A3A01050861).

Conflicts of Interest: The authors declare no conflict of interest.

References

1. Sakaguchi, K. On a certain univalent mapping. *J. Math. Soc. Jpn.* **1959**, *2*, 72–75. [CrossRef]
2. Dziok, J. On Janowski harmonic functions. *J. Appl. Anal.* **2015**, *21*, 99–107. [CrossRef]
3. Janowski, W. Some extremal problems for certain families of analytic functions I. *Ann. Polon. Math.* **1973**, *28*, 297–326. [CrossRef]
4. Jahangiri, J.M. Harmonic functions starlike in the unit disk. *J. Math. Anal. Appl.* **1999**, *235*, 470–477. [CrossRef]
5. Ahuja, O.P. Planar Harmonic Univalent and Related Mappings. *J. Inequl. Pure Appl. Math.* **2005**, *6*, 122.
6. Ahuja, O.P.; Jahangiri, J.M. Sakaguchi-type harmonic univalent functions. *Sci. Math. Jpn.* **2004**, *59*, 163–168.
7. Al Amoush, A.G.; Darus, M. Harmonic starlike functions with respect to symmetric points. *Matematika* **2016**, *32*, 121–131.
8. Al-Khal, R.A.; Al-Kharsani, H.A. Salagean-type harmonic univalent functions with respect to symmetric points. *Aust. J. Math. Anal. Appl.* **2007**, *4*, 6.
9. Janteng, A.; Halim, S.A. Properties of harmonic functions which are starlike of complex order with respect to symmetric points. *Int. J. Contemp. Math. Sci.* **2009**, *4*, 275–281.
10. Ruscheweyh, S. *Convolutions in Geometric Function Theory—Seminaire de Mathematiques Superieures*; Gaetan Morin Editeur Ltee: Boucherville, QC, Canada, 1982.
11. Lewy, H. On the non-vanishing of the Jacobian in certain one-to-one mappings. *Bull. Am. Math. Soc.* **1936**, *42*, 689–692. [CrossRef]
12. Silverman, H. Harmonic univalent functions with negative coefficients. *J. Math. Anal. Appl.* **1998**, *220*, 283–289. [CrossRef]
13. Krein, M.; Milman, D. On the extreme points of regularly convex sets. *Stud. Math.* **1940**, *9*, 133–138. [CrossRef]
14. Hallenbeck, D.J.; MacGregor, T.H. *Linear Problems and Convexity Techniques in Geometric Function Theory*; Pitman Advanced Publishing Program: Boston, MA, USA; Pitman, NJ, USA, 1984.
15. Montel, P. Sur les families de functions analytiques qui admettent des valeurs exceptionelles dans un domaine. *Annales Scientifiques de l'École Normale Supérieure* **1912**, *23*, 487–535. [CrossRef]
16. Littlewood, J.E. On inequalities in theory of functions. *Proc. Lond. Math. Soc.* **1925**, *23*, 481–519. [CrossRef]

© 2019 by the authors. Licensee MDPI, Basel, Switzerland. This article is an open access article distributed under the terms and conditions of the Creative Commons Attribution (CC BY) license (http://creativecommons.org/licenses/by/4.0/).

Article
Oscillation Results for Higher Order Differential Equations

Choonkil Park [1,*,†], Osama Moaaz [2,†] and Omar Bazighifan [2,†]

1 Research Institute for Natural Sciences, Hanyang University, Seoul 04763, Korea
2 Department of Mathematics, Faculty of Science, Mansoura University, Mansoura 35516, Egypt; o_moaaz@mans.edu.eg (O.M.); o.bazighifan@gmail.com (O.B.)
* Correspondence: baak@hanyang.ac.kr
† These authors contributed equally to this work.

Received: 22 December 2019; Accepted: 25 January 2020; Published: 3 February 2020

Abstract: The objective of our research was to study asymptotic properties of the class of higher order differential equations with a p-Laplacian-like operator. Our results supplement and improve some known results obtained in the literature. An illustrative example is provided.

Keywords: ocillation; higher-order; differential equations; p-Laplacian equations

1. Introduction

In this work, we are concerned with oscillations of higher-order differential equations with a p-Laplacian like operator of the form

$$\left(r(t)\left|\left(y^{(n-1)}(t)\right)\right|^{p-2} y^{(n-1)}(t)\right)' + q(t)|y(\tau(t))|^{p-2} y(\tau(t)) = 0. \tag{1}$$

We assume that $p > 1$ is a constant, $r \in C^1([t_0, \infty), \mathbb{R})$, $r(t) > 0$, $q, \tau \in C([t_0, \infty), \mathbb{R})$, $q > 0$, $\tau(t) \leq t$, $\lim_{t \to \infty} \tau(t) = \infty$ and the condition

$$\eta(t_0) = \infty, \tag{2}$$

where

$$\eta(t) := \int_t^\infty \frac{ds}{r^{1/(p-1)}(s)}.$$

By a solution of (1) we mean a function $y \in C^{n-1}[T_y, \infty)$, $T_y \geq t_0$, which has the property $r(t)\left|\left(y^{(n-1)}(t)\right)\right|^{p-2} y^{(n-1)}(t) \in C^1[T_y, \infty)$, and satisfies (1) on $[T_y, \infty)$. We consider only those solutions y of (1) which satisfy $\sup\{|y(t)| : t \geq T\} > 0$, for all $T > T_y$. A solution of (1) is called oscillatory if it has arbitrarily large number of zeros on $[T_y, \infty)$, and otherwise it is called to be nonoscillatory; (1) is said to be oscillatory if all its solutions are oscillatory.

In recent decades, there has been a lot of research concerning the oscillation of solutions of various classes of differential equations; see [1–24].

It is interesting to study Equation (1) since the p-Laplace differential equations have applications in continuum mechanics [14,25]. In the following, we briefly review some important oscillation criteria obtained for higher-order equations, which can be seen as a motivation for this paper.

Elabbasy et al. [26] proved that the equation

$$\left(r(t)\left|\left(y^{(n-1)}(t)\right)\right|^{p-2} y^{(n-1)}(t)\right)' + q(t) f(y(\tau(t))) = 0,$$

is oscillatory, under the conditions
$$\int_{t_0}^{\infty} \frac{1}{r^{p-1}(t)} dt = \infty;$$

additionally,
$$\int_{\ell_0}^{\infty} \left(\psi(s) - \frac{1}{p^p} \phi^p(s) \frac{((n-1)!)^{p-1} \rho(s) a(s)}{((p-1) \mu s^{n-1})^{p-1}} - \frac{(p-1)\rho(s)}{a^{1/(p-1)}(s) \eta^p(s)} \right) ds = +\infty,$$

for some constant $\mu \in (0,1)$ and
$$\int_{\ell_0}^{\infty} kq(s) \frac{\tau(s)^{p-1}}{s^{p-1}} ds = \infty.$$

Agarwal et al. [2] studied the oscillation of the higher-order nonlinear delay differential equation
$$\left[\left| y^{(n-1)}(t) \right|^{\alpha-1} y^{(n-1)}(t) \right]' + q(t) |y(\tau(t))|^{\alpha-1} y(\tau(t)) = 0.$$

where α is a positive real number. In [27], Zhang et al. studied the asymptotic properties of the solutions of equation
$$\left[r(t) \left(y^{(n-1)}(t) \right)^{\alpha} \right]' + q(t) y^{\beta}(\tau(t)) = 0, \quad t \geq t_0.$$

where α and β are ratios of odd positive integers, $\beta \leq \alpha$ and
$$\int_{t_0}^{\infty} r^{-1/\alpha}(s) ds < \infty. \qquad (3)$$

In this work, by using the Riccati transformations, the integral averaging technique and comparison principles, we establish a new oscillation criterion for a class of higher-order neutral delay differential Equations (1). This theorem complements and improves results reported in [26]. An illustrative example is provided.

In the sequel, all occurring functional inequalities are assumed to hold eventually; that is, they are satisfied for all t large enough.

2. Main Results

In this section, we establish some oscillation criteria for Equation (1). For convenience, we denote that $F_+(t) := \max\{0, F(t)\}$,
$$B(t) := \frac{1}{(n-4)!} \int_t^{\infty} (\theta - t)^{n-4} \left(\frac{\int_{\theta}^{\infty} q(s) \left(\frac{\tau(s)}{s} \right)^{p-1} ds}{r(\theta)} \right)^{1/(p-1)} d\theta$$

and
$$D(s) := \frac{r(s) \delta(s) |h(t,s)|^p}{p^p \left[H(t,s) A(s) \mu \frac{s^{n-2}}{(n-2)!} \right]^{p-1}}.$$

We begin with the following lemmas.

Lemma 1 (Agarwal [1]). *Let $y(t) \in C^m[t_0, \infty)$ be of constant sign and $y^{(m)}(t) \neq 0$ on $[t_0, \infty)$ which satisfies $y(t) y^{(m)}(t) \leq 0$. Then,*
(I) *There exists a $t_1 \geq t_0$ such that the functions $y^{(i)}(t)$, $i = 1, 2, ..., m-1$ are of constant sign on $[t_0, \infty)$;*

(**II**) There exists a number $k \in \{1, 3, 5, ..., m-1\}$ when m is even, $k \in \{0, 2, 4, ..., m-1\}$ when m is odd, such that, for $t \geq t_1$,
$$y(t) y^{(i)}(t) > 0,$$
for all $i = 0, 1, ..., k$ and
$$(-1)^{m+i+1} y(t) y^{(i)}(t) > 0,$$
for all $i = k+1, ..., m$.

Lemma 2 (Kiguradze [15]). *If the function y satisfies $y^{(j)} > 0$ for all $j = 0, 1, ..., m$, and $y^{(m+1)} < 0$, then*
$$\frac{m!}{t^m} y(t) - \frac{(m-1)!}{t^{m-1}} y'(t) \geq 0.$$

Lemma 3 (Bazighifan [7]). *Let $h \in C^m([t_0, \infty), (0, \infty))$. Suppose that $h^{(m)}(t)$ is of a fixed sign, on $[t_0, \infty)$, $h^{(m)}(t)$ not identically zero, and that there exists a $t_1 \geq t_0$ such that, for all $t \geq t_1$,*
$$h^{(m-1)}(t) h^{(m)}(t) \leq 0.$$
If we have $\lim_{t \to \infty} h(t) \neq 0$, then there exists $t_\lambda \geq t_0$ such that
$$h(t) \geq \frac{\lambda}{(m-1)!} t^{m-1} \left| h^{(m-1)}(t) \right|,$$
for every $\lambda \in (0, 1)$ and $t \geq t_\lambda$.

Lemma 4. *Let $n \geq 4$ be even, and assume that y is an eventually positive solution of Equation (1). If (2) holds, then there exists two possible cases for $t \geq t_1$, where $t_1 \geq t_0$ is sufficiently large:*

(C_1) $y'(t) > 0, y''(t) > 0, y^{(n-1)}(t) > 0, y^{(n)}(t) < 0,$
(C_2) $y^{(j)}(t) > 0, y^{(j+1)}(t) < 0$ for all odd integer
       $j \in \{1, 2, ..., n-3\}, y^{(n-1)}(t) > 0, y^{(n)}(t) < 0.$

Proof. Let y be an eventually positive solution of Equation (1). By virtue of (1), we get
$$\left(r(t) \left| \left(y^{(n-1)}(t) \right) \right|^{p-2} y^{(n-1)}(t) \right)' < 0. \tag{4}$$

From ([11] Lemma 4), we have that $y^{(n-1)}(t) > 0$ eventually. Then, we can write (4) in the from
$$\left(r(t) \left(y^{(n-1)}(t) \right)^{p-1} \right)' < 0,$$
which gives
$$r'(t) \left(y^{(n-1)}(t) \right)^{p-1} + r(t)(p-1) \left(y^{(n-1)}(t) \right)^{p-2} y^{(n)}(t) < 0.$$

Thus, $y^{(n)}(t) < 0$ eventually. Thus, by Lemma 1, we have two possible cases (C_1) and (C_2). This completes the proof. □

Lemma 5. *Let y be an eventually positive solution of Equation (1) and assume that Case (C_1) holds. If*
$$\omega(t) := \delta(t) \left(\frac{r(t) \left| \left(y^{(n-1)}(t) \right) \right|^{p-1}}{y^{p-1}(t)} \right), \tag{5}$$

where $\delta \in C^1\left([t_0, \infty), (0, \infty)\right)$, then

$$\omega'(t) \leq \frac{\delta'_+(t)}{\delta(t)}\omega(t) - \delta(t) q(t) \left(\frac{\tau^{n-1}(t)}{t^{n-1}}\right)^{p-1} - \frac{(p-1)\mu t^{n-2}}{(n-2)!\left(\delta(t) r(t)\right)^{1/(p-1)}}\omega^{p/(p-1)}(t). \quad (6)$$

Proof. Let y be an eventually positive solution of Equation (1) and assume that Case (C_1) holds. From the definition of ω, we see that $\omega(t) > 0$ for $t \geq t_1$, and

$$\omega'(t) \leq \delta'(t) \frac{r(t)\left|\left(y^{(n-1)}(t)\right)\right|^{p-1}}{y^{p-1}(t)} + \delta(t) \frac{\left(r(t)\left|\left(y^{(n-1)}(t)\right)\right|^{p-1}\right)'}{y^{p-1}(t)}$$
$$-\delta(t) \frac{(p-1) y'(t) r(t) \left|\left(y^{(n-1)}(t)\right)\right|^{p-1}}{y^p(t)}.$$

Using Lemma 3 with $m = n-1$, $h(t) = y'(t)$, we get

$$y'(t) \geq \frac{\mu}{(n-2)!} t^{n-2} y^{(n-1)}(t), \quad (7)$$

for every constant $\mu \in (0, 1)$. From (5) and (7), we obtain

$$\omega'(t) \leq \delta'(t) \frac{r(t)\left|\left(y^{(n-1)}(t)\right)\right|^{p-1}}{y^{p-1}(t)} + \delta(t) \frac{\left(r(t)\left|\left(y^{(n-1)}(t)\right)\right|^{p-1}\right)'}{y^{p-1}(t)}$$
$$-\delta(t) \frac{(p-1)\mu t^{n-2}}{(n-2)!} \frac{r(t)\left|\left(y^{(n-1)}(t)\right)\right|^p}{y^p(t)}. \quad (8)$$

By Lemma 2, we have

$$\frac{y(t)}{y'(t)} \geq \frac{t}{n-1}.$$

Integrating this inequality from $\tau(t)$ to t, we obtain

$$\frac{y(\tau(t))}{y(t)} \geq \frac{\tau^{n-1}(t)}{t^{n-1}}. \quad (9)$$

Combining (1) and (8), we get

$$\omega'(t) \leq \delta'(t) \frac{r(t)\left|\left(y^{(n-1)}(t)\right)\right|^{p-1}}{y^{p-1}(t)} - \delta(t) \frac{q(t)\left(y^{(p-1)}(\tau(t))\right)}{y^{p-1}(t)}$$
$$-\delta(t) \frac{(p-1)\mu t^{n-2}}{(n-2)!} \frac{r(t)\left|\left(y^{(n-1)}(t)\right)\right|^p}{y^p(t)}. \quad (10)$$

From (9) and (10), we obtain

$$\omega'(t) \leq \frac{\delta'_+(t)}{\delta(t)}\omega(t) - \delta(t) q(t) \left(\frac{\tau^{n-1}(t)}{t^{n-1}}\right)^{p-1} - \frac{(p-1)\mu t^{n-2}}{(n-2)!\left(\delta(t) r(t)\right)^{1/(p-1)}}\omega^{p/(p-1)}(t). \quad (11)$$

It follows from (11) that

$$\delta(t) q(t) \left(\frac{\tau^{n-1}(t)}{t^{n-1}}\right)^{p-1} \leq \frac{\delta'_+(t)}{\delta(t)}\omega(t) - \omega'(t) - \frac{(p-1)\mu t^{n-2}}{(n-2)!\left(\delta(t) r(t)\right)^{1/(p-1)}}\omega^{p/(p-1)}(t).$$

This completes the proof. □

Lemma 6. Let y be an eventually positive solution of Equation (1) and assume that Case (C_2) holds. If

$$\psi(t) := \sigma(t) \frac{y'(t)}{y(t)}, \tag{12}$$

where $\sigma \in C^1([t_0, \infty), (0, \infty))$, then

$$\sigma(t) B(t) \leq -\psi'(t) + \frac{\sigma'(t)}{\sigma(t)} \psi(t) - \frac{1}{\sigma(t)} \psi^2(t). \tag{13}$$

Proof. Let y be an eventually positive solution of Equation (1) and assume that Case (C_2) holds. Using Lemma 2, we obtain

$$y(t) \geq t y'(t).$$

Thus we find that y/t is nonincreasing, and hence

$$y(\tau(t)) \geq y(t) \frac{\tau(t)}{t}. \tag{14}$$

Since $y > 0$, (1) becomes

$$\left(r(t) \left(y^{(n-1)}(t) \right)^{p-1} \right)' + q(t) y^{p-1}(\tau(t)) = 0.$$

Integrating that equation from t to ∞, we see that

$$\lim_{t \to \infty} \left(r(t) \left(y^{(n-1)}(t) \right)^{p-1} \right) - r(t) \left(y^{(n-1)}(t) \right)^{p-1} + \int_t^\infty q(s) y^{p-2}(\tau(s)) = 0. \tag{15}$$

Since the function $r \left(y^{(n-1)} \right)^{p-1}$ is positive $\left[r > 0 \text{ and } y^{(n-1)} > 0 \right]$ and nonincreasing $\left(\left(r \left(y^{(n-1)} \right)^{p-1} \right)' < 0 \right)$, there exists a $t_2 \geq t_0$ such that $r \left(y^{(n-1)} \right)^{p-1}$ is bounded above for all $t \geq t_2$, and so $\lim_{t \to \infty} \left(r(t) \left(y^{(n-1)}(t) \right)^{p-1} \right) = c \geq 0$. Then, from (15), we obtain

$$-r(t) \left(y^{(n-1)}(t) \right)^{p-1} + \int_t^\infty q(s) y^{p-2}(\tau(s)) \leq -c \leq 0.$$

From (14), we obtain

$$-r(t) \left(y^{(n-1)}(t) \right)^{p-1} + \int_t^\infty q(s) y(s)^{p-1} \frac{\tau(s)^{p-1}}{s^{p-1}} ds \leq 0.$$

It follows from $y'(t) > 0$ that

$$-y^{(n-1)}(t) + \frac{y(t)}{r^{1/(p-1)}(t)} \left(\int_t^\infty q(s) \left(\frac{\tau(s)}{s} \right)^{p-1} ds \right)^{1/(p-1)} \leq 0.$$

Integrating the above inequality from t to ∞ for a total of $(n-3)$ times, we get

$$y''(t) + \frac{\int_t^\infty (\theta - t)^{n-4} \left(\frac{\int_\theta^\infty q(s) \left(\frac{\tau(s)}{s} \right)^{p-1} ds}{r(\theta)} \right)^{1/(p-1)} d\theta}{(n-4)!} y(t) \leq 0. \tag{16}$$

From the definition of $\psi(t)$, we see that $\psi(t) > 0$ for $t \geq t_1$, and

$$\psi'(t) = \sigma'(t) \frac{y'(t)}{y(t)} + \sigma(t) \frac{y''(t) y(t) - (y'(t))^2}{y^2(t)}. \tag{17}$$

It follows from (16) and (17) that

$$\sigma(t) B(t) \leq -\psi'(t) + \frac{\sigma'(t)}{\sigma(t)} \psi(t) - \frac{1}{\sigma(t)} \psi^2(t).$$

This completes the proof. □

Definition 1. Let

$$D = \{(t,s) \in \mathbb{R}^2 : t \geq s \geq t_0\} \text{ and } D_0 = \{(t,s) \in \mathbb{R}^2 : t > s \geq t_0\}.$$

We say that a function $H \in C(D, \mathbb{R})$ belongs to the class $\Re$ if
(i_1) $H(t,t) = 0$ for $t \geq t_0$, $H(t,s) > 0$, $(t,s) \in D_0$.
(i_2) H has a nonpositive continuous partial derivative $\partial H / \partial s$ on D_0 with respect to the second variable.

Theorem 1. Let $n \geq 4$ be even. Assume that there exist functions $H, H_* \in \Re$, $\delta, A, \sigma, A_* \in C^1([t_0, \infty), (0, \infty))$ and $h, h_* \in C(D_0, \mathbb{R})$ such that

$$-\frac{\partial}{\partial s}(H(t,s) A(s)) = H(t,s) A(s) \frac{\delta'(t)}{\delta(t)} + h(t,s). \tag{18}$$

and

$$-\frac{\partial}{\partial s}(H_*(t,s) A_*(s)) = H_*(t,s) A_*(s) \frac{\sigma'(t)}{\sigma(t)} + h_*(t,s). \tag{19}$$

If

$$\limsup_{t \to \infty} \frac{1}{H(t,t_0)} \int_{t_0}^{t} \left[H(t,s) A(s) \delta(s) q(s) \left(\frac{\tau^{n-1}(s)}{s^{n-1}} \right)^{p-1} - D(s) \right] ds = \infty, \tag{20}$$

for some constant $\mu \in (0,1)$ and

$$\limsup_{t \to \infty} \frac{1}{H_*(t,t_0)} \int_{t_0}^{t} \left(H_*(t,s) A_*(s) \sigma(s) B(s) - \frac{\sigma(s) |h_*(t,s)|^2}{4 H_*(t,s) A_*(s)} \right) ds = \infty, \tag{21}$$

then every solution of (1) is oscillatory.

Proof. Let y be a nonoscillatory solution of Equation (1) on the interval $[t_0, \infty)$. Without loss of generality, we can assume that y is an eventually positive. By Lemma 4, there exist two possible cases for $t \geq t_1$, where $t_1 \geq t_0$ is sufficiently large.

Assume that (C_1) holds. From Lemma 5, we get that (6) holds. Multiplying (6) by $H(t,s) A(s)$ and integrating the resulting inequality from t_1 to t, we have

$$\int_{t_1}^{t} H(t,s) A(s) \delta(s) q(s) \left(\frac{\tau^{n-1}(s)}{s^{n-1}} \right)^{p-1} ds$$

$$\leq - \int_{t_1}^{t} H(t,s) A(s) \omega'(s) ds + \int_{t_1}^{t} H(t,s) A(s) \frac{\delta'(s)}{\delta(s)} \omega(s) ds$$

$$- \int_{t_1}^{t} H(t,s) A(s) \frac{(p-1) \mu s^{n-2}}{(n-2)! (\delta(s) r(s))^{1/(p-1)}} \omega^{p/(p-1)}(s) ds$$

Thus

$$\int_{t_1}^{t} H(t,s) A(s) \delta(s) q(s) \left(\frac{\tau^{n-1}(s)}{s^{n-1}}\right)^{p-1} ds$$
$$\leq H(t,t_1) A(t_1) \omega(t_1) - \int_{t_1}^{t} \left(-\frac{\partial}{\partial s}(H(t,s) A(s)) - H(t,s) A(s) \frac{\delta'(t)}{\delta(t)}\right) \omega(s) ds$$
$$- \int_{t_1}^{t} H(t,s) A(s) \frac{(p-1) \mu s^{n-2}}{(n-2)!(\delta(s) r(s))^{1/(p-1)}} \omega^{p/(p-1)}(s) ds$$

This implies

$$\int_{t_1}^{t} H(t,s) A(s) \delta(s) q(s) \left(\frac{\tau^{n-1}(s)}{s^{n-1}}\right)^{p-1} ds$$
$$\leq H(t,t_1) A(t_1) \omega(t_1) + \int_{t_1}^{t} |h(t,s)| \omega(s) d(s) \qquad (22)$$
$$- \int_{t_1}^{t} H(t,s) A(s) \frac{(p-1) \mu s^{n-2}}{(n-2)!(\delta(s) r(s))^{1/(p-1)}} \omega^{p/(p-1)}(s) ds.$$

Using the inequality

$$\beta U V^{\beta-1} - U^\beta \leq (\beta - 1) V^\beta, \quad \beta > 1, \ U \geq 0 \text{ and } V \geq 0, \qquad (23)$$

with $\beta = p/(p-1)$,

$$U = \left((p-1) H(t,s) A(s) \frac{\mu s^{n-2}}{(n-2)!}\right)^{(p-1)/p} \frac{\omega(s)}{(\delta(s) r(s))^{1/p}}$$

and

$$V = \left(\frac{p-1}{p}\right)^{p-1} |h(t,s)|^{p-1} \left(\frac{\delta(s) r(s)}{\left((p-1) H(t,s) A(s) \frac{\mu s^{n-2}}{(n-2)!}\right)^{p-1}}\right)^{(p-1)/p},$$

we get

$$|h(t,s)| \omega(s) - H(t,s) A(s) \frac{(p-1) \mu s^{n-2}}{(n-2)!(\delta(s) r(s))^{1/(p-1)}} \omega^{p/(p-1)}$$
$$\leq \frac{\delta(s) r(s)}{\left(H(t,s) A(s) \frac{\mu s^{n-2}}{(n-2)!}\right)^{p-1}} \left(\frac{|h(t,s)|}{p}\right)^p,$$

which with (23) gives

$$\int_{t_1}^{t} \left(H(t,s) A(s) \delta(s) q(s) \left(\frac{\tau^{n-1}(s)}{s^{n-1}}\right)^{p-1} - D(s)\right) ds \leq H(t,t_1) A(t_1) \omega(t_1)$$
$$\leq H(t,t_0) A(t_1) \omega(t_1).$$

Then

$$\frac{1}{H(t,t_0)} \int_{t_0}^{t} \left(H(t,s) A(s) \delta(s) q(s) \left(\frac{\tau^{n-1}(s)}{s^{n-1}}\right)^{p-1} - D(s) \right) ds$$

$$\leq A(t_1) \omega(t_1) + \int_{t_0}^{t_1} A(s) \delta(s) q(s) \left(\frac{\tau^{n-1}(s)}{s^{n-1}}\right)^{p-1} ds$$

$$< \infty,$$

for some $\mu \in (0, 1)$, which contradicts (20).

Assume that Case (C_2) holds. From Lemma 6, we get that (13) holds. Multiplying (13) by $H_*(t,s) A_*(s)$, and integrating the resulting inequality from t_1 to t, we have

$$\int_{t_1}^{t} H_*(t,s) A_*(s) \sigma(s) B(s) ds \leq -\int_{t_1}^{t} H_*(t,s) A_*(s) \psi'(s) ds + \int_{t_1}^{t} H_*(t,s) A_*(s) \frac{\sigma'(s)}{\sigma(s)} \psi(s) ds$$

$$- \int_{t_1}^{t} \frac{H_*(t,s) A_*(s)}{\sigma(s)} \psi^2(s) ds$$

$$= H_*(t,t_1) A_*(t_1) \psi(t_1) - \int_{t_1}^{t} \frac{H_*(t,s) A_*(s)}{\sigma(s)} \psi^2(s) ds$$

$$- \int_{t_1}^{t} \left(-\frac{\partial}{\partial s} (H_*(t,s) A_*(s)) - H_*(t,s) A_*(s) \frac{\sigma'(t)}{\sigma(t)} \right) \psi(s) ds.$$

Then

$$\int_{t_1}^{t} H_*(t,s) A_*(s) \sigma(s) B(s) ds \leq H_*(t,t_1) A_*(t_1) \psi(t_1) + \int_{t_1}^{t} |h_*(t,s)| \psi(s) d(s)$$

$$- \int_{t_1}^{t} \frac{H_*(t,s) A_*(s)}{\sigma(s)} \psi^2(s) ds.$$

Hence we have

$$\int_{t_1}^{t} \left(H_*(t,s) A_*(s) \sigma(s) B(s) - \frac{\sigma(s) |h_*(t,s)|^2}{4 H_*(t,s) A_*} \right) ds \leq H_*(t,t_1) A_*(t_1) \psi(t_1)$$

$$\leq H_*(t,t_0) A_*(t_1) \psi(t_1).$$

This implies

$$\frac{1}{H_*(t,t_0)} \int_{t_0}^{t} \left(H_*(t,s) A_*(s) \sigma(s) B(s) - \frac{\sigma(s) |h_*(t,s)|^2}{4 H_*(t,s) A_*} \right) ds$$

$$\leq A_*(t_1) \psi(t_1) + \int_{t_0}^{t} A_*(s) \sigma(s) B(s) ds < \infty$$

which contradicts (21). Therefore, every solution of (1) is oscillatory. □

In the next theorem, we establish new oscillation results for Equation (1) by using the comparison technique with the first-order differential inequality:

Theorem 2. *Let $n \geq 2$ be even and $r'(t) > 0$. Assume that for some constant $\lambda \in (0,1)$, the differential equation*

$$\varphi'(t) + \frac{q(t)}{r(\tau(t))} \left(\frac{\lambda \tau^{n-1}(t)}{(n-1)!} \right)^{p-1} \varphi(\tau(t)) = 0 \qquad (24)$$

is oscillatory. Then every solution of (1) is oscillatory.

Proof. Let (1) have a nonoscillatory solution y. Without loss of generality, we can assume that $y(t) > 0$ for $t \geq t_1$, where $t_1 \geq t_0$ is sufficiently large. Since $r'(t) > 0$, we have

$$y'(t) > 0, \ y^{(n-1)}(t) > 0 \text{ and } y^{(n)}(t) < 0. \tag{25}$$

From Lemma 3, we get

$$y(t) \geq \frac{\lambda t^{n-1}}{(n-1)!r^{1/p-1}(t)} r^{1/p-1}(t) y^{(n-1)}(t), \tag{26}$$

for every $\lambda \in (0,1)$. Thus, if we set

$$\varphi(t) = r(t) \left[y^{(n-1)}(t) \right]^{p-1} > 0,$$

then we see that φ is a positive solution of the inequality

$$\varphi'(t) + \frac{q(t)}{r(\tau(t))} \left(\frac{\lambda \tau^{n-1}(t)}{(n-1)!} \right)^{p-1} \varphi(\tau(t)) \leq 0. \tag{27}$$

From [22] (Theorem 1), we conclude that the corresponding Equation (24) also has a positive solution, which is a contradiction.

Theorem 2 is proved. □

Corollary 1. *Assume that (2) holds and let $n \geq 2$ be even. If*

$$\lim_{t \to \infty} \int_{\tau(t)}^{t} \frac{q(s)}{r(\tau(s))} \left(\tau^{n-1}(s) \right)^{p-1} ds > \frac{((n-1)!)^{p-1}}{e}, \tag{28}$$

then every solution of (1) is oscillatory.

Next, we give the following example to illustrate our main results.

Example 1. *Consider the equation*

$$y^{(4)}(t) + \frac{\gamma}{t^4} y \left(\frac{9}{10} t \right) = 0, \ t \geq 1, \tag{29}$$

where $\gamma > 0$ is a constant. We note that $n = 4$, $r(t) = 1$, $p = 2$, $\tau(t) = 9t/10$ and $q(t) = \gamma/t^4$. If we set $H(t,s) = H_*(t,s) = (t-s)^2$, $A(s) = A_*(s) = 1$, $\delta(s) = t^3$, $\sigma(s) = t$, $h(t,s) = (t-s)(5-3ts^{-1})$ and $h_*(t,s) = (t-s)(3-ts^{-1})$ then we get

$$\eta(s) = \int_{t_0}^{\infty} \frac{1}{r^{1/(p-1)}(s)} ds = \infty$$

and

$$B(t) = \frac{1}{(n-4)!} \int_{t}^{\infty} (\theta - t)^{n-4} \left(\frac{\int_{\theta}^{\infty} q(s) \left(\frac{\tau(s)}{s} \right)^{p-1} ds}{r(\theta)} \right)^{1/(p-1)} d\theta$$

$$= 3\gamma / \left(20t^2 \right).$$

Hence conditions (20) and (21) become

$$\limsup_{t\to\infty} \frac{1}{H(t,t_0)} \int_{t_0}^{t} \left(H(t,s) A(s) \delta(s) q(s) \left(\frac{\tau^{n-1}(s)}{s^{n-1}}\right)^{p-1} - D(s) \right) ds$$

$$= \limsup_{t\to\infty} \frac{1}{(t-1)^2} \int_{1}^{t} \left[\frac{729\gamma}{1000} t^2 s^{-1} + \frac{729\gamma}{1000} s - \frac{729\gamma}{500} t - \frac{s}{2\mu}\left(25 + 9t^2 s^{-2} - 30ts^{-1}\right) \right] ds$$

$$= \infty \quad (\text{if } \gamma > 500/81)$$

and

$$\limsup_{t\to\infty} \frac{1}{H_*(t,t_0)} \int_{t_0}^{t} \left(H_*(t,s) A_*(s) \sigma(s) B(s) - \frac{\sigma(s)|h_*(t,s)|^2}{4H_*(t,s) A_*(s)} \right) ds$$

$$= \limsup_{t\to\infty} \frac{1}{(t-1)^2} \int_{1}^{t} \left[\frac{3\gamma}{20} t^2 s^{-1} + \frac{3\gamma}{20} s - \frac{3\gamma}{10} t - \frac{s}{4}\left(9 - 630ts^{-1} + t^2 s^{-2}\right) \right] ds$$

$$= \infty \quad (\text{if } \gamma > 5/3).$$

Thus, by Theorem 1, every solution of Equation (29) is oscillatory if $\gamma > 500/81$.

3. Conclusions

In this work, we have discussed the oscillation of the higher-order differential equation with a p-Laplacian-like operator and we proved that Equation (1) is oscillatory by using the following methods:

1. The Riccati transformation technique.
2. Comparison principles.
3. The Integral averaging technique.

Additionally, in future work we could try to get some oscillation criteria of Equation (1) under the condition $\int_{t_0}^{\infty} \frac{1}{r^{1/(p-1)}(t)} dt < \infty$. Thus, we would discuss the following two cases:

(C_1) $y(t) > 0$, $y^{(n-1)}(t) > 0$, $y^{(n)}(t) < 0$,
(C_2) $y(t) > 0$, $y^{(n-2)}(t) > 0$, $y^{(n-1)}(t) < 0$.

Author Contributions: The authors claim to have contributed equally and significantly in this paper. All authors have read and agreed to the published version of the manuscript.

Funding: The authors received no direct funding for this work.

Acknowledgments: The authors thank the reviewers for for their useful comments, which led to the improvement of the content of the paper.

Conflicts of Interest: The authors declare no conflict of interest.

References

1. Agarwal, R.; Grace, S.; O'Regan, D. *Oscillation Theory for Difference and Functional Differential Equations*; Kluwer Acad. Publ.: Dordrecht, The Netherlands, 2000.
2. Agarwal, R.; Grace, S.; O'Regan, D. Oscillation criteria for certain nth order differential equations with deviating arguments. *J. Math. Appl. Anal.* **2001**, *262*, 601–622. [CrossRef]
3. Agarwal, R.; Shieh, S.L.; Yeh, C.C. Oscillation criteria for second order retarde ddifferential equations. *Math. Comput. Model.* **1997**, *26*, 1–11. [CrossRef]
4. Agarwal, R.P.; Zhang, C.; Li, T. Some remarks on oscillation of second order neutral differential equations. *Appl. Math. Compt.* **2016**, *274*, 178–181. [CrossRef]

5. Baculikova, B.; Dzurina, J.; Graef, J.R. On the oscillation of higher-order delay differential equations. *Math. Slovaca* **2012**, *187*, 387–400. [CrossRef]
6. Bazighifan, O.; Cesarano, C. Some New Oscillation Criteria for Second-Order Neutral Differential Equations with Delayed Arguments. *Mathematics* **2019**, *7*, 619. [CrossRef]
7. Bazighifan, O.; Elabbasy, E.M.; Moaaz, O. Oscillation of higher-order differential equations with distributed delay. *J. Inequal. Appl.* **2019**, *55*, 1–9. [CrossRef]
8. Chatzarakis, G.E.; Elabbasy, E.M.; Bazighifan, O. An oscillation criterion in 4th-order neutral differential equations with a continuously distributed delay. *Adv. Difference Equ.* **2019**, *336*, 1–9.
9. Cesarano, C.; Pinelas, S.; Al-Showaikh, F.; Bazighifan, O. Asymptotic Properties of Solutions of Fourth-Order Delay Differential Equations. *Symmetry* **2019**, *11*, 628. [CrossRef]
10. Cesarano, C.; Bazighifan, O. Oscillation of fourth-order functional differential equations with distributed delay. *Axioms* **2019**, *7*, 61. [CrossRef]
11. Cesarano, C.; Bazighifan, O. Qualitative behavior of solutions of second order differential equations. *Symmetry* **2019**, *11*, 777. [CrossRef]
12. Grace, S.; Dzurina, J.; Jadlovska, I.; Li, T. On the oscillation of fourth order delay differential equations. *Adv. Differ. Equ.* **2019**, *118*, 1–15. [CrossRef]
13. Gyori, I.; Ladas, G. *Oscillation Theory of Delay Differential Equations with Applications*; Clarendon Press: Oxford, UK, 1991.
14. Hale, J.K. *Theory of Functional Differential Equations*; Springer: New York, NY, USA, 1977.
15. Kiguradze, I.; Chanturia, T. *Asymptotic Properties of Solutions of Nonautonomous Ordinary Differential Equations*; Kluwer Acad. Publ.: Drodrcht, The Netherlands, 1993.
16. Ladde, G.; Lakshmikantham, V.; Zhang, B. *Oscillation Theory of Differential Equations with Deviating Arguments*; Marcel Dekker: New York, NY, USA, 1987.
17. Li, T.; Baculikova, B.; Dzurina, J.; Zhang, C. Oscillation of fourth order neutral differential equations with p-Laplacian like operators. *Bound. Value Probl.* **2014**, *56*, 41–58. [CrossRef]
18. Moaaz, O. New criteria for oscillation of nonlinear neutral differential equations. *Adv. Differ. Equ.* **2019**, *2019*, 484. [CrossRef]
19. Moaaz, O.; Elabbasy, E.M.; Bazighifan, O. On the asymptotic behavior of fourth-order functional differential equations. *Adv. Differ. Equ.* **2017**, *2017*, 261. [CrossRef]
20. Moaaz, O.; Elabbasy, E.M.; Muhib, A. Some new oscillation results for fourth-order neutral differential equations. *Adv. Differ. Equ.* **2019**, *2019*, 297. [CrossRef]
21. Moaaz, O.; Elabbasy, E.M.; Shaaban, E. Oscillation criteria for a class of third order damped differential equations. *Arab J. Math. Sci.* **2018**, *24*, 16–30. [CrossRef]
22. Philos, C. On the existence of nonoscillatory solutions tending to zero at ∞ for differential equations with positive delay. *Arch. Math. (Basel)* **1981**, *36*, 168–178. [CrossRef]
23. Rehak, P. How the constants in Hille–Nehari theorems depend on time scales. *Adv. Differ. Equ.* **2006**, *2006*, 1–15. [CrossRef]
24. Zhang, C.; Agarwal, R.P.; Bohner, M.; Li, T. New results for oscillatory behavior of even-order half-linear delay differential equations. *Appl. Math. Lett.* **2013**, *26*, 179–183. [CrossRef]
25. Aronsson, G.; Janfalk, U. On Hele-Shaw flow of power-law fluids. *Eur. J. Appl. Math.* **1992**, *3*, 343–366. [CrossRef]
26. Elabbasy, E.M.; Cesarano, C.; Bazighifan, O.; Moaaz, O. Asymptotic and oscillatory behavior of solutions of a class of higher-order differential equations. *Symmetry* **2019**, *11*, 1434. [CrossRef]
27. Zhang, C.; Li, T.; Suna, B.; Thandapani, E. On the oscillation of higher-order half-linear delay differential equations. *Appl. Math. Lett.* **2011**, *24*, 1618–1621. [CrossRef]

© 2020 by the authors. Licensee MDPI, Basel, Switzerland. This article is an open access article distributed under the terms and conditions of the Creative Commons Attribution (CC BY) license (http://creativecommons.org/licenses/by/4.0/).

Article

Stability of Equilibria of Rumor Spreading Model under Stochastic Perturbations

Leonid Shaikhet

Department of Mathematics, Ariel University, Ariel 40700, Israel; leonid.shaikhet@usa.net

Received: 2 February 2020; Accepted: 11 February 2020; Published: 18 February 2020

Abstract: The known mathematical model of rumor spreading, which is described by a system of four nonlinear differential equations and is very popular in research, is considered. It is supposed that the considered model is influenced by stochastic perturbations that are of the type of white noise and are proportional to the deviation of the system state from its equilibrium point. Sufficient conditions of stability in probability for each from the five equilibria of the considered model are obtained by virtue of the Routh–Hurwitz criterion and the method of linear matrix inequalities (LMIs). The obtained results are illustrated by numerical analysis of appropriate LMIs and numerical simulations of solutions of the considered system of stochastic differential equations. The research method can also be used in other applications for similar nonlinear models with the order of nonlinearity higher than one.

Keywords: rumor spreading model; white noise; stochastic differential equations; asymptotic mean square stability; stability in probability; linear matrix inequality

1. Introduction

There are two classes of mathematical models of the type of epidemics: medical epidemics (see, for instance, the so-called SIR-epidemic model [1–3]) and different social epidemics (see, for instance, the alcohol consumption model [4] or the model of obesity epidemic [5]). During the last two decades, the rumor spreading model, that is an epidemic of the social type too, is extremely popular in research (see, [6–29]). Following [26], we will consider the rumor spreading model (the so-called I2SR-model) in the form

$$\begin{aligned}
\dot{I}(t) &= p - \lambda_1 I(t) S_1(t) - \lambda_2 I(t) S_2(t) - q I(t), \\
\dot{S}_1(t) &= \lambda_1 I(t) S_1(t) + \alpha S_2(t) - \delta_1 S_1(t) R(t) - q S_1(t), \\
\dot{S}_2(t) &= \lambda_2 I(t) S_2(t) - \alpha S_2(t) - \delta_2 S_2(t) R(t) - q S_2(t), \\
\dot{R}(t) &= \delta_1 S_1(t) R(t) + \delta_2 S_2(t) R(t) - q R(t),
\end{aligned} \quad (1)$$

where $I(t), S_1(t), S_2(t), R(t)$ are respectively the density of ignorants, the low rate of active spreaders, the high rate of active spreaders and stiflers at time t, $p, q, \alpha, \delta_1, \delta_2, \lambda_1, \lambda_2$ are positive parameters.

Please note that the sense of the parameters $p, q, \alpha, \delta_1, \delta_2, \lambda_1, \lambda_2$ that are used in the rumor spreading model (1) are described in [26]. We will consider the system (1) as a mathematical object and show how stability of nonlinear mathematical models of the similar type can be investigated under influence of stochastic perturbations. In particular, we will consider here the simple parameters λ_i and δ_i unlike from [26], where these parameters are considered in the form of the product of two parameters: $\lambda_i \bar{k}$ and $\delta_i \bar{k}$, $i = 1, 2$. We will not suppose in the general case as it is made in [26] that $p = q$ and $\delta_1 = \delta_2$. We will correct also some errors and inaccuracies which there are in [26]. For example, in [26] it is supposed that $\lambda_2 > \lambda_1$ (p. 856) but in the numerical examples the following values are used: $\lambda_1 = 0.05$ and $\lambda_2 = 0.007$ or $\lambda_2 = 0.003$ (p. 862), all equilibria and stability conditions are obtained under the assumption $\delta_1 = \delta_2 = \delta$ (p. 857) but in the numerical examples one can see $\delta_1 = 0.007$ and $\delta_2 = 0.59$ (p. 862) or $\delta_2 = 0.009$ (p. 863) and so on.

The purpose of the proposed research is to calculate of equilibria of the system (1) and to obtain stability conditions for each from these equilibria under assumption that the system is exposed to stochastic perturbations. Sufficient conditions of stability in probability for each from the five equilibria of the considered model are obtained by virtue of the Routh–Hurwitz criterion [30] and the method of linear matrix inequalities (LMIs) [31,32]. The proposed research method can be used for a lot of other similar nonlinear models in different applications.

2. Equilibria of the Model

Equilibria $E = (I^*, S_1^*, S_2^*, R^*)$ of the model (1) are defined by the system of algebraic equations

$$\begin{cases} (\lambda_1 S_1 + \lambda_2 S_2 + q)I = p, \\ (\delta_1 R - \lambda_1 I + q)S_1 = \alpha S_2, \\ (\delta_2 R - \lambda_2 I + \alpha + q)S_2 = 0, \\ (\delta_1 S_1 + \delta_2 S_2 - q)R = 0, \end{cases} \quad (2)$$

that follows from (1) by the condition that $I(t), S_1(t), S_2(t), R(t)$ are constants.

Please note that the solution of the system (2) is not unique. Solving the system (2) gives the following five equilibria $E_i = (I_i^*, S_{1i}^*, S_{2i}^*, R_i^*), i = 0, ..., 4$, where (see Appendix A.1)

$$\begin{aligned}
E_0 &= (I_0^*, 0, 0, 0), \quad I_0^* = \frac{p}{q}; \\
E_1 &= (I_1^*, S_{11}^*, 0, 0), \quad I_1^* = \frac{q}{\lambda_1}, \quad S_{11}^* = \frac{p}{q} - \frac{q}{\lambda_1}; \\
E_2 &= (I_2^*, S_{12}^*, 0, R_2^*), \quad I_2^* = \frac{p\delta_1}{q(\delta_1 + \lambda_1)}, \quad S_{12}^* = \frac{q}{\delta_1}, \quad R_2^* = \frac{p\lambda_1}{q(\delta_1 + \lambda_1)} - \frac{q}{\delta_1}; \\
E_3 &= (I_3^*, S_{13}^*, S_{23}^*, 0), \\
I_3^* &= \frac{\alpha + q}{\lambda_2}, \quad S_{13}^* = \frac{\alpha(p\lambda_2 - q(\alpha + q))}{q(\lambda_2 - \lambda_1)(\alpha + q)}, \quad S_{23}^* = \frac{(q(\lambda_2 - \lambda_1) - \alpha\lambda_1)(p\lambda_2 - q(\alpha + q))}{\lambda_2 q(\lambda_2 - \lambda_1)(\alpha + q)}; \\
E_4 &= (I_4^*, S_{14}^*, S_{24}^*, R_4^*), \\
&\text{if } (\delta_2 - \delta_1)(\lambda_2\delta_1 - \lambda_1\delta_2) \neq 0 \text{ then } S_{14}^* \text{ is a positive root of the quadratic equation} \\
&S_1^2 - \nu_1 S_1 + \nu_2 = 0, \quad \nu_1 = \frac{q\alpha + p\delta_2}{q(\delta_2 - \delta_1)} + \frac{q(\lambda_2 + \delta_2)}{\lambda_2\delta_1 - \lambda_1\delta_2}, \quad \nu_2 = \frac{\alpha q(\lambda_2 + \delta_2)}{(\delta_2 - \delta_1)(\lambda_2\delta_1 - \lambda_1\delta_2)}, \\
S_{14}^* &= \begin{cases} \frac{\alpha q^2(\delta + \lambda_2)}{\delta(\lambda_2 - \lambda_1)(q\alpha + p\delta)} & \text{if } \delta_2 = \delta_1 = \delta, \quad \lambda_2 > \lambda_1, \\ \frac{\alpha}{\delta_2 - \delta_1} & \text{if } \lambda_2\delta_1 = \lambda_1\delta_2, \quad \delta_2 > \delta_1, \end{cases} \\
S_{24}^* &= \frac{1}{\delta_2}(q - \delta_1 S_{14}^*), \quad I_4^* = \frac{p}{\lambda_1 S_{14}^* + \lambda_2 S_{24}^* + q}, \quad R_4^* = \frac{\lambda_2 I_4^* - \alpha - q}{\delta_2}, \quad S_{14}^* < \frac{q}{\delta_1}, \quad I_4^* > \frac{\alpha + q}{\lambda_2}.
\end{aligned} \quad (3)$$

It is supposed that all nonzero elements of all equilibria are positive.

Putting $N(t) = I(t) + S_1(T) + S_2(t) + R(t)$ and summing all equations of the system (1), we obtain

$$\dot{N}(t) = p - qN(t), \quad N(t) = \left(N(0) - \frac{p}{q}\right)e^{-qt} + \frac{p}{q}, \quad \lim_{t\to\infty} N(t) = \frac{p}{q}. \quad (4)$$

In accordance with (4) for all equilibria we have

$$N^* = I_i^* + S_{1i}^* + S_{2i}^* + R_i^* = \frac{p}{q}, \quad i = 0, ..., 4. \quad (5)$$

3. Stochastic Perturbations, Centralization, and Linearization

Let us suppose that the system (1) is exposed to stochastic perturbations which are directly proportional to the deviation of the system (1) state $(I(t), S_1(t), S_2(t), R(t))$ from the equilibrium (I^*, S_1^*, S_2^*, R^*) and are of the type of white noise $(\dot{w}_0(t), \dot{w}_1(t), \dot{w}_2(t), \dot{w}_3(t))$, where $(w_0(t), w_1(t), w_2(t), w_3(t))$ are mutually independent standard Wiener processes. Therefore, we obtain the following system of the Ito stochastic differential equations [33]

$$\begin{aligned}
\dot{I}(t) &= p - \lambda_1 I(t)S_1(t) - \lambda_2 I(t)S_2(t) - qI(t) + \sigma_0(I(t) - I^*)\dot{w}_0(t), \\
\dot{S}_1(t) &= \lambda_1 I(t)S_1(t) + \alpha S_2(t) - \delta_1 S_1(t)R(t) - qS_1(t) + \sigma_1(S_1(t) - S_1^*)\dot{w}_1(t), \\
\dot{S}_2(t) &= \lambda_2 I(t)S_2(t) - \alpha S_2(t) - \delta_2 S_2(t)R(t) - qS_2(t) + \sigma_2(S_2(t) - S_2^*)\dot{w}_2(t), \\
\dot{R}(t) &= \delta_1 S_1(t)R(t) + \delta_2 S_2(t)R(t) - qR(t) + \sigma_3(R(t) - R^*)\dot{w}_3(t).
\end{aligned} \quad (6)$$

Please note that the equilibrium (I^*, S_1^*, S_2^*, R^*) of the deterministic system (1) is also a solution of the system with stochastic perturbations (6).

Let (I^*, S_1^*, S_2^*, R^*) be one of the equilibria of the system (1). Putting in (6) $I(t) = y_0(t) + I^*$, $S_1(t) = y_1(t) + S_1^*$, $S_2(t) = y_2(t) + S_2^*$, $R(t) = y_3(t) + R^*$, we obtain

$$\begin{aligned}
\dot{y}_0(t) &= p - (y_0(t) + I^*)[\lambda_1(y_1(t) + S_1^*) + \lambda_2(y_2(t) + S_2^*) + q] + \sigma_0 y_0(t)\dot{w}_0(t), \\
\dot{y}_1(t) &= (y_1(t) + S_1^*)[\lambda_1(y_0(t) + I^*) - \delta_1(y_3(t) + R^*) - q] + \alpha(y_2(t) + S_2^*) + \sigma_1 y_1(t)\dot{w}_1(t), \\
\dot{y}_2(t) &= (y_2(t) + S_2^*)[\lambda_2(y_0(t) + I^*) - \delta_2(y_3(t) + R^*) - \alpha - q] + \sigma_2 y_2(t)\dot{w}_2(t), \\
\dot{y}_3(t) &= (y_3(t) + R^*)[\delta_1(y_1(t) + S_1^*) + \delta_2(y_2(t) + S_2^*) - q] + \sigma_3 y_3(t)\dot{w}_3(t).
\end{aligned} \quad (7)$$

It is clear that stability of the zero solution of the system (7) is equivalent to stability of the equilibrium (I^*, S_1^*, S_2^*, R^*) of the system (6).

Removing from the system (7) nonlinear terms and using the system for equilibria (2) we obtain the linear part of the system (7)

$$\begin{aligned}
\dot{z}_0(t) &= -p(I^*)^{-1}z_0(t) - \lambda_1 I^* z_1(t) - \lambda_2 I^* z_2(t) + \sigma_0 z_0(t)\dot{w}_0(t), \\
\dot{z}_1(t) &= \lambda_1 S_1^* z_0(t) - (q + \delta_1 R^* - \lambda_1 I^*)z_1(t) + \alpha z_2(t) - \delta_1 S_1^* z_3(t) + \sigma_1 z_1(t)\dot{w}_1(t), \\
\dot{z}_2(t) &= \lambda_2 S_2^* z_0(t) - (\alpha + q - \lambda_2 I^* + \delta_2 R^*)z_2(t) - \delta_2 S_2^* z_3(t) + \sigma_2 z_2(t)\dot{w}_2(t), \\
\dot{z}_3(t) &= \delta_1 R^* z_1(t) + \delta_2 R^* z_2(t) - (q - \delta_1 S_1^* - \delta_2 S_2^*)z_3(t) + \sigma_3 z_3(t)\dot{w}_3(t).
\end{aligned} \quad (8)$$

Let us present the system (8) in the matrix form

$$dz(t) = Az(t)dt + C(z(t))dw(t), \quad (9)$$

where $z(t) = (z_0(t), z_1(t), z_2(t), z_3(t))'$, $w(t) = (w_0(t), w_1(t), w_2(t), w_3(t))'$, $C(z(t)) = diag(\sigma_0 z_0(t), ..., \sigma_3 z_3(t))$,

$$A = \begin{bmatrix} -p(I^*)^{-1} & -\lambda_1 I^* & -\lambda_2 I^* & 0 \\ \lambda_1 S_1^* & -(q + \delta_1 R^* - \lambda_1 I^*) & \alpha & -\delta_1 S_1^* \\ \lambda_2 S_2^* & 0 & -(\alpha + q - \lambda_2 I^* + \delta_2 R^*) & -\delta_2 S_2^* \\ 0 & \delta_1 R^* & \delta_2 R^* & -(q - \delta_1 S_1^* - \delta_2 S_2^*) \end{bmatrix}. \quad (10)$$

Remark 1. *The order of nonlinearity of the nonlinear system (7) is higher than one. For systems of such type a sufficient condition for asymptotic mean square stability of the zero solution of its linear part (9) provides stability in probability of the zero solution of the initial nonlinear system (7) [30]. Therefore, a sufficient condition for asymptotic mean square stability of the zero solution of the linear Equation (9) provides stability in probability of the equilibrium (I^*, S_1^*, S_2^*, R^*) of the initial system (6).*

Following Remark 1, below we will have sufficient conditions for asymptotic mean square stability of the zero solution of the linear Equation (9) for each from the equilibria (3).

4. Stability of the Equilibria

Consider some definitions and statements that will be used below [30].

Definition 1. *The zero solution of the system (7) is called stable in probability if for any $\varepsilon_1 > 0$ and $\varepsilon_2 > 0$ there exists $\delta > 0$ such that the solution $y(t) = (y_0(t), y_1(t), y_2(t), y_3(t))'$ of the system (7) satisfies the condition $\mathbf{P}\{\sup_{t \geq 0} |y(t)| > \varepsilon_1\} < \varepsilon_2$ provided that $\mathbf{P}\{|y(0)| < \delta\} = 1$.*

Definition 2. *The zero solution of the system (9) is called:*

- *mean square stable if for each $\varepsilon > 0$ there exists a $\delta > 0$ such that $\mathbf{E}|z(t)|^2 < \varepsilon$, $t \geq 0$, provided that $\mathbf{E}|z(0)|^2 < \delta$;*
- *asymptotically mean square stable if it is mean square stable and the solution $z(t)$ of Equation (9) satisfies the condition $\lim_{t \to \infty} \mathbf{E}|z(t)|^2 = 0$ provided that $\mathbf{E}|z(0)|^2 < \infty$.*

The generator L of the Ito stochastic differential Equation (9) is defined on the functions $V(t,z)$ which have one continuous derivative with respect to t (V_t), two continuous derivatives (∇V and $\nabla^2 V$) with respect to z and has the form [30,33]

$$LV(t,z(t)) = V_t(t,z(t)) + \nabla V'(t,z(t))Az(t) + \frac{1}{2}Tr[C(z(t))\nabla^2 V(t,z(t))C(z(t))]. \tag{11}$$

Theorem 1. *Let there exist a function $V(t,z)$ with continuous derivatives V_t, ∇V, $\nabla^2 V$, positive constants c_1, c_2, c_3, such that the following conditions hold:*

$$\mathbf{E}V(t,z(t)) \geq c_1 \mathbf{E}|z(t)|^2, \quad \mathbf{E}V(0,z(0)) \leq c_2 \mathbf{E}|z(0)|^2, \quad \mathbf{E}LV(t,z(t)) \leq -c_3 \mathbf{E}|z(t)|^2.$$

Then the zero solution of Equation (9) is asymptotically mean square stable.

Lemma 1. *Let there exist a positive definite matrix $P = \|p_{ij}\|$ $(i,j = 1,2,3,4)$ such that the matrix (10) with the equilibrium (I^*, S_1^*, S_2^*, R^*) satisfies the linear matrix inequality (LMI)*

$$PA + A'P + P_\sigma < 0, \qquad P_\sigma = \mathrm{diag}\{p_{11}\sigma_0^2, ..., p_{44}\sigma_3^2\}. \tag{12}$$

Then the equilibrium (I^*, S_1^*, S_2^*, R^*) of the system (6) is stable in probability.

Proof. For the function $V(t,z) = z'Pz$ from (11) and LMI (12) for some $c > 0$ we have

$$\begin{aligned}LV(t,z(t)) &= 2z'(t)PAz(t) + Tr[C(z(t))PC(z(t))] \\ &= z'(t)(PA + A'P + P_\sigma)z(t) \leq -c|z(t)|^2.\end{aligned}$$

From Theorem 1 it follows that the zero solution of Equation (9) is asymptotically mean square stable. Via Remark 1 one can conclude that the equilibrium (I^*, S_1^*, S_2^*, R^*) of the system (6) is stable in probability. The proof is completed. □

Note to satisfy the LMI (12) the matrix A must be the Hurwitz matrix [30,31].

Definition 3. *The trace of the $k-$th order of a $n \times n$-matrix $A = \|a_{ij}\|$ is defined as follows:*

$$T_k = \sum_{1 \leq i_1 < ... < i_k \leq n} \begin{vmatrix} a_{i_1 i_1} & ... & a_{i_1 i_k} \\ ... & ... & ... \\ a_{i_k i_1} & ... & a_{i_k i_k} \end{vmatrix}, \quad k = 1, ..., n.$$

Here, in particular, $T_1 = Tr(A)$, $T_n = \det(A)$, $T_{n-1} = \sum_{i=1}^{n} A_{ii}$, where A_{ii} is the algebraic complement of the diagonal element a_{ii} of the matrix A.

Lemma 2. *[30,31] Let T_k, $k = 1,2,3,4$, be the trace of the k-th order of a 4×4-matrix A. The matrix A is the Hurwitz matrix if and only if*

$$T_1 < 0, \quad T_1 T_2 < T_3 < 0, \quad 0 < T_1^2 T_4 < (T_1 T_2 - T_3) T_3. \tag{13}$$

A 3×3-matrix A is the Hurwitz matrix if and only if first two conditions (13) hold.

In general, the LMI (12) for each equilibrium (3) must be numerically investigated via MATLAB. However, in some particular cases this process can be simplified and analytical conditions can be obtained. Below it is shown in investigation of stability of the equilibria (3).

4.1. Stability of the Equilibrium $E_0 = (\frac{p}{q}, 0, 0, 0)$

Theorem 2. *If*

$$\frac{1}{\lambda_1} > \frac{p}{q^2}, \quad \frac{1}{\lambda_2}\left(1 + \frac{\alpha}{q}\right) > \frac{p}{q^2}, \tag{14}$$

and

$$\sigma_0^2 < 2q, \quad \sigma_1^2 < 2\left(q - \lambda_1 \frac{p}{q}\right), \quad \sigma_2^2 < 2\left(\alpha + q - \lambda_2 \frac{p}{q}\right), \quad \sigma_3^2 < 2q, \tag{15}$$

then the equilibrium E_0 is stable in probability.

Proof. For the equilibrium $E_0 = \left(\frac{p}{q}, 0, 0, 0\right)$ the system (8) takes the form

$$\begin{aligned}
\dot{z}_0(t) &= -qz_0(t) - \lambda_1 pq^{-1} z_1(t) - \lambda_2 pq^{-1} z_2(t) + \sigma_0 z_0(t)\dot{w}_0(t), \\
\dot{z}_1(t) &= -(q - \lambda_1 pq^{-1})z_1(t) + \alpha z_2(t) + \sigma_1 z_1(t)\dot{w}_1(t), \\
\dot{z}_2(t) &= -(\alpha + q - \lambda_2 pq^{-1})z_2(t) + \sigma_2 z_2(t)\dot{w}_2(t), \\
\dot{z}_3(t) &= -qz_3(t) + \sigma_3 z_3(t)\dot{w}_3(t).
\end{aligned} \tag{16}$$

The conditions (14) provide negativity of the coefficients before $z_1(t)$ and $z_2(t)$ in the second and the third equations (16). It is known [30] that the last two inequalities (15) are the necessary and sufficient conditions for asymptotic mean square stability of the zero solutions of the last two equations in (16) which do not depend on $z_0(t)$ and $z_1(t)$ and can be considered separately. Since $\lim_{t \to \infty} Ez_2^2(t) = 0$ then the system of first two Equation (16) for $z_0(t)$ and $z_1(t)$ can be considered without the process $z_2(t)$, i.e.,

$$\begin{aligned}
\dot{z}_0(t) &= -qz_0(t) - \lambda_1 pq^{-1} z_1(t) + \sigma_0 z_0(t)\dot{w}_0(t), \\
\dot{z}_1(t) &= -(q - \lambda_1 pq^{-1})z_1(t) + \sigma_1 z_1(t)\dot{w}_1(t).
\end{aligned} \tag{17}$$

Via Remark A2 (see Appendix A.2) the first two inequalities (15) are sufficient for asymptotic mean square stability of the zero solution of the system (17). Therefore, the conditions (14), (15) provide asymptotic mean square stability of the zero solution of the system (16) and via Remark 1 stability in probability of the equilibrium E_0 of the system (6). The proof is completed. □

Remark 2. *One can check that by the conditions (14) and (15) the matrix*

$$A = \begin{bmatrix} -q & -\lambda_1 pq^{-1} & -\lambda_2 pq^{-1} & 0 \\ 0 & -(q - \lambda_1 pq^{-1}) & \alpha & 0 \\ 0 & 0 & -(\alpha + q - \lambda_2 pq^{-1}) & 0 \\ 0 & 0 & 0 & -q \end{bmatrix} \tag{18}$$

of the system (16) satisfies the conditions (13).

Example 1. *Put*

$$\alpha = 0.4, \quad \lambda_1 = 0.5, \quad \lambda_2 = 0.7, \quad \delta_1 = \delta_2 = 0.2, \quad p = 0.8, \quad q = 0.7,$$
$$\sigma_0 = 1.18, \quad \sigma_1 = 0.50, \quad \sigma_2 = 0.77, \quad \sigma_3 = 1.18. \quad (19)$$

By these values of the parameters the conditions (14) and (15) hold:

$$\frac{1}{\lambda_1} = 2 > \frac{p}{q^2} = 1.63, \quad \frac{1}{\lambda_2}\left(1+\frac{\alpha}{q}\right) = 2.24 > \frac{p}{q^2} = 1.63,$$
$$\sigma_0^2 = 1.3924 < 2q = 1.4, \quad \sigma_1^2 = 0.25 < 2\left(q - \lambda_1\frac{p}{q}\right) = 0.257,$$
$$\sigma_2^2 = 0.5929 < 2\left(\alpha + q - \lambda_2\frac{p}{q}\right) = 0.6, \quad \sigma_3^2 = 1.3924 < 2q = 1.4.$$

Using MATLAB it was shown that by the values of the parameters (19) the matrix (18) satisfies the LMI (12). The conditions (13) with

$$T_1 = -1.8286 < 0, \quad T_2 = 1.1286 > 0, \quad T_3 = -0.2640 < 0, \quad T_4 = 0.0189 > 0,$$
$$T_3 - T_1 T_2 = 1.7997 > 0, \quad (T_1 T_2 - T_3)T_3 - T_1^2 T_4 = 0.4119 > 0,$$

hold too. Therefore, the equilibrium E_0 is stable in probability.

In Figure 1 one can see 30 trajectories of the system (6) solution for the equilibrium E_0 with the initial condition $I(0) = 1.7$, $S_1(0) = 0.9$, $S_2(0) = 0.7$, $R(0) = 0.5$: all trajectories $I(t)$ (yellow), $S_1(t)$ (green), $S_2(t)$ (blue), $R(t)$ (red) converge to the equilibrium $E_0 = (I^*, S_1^*, S_2^*, R^*) = (1.1429, 0, 0, 0)$.

Figure 1. 30 trajectories of the system (6) solution with the initial condition $I(0) = 1.7$, $S_1(0) = 0.9$, $S_2(0) = 0.7$, $R(0) = 0.5$: all trajectories $I(t)$ (yellow), $S_1(t)$ (green), $S_2(t)$ (blue), $R(t)$ (red) converge to the equilibrium $E_0 = (I^*, S_1^*, S_2^*, R^*) = (1.1429, 0, 0, 0)$.

4.2. Stability of the Equilibrium $E_1 = \left(\frac{q}{\lambda_1}, \frac{p}{q} - \frac{q}{\lambda_1}, 0, 0\right)$

Theorem 3. *If*

$$\frac{1}{\lambda_1} + \frac{1}{\delta_1} > \frac{p}{q^2} > \frac{1}{\lambda_1}, \quad \frac{1}{\lambda_2}\left(1+\frac{\alpha}{q}\right) > \frac{1}{\lambda_1}, \quad (20)$$

and

$$\sigma_0^2 < \frac{2p\lambda_1}{q}, \quad \sigma_1^2 < \frac{2p\lambda_1}{q + \left[\frac{q}{p\lambda_1}\left(1 - \frac{q^2}{p\lambda_1}\right)\right]^{-1}}, \quad \sigma_2^2 < 2\left(\alpha + q - q\frac{\lambda_2}{\lambda_1}\right), \quad \sigma_3^2 < 2\left(q - \delta_1\left(\frac{p}{q} - \frac{q}{\lambda_1}\right)\right), \quad (21)$$

then the equilibrium E_1 is stable in probability.

Proof. For the equilibrium $E_1 = \left(\frac{q}{\lambda_1}, \frac{p}{q} - \frac{q}{\lambda_1}, 0, 0\right)$ the system (8) takes the form

$$\begin{aligned}
\dot{z}_0(t) &= -pq^{-1}\lambda_1 z_0(t) - q z_1(t) - q\lambda_2\lambda_1^{-1} z_2(t) + \sigma_0 z_0(t)\dot{w}_0(t), \\
\dot{z}_1(t) &= \lambda_1\left(\frac{p}{q} - \frac{q}{\lambda_1}\right) z_0(t) + \alpha z_2(t) - \delta_1\left(\frac{p}{q} - \frac{q}{\lambda_1}\right) z_3(t) + \sigma_1 z_1(t)\dot{w}_1(t), \\
\dot{z}_2(t) &= -(\alpha + q - q\lambda_2\lambda_1^{-1}) z_2(t) + \sigma_2 z_2(t)\dot{w}_2(t), \\
\dot{z}_3(t) &= -\left(q - \delta_1\left(\frac{p}{q} - \frac{q}{\lambda_1}\right)\right) z_3(t) + \sigma_3 z_3(t)\dot{w}_3(t).
\end{aligned} \qquad (22)$$

The conditions (20) provide positivity of the nonzero component of the equilibrium E_1 and negativity of the coefficients before $z_2(t)$ and $z_3(t)$ in the last two equations (22). The last two inequalities (21) are the necessary and sufficient conditions for asymptotic mean square stability of the zero solutions of last two equations in (22) [30] which do not depend on $z_0(t)$ and $z_1(t)$ and can be considered separately. Since $\lim_{t\to\infty} Ez_2^2(t) = 0$ and $\lim_{t\to\infty} Ez_3^2(t) = 0$ then the system of first two Equation (22) for $z_0(t)$ and $z_1(t)$ can be considered without the processes $z_2(t), z_3(t)$, i.e.,

$$\begin{aligned}
\dot{z}_0(t) &= -pq^{-1}\lambda_1 z_0(t) - q z_1(t) + \sigma_0 z_0(t)\dot{w}_0(t), \\
\dot{z}_1(t) &= \lambda_1\left(\frac{p}{q} - \frac{q}{\lambda_1}\right) z_0(t) + \sigma_1 z_1(t)\dot{w}_1(t).
\end{aligned} \qquad (23)$$

Via Remark A2 (see Appendix A.2) first two inequalities (21) are sufficient for asymptotic mean square stability of the zero solution of the system (23). Therefore, the conditions (20) and (21) provide asymptotic mean square stability of the zero solution of the system (22) and via Remark 1 stability in probability of the equilibrium E_1 of the system (6). The proof is completed. □

Remark 3. One can check that by the conditions (20) and (21) the matrix

$$A = \begin{bmatrix} -pq^{-1}\lambda_1 & -q & -q\lambda_2\lambda_1^{-1} & 0 \\ \lambda_1\left(\frac{p}{q} - \frac{q}{\lambda_1}\right) & 0 & \alpha & -\delta_1\left(\frac{p}{q} - \frac{q}{\lambda_1}\right) \\ 0 & 0 & -(\alpha + q - q\lambda_2\lambda_1^{-1}) & 0 \\ 0 & 0 & 0 & -\left(q - \delta_1\left(\frac{p}{q} - \frac{q}{\lambda_1}\right)\right) \end{bmatrix} \qquad (24)$$

of the system (22) satisfies the conditions (13).

Example 2. *Put*

$$\begin{aligned}
&\alpha = 0.4, \quad \lambda_1 = 0.65, \quad \lambda_2 = 0.75, \quad \delta_1 = \delta_2 = 0.2, \quad p = 0.9, \quad q = 0.7, \\
&\sigma_0 = 1.01, \quad \sigma_1 = 0.41, \quad \sigma_2 = 0.76, \quad \sigma_3 = 1.14.
\end{aligned} \qquad (25)$$

By these values of the parameters the conditions (20) and (21) hold:

$$\frac{1}{\lambda_1} + \frac{1}{\delta_1} = 6.538 > \frac{p}{q^2} = 1.837 > \frac{1}{\lambda_1} = 1.538, \quad \frac{1}{\lambda_2}\left(1 + \frac{\alpha}{q}\right) = 2.095 > \frac{1}{\lambda_1} = 1.538,$$

$$\sigma_0^2 = 1.0201 < \frac{2p\lambda_1}{q} < 1.671, \quad \sigma_1^2 = 0.1681 < \frac{2p\lambda_1}{q + \left[\frac{q}{p\lambda_1}\left(1 - \frac{q^2}{p\lambda_1}\right)\right]^{-1}} = 0.2001,$$

$$\sigma_2^2 = 0.5776 < 2\left(\alpha + q - q\frac{\lambda_2}{\lambda_1}\right) = 0.5846, \quad \sigma_3^2 = 1.2996 < 2\left(q - \delta_1\left(\frac{p}{q} - \frac{q}{\lambda_1}\right)\right) = 1.3165.$$

Using MATLAB it was shown that by the values of the parameters (25) the matrix (24) satisfies the LMI (12), the conditions (13) with

$$T_1 = -1.7863 < 0, \quad T_2 = 1.0818 > 0, \quad T_3 = -0.2511 < 0, \quad T_4 = 0.0183 > 0,$$
$$T_3 - T_1 T_2 = 1.6813 > 0, \quad (T_1 T_2 - T_3) T_3 - T_1^2 T_4 = 0.3638 > 0,$$

hold too. Therefore, the equilibrium E_1 is stable in probability.

In Figure 2 one can see 30 trajectories of the system (6) solution for the equilibrium E_1 with the initial condition $I(0) = 1.7$, $S_1(0) = 0.9$, $S_2(0) = 0.7$, $R(0) = 0.5$: all trajectories $I(t)$ (yellow), $S_1(t)$ (green), $S_2(t)$ (blue), $R(t)$ (red) converge to the equilibrium $E_1 = (I^*, S_1^*, S_2^*, R^*) = (1.0769, 0.2088, 0, 0)$.

Figure 2. 30 trajectories of the system (6) solution with the initial condition $I(0) = 1.7$, $S_1(0) = 0.9$, $S_2(0) = 0.7$, $R(0) = 0.5$: all trajectories $I(t)$ (yellow), $S_1(t)$ (green), $S_2(t)$ (blue), $R(t)$ (red) converge to the equilibrium $E_1 = (I^*, S_1^*, S_2^*, R^*) = (1.0769, 0.2088, 0, 0)$.

4.3. Stability of the Equilibrium $E_2 = (I_2^*, S_{12}^*, 0, R_2^*)$

For the equilibrium E_2 the system (8) takes the form

$$\begin{aligned}
\dot{z}_0(t) &= -q(1 + \lambda_1 \delta_1^{-1}) z_0(t) - \lambda_1 I_2^* z_1(t) - \lambda_2 I_2^* z_2(t) + \sigma_0 z_0(t) \dot{w}_0(t), \\
\dot{z}_1(t) &= q\lambda_1 \delta_1^{-1} z_0(t) - (q + \delta_1 R_2^* - \lambda_1 I_2^*) z_1(t) + \alpha z_2(t) - q z_3(t) + \sigma_1 z_1(t) \dot{w}_1(t), \\
\dot{z}_2(t) &= -(\alpha + q - \lambda_2 I_2^* + \delta_2 R_2^*) z_2(t) + \sigma_2 z_2(t) \dot{w}_2(t), \\
\dot{z}_3(t) &= \delta_1 R_2^* z_1(t) + \delta_2 R_2^* z_2(t) + \sigma_3 z_3(t) \dot{w}_3(t),
\end{aligned} \quad (26)$$

where I_2^* and R_2^* are defined in (3).

Lemma 3. *If*

$$\frac{p}{q^2} > \frac{1}{\lambda_1} + \frac{1}{\delta_1}, \quad 1 + \frac{\alpha}{q} > \frac{p(\lambda_2 \delta_1 - \lambda_1 \delta_2)}{q^2(\delta_1 + \lambda_1)} + \frac{\delta_2}{\delta_1}, \qquad (27)$$

then the matrix

$$A = \begin{bmatrix} -q(1 + \lambda_1 \delta_1^{-1}) & -\lambda_1 I_2^* & -\lambda_2 I_2^* & 0 \\ q\lambda_1 \delta_1^{-1} & 0 & \alpha & -q \\ 0 & 0 & -(\alpha + q - \lambda_2 I_2^* + \delta_2 R_2^*) & 0 \\ 0 & \delta_1 R_2^* & \delta_2 R_2^* & 0 \end{bmatrix} \qquad (28)$$

of the system (26) *is the Hurwitz matrix.*

Proof. The first and the second conditions (27) provide respectively a positivity of R_2^* and a negativity of the coefficient before $z_2(t)$ in the third equation of the system (26). Please note that the inequality

$$\sigma_2^2 < 2q\left(1 + \frac{\alpha}{q} - \frac{p(\lambda_2 \delta_1 - \lambda_1 \delta_2)}{q^2(\delta_1 + \lambda_1)} - \frac{\delta_2}{\delta_1}\right) \qquad (29)$$

is the necessary and sufficient condition for asymptotic mean square stability of the zero solution of the equation for $z_2(t)$ of the system (26). Therefore, by this condition $\lim_{t \to \infty} \mathbf{E}z_2^2(t) = 0$ it is enough to show that the matrix

$$A = \begin{bmatrix} -q(1 + \lambda_1 \delta_1^{-1}) & -\lambda_1 I_2^* & 0 \\ q\lambda_1 \delta_1^{-1} & 0 & -q \\ 0 & \delta_1 R_2^* & 0 \end{bmatrix} \qquad (30)$$

is the Hurwitz matrix. Really, for the matrix (30) we have

$$T_1 = -q(1 + \lambda_1 \delta_1^{-1}) < 0, \quad T_2 = q\lambda_1^2 \delta_1^{-1} I_2^* + q\delta_1 R_2^* > 0, \quad T_3 = -q^2 \delta_1 R_2^* (1 + \lambda_1 \delta_1^{-1}) < 0,$$

and

$$T_1 T_2 = (-q(1 + \lambda_1 \delta_1^{-1}))(q\lambda_1^2 \delta_1^{-1} I_2^* + q\delta_1 R_2^*)$$
$$< -q^2 \delta_1 R_2^* (1 + \lambda_1 \delta_1^{-1}) = T_3.$$

Therefore, the matrix (30) is the Hurwitz matrix. Therefore the matrix (28) is the Hurwitz matrix too. The proof is completed. □

Corollary 1. *If the conditions* (27) *and* (29) *hold then for small enough* $\sigma_0^2, \sigma_1^2, \sigma_3^2$ *the LMI* (12) *holds. It means that the zero solution of the linear system* (26) *is asymptotically mean square stable and therefore (Remark* 1) *the equilibrium* E_2 *is stable in probability.*

Example 3. *Put*

$$\alpha = 0.4, \quad \lambda_1 = 1, \quad \lambda_2 = 1.3, \quad \delta_1 = 0.5, \quad \delta_2 = 0.7, \quad p = 0.9, \quad q = 0.5,$$
$$\sigma_0 = 0.55, \quad \sigma_1 = 0.30, \quad \sigma_2 = 0.72, \quad \sigma_3 = 0.44. \qquad (31)$$

By these values of the parameters the conditions (27) *and* (29) *hold:*

$$\frac{p}{q^2} = 3.6 > \frac{1}{\lambda_1} + \frac{1}{\delta_1} = 3, \quad 1 + \frac{\alpha}{q} = 1.8 > \frac{p(\lambda_2 \delta_1 - \lambda_1 \delta_2)}{q^2(\delta_1 + \lambda_1)} + \frac{\delta_2}{\delta_1} = 1.28,$$
$$\sigma_2^2 = 0.5184 < 2q\left(1 + \frac{\alpha}{q} - \frac{p(\lambda_2 \delta_1 - \lambda_1 \delta_2)}{q^2(\delta_1 + \lambda_1)} - \frac{\delta_2}{\delta_1}\right) = 0.52.$$

Using MATLAB it was shown that by the values of the parameters (31) *the matrix* (28) *satisfies the LMI* (12), *via Lemma* 3 *the conditions* (13) *hold too. Therefore, the equilibrium* E_2 *is stable in probability.*

In Figure 3 one can see 30 trajectories of the system (6) solution for the equilibrium E_2 with the initial condition $I(0) = 1.7$, $S_1(0) = 0.9$, $S_2(0) = 0.7$, $R(0) = 0.5$: all trajectories $I(t)$ (yellow), $S_1(t)$ (green), $S_2(t)$ (blue), $R(t)$ (red) converge to the equilibrium $E_2 = (I^*, S_1^*, S_2^*, R^*) = (0.6, 1, 0, 0.2)$. In accordance with (5) $I^* + S_1^* + S_2^* + R^* = pq^{-1} = 1.8$.

Figure 3. 30 trajectories of the system (6) solution with the initial condition $I(0) = 1.7$, $S_1(0) = 0.9$, $S_2(0) = 0.7$, $R(0) = 0.5$: all trajectories $I(t)$ (yellow), $S_1(t)$ (green), $S_2(t)$ (blue), $R(t)$ (red) converge to the equilibrium $E_2 = (I^*, S_1^*, S_2^*, R^*) = (0.6, 1, 0, 0.2)$.

4.4. Stability of the Equilibrium $E_3 = (I_3^*, S_{13}^*, S_{23}^*, 0)$

For the equilibrium E_3 the system (8) takes the form

$$\begin{aligned}
\dot{z}_0(t) &= -p\lambda_2(\alpha+q)^{-1}z_0(t) - \lambda_1\lambda_2^{-1}(\alpha+q)z_1(t) - (\alpha+q)z_2(t) + \sigma_0 z_0(t)\dot{w}_0(t), \\
\dot{z}_1(t) &= \lambda_1 S_{13}^* z_0(t) - (q - \lambda_1\lambda_2^{-1}(\alpha+q))z_1(t) + \alpha z_2(t) - \delta_1 S_{13}^* z_3(t) + \sigma_1 z_1(t)\dot{w}_1(t), \\
\dot{z}_2(t) &= \lambda_2 S_{23}^* z_0(t) - \delta_2 S_{23}^* z_3(t) + \sigma_2 z_2(t)\dot{w}_2(t), \\
\dot{z}_3(t) &= -(q - \delta_1 S_{13}^* - \delta_2 S_{23}^*)z_3(t) + \sigma_3 z_3(t)\dot{w}_3(t),
\end{aligned} \qquad (32)$$

where S_{13}^*, S_{23}^* are defined in (3).

Lemma 4. *If*

$$\frac{p}{q^2} > \frac{1}{\lambda_2}\left(1 + \frac{\alpha}{q}\right), \qquad \frac{1}{\lambda_1} > \frac{1}{\lambda_2}\left(1 + \frac{\alpha}{q}\right), \qquad q > \delta_1 S_{13}^* + \delta_2 S_{23}^*, \qquad (33)$$

then the matrix

$$A = \begin{bmatrix} -p\lambda_2(\alpha+q)^{-1} & -\lambda_1\lambda_2^{-1}(\alpha+q) & -(\alpha+q) & 0 \\ \lambda_1 S_{13}^* & -(q - \lambda_1\lambda_2^{-1}(\alpha+q)) & \alpha & -\delta_1 S_{13}^* \\ \lambda_2 S_{23}^* & 0 & 0 & -\delta_2 S_{23}^* \\ 0 & 0 & 0 & -(q - \delta_1 S_{13}^* - \delta_2 S_{23}^*) \end{bmatrix} \qquad (34)$$

of the system (26) *is the Hurwitz matrix.*

Proof. The conditions (33) provide a positivity of S_{13}^* and S_{23}^* and a negativity of the diagonal elements of the matrix (34). Please note that the inequality

$$\sigma_3^2 < 2(q - \delta_1 S_{13}^* - \delta_2 S_{23}^*) \tag{35}$$

is the necessary and sufficient condition for asymptotic mean square stability of the zero solution of the equation for $z_3(t)$ of the system (32). Therefore, by this condition $\lim_{t\to\infty} \mathbf{E} z_3^2(t) = 0$ and it is enough to show that the matrix

$$A = \begin{bmatrix} -p\lambda_2(\alpha+q)^{-1} & -\lambda_1\lambda_2^{-1}(\alpha+q) & -(\alpha+q) \\ \lambda_1 S_{13}^* & -(q-\lambda_1\lambda_2^{-1}(\alpha+q)) & \alpha \\ \lambda_2 S_{23}^* & 0 & 0 \end{bmatrix} \tag{36}$$

with

$$T_1 = -p\lambda_2(\alpha+q)^{-1} - (q - \lambda_1\lambda_2^{-1}(\alpha+q)) < 0,$$
$$T_2 = p(\alpha+q)^{-1}(\lambda_2 q - \lambda_1(\alpha+q)) + (\alpha+q)[\lambda_1^2 \lambda_2^{-1} S_{13}^* + \lambda_2 S_{23}^*] > 0,$$
$$T_3 = -q(\alpha+q)(\lambda_2 - \lambda_1) S_{23}^* < 0,$$

is the Hurwitz matrix, i.e., $T_1 T_2 < T_3$. The proof is completed. □

Corollary 2. *If the conditions (33) and (35) hold then for small enough $\sigma_0^2, \sigma_1^2, \sigma_2^2$ the LMI (12) holds. It means that the zero solution of the linear system (32) is asymptotically mean square stable and therefore (Remark 1) the equilibrium E_3 is stable in probability.*

Example 4. *Put*

$$\alpha = 0.8, \quad \lambda_1 = 0.3, \quad \lambda_2 = 0.9, \quad \delta_1 = 0.8, \quad \delta_2 = 0.7, \quad p = 1.2, \quad q = 0.6,$$
$$\sigma_0 = 0.91, \quad \sigma_1 = 0.50, \quad \sigma_2 = 0.40, \quad \sigma_3 = 0.70. \tag{37}$$

By these values of the parameters the conditions (33) and (35) hold:

$$\frac{p}{q^2} = 3.33 > \frac{1}{\lambda_2}\left(1 + \frac{\alpha}{q}\right) = 2.59, \quad \frac{1}{\lambda_1} = 3.33 > \frac{1}{\lambda_2}\left(1 + \frac{\alpha}{q}\right) = 2.59,$$
$$q = 0.6 > \delta_1 S_{13}^* + \delta_2 S_{23}^* = 0.1715, \quad \sigma_3^2 = 0.49 < 2(q - \delta_1 S_{13}^* - \delta_2 S_{23}^*) = 0.5015.$$

Using MATLAB it was shown that by the values of the parameters (37) the matrix (34) satisfies the LMI (12), for the matrix (36) $T_1 = -0.9048 < 0$, $T_2 = 0.2362 > 0$, $T_3 = -0.0320 < 0$, $T_3 - T_1 T_2 = 0.1817 > 0$, the conditions (13) hold too. Therefore, the equilibrium E_3 is stable in probability.

In Figure 4 one can see 30 trajectories of the system (6) solution for the equilibrium E_3 with the initial condition $I(0) = 1.9$, $S_1(0) = 0.8$, $S_2(0) = 0.4$, $R(0) = 0.4$: all trajectories $I(t)$ (yellow), $S_1(t)$ (green), $S_2(t)$ (blue), $R(t)$ (red) converge to the equilibrium $E_3 = (I^*, S_1^*, S_2^*, R^*) = (1.5556, 0.3810, 0.0634, 0)$. In accordance with (5) $I^* + S_1^* + S_2^* + R^* = pq^{-1} = 2$.

Figure 4. 30 trajectories of the system (6) solution with the initial condition $I(0) = 1.9$, $S_1(0) = 0.8$, $S_2(0) = 0.4$, $R(0) = 0.4$: all trajectories $I(t)$ (yellow), $S_1(t)$ (green), $S_2(t)$ (blue), $R(t)$ (red) converge to the equilibrium $E_3 = (I^*, S_1^*, S_2^*, R^*) = (1.5556, 0.3810, 0.0634, 0)$.

4.5. Stability of the Equilibrium $E_4 = (I_4^*, S_{14}^*, S_{24}^*, R_4^*)$

For the equilibrium E_4 the system (8) by virtue of (2) takes the form

$$\begin{aligned}
\dot{z}_0(t) &= -p(I_4^*)^{-1}z_0(t) - \lambda_1 I_4^* z_1(t) - \lambda_2 I_4^* z_2(t) + \sigma_0 z_0(t)\dot{w}_0(t), \\
\dot{z}_1(t) &= \lambda_1 S_{14}^* z_0(t) - \alpha S_{24}^*(S_{14}^*)^{-1} z_1(t) + \alpha z_2(t) - \delta_1 S_{14}^* z_3(t) + \sigma_1 z_1(t)\dot{w}_1(t), \\
\dot{z}_2(t) &= \lambda_2 S_{24}^* z_0(t) - \delta_2 S_{24}^* z_3(t) + \sigma_2 z_2(t)\dot{w}_2(t), \\
\dot{z}_3(t) &= \delta_1 R_4^* z_1(t) + \delta_2 R_4^* z_2(t) + \sigma_3 z_3(t)\dot{w}_3(t),
\end{aligned} \tag{38}$$

where $I_4^*, S_{14}^*, S_{24}^*, R_4^*$ are defined in (3).

Let us show that the matrix

$$A = \begin{bmatrix} -p(I_4^*)^{-1} & -\lambda_1 I_4^* & -\lambda_2 I_4^* & 0 \\ \lambda_1 S_{14}^* & -\alpha S_{24}^*(S_{14}^*)^{-1} & \alpha & -\delta_1 S_{14}^* \\ \lambda_2 S_{24}^* & 0 & 0 & -\delta_2 S_{24}^* \\ 0 & \delta_1 R_4^* & \delta_2 R_4^* & 0 \end{bmatrix} \tag{39}$$

of the system (38) is the Hurwitz matrix. Really, the conditions (13) for the matrix (39) hold with

$$\begin{aligned}
T_1 &= -p(I_4^*)^{-1} - \alpha S_{24}^*(S_{14}^*)^{-1} < 0, \\
T_2 &= \alpha p S_{24}^*(I_4^* S_{14}^*)^{-1} + I_4^*(\lambda_1^2 S_{14}^* + \lambda_2^2 S_{24}^*) + R_4^*(\delta_1^2 S_{14}^* + \delta_2^2 S_{24}^*) > 0, \\
T_3 &= -\alpha S_{24}^*(\lambda_1 \lambda_2 I_4^* + \delta_1 \delta_2 R_4^*) - p(I_4^*)^{-1} R_4^*(\delta_1^2 S_{14}^* + \delta_2^2 S_{24}^*) - \alpha(S_{24}^*)^2(S_{14}^*)^{-1}(\lambda_2^2 I_4^* + \delta_2^2 R_4^*) < 0, \\
T_4 &= \alpha p R_4^*(I_4^* S_{14}^*)^{-1}(\delta_1^2(S_{14}^*)^2 + \delta_2^2(S_{24}^*)^2) + (\lambda_2 \delta_1 - \lambda_1 \delta_2)^2 I_4^* S_{14}^* S_{24}^* R_4^* > 0.
\end{aligned}$$

Example 5. *Put*

$$\begin{aligned} \alpha &= 0.4, \quad \lambda_1 = 0.3, \quad \lambda_2 = 0.9, \quad \delta_1 = \delta_2 = 0.8, \quad p = 1.2, \quad q = 0.6, \\ \sigma_0 &= 0.51, \quad \sigma_1 = 0.51, \quad \sigma_2 = 0.55, \quad \sigma_3 = 0.34. \end{aligned} \tag{40}$$

Using MATLAB it was shown that by the values of the parameters (40) *the matrix* (39) *satisfies the LMI* (12), *the conditions* (13) *hold too:* $T_1 = -1.3259 < 0$, $T_2 = 0.7020 > 0$, $T_3 = -0.1828 < 0$, $T_4 = 0.0187 > 0$,

$T_3 - T_1T_2 = 0.7479 > 0$, $(T_1T_2 - T_3)T_3 - T_1^2T_4 = 0.1039 > 0$. Therefore, the equilibrium E_4 is stable in probability.

In Figure 5 one can see 30 trajectories of the system (6) solution for the equilibrium E_4 with the initial condition $I(0) = 1.9$, $S_1(0) = 0.8$, $S_2(0) = 0.7$, $R(0) = 0.4$: all trajectories $I(t)$ (yellow), $S_1(t)$ (green), $S_2(t)$ (blue), $R(t)$ (red) converge to the equilibrium $E_4 = (I^*, S_1^*, S_2^*, R^*) = (1.1765, 0.4250, 0.3250, 0.0735)$. In accordance with (5) $I^* + S_1^* + S_2^* + R^* = pq^{-1} = 2$.

Please note that decreasing δ_2 from $\delta_2 = 0.8$ to $\delta_2 = 0.7$, we obtain that S_1^* unlike from the previous case is calculated via quadratic equation (see (3)). By that with the same values of all other parameters the equilibrium E_4 a bit changed $E_4 = (I^*, S_1^*, S_2^*, R^*) = (1.1534, 0.4543, 0.3379, 0.0544)$, $I^* + S_1^* + S_2^* + R^* = pq^{-1} = 2$, but remains stable in probability and the conditions (13) hold with $T_1 = -1.3379 < 0$, $T_2 = 0.6971 > 0$, $T_3 = -0.1686 < 0$, $T_4 = 0.0143 > 0$, $T_3 - T_1T_2 = 0.7641 > 0$, $(T_1T_2 - T_3)T_3 - T_1^2T_4 = 0.1033 > 0$.

Figure 5. 30 trajectories of the system (6) solution with the initial condition $I(0) = 1.9$, $S_1(0) = 0.8$, $S_2(0) = 0.7$, $R(0) = 0.4$: all trajectories $I(t)$ (yellow), $S_1(t)$ (green), $S_2(t)$ (blue), $R(t)$ (red) converge to the equilibrium $E_4 = (I^*, S_1^*, S_2^*, R^*) = (1.1765, 0.4250, 0.3250, 0.0735)$.

5. Conclusions

In this paper, it is shown how the dynamics of the very popular I2SR rumor spreading model can be investigated under stochastic perturbations. It is shown that for some equilibria of the considered model it is possible to get conditions for stability in probability in an analytical form, for other equilibria stability condition can be obtained numerically by an appropriate linear matrix inequality via MATLAB.

The proposed way of research can be used for more detail investigation both the considered I2SR rumor spreading model and also all other known type of rumor spreading models [6–29].

Besides, this research method can be used for a detailed investigation of many other nonlinear mathematical models (with the order of nonlinearity higher than one) in different other applications. In particular, the proposed method can be used for systems with exponential nonlinearity [34,35], together with stochastic perturbations of the type of white noise other types of perturbations can be used, for instance, perturbations of the type of Poisson's jumps [35], the method does not depend on the dimension of the considered system and can be used for systems of more than four equations.

Funding: This research received no external funding.

Conflicts of Interest: The author declares no conflict of interest.

Appendix A

Appendix A.1. Equilibria of the System (1)

The equilibria $E_0, \ldots, E_3$ of the system (1) are obtained from the system (2) quite simply (see (3)). To get the equilibrium E_4 note that from the second and the third equations of the system (2) we obtain

$$R = \frac{1}{\delta_1}(\alpha S_2 S_1^{-1} - q + \lambda_1 I) = \frac{1}{\delta_2}(\lambda_2 I - (\alpha + q)).$$

From this and the first equation of the system (2) we have

$$I = \frac{(\alpha + q)\delta_1 + (\alpha S_2 S_1^{-1} - q)\delta_2}{\lambda_2 \delta_1 - \lambda_1 \delta_2} = \frac{p}{\lambda_1 S_1 + \lambda_2 S_2 + q},$$

and therefore

$$((\alpha + q)\delta_1 + (\alpha S_2 S_1^{-1} - q)\delta_2)(\lambda_1 S_1 + \lambda_2 S_2 + q) = p(\lambda_2 \delta_1 - \lambda_1 \delta_2). \tag{A1}$$

From the last equation of the system (2) and $R^* \neq 0$ it follows that

$$S_2 S_1^{-1} = \frac{1}{\delta_2}\left(\frac{q}{S_1} - \delta_1\right). \tag{A2}$$

Substituting (A2) into (A1) we obtain the equation for S_1

$$\left((\alpha + q)\delta_1 + \alpha\left(\frac{q}{S_1} - \delta_1\right) - q\delta_2\right)\left(\lambda_1 S_1 + \frac{\lambda_2}{\delta_2}(q - \delta_1 S_1) + q\right) = p(\lambda_2 \delta_1 - \lambda_1 \delta_2),$$

$$q(\alpha - (\delta_2 - \delta_1)S_1)\left(q(\lambda_2 + \delta_2) - (\lambda_2 \delta_1 - \lambda_1 \delta_2)S_1\right) = p\delta_2(\lambda_2 \delta_1 - \lambda_1 \delta_2)S_1,$$

$$q(\delta_2 - \delta_1)(\lambda_2 \delta_1 - \lambda_1 \delta_2)S_1^2 - [(q\alpha + p\delta_2)(\lambda_2 \delta_1 - \lambda_1 \delta_2) + q^2(\delta_2 - \delta_1)(\lambda_2 + \delta_2)]S_1 + \alpha q^2(\lambda_2 + \delta_2) = 0.$$

If $(\delta_2 - \delta_1)(\lambda_2 \delta_1 - \lambda_1 \delta_2) \neq 0$ then S_1 is a positive root of the quadratic equation $S_1^2 - \nu_1 S_1 + \nu_2 = 0$, where

$$\nu_1 = \frac{q\alpha + p\delta_2}{q(\delta_2 - \delta_1)} + \frac{q(\lambda_2 + \delta_2)}{\lambda_2 \delta_1 - \lambda_1 \delta_2}, \quad \nu_2 = \frac{\alpha q(\lambda_2 + \delta_2)}{(\delta_2 - \delta_1)(\lambda_2 \delta_1 - \lambda_1 \delta_2)}.$$

Remark A1. *Please note that a positive root of the quadratic equation $S_1^2 - \nu_1 S_1 + \nu_2 = 0$ may not exist, for instance, if $\nu_1 < 0$ and $\nu_2 > 0$. In this case a positive equilibrium E_4 does not exist too. On the other hand for some values of the parameters the quadratic equation $S_1^2 - \nu_1 S_1 + \nu_2 = 0$ may have two positive roots, for instance, if $\nu_1 > 0$ and $0 < 4\nu_2 < \nu_1^2$: $S_1^* = \frac{1}{2}(\nu_1 \pm \sqrt{\nu_1^2 - 4\nu_2})$. In this case there are two equilibria of the type of E_4.*

If $\delta_1 = \delta_2 = \delta, \lambda_2 > \lambda_1$ then $S_1^* = \frac{\alpha q^2(\delta + \lambda_2)}{\delta(\lambda_2 - \lambda_1)(q\alpha + p\delta)}$. If $\lambda_2 \delta_1 = \lambda_1 \delta_2, \delta_2 > \delta_1$ then $S_1^* = \frac{\alpha}{\delta_2 - \delta_1}$.
If S_1^* is defined then via (2)

$$S_2^* = \frac{1}{\delta_2}(q - \delta_1 S_1^*), \quad I^* = \frac{p}{\lambda_1 S_1^* + \lambda_2 S_2^* + q}, \quad R^* = \frac{\lambda_2 I^* - \alpha - q}{\delta_2}.$$

For positivity of the equilibrium E_4 must be $S_1^* < \frac{q}{\delta_1}$ and $I^* > \frac{\alpha + q}{\lambda_2}$.

Appendix A.2. Stability of the System of Two Stochastic Differential Equations

Consider the system of two stochastic differential equations

$$\dot{x}_1(t) = a_{11}x_1(t) + a_{12}x_2(t) + \sigma_1 x_1(t)\dot{w}_1(t),$$
$$\dot{x}_2(t) = a_{21}x_1(t) + a_{22}x_2(t) + \sigma_1 x_2(t)\dot{w}_2(t),$$
(A3)

where a_{ij}, σ_i, $i,j = 1,2$, are constants, $w_1(t)$ and $w_2(t)$ are mutually independent standard Wiener processes [30,33].

Lemma A1. *[30] Put $A = \|a_{ij}\|$, $i,j = 1,2$, $A_i = \det(A) + a_{ii}^2$, $\mu_i = \frac{1}{2}\sigma_i^2$, $i = 1,2$, and suppose that the following conditions hold*

$$Tr(A) = a_{11} + a_{22} < 0, \quad \det(A) = a_{11}a_{22} - a_{12}a_{21} > 0,$$
$$\mu_1 < \frac{|Tr(A)|\det(A)}{A_2}, \quad \mu_2 < \frac{|Tr(A)|\det(A) - A_2\mu_1}{A_1 - |Tr(A)|\mu_1}.$$
(A4)

Then the zero solution of the system (A3) is asymptotically mean square stable.

Remark A2. *Please note that if $a_{12}a_{21} = 0$ then the last two conditions in (A4) take the form $\mu_1 < -a_{11}$, $\mu_2 < -a_{22}$.*

References

1. Beretta, E.; Kolmanovskii, V.; Shaikhet, L. Stability of epidemic model with time delays influenced by stochastic perturbations. *Math. Comput. Simul.* **1998**, *45*, 269–277. [CrossRef]
2. Magal, P.; Webb, G. The parameter identification problem for SIR epidemic models: Identifying unreported cases. *J. Math. Biol.* **2018**, *77*, 1629–1648. [CrossRef] [PubMed]
3. Kiouach, D.; Sabbar, Y. Stability and Threshold of a Stochastic SIRS Epidemic Model with Vertical Transmission and Transfer from Infectious to Susceptible Individuals. *Discret. Dyn. Nat. Soc.* **2018**, *2018*, 7570296. [CrossRef]
4. Santonja, F.-J.; Shaikhet, L. Analysing social epidemics by delayed stochastic models. *Discret. Dyn. Nat. Soc.* **2012**, *2012*, 530472. [CrossRef]
5. Santonja, F.-J.; Shaikhet, L. Probabilistic stability analysis of social obesity epidemic by a delayed stochastic model. *Nonlinear Anal. Real World Appl.* **2014**, *17*, 114–125. [CrossRef]
6. Dietz, K. Epidemics and Rumours: A Survey. *J. R. Stat. Soc. Ser. A Gener.* **1967**, *130*, 505–528. [CrossRef]
7. Zanette, D.H. Dynamics of rumor propagation on small-world networks. *Phys. Rev. E* **2002**, *65*, 041908. [CrossRef]
8. Galam, S. Modelling rumors: The no plane pentagon french hoax case. *Physica A* **2003**, *320*, 571–580. [CrossRef]
9. Moreno, Y.; Nekovee, M.; Pacheco, A.F. Dynamics of rumor spreading in complex networks. *Phys. Rev. E* **2004**, *69*, 066130. [CrossRef]
10. Nekovee, M.; Moreno, Y.; Bianconi, G.; Marsili, M. Theory of rumour spreading in complex social networks. *Physica A* **2007**, *374*, 457–470. [CrossRef]
11. Kawachi, K. Deterministic models for rumor transmission. *Nonlinear Anal. Real World Appl.* **2008**, *9*, 1989–2028. [CrossRef]
12. Zhang, Z.; Zhang, Z. An interplay model for rumour spreading and emergency development. *Physica A* **2009**, *388*, 4159–4166. [CrossRef]
13. Roshani, F.; Naimi, Y. Effects of degree-biased transmission rate and nonlinear infectivity on rumor spreading in complex social networks. *Phys. Rev. E* **2012**, *85*, 036109. [CrossRef]
14. Zhao, L.J.; Wang, J.J.; Chen, Y.C.; Wang, Q.; Cheng, J.J.; Cui, H.X. SIHR rumor spreading model in social networks. *Physica A* **2012**, *391*, 2444–2453. [CrossRef]
15. Wang, Y.; Yang, X.; Han, Y.; Wang, X. Rumor spreading model with trust mechanism in complex social networks. *Commun. Theor. Phys.* **2013**, *59*, 510–516. [CrossRef]

16. Zhao, L.; Qiu, X.; Wang, X.; Wang, J. Rumor spreading model considering forgetting and remembering mechanisms in homogeneous networks. *Physica A* **2013**, *392*, 987–994. [CrossRef]
17. Wang, J.; Zhao, L.; Huang, R. SIRaRu rumor spreading model in complex networks. *Physica A* **2014**, *398*, 43–55. [CrossRef]
18. Wang, J.; Zhao, L.; Huang, R. 2SI2R rumor spreading model in homogeneous networks. *Physica A* **2014**, *413*, 153–161. [CrossRef]
19. Zan, Y.; Wu, J.; Li, P.; Yu, Q. SICR rumor spreading model in complex networks: Counterattack and self-resistance. *Physica A* **2014**, *405*, 159–170. [CrossRef]
20. Ji, K.; Liu, J.; Xiang, G. Anti-rumor dynamics and emergence of the timing threshold on complex network. *Physica A* **2014**, *411*, 87–94. [CrossRef]
21. Afassinou, K. Analysis of the impact of education rate on the rumor spreading mechanism. *Physica A* **2014**, *414*, 43–52. [CrossRef]
22. Xia, L.L.; Jiang, G.P.; Song, B.; Song, Y.R. Rumor spreading model considering hesitating mechanism in complex social networks. *Physica A* **2015**, *437*, 295–303. [CrossRef]
23. Zhang, N.; Huang, H.; Duarte, M.; Zhang, J. Risk analysis for rumor propagation in metropolises based on improved 8-state ICSAR model and dynamic personal activity trajectories. *Physica A* **2016**, *451*, 403–419. [CrossRef]
24. Wan, C.; Li, T.; Wang, Y.; Liu, X. Rumor Spreading of a SICS Model on complex social networks with counter mechanism. *Open Access Libr. J.* **2016**, *3*, 1–11. [CrossRef]
25. Jie, R.; Qiao, J.; Xu, G.; Meng, Y. A study on the interaction between two rumors in homogeneous complex networks under symmetric conditions. *Physica A* **2016**, *454*, 129–142. [CrossRef]
26. Huo, L.; Wang, L.; Song, N.; Ma, C.; He, B. Rumor spreading model considering the activity of spreaders in the homogeneous network. *Physica A* **2017**, *468*, 855–865. [CrossRef]
27. Liu, Q.M.; Li, T.; Su, M.C. The analysis of an SEIR rumor propagation model on heterogeneous network. *Physica A* **2017**, *469*, 372–380. [CrossRef]
28. Zhu, L.; Wang, Y.G. Rumor spreading model with noise interference in complex social networks. *Physica A* **2017**, *469*, 750–760. [CrossRef]
29. Zhang, Y.; Su, Y.; Li, W.; Liu, H. Rumor and authoritative information propagation model considering super spreading in complex social networks. *Physica A* **2018**, *506*, 395–411. [CrossRef]
30. Shaikhet, L. *Lyapunov Functionals and Stability of Stochastic Functional Differential Equations*; Springer Science & Business Media: Berlin, Germany, 2013.
31. Shaikhet, L.; Bunimovich-Mendrazitsky, S. Stability Analysis of Delayed Immune Response BCG Infection in Bladder Cancer Treatment Model by Stochastic Perturbations. *Comput. Math. Methods Med.* **2018**, *2018*, 9653873. [CrossRef] [PubMed]
32. Fridman, E.; Shaikhet, L. Simple LMIs for stability of stochastic systems with delay term given by Stieltjes integral or with stabilizing delay. *Syst. Control Lett.* **2019**, *124*, 83–91. [CrossRef]
33. Gikhman, I.I.; Skorokhod, A.V. *Stochastic Differential Equations*; Springer: Berlin/Heidelberg, Germany, 1972.
34. Shaikhet, L. Stability of the zero and positive equilibria of two connected neoclassical growth models under stochastic perturbations. *Commun. Nonlinear Sci. Numer. Simul.* **2019**, *68*, 86–93. [CrossRef]
35. Shaikhet, L. Stability of the neoclassical growth model under perturbations of the type of Poisson's jumps: Analytical and numerical analysis. *Commun. Nonlinear Sci. Numer. Simul.* **2019**, *72*, 78–87. [CrossRef]

© 2020 by the authors. Licensee MDPI, Basel, Switzerland. This article is an open access article distributed under the terms and conditions of the Creative Commons Attribution (CC BY) license (http://creativecommons.org/licenses/by/4.0/).

Article

Dynamics of HIV-TB Co-Infection Model

Nita H Shah [1],*, Nisha Sheoran [1],* and Yash Shah [2]

[1] Department of Mathematics, Gujarat University, Ahmedabad 380009, Gujarat, India
[2] GCS Medical College, Ahmedabad 380054, Gujarat, India; yashshah95aries@gmail.com
* Correspondence: nitahshah@gmail.com (N.H.S.); sheorannisha@gmail.com (N.S.)

Received: 3 February 2020; Accepted: 5 March 2020; Published: 11 March 2020

Abstract: According to World Health Organization (WHO), the population suffering from human immunodeficiency virus (HIV) infection over a period of time may suffer from TB infection which increases the death rate. There is no cure for acquired immunodeficiency syndrome (AIDS) to date but antiretrovirals (ARVs) can slow down the progression of disease as well as prevent secondary infections or complications. This is considered as a medication in this paper. This scenario of HIV-TB co-infection is modeled using a system of non-linear differential equations. This model considers HIV-infected individual as the initial stage. Four equilibrium points are found. Reproduction number R_0 is calculated. If $R_0 > 1$ disease persists uniformly, with reference to the reproduction number, backward bifurcation is computed for pre-AIDS (latent) stage. Global stability is established for the equilibrium points where there is no Pre-AIDS TB class, point without co-infection and for the endemic point. Numerical simulation is carried out to validate the data. Sensitivity analysis is carried out to determine the importance of model parameters in the disease dynamics.

Keywords: Co-infection of HIV-TB; equilibrium point; reproduction number; stability analysis; backward bifurcation

1. Introduction

In the public health sector, human immunodeficiency virus (HIV) continues to be the major health threat globally, having claimed more than 32 million lives to date [1]. There were approximately 37.9 million people living with HIV at the end of 2018 [1]. The human immunodeficiency virus (HIV) is a virus that spread through certain body fluids, attacking the body's immune system, specifically the CD4 cells. The immune function is typically measured by CD4 cell count. Over time, HIV can destroy so many of these cells that the body can't fight against infections and diseases, which paves the way for many opportunistic diseases. One such disease is tuberculosis (TB). It is a contagious disease caused by bacteria called Mycobacterium tuberculosis. The bacteria mostly attack the lungs, but can also damage other parts of the body. The population living with HIV are 15–22 times more likely to develop TB [2]. It is the most commonly occurring illness among HIV-infected individuals, including among those taking antiretroviral treatment (ART). This interaction explains the fact that HIV and TB co-infection is a deadly human syndemic, where syndemic refers to the convergence of two or more diseases that exacerbate the burden of the disease [3]. For the treatment of HIV, HIV drugs called antiretrovirals (ARVs) are advised. ART reduces the risk of TB infection in people living with HIV by 65% [4]. It plays a significant role in preventing TB.

Mathematical modelling has enhanced understanding of disease dynamics. The first compartmental model was given by Kermack and McKendrick [5]. Some basic papers like [6,7] have constructed mathematical models by formulating non-linear differential equation for their respective models and have worked out the critical point/equilibrium points of the respective system and various related properties. In some the related research, many authors have worked out various types of HIV-TB co-infection model. Kirschner et al. [8] developed a model for HIV-1 and TB coinfection inside a host. This was the first attempt to understand how TB affects the dynamics of HIV-infected individuals. TB is known to be the common serious opportunistic infection occurring in HIV individuals and it occurs in more than 50% of the acquired immunodeficiency syndrome (AIDS) cases in developing countries. Naresh et al. [9] developed a simple nonlinear mathematical model dividing the population into four sub-classes, namely the susceptible, TB-infective, HIV-infective and AIDS patients. The treatment class in the HIV-AIDS co-infection model was first introduced by Huo et al. [10], however, Bhunu et al. [11] in his co-infection model considered all aspects of TB and HIV transmission dynamics with both HIV and TB treatment. This paper incorporated ARTs for AIDS cases and studied its implication on TB. However, the author did not consider the case where individual co-infected with HIV-TB can effectively recover from TB infection. Another HIV-TB co-infection model was formulated by Roeger et al. [12], assuming TB-infected individuals in the active stage of disease to be sexually inactive. Singh et al. [13] studied the transmission dynamics of the HIV/AIDS epidemic model considering three different latent stages based on treatment. Torres et al. [14], in his model, incorporates both TB and AIDS treatment for individuals suffering with either or both disease.

The model formulated in this paper considers the susceptible class to be HIV-infected. The paper is organized as follows. The model is formulated and its description is given in Section 2. Calculation of reproduction number and uniform persistence of the disease is shown in Section 2.3. In Section 2.4, global stability for all the equilibrium points is done. In Section 3, backward bifurcation is established. The sensitivity of reproduction number is done in Section 3.1. Section 3.2 presents a numerical simulation. The paper concludes in Section 4.

2. Mathematical Model

We begin with seven mutually exclusive compartmental models showing HIV-TB co-infection. In this model, the human population is divided into sub-populations as follows: acute HIV-infected individuals (H), co-infected with HIV-TB (H_{TB}), Pre-AIDS stage(P_A), infected individuals undergoing any type of treatment say ARV's and any TB treatment (M), Pre-AIDS stage with TB disease (P_{ATB}), HIV-infected individuals showing clinical AIDS symptoms (A), HIV-infected individuals with AIDS symptoms coinfected with TB disease (A_{TB}).

The notations and parametric values assumed in the paper for the study of dynamical system of HIV-TB co-infection model is tabulated in Table 1.

Table 1. Parametric definitions and its values.

Notations	Description	Parametric Values
$N(t)$	Number of individuals at any instant of time	100
B	Birth rate	0.2
β_1	Rate at which co-infection occurs	0.45
β_2	Rate at which HIV-infected individuals reaches pre-AIDS stage	0.48
β_3	Rate at which HIV-infected individuals opt for medication	0.31
β_4	Rate at which co-infected individual goes for medication	0.1
β_5	Rate at which co-infection (HIV-TB) individual joins pre-AIDS TB stage	0.037
β_6	Rate at which pre-AIDS infectives opt for medication	0.25
β_7	Rate at which pre-AIDS TB infectives undergo medication	0.15
β_8	Rate at which pre-AIDS infected individuals join pre-AIDS TB class	0.8
β_9	Rate at which pre-AIDS suffer from full-blown AIDS	0.3
β_{10}	Rate at which Pre-AIDS TB infectives joins full-blown AIDS TB class	0.001
β_{11}	Rate at which treated infectives move to AIDS class	0.78
β_{12}	Rate at which individuals with full-blown AIDS suffer from TB	0.35
μ	Natural death rate	0.002
μ_D	Death rate due to AIDS	0.6
μ_{DTB}	Death rate due to co-infection	0.52

In this paper, the susceptible class is considered to be HIV-infected (acute HIV infection). This class is increased by recruitments of newly HIV-infected individuals at the rate B. All the individuals in their respective compartments suffer from natural death at the constant rate μ. Individuals undergoing medication (treatment) through ARTs lower the rate of progression from HIV disease to AIDS, as HIV can never be cured.

Here, the individuals infected with HIV develop a very weak immune system, which means they are likely to get infected by many opportunistic diseases. As TB is considered to be one of the most commonly occurring disease among HIV patients [15], the individual infected with HIV gets TB disease moving towards H_{TB} by rate β_1. The HIV-infected individuals are also assumed to progress to the asymptotic pre-AIDS class (P_A) at the rate β_2. The HIV-infected and co-infected individuals undergoing treatment move to class M with the rates β_3 and β_4, respectively. Similarly, individuals with a co-infection of HIV-TB move towards P_{ATB} with the rate β_5. Individuals showing symptoms of AIDS (P_A) suffer from full-blown AIDS, joining A, at the rate β_9, and they are more likely to develop TB, progressing to class P_{ATB} with the rate β_8. P_A class individuals undergoing ARTs treatment (anti-retroviral therapy) join M at the rate β_6. Individuals in P_{ATB} are treated for TB at the constant rate β_7, joining M, and some of them can also develop full-blown AIDS, moving to A_{TB} class with the constant rate β_{10}. Treated individuals, recovered from TB but still with HIV infection (as it cannot be cured) move to full-blown AIDS (A) with the constant rate β_{11}. Individuals suffering with AIDS have such a badly damaged immune system that they get an increasing number of severe illnesses (here, TB) and hence move towards full-blown AIDS-TB class with the constant rate β_{12}. The death rates μ_D and μ_{DTB} are considered as deaths due to individuals infected with AIDS and AIDS-TB, respectively.

2.1. HIV-TB Co-infection Model

Considering the aforementioned assumptions and Figure 1 gives rise to the following set of non-linear differential equations for the HIV-TB co-infection model:

$$\begin{aligned}
\frac{dH}{dt} &= B - \beta_1 H H_{TB} - (\beta_2 + \beta_3 + \mu)H \\
\frac{dH_{TB}}{dt} &= \beta_1 H H_{TB} - \beta_5 H_{TB} P_{ATB} - (\mu + \beta_4) H_{TB} \\
\frac{dP_A}{dt} &= \beta_2 H - \beta_8 P_A P_{ATB} - (\mu + \beta_6 + \beta_9) P_A \\
\frac{dM}{dt} &= \beta_3 H + \beta_4 H_{TB} + \beta_6 P_A + \beta_7 P_{ATB} - (\mu + \beta_{11}) M \\
\frac{dP_{ATB}}{dt} &= \beta_5 H_{TB} P_{ATB} + \beta_8 P_A P_{ATB} - (\mu + \beta_7 + \beta_{10}) P_{ATB} \\
\frac{dA}{dt} &= \beta_9 P_A + \beta_{11} M - (\mu + \mu_D + \beta_{12}) A \\
\frac{dA_{TB}}{dt} &= \beta_{10} P_{ATB} + \beta_{12} A - (\mu + \mu_{DTB}) A_{TB}
\end{aligned} \qquad (1)$$

where $N(t) = H(t) + H_{TB}(t) + P_A(t) + M(t) + P_{ATB}(t) + A(t) + A_{TB}(t)$.

The system satisfies the conditions:

$$H(t) \geq 0, H_{TB}(t) \geq 0, P_A(t) \geq 0, M(t) \geq 0, P_{ATB}(t) \geq 0, A(t) \geq 0, A_{TB}(t) \geq 0$$

Adding the above set of differential equations, we get,

$$\begin{aligned}
\frac{dN(t)}{dt} &= B - \mu(H + H_{TB} + P_A + M + P_{ATB} + A + A_{TB}) - \mu_D A - \mu_{DTB} A_{TB} \\
&\leq B - \mu(H + H_{TB} + P_A + M + P_{ATB} + A + A_{TB})
\end{aligned}$$

Hence, $\frac{dN(t)}{dt} \leq B - \mu N$, so that $\limsup_{t\to\infty} N \leq \frac{B}{\mu}$

The feasible region for the system is defined as

$$\Lambda = \left\{ (H, H_{TB}, P_A, M, P_{ATB}, A, A_{TB}) : 0 \leq H + H_{TB} + P_A + M + P_{ATB} + A + A_{TB} \leq \frac{B}{\mu} \right\}$$

We assume $L_1 = \beta_2 + \beta_3$, $L_2 = \beta_6 + \beta_9$, $L_3 = \beta_7 + \beta_{10}$. The modified system is

$$\begin{aligned}
\frac{dH}{dt} &= B - \beta_1 H H_{TB} - (L_1 + \mu) H \\
\frac{dH_{TB}}{dt} &= \beta_1 H H_{TB} - \beta_5 H_{TB} P_{ATB} - (\mu + \beta_4) H_{TB} \\
\frac{dP_A}{dt} &= \beta_2 H - \beta_8 P_A P_{ATB} - (L_2 + \mu) P_A \\
\frac{dM}{dt} &= \beta_3 H + \beta_4 H_{TB} + \beta_6 P_A + \beta_7 P_{ATB} - (\mu + \beta_{11}) M \\
\frac{dP_{ATB}}{dt} &= \beta_5 H_{TB} P_{ATB} + \beta_8 P_A P_{ATB} - (L_3 + \mu) P_{ATB} \\
\frac{dA}{dt} &= \beta_9 P_A + \beta_{11} M - (\mu + \mu_D + \beta_{12}) A \\
\frac{dA_{TB}}{dt} &= \beta_{10} P_{ATB} + \beta_{12} A - (\mu + \mu_{DTB}) A_{TB}
\end{aligned} \qquad (2)$$

System (1) and (2) are equivalent, hence Λ is also the feasible region for system (2).

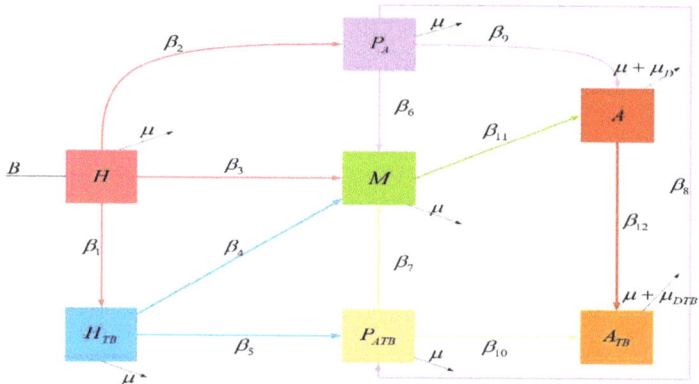

Figure 1. Transmission of individuals in different compartments.

2.2. Equilibrium Solutions

Equating $\frac{dH}{dt} = \frac{dH_{TB}}{dt} = \frac{dP_A}{dt} = \frac{dM}{dt} = \frac{dP_{ATB}}{dt} = \frac{dA}{dt} = \frac{dA_{TB}}{dt} = 0$ and solving for the compartments following are the equilibria:

1. $E_1(H_1, 0, P_{A_1}, M_1, 0, A_1, A_{TB_1})$

$$H_1 = \frac{B}{L_1+\mu}, H_{TB_1} = 0, P_{A_1} = \frac{B\beta_2}{(L_1+\mu)(L_2+\mu)}, M_1 = \frac{B(L_2\beta_3 + \beta_2\beta_6 + \beta_3\mu)}{(L_1+\mu)(L_2+\mu)(\beta_{11}+\mu)},$$
$$P_{ATB_1} = 0, A_1 = \frac{B(\beta_{11}(L_2\beta_3 + \beta_2\beta_6 + \beta_3\mu) + \beta_2\beta_9(\beta_{11}+\mu))}{(L_1+\mu)(L_2+\mu)(\beta_{11}+\mu)(\mu+\mu_D+\beta_{12})},$$
$$A_{TB_1} = \frac{B\beta_{12}(\beta_{11}(L_2\beta_3 + \beta_2\beta_6 + \beta_3\mu) + \beta_2\beta_9(\beta_{11}+\mu))}{(L_1+\mu)(L_2+\mu)(\beta_{11}+\mu)(\mu+\mu_D+\beta_{12})(\mu+\mu_D)}.$$

2. $E_2(H_2, 0, P_{A_2}, M_2, P_{ATB_2}, A_2, A_{TB_2})$

$$H_2 = \frac{B}{L_1+\mu}, H_{TB_2} = 0, P_{A_2} = \frac{L_3+\mu}{\beta_8}, P_{ATB_2} = \frac{B\beta_2\beta_8 - (L_1+\mu)(L_2+\mu)(L_3+\mu)}{\beta_8(L_1+\mu)(L_3+\mu)},$$
$$M_2 = \frac{B\beta_8(L_3\beta_3 + \beta_2\beta_7 + \beta_3\mu) + (L_1+\mu)(L_3+\mu)(\beta_6(L_3+\mu) - \beta_7(L_2+\mu))}{\beta_8(L_1+\mu)(L_3+\mu)(\beta_{11}+\mu)},$$
$$A_2 = \frac{\{B\beta_8\beta_{11}(L_3\beta_3 + \beta_2\beta_7 + \beta_3\mu) + (L_3+\mu)^2(L_1+\mu)(\beta_9(\beta_{11}+\mu) + \beta_6\beta_{11}) - (L_1+\mu)(L_2+\mu)(L_3+\mu)\beta_7\beta_{11}\}}{\beta_8(L_1+\mu)(L_3+\mu)(\mu+\mu_D+\beta_{12})},$$
$$A_{TB_2} = \frac{\begin{array}{c}B\beta_8\beta_{11}\beta_{12}(\beta_2\beta_7 + L_3\beta_3 + \mu\beta_3) + (\beta_{11}+\mu)\mu_D B\beta_2\beta_8\beta_{10} - (L_1+\mu)(L_2+\mu) \\ (L_3+\mu)(\beta_7\beta_{11}\beta_{12} + (\beta_{11}+\mu)\mu_D\beta_{10}) + (\beta_{12}+\mu)(\beta_{11}+\mu)B\beta_2\beta_8\beta_{10} + (L_1+\mu) \\ (L_3+\mu)(\beta_6\beta_{11} + (\beta_{11}+\mu)\beta_9\beta_{12}) - (L_1+\mu)(L_2+\mu)(L_3+\mu)(\beta_{11}+\mu)(\beta_{12}+\mu)\end{array}}{\beta_8(L_1+\mu)(L_3+\mu)(\beta_{11}+\mu)(\mu+\mu_{DTB})(\mu+\mu_D+\beta_{12})}$$

3. $E_3(H_3, H_{TB_3}, P_{A_3}, M_3, 0, A_3, A_{TB_3})$

$$H_3 = \frac{\beta_4+\mu}{\beta_1}, H_{TB_3} = \frac{B\beta_1-(L_1+\mu)(\beta_4+\mu)}{\beta_1(\beta_4+\mu)}, P_{A_3} = \frac{\beta_2(\beta_4+\mu)}{\beta_1(L_2+\mu)},$$

$$M_3 = \frac{(L_2+\mu)(B\beta_1\beta_4-(\beta_4+\mu)(L_1\beta_4-\beta_3\beta_4-\beta_3\mu-\beta_4\mu))+\beta_2\beta_6(\beta_4+\mu)^2}{\beta_1(L_2+\mu)(\beta_4+\mu)(\beta_{11}+\mu)},$$

$$P_{ATB_3} = 0, A_3 = \frac{\{\beta_4\beta_{11}(B\beta_1-L_1(\beta_4+\mu))(L_2+\mu)+(\beta_4+\mu)(L_2\beta_{11}(\beta_3\beta_4+\beta_3\mu-\beta_4\mu)\\-\beta_4\beta_{11}\mu^2)+(\beta_4+\mu)^2(\beta_2\beta_9(\beta_{11}+\mu)+\beta_2\beta_6\beta_{11}+\beta_3\beta_{11}\mu)\}}{\beta_1(L_2+\mu)(\beta_4+\mu)(\beta_{11}+\mu)(\mu+\mu_D+\beta_{12})}$$

$$A_{TB_3} = \frac{\{\beta_{12}(\beta_4\beta_{11}(B\beta_1-L_1(\beta_4+\mu))(L_2+\mu)+(\beta_4+\mu)(L_2\beta_{11}(\beta_3\beta_4+\beta_3\mu-\beta_4\mu)\\-\beta_4\beta_{11}\mu^2))+(\beta_4+\mu)^2(\beta_2\beta_9(\beta_{11}+\mu)+\beta_2\beta_6\beta_{11}+\beta_3\beta_{11}\mu)\}}{\beta_1(L_2+\mu)(\beta_4+\mu)(\beta_{11}+\mu)(\mu+\mu_D+\beta_{12})(\mu+\mu_{DTB})}$$

4. Endemic Equilibrium point $E^*(H^*, H_{TB}^*, P_A^*, M^*, P_{ATB}^*, A^*, A_{TB}^*)$

$$H^* = \frac{r(\beta_8(\beta_4+\mu)-\beta_5(L_2+\mu))+B\beta_5}{\beta_5(L_1+\mu)+\beta_1(L_3+\mu)-\beta_2\beta_5}, H_{TB}^* = \frac{-\beta_8 r+L_3+\mu}{\beta_5}, P_A^* = r,$$

$$P_{ATB}^* = \frac{r\beta_1(\beta_8(\beta_4+\mu)-\beta_5(L_2+\mu))+\beta_5(B\beta_1\\-(L_1+\mu)(\beta_4+\mu)+\beta_2(\beta_4+\mu))-\beta_1(L_3+\mu)(\beta_4+\mu)}{\beta_5(L_1\beta_5+L_3\beta_1+\beta_1\mu-\beta_2\beta_5+\beta_5\mu)},$$

$$M^* = \frac{r[(\beta_5\beta_6-\beta_4\beta_8)[\beta_5(L_1+\mu)+\beta_1(L_3+\mu)-\beta_2\beta_5]+(\beta_3\beta_5+\beta_7\beta_{11})(\beta_8(\beta_4+\mu)\\-\beta_5(L_2+\mu))]+\beta_5\beta_7[(B\beta_1-(L_1+\mu)(\beta_4+\mu))+\beta_2(\beta_4+\mu)]+\beta_4\beta_5(L_1+\mu)\\(L_3+\mu)+(L_3+\mu)\beta_1(\beta_4(L_3+\mu)-\beta_7(\beta_4+\mu))-\beta_2\beta_4\beta_5(L_3+\mu)+B\beta_3\beta_5^2}{\beta_5(\beta_{11}+\mu)(\beta_5(L_1+\mu)+\beta_1(L_3+\mu)-\beta_2\beta_5)},$$

$$A^* = \frac{r(\beta_{11}(\beta_1\beta_7+\beta_3\beta_5)(\beta_8(\beta_4+\mu)-\beta_5(L_2+\mu))+(\beta_1(L_3+\mu)+\beta_5(L_1+\mu)-\beta_2\beta_5)\\((\beta_{11}+\mu)\beta_5\beta_9+(\beta_5\beta_6-\beta_4\beta_8)\beta_{11}))+\beta_{11}(\beta_5\beta_7(B\beta_1-(L_1+\mu)(\beta_4+\mu))+\beta_2(\beta_4+\mu))\\+\beta_4\beta_5(L_1+\mu)(L_3+\mu)+\beta_1(L_3+\mu)(\beta_4(L_3+\mu)-\beta_7(\beta_4+\mu))-\beta_2\beta_4\beta_5(L_3+\mu)+B\beta_3\beta_5^2}{\beta_5(\mu+\beta_{11})(\mu+\mu_D+\beta_{12})(\beta_5(L_1+\mu)+\beta_1(L_3+\mu)-\beta_2\beta_5)},$$

$$A_{TB}^* = \frac{r[(\beta_8(\beta_4+\mu)-\beta_5(L_2+\mu))((\beta_{11}+\mu)\beta_1\beta_{10}(\mu+\mu_D+\beta_{12})+\beta_{11}\beta_{12}(\beta_3\beta_5+\beta_1\beta_7))\\+(\beta_5(L_1+\mu)+\beta_1(L_3+\mu)-\beta_2\beta_5)\beta_{12}((\beta_4+\mu)\beta_5\beta_9+(\beta_5\beta_6-\beta_4\beta_8)\beta_{11})]+\mu_D\beta_{10}\\(\beta_{11}+\mu)(\beta_4+\mu)(\beta_2\beta_5-\beta_1(L_3+\mu)-\beta_5(L_1+\mu))+\mu_D B\beta_1\beta_5\beta_{10}(\beta_{11}+\mu)+\beta_7\beta_{11}\beta_{12}\\(\beta_4+\mu)(\beta_2\beta_5-\beta_5\mu-L_3\beta_1)-\beta_{10}(\beta_{12}+\mu)(\beta_{11}+\mu)^2(\beta_5(L_1+\mu)+\beta_1(L_3+\mu))+\beta_4\\\beta_{11}\beta_{12}(L_3+\mu)(\beta_5(L_1+\mu)+\beta_1(L_3+\mu)-\beta_2\beta_5)+\beta_5\beta_{10}(\beta_{11}+\mu)(\beta_{12}+\mu)(\beta_2(\beta_4+\mu)\\+B\beta_1)-\beta_1\beta_7\beta_{11}\beta_{12}(\beta_5+\mu)(\beta_4+\mu)+B\beta_5\beta_{11}\beta_{12}(\beta_1\beta_7+\beta_3\beta_5)}{\beta_5(\mu+\beta_{11})(\mu+\mu_{DTB})(\mu+\mu_D+\beta_{12})(\beta_5(L_1+\mu)+\beta_1(L_3+\mu)-\beta_2\beta_5)},$$

where, $r = rootof((\beta_1\beta_8(\beta_5(L_2+\mu)-\beta_8(\beta_4+\mu)))Z^2+(-\beta_5\beta_8(B\beta_1-(L_1+\mu)(\beta_4+\mu))\\-\beta_5^2(L_1+\mu)(L_2+\mu)+\beta_1(L_3+\mu)(\beta_8(\beta_4+\mu)-\beta_5(L_2+\mu)))Z+B\beta_2\beta_5^2$.

2.3. Reproduction Number

The reproduction number measures the expected number of secondary infected individuals produced due to an infected individual during the entire death period in an uninfected population.

In this paper, reproduction number R_0 is defined as the number of infected individuals due to an AIDS- or TB-infected individual in the HIV infected-population. It is calculated using next-generation matrix method [16] and is defined as the spectral radius of FV^{-1} at E_1.

$$\text{where, } F = \begin{bmatrix} \beta_1 H & 0 & 0 & 0 & 0 & 0 & \beta_1 H_{TB} \\ 0 & 0 & 0 & 0 & 0 & 0 & 0 \\ 0 & 0 & 0 & 0 & 0 & 0 & 0 \\ \beta_5 P_{ATB} & \beta_8 P_{ATB} & 0 & \beta_5 H_{TB}+\beta_8 P_A & 0 & 0 & 0 \\ 0 & 0 & 0 & 0 & 0 & 0 & 0 \\ 0 & 0 & 0 & 0 & 0 & 0 & 0 \\ 0 & 0 & 0 & 0 & 0 & 0 & 0 \end{bmatrix}$$

$$V = \begin{bmatrix} \beta_4 + \beta_5 P_{ATB} + \mu & 0 & 0 & \beta_5 H_{TB} & 0 & 0 & 0 \\ 0 & \beta_8 P_{ATB} + L_2 + \mu & 0 & \beta_8 P_A & 0 & 0 & -\beta_2 \\ -\beta_4 & -\beta_6 & \beta_{11} + \mu & -\beta_7 & 0 & 0 & -\beta_3 \\ 0 & 0 & 0 & L_3 + \mu & 0 & 0 & 0 \\ 0 & -\beta_9 & -\beta_{11} & 0 & \beta_{12} + \mu_D + \mu & 0 & 0 \\ 0 & 0 & 0 & -\beta_{10} & -\beta_{12} & \mu + \mu_{DTB} & 0 \\ \beta_1 H & 0 & 0 & 0 & 0 & 0 & \beta_1 H_{TB} + L_1 + \mu \end{bmatrix}$$

The dominant eigenvalue of FV^{-1} at E_1 is $R_0 = \frac{B\beta_2\beta_8}{(L_1+\mu)(L_2+\mu)(L_3+\mu)} + \frac{\beta_1 B}{(L_1+\mu)(\beta_4+\mu)}$.

2.4. Persistence of Disease

Now, uniform persistence for the system (1) is constructed. The model system (1) is said to be uniformly persistent if there is a constant f, such that any solution $(H(t), H_{TB}(t), P_A(t), M(t), P_{ATB}(t), A(t), A_{TB}(t))$ satisfies [17,18].

$$\liminf_{t \to \infty} H(t) > f, \; \liminf_{t \to \infty} H_{TB}(t) > f, \; \liminf_{t \to \infty} P_A(t) > f, \; \liminf_{t \to \infty} M(t) > f,$$
$$\liminf_{t \to \infty} P_{ATB}(t) > f, \; \liminf_{t \to \infty} A(t) > f, \; \liminf_{t \to \infty} A_{TB}(t) > f.$$

Provided that $(H(0), H_{TB}(0), P_A(0), M(0), P_{ATB}(0), A(0), A_{TB}(0)) \in \Lambda$

Theorem 1. *The model (1) is uniformly persistent in Λ only if $R_0 > 1$.*

2.5. Stability Analysis

In this section, global stability is studied for all the equilibrium points obtained.

Theorem 2. *Global Stability of $E_1(H_1, 0, P_{A_1}, M_1, 0, A_1, A_{TB_1})$*

The system (2) of the model can be written as

$$\frac{dX_1}{dt} = F_1(X_1, Z_1) \tag{3}$$

$$\frac{dZ_1}{dt} = G_1(X_1, Z_1), G_1(X_1, 0) = 0 \tag{4}$$

where $X_1 = (H, P_A, M, A, A_{TB})$ and $Z_1 = (H_{TB}, P_{ATB})$. According to this notation, equilibrium point is denoted by $E_1 = (X_1', 0)$, where $X_1' = (H_1, 0, P_{A_1}, M_1, 0, A_1, A_{TB_1})$.

By the Castillo Chavez method, the following two condition ensure the global stability of the given equilibrium point:

P.1 For $\frac{dX_1}{dt} = F_1(X_1, 0)$, E_1 is globally asymptotically stable.
P.2 $G_1(X_1, Z_1) = AZ_1 - \hat{G}_1(X_1, Z_1)$, where $\hat{G}_1(X_1, Z_1) \geq 0$ for $(X_1, Z_1) \in \Lambda$.

where $A = D_{Z_1} G_1(X_1, 0)$ is a M-matrix (matrix with non-negative off diagonal elements) and Λ is the region defined above. We have,

$$F_1(X_1, 0) = \begin{bmatrix} B - (\beta_2 + \beta_3 + \mu)H \\ \beta_2 H - (\mu + \beta_6 + \beta_9)P_A \\ \beta_3 H + \beta_6 P_A - (\mu + \beta_{11})M \\ \beta_9 P_A + \beta_{11} M - (\mu + \mu_D + \beta_{12})A \\ \beta_{12} A - (\mu + \mu_{DTB})A_{TB} \end{bmatrix}$$

$$G_1(X_1, Z_1) = \begin{bmatrix} \beta_1 H H_{TB} - \beta_5 H_{TB} P_{ATB} - (\mu + \beta_4) H_{TB} \\ \beta_5 H_{TB} P_{ATB} + \beta_8 P_A P_{ATB} - (\mu + \beta_7 + \beta_{10}) P_{ATB} \end{bmatrix} \text{ and } G_1(X_1, 0) = 0, \text{ thus}$$

$$A = D_{Z_1} G_1(X_1', 0) = \begin{bmatrix} \beta_1 H - (\mu + \beta_4) & 0 \\ 0 & \beta_8 P_A - (\mu + \beta_7 + \beta_{10}) \end{bmatrix}$$

$$\hat{G}(X_1, Z_1) = \begin{bmatrix} \beta_5 H_{TB} P_{ATB} \\ -\beta_5 H_{TB} P_{ATB} \end{bmatrix} \tag{5}$$

From Equation (5), the condition P.2 is not satisfied, since $\hat{G}_1(X_1, Z_1) \geq 0$ is not true. Therefore, the equilibrium point E_1 may not be globally stable. Here, since disease (HIV-AIDS) persists at this point, it will not be globally stable. Following [19], the backward bifurcation occurs at $R_0 = 1$.

Theorem 3. *Global Stability of $E_2(H_2, 0, P_{A_2}, M_2, P_{ATB_2}, A_2, A_{TB_2})$*

The system (2) of the model can be written as

$$\frac{dX_2}{dt} = F_2(X_2, Z_2) \tag{6}$$

$$\frac{dZ_2}{dt} = G_2(X_2, Z_2), \; G_2(X_2, 0) = 0 \tag{7}$$

where $X_2 = (H, P_A, M, P_{ATB}, A, A_{TB})$ and $Z_2 = (H_{TB})$. According to this notation, the equilibrium point is denoted by $E_2 = (X_2', 0)$, where $X_2' = (H_2, 0, P_{A_2}, M_2, P_{ATB_2}, A_2, A_{TB_2})$.

Using the Castillo Chavez method [20], the following two condition ensure the global stability of the given equilibrium point:

P.3 For $\frac{dX_2}{dt} = F_2(X_2, 0)$, E_2 is globally asymptotically stable.
P.4 $G_2(X_2, Z_2) = BZ_2 - \hat{G}_2(X_2, Z_2)$, where $\hat{G}(X_2, Z_2) \geq 0$ for $(X_2, Z_2) \in \Lambda$.

where $B = D_{Z_2} G_2(X_2, 0)$ is an M-matrix (matrix with non-negative off diagonal elements) and Λ is the region defined above.

The equilibrium point $E_2(H_2, 0, P_{A_2}, M_2, P_{ATB_2}, A_2, A_{TB_2})$ is the globally asymptotically stable equilibrium of the system (P.3)–(P.4)

we have $F_2(X_2, 0) = \begin{bmatrix} B - (\beta_2 + \beta_3 + \mu) H \\ \beta_2 H - \beta_8 P_A P_{ATB} - (\mu + \beta_6 + \beta_9) P_A \\ \beta_3 H + \beta_6 P_A + \beta_7 P_{ATB} - (\mu + \beta_{11}) M \\ \beta_8 P_A P_{ATB} - (\mu + \beta_7 + \beta_{10}) P_{ATB} \\ \beta_9 P_A + \beta_{11} M - (\mu + \mu_D + \beta_{12}) A \\ \beta_{10} P_{ATB} + \beta_{12} A - (\mu + \mu_{DTB}) A_{TB} \end{bmatrix}$

The eigenvalues of the characteristic polynomial of its Jacobian matrix are given as

$$\lambda_1 = -(\mu + \beta_2 + \beta_3), \lambda_2 = -(\mu + \beta_{11}), \lambda_3 = -(\mu + \mu_{DTB}), \lambda_4 = -(\mu + \mu_D + \beta_{12}),$$

$$\lambda_5 = -\tfrac{1}{2}((\beta_8 P_{ATB} + \beta_6 + \beta_7 + \beta_9 + \beta_{10} + 2\mu - \beta_8 P_A)$$
$$- \sqrt{\beta_8^2 (P_A - P_{ATB})^2 + (\beta_6 - \beta_7 + \beta_9 - \beta_{10})^2 + 2\beta_8 (P_A + P_{ATB})(\beta_6 - \beta_7 + \beta_9 - \beta_{10})}),$$

$$\lambda_6 = -\tfrac{1}{2}((\beta_8 P_{ATB} + \beta_6 + \beta_7 + \beta_9 + \beta_{10} + 2\mu - \beta_8 P_A)$$
$$+ \sqrt{\beta_8^2 (P_A - P_{ATB})^2 + (\beta_6 - \beta_7 + \beta_9 - \beta_{10})^2 + 2\beta_8 (P_A + P_{ATB})(\beta_6 - \beta_7 + \beta_9 - \beta_{10})})$$

Here, λ_5, λ_6 have a negative real part if $(\beta_6 + \beta_9 + \mu)\beta_8 P_A < (\beta_7 + \beta_{10} + \mu)(\beta_6 + \beta_9 + \mu + \beta_8 P_{ATB})$. Hence, by Routh–Hurwitz criterion, the system is globally asymptotically stable.
Next,

$$G_2(X_2, Z_2) = (\beta_1 H_2 + \beta_5 P_{ATB_2} - (\mu + \beta_4))H_{TB} - (\beta_1(H_2 - H)H_{TB} + \beta_5(P_{ATB_2} - P_{ATB})H_{TB})$$
$$= BH_{TB} - \hat{G}_2(X_2, Z_2)$$

Here, $\hat{G}_2(X_2, Z_2) \geq 0$, hence the conditions of P.3 and P.4 are satisfied. Hence, by Castillo Chavez the system is globally stable.

Theorem 4. *Global stability of* $E_3(H_3, H_{TB_3}, P_{A_3}, M_3, 0, A_3, A_{TB_3})$

The system (1) of the model can be written as

$$\frac{dX_3}{dt} = F_3(X_3, Z_3) \tag{8}$$

$$\frac{dZ_3}{dt} = G_3(X_3, Z_3), \; G_3(X_3, 0) = 0 \tag{9}$$

where $X_3 = (H, H_{TB}, P_A, M, A, A_{TB})$ and $Z_3 = (P_{ATB})$. According to this notation, the equilibrium point is denoted by $E_3 = (X_3', 0)$, where $X_3' = (H_3, H_{TB_3}, P_{A_3}, M_3, 0, A_3, A_{TB_3})$.

The following two conditions ensure the global stability of this equilibrium point

P.5 For $\frac{dX_3}{dt} = F_3(X_3, 0)$, E_3 is globally asymptotically stable.
P.6 $G_3(X_3, Z_3) = CZ_3 - \hat{G}_3(X_3, Z_3)$, where $\hat{G}_3(X_3, Z_3) \geq 0$ for $(X_3, Z_3) \in \Lambda$

where $C = D_{Z_3} G_3(X_3, 0)$ is an M-matrix (matrix with non-negative off diagonal elements) and Λ is the region defined above.

The equilibrium point $E_3(H_3, H_{TB_3}, P_{A_3}, M_3, 0, A_3, A_{TB_3})$ is the globally asymptotically stable equilibrium of the system (P.5)–(P.6)

we have $F_3(X_3, 0) = \begin{bmatrix} B - \beta_1 H H_{TB} - (\beta_2 + \beta_3 + \mu)H \\ \beta_1 H H_{TB} - (\mu + \beta_4)H_{TB} \\ \beta_2 H - (\mu + \beta_6 + \beta_9)P_A \\ \beta_3 H + \beta_4 H_{TB} + \beta_6 P_A - (\mu + \beta_{11})M \\ \beta_9 P_A + \beta_{11} M - (\mu + \mu_D + \beta_{12})A \\ \beta_{12} A - (\mu + \mu_{DTB})A_{TB} \end{bmatrix}$

The eigenvalues of the characteristic polynomial of its Jacobian matrix are given as

$$\lambda_1 = -(\mu + \beta_6 + \beta_9), \lambda_2 = -(\mu + \beta_{11}), \lambda_3 = -(\mu + \mu_{DTB}),$$
$$\lambda_4 = -(\mu + \mu_D + \beta_{12}),$$
$$\lambda_5 = -\tfrac{1}{2}((\beta_1 H_{TB} + \beta_2 + \beta_3 + \beta_4 + 2\mu - \beta_1 H)$$
$$- \sqrt{\beta_1^2(H - H_{TB})^2 + (\beta_2 + \beta_3 - \beta_4)^2 + 2\beta_1(H + H_{TB})(\beta_2 + \beta_3 - \beta_4)})$$
$$\lambda_6 = -\tfrac{1}{2}((\beta_1 H_{TB} + \beta_2 + \beta_3 + \beta_4 + 2\mu - \beta_1 H)$$
$$+ \sqrt{\beta_1^2(H - H_{TB})^2 + (\beta_2 + \beta_3 - \beta_4)^2 + 2\beta_1(H + H_{TB})(\beta_2 + \beta_3 - \beta_4)})$$

here, λ_5, λ_6 have negative real part if $(\beta_2 + \beta_3 + \mu)\beta_1 H < (\beta_4 + \mu)(\beta_2 + \beta_3 + \mu + H_{TB}\beta_1)$. Hence, by Routh–Hurwitz criterion, the system is globally stable.
Next,

$$G_3(X_3, Z_3) = (\beta_5 H_{TB} + \beta_8 P_A - (\mu + \beta_7 + \beta_{10}))P_{ATB} - (\beta_5(H_{TB_3} - H_{TB})P_{ATB} + \beta_8(P_{A_3} - P_A)P_{ATB})$$
$$= CP_{ATB} - \hat{G}_3(X_3, Z_3)$$

Here, $\hat{G}_3(X_3, Z_3) \geq 0$, hence the conditions P.5 and P.6 are satisfied. Therefore, by Castillo Chavez the system is globally stable.

Theorem 5. *The endemic equilibrium point $E^*(H^*, H^*_{TB}, P^*_A, M^*, P^*_{ATB}, A^*, A^*_{TB})$ is globally asymptotically stable.*

Proof. Let us assume Lyapunov function

$$L^*(t) = \tfrac{1}{2}[(H - H^*) + (H_{TB} - H^*_{TB}) + (P_A - P^*_A) + (M - M^*) + (P_{ATB} - P^*_{ATB}) + (A - A^*) + (A_{TB} - A^*_{TB})]^2$$

$$\begin{aligned}
\frac{dL^*}{dt} &= [(H - H^*) + (H_{TB} - H^*_{TB}) + (P_A - P^*_A) + (M - M^*) + (P_{ATB} - P^*_{ATB}) + (A - A^*) + (A_{TB} - A^*_{TB})] \\
&\quad [H' + H'_{TB} + P'_A + M' + P'_{ATB} + A' + A'_{ATB}] \\
&= [(H - H^*) + (H_{TB} - H^*_{TB}) + (P_A - P^*_A) + (M - M^*) + (P_{ATB} - P^*_{ATB}) + (A - A^*) + (A_{TB} - A^*_{TB})] \\
&\quad [B - \mu(H + H_{TB} + P_A + M + P_{ATB} + A + A_{TB}) - \mu_D A - \mu_{DTB} A_{TB}] \\
&= -[(H - H^*) + (H_{TB} - H^*_{TB}) + (P_A - P^*_A) + (M - M^*) + (P_{ATB} - P^*_{ATB}) + (A - A^*) + (A_{TB} - A^*_{TB})] \\
&\quad [\mu((H - H^*) + (H_{TB} - H^*_{TB}) + (P_A - P^*_A) + (M - M^*) + (P_{ATB} - P^*_{ATB}) + (A - A^*) + (A_{TB} - A^*_{TB}))] \\
&= -\mu[(H - H^*) + (H_{TB} - H^*_{TB}) + (P_A - P^*_A) + (M - M^*) + (P_{ATB} - P^*_{ATB}) + (A - A^*) + (A_{TB} - A^*_{TB})]^2 \\
&\leq 0
\end{aligned}$$

where $B = \mu(H^* + H^*_{TB} + P^*_A + M^* + P^*_{ATB} + A^* + A^*_{TB}) + \mu_D A + \mu_{DTB} A_{TB}$ □

Here, $\frac{dL^*}{dt} \leq 0$. Hence, by LaSalle Invariance principle [21] the endemic equilibrium point is globally asymptotically stable.

3. Backward Bifurcation

If the reproduction number $R_0 > 1$, then $P_A > 0$, the system (1) exhibits a unique positive solution E^*. Now, on solving system (2), we have

$$F(P^*_A) = b_2 P^{*2}_A + b_1 P^*_A + b_0 \tag{10}$$

where,

$$b_2 = \beta_1 \beta_8 (\beta_5(L_2 + \mu) - \beta_8(\beta_4 + \mu))$$
$$b_1 = \beta_8(\mu + \beta_4)[\beta_1(L_1 + \mu) + \beta_5(L_3 + \mu)] - \beta_1 \beta_5[(L_2 + \mu)(L_3 + \mu) + B\beta_8] - \beta_5^2(L_1 + \mu)(L_2 + \mu)$$
$$b_0 = B \beta_1 \beta_5^2$$

Here, the coefficient $b_2 < 0$, and b_0 depends on the value of R_0. If $R_0 < 1$, then b_0 is positive and if $R_0 > 1$, then b_0 is negative. For $R_0 > 1$, Equation (10) has two roots, positive and negative.

For $b_1 > 0$, the system has endemic equilibria continuously depending on R_0; this shows that there exists an interval for R_0, which has two positive equilibria as follows:

$$I_1 = \frac{-b_1 - \sqrt{b_1^2 - 4 b_2 b_0}}{2 b_2}, \quad I_2 = \frac{-b_1 + \sqrt{b_1^2 - 4 b_2 b_0}}{2 b_2}$$

For Backward Bifurcation, setting $b_1^2 - 4 b_2 b_0 = 0$ and solving for critical points of R_0 gives

$$R_C = 1 - \frac{[\beta_8(\mu + \beta_4)[\beta_1(L_1 + \mu) + \beta_5(L_3 + \mu)] - \beta_1 \beta_5[(L_2 + \mu)(L_3 + \mu) + B\beta_8] - \beta_5^2(L_1 + \mu)(L_2 + \mu)]^2((L_1 + \mu)(L_2 + \mu)(L_3 + \mu) - B\beta_1 \beta_8)}{4(\beta_1 \beta_8(\beta_5(L_2 + \mu) - \beta_8(\beta_4 + \mu))) B \beta_2 \beta_5^2 (L_1 + \mu)(L_2 + \mu)(L_3 + \mu)}$$

If $R_C < R_0$, then, equivalently, $b_1^2 - 4 b_2 b_0 > 0$ and backward bifurcation occur for the points of R_0, such that $R_C < R_0 < 1$ [22], as shown in the above Figure 2. Here, $R_C = 0.95$ is the critical value after which co-infection attains stability.

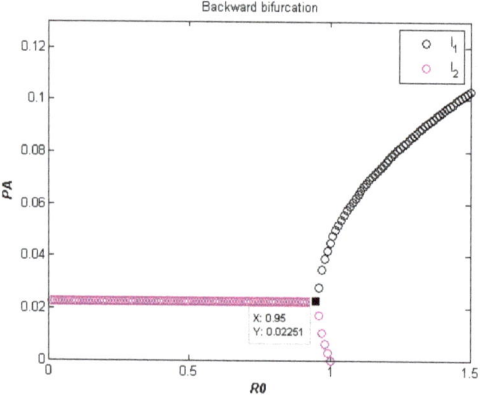

Figure 2. Pre-AIDS class at equilibria versus R_0.

3.1. Sensitivity Analysis of R_0

In this section, sensitivity indices of R_0 with respect to different parameters are calculated as shown in Table 2, using the formula $\gamma_\alpha^{R_0} = \frac{\partial R_0}{\partial \alpha} \cdot \frac{\alpha}{R_0}$, where α is the model parameter. These indices show how crucial each parameter is to disease transmission.

Table 2. Effect of Parameters on Sensitivity.

Parameter	Value	Observation
B	1	The transmission rate of HIV is directly proportional to birth rate.
β_1	0.4925	The transmission rate of co-infection occurs at 49%.
β_2	0.9013	Among HIV infectives, around 90% of them join the pre-AIDS stage.
β_3	0.6084	Individuals moving toward medication can be increased further by creating awareness programs.
β_4	0.5172	
β_6	0.77	77% of individuals in pre-AIDS class opt for medication.
β_7	0.5022	From the pre-AIDS, class 50% of individuals undergo medication for TB disease.
β_8	0.5075	Transmission occurs at the rate of 50% from the pre-AIDS class to pre-AIDS TB.
β_9	0.724	The number of individuals in pre-AIDS class suffering from AIDS can be reduced if they take treatment while in pre-AIDS class.
β_{10}	0.9967	The transmission rate of individuals from pre-AIDS TB stage to AIDS TB stage highly effects the sensitivity of R_0.
μ	0.9793	Natural death rate cannot be removed completely even if the treatment is opted for in initial stage.

The other parameters $\beta_5, \beta_{11}, \beta_{12}, \mu_D, \mu_{DTB}$ do not have any impact on the sensitivity of reproduction number.

3.2. Numerical Simulation

From Figure 3 it can be observed that about 34% of the total HIV-infected population gets TB infection within 15 months. Approximately 30% of individuals infected with HIV go for treatment in 27 months. Co-infected individuals undergo treatment for TB in 11 months. Within 26 months, approximately 31% of HIV-infectives proceed to next stage, i.e., AIDS. About 22% of pre-AIDS infectives get TB infection and join pre-AIDS TB in 20 months. Individuals in the pre-AIDS class initially undergo medication then, due to ignorance or any other social reason, individuals leave the compartment and, after some time, joins the medication class again. Between approximately 1.7 and 4 months, individuals in the pre-AIDS class (not taking any kind of medication) suffer from AIDS, whereas those undergoing treatment get infected by AIDS within 28 months. This shows that medication is helpful. Even though it does not help the complete elimination of disease, the rate of disease spread can be controlled.

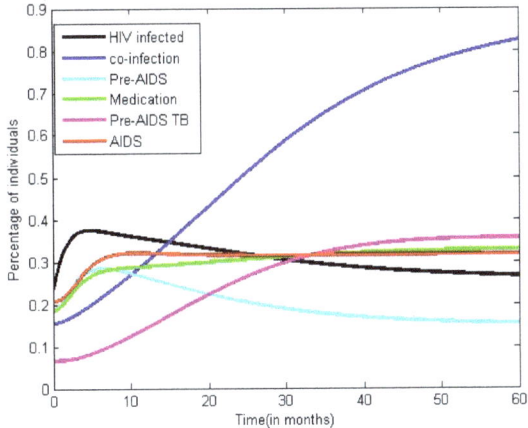

Figure 3. Transmission of HIV–TB co-infection.

From Figure 4, we can conclude that individuals in Pre-AIDS class for a longer duration get AIDS at faster rate than the individuals who have just joined the pre-AIDS class.

Figure 4. Intensity of pre-AIDS class versus AIDS class.

Figure 5 indicates that individuals suffering from HIV suffer from TB also, and both the compartments stabilize after some time. Figure 6 shows that individuals in the pre-AIDS class also suffer from TB. The trajectory is stable.

Figure 5. Behavior of H v/s H_{TB}.

Figure 6. Trajectory of P_{ATB} and P_A.

Figure 7 shows the stability of the respective compartments.

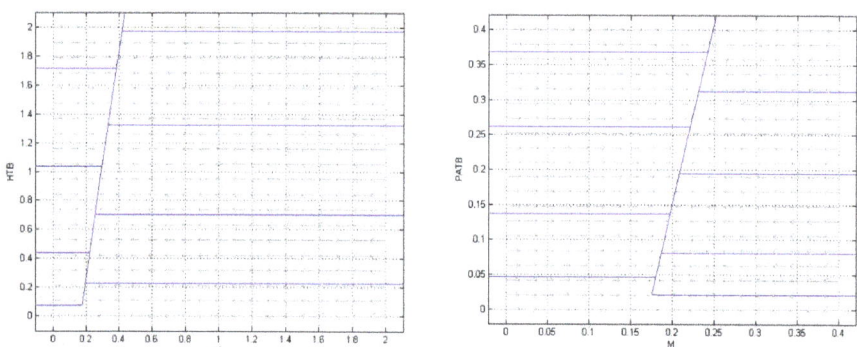

Figure 7. Phase transition plot of M with H_{TB} and P_{ATB}.

Figure 8 shows that the newly HIV-TB infected individuals and individuals in Pre-AIDS TB class will oscillate cyclically. Here, neither compartment will die out completely nor they will grow indefinitely.

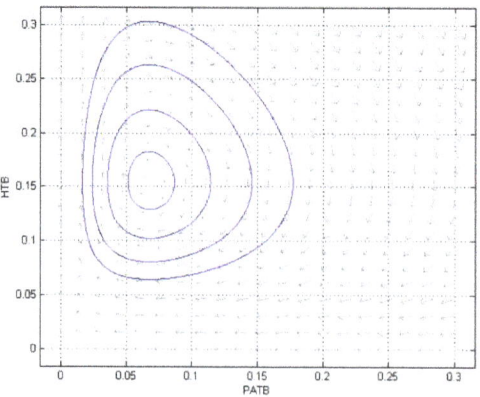

Figure 8. Phase transition of co-infection and pre-stage of co-infection.

From Figure 9, it can be observed that, of the total population, 21% are HIV infected and 13% are HIV-TB infected, whereas the percentage of individuals in pre-AIDS and pre-AIDS TB stage is 16% and 6%, respectively. A total of 15% of the population undergoes treatment for both diseases. Since HIV-AIDS is not curable, even after taking treatment, 17% of cases lead to AIDS infections and 12% are infected by TB, moving towards the AIDS TB class.

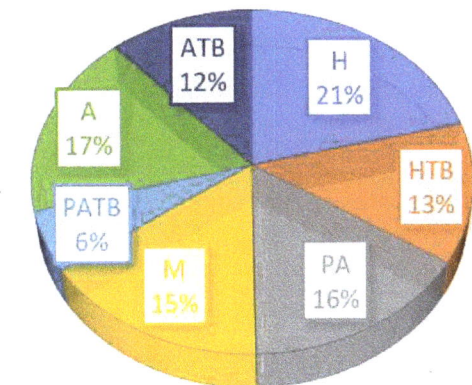

Figure 9. Percentage wise distribution of different compartments.

4. Conclusions

In this paper, a mathematical model of HIV-TB considering the HIV-infected population is studied. Using data tabulated in Table 1, we have $R_0 = 2.262 > 1$, which shows the persistence of the disease in the society. HIV-AIDS cannot be eradicated completely from the infected population. Next, global stability is shown for the equilibrium points where there is no co-infection, and instances when there is no individual in the pre-AIDS TB class are shown using Castillo Chavez method. The equilibrium points where there is no co-infection and no individual in the pre-AIDS TB class proved to be globally unstable and is said to exhibit bifurcation. The endemic point is proven to be globally stable using Lyapunov function. Backward bifurcation analysis is studied, which indicates that a minimum of 95% of individuals join the pre-AIDS class. Numerical simulation is done to validate the model, which concludes that the medication plays a vital role in controlling disease spread. Here, we can observe that if treatment is provided at the initial stage of disease, its further progression can be prevented, and survival of individuals can be extended. The value of the reproduction number is highly affected by the rate at which individuals join the AIDS TB class. The pie-chart exhibits the distribution of the population in various compartments in the model.

Author Contributions: Conceptualization, N.H.S.; Formal analysis, N.S.; Writing—original draft, Y.S. All authors have read and agreed to the published version of the manuscript.

Funding: The authors thank DST-FIST file #MSI-097 for technical support to the department (only technical support to department).

Acknowledgments: The paper is prepared under the guidance of Nita H. shah.

Conflicts of Interest: The authors declare no conflict of interest.

References

1. Available online: https://www.who.int/news-room/fact-sheets/detail/hiv-aids (accessed on 3 January 2020).
2. Available online: https://www.who.int/tb/areas-of-work/tb-hiv/en/ (accessed on 3 January 2020).
3. Kwan, C.K.; Ernst, J.D. HIV and tuberculosis: A deadly human syndemic. *Clin. Microbiol. Rev.* **2011**, *24*, 351–376. [CrossRef] [PubMed]

4. Available online: https://www.avert.org/professionals/hiv-programming/hiv-tb-coinfection (accessed on 3 January 2020).
5. Kermack, W.O.; McKendrick, A.G. A contribution to the mathematical theory of epidemics. *Proc. R. Soc. Lond. A Math. Phys. Char.* **1927**, *115*, 700–721.
6. Cattani, C.; Ciancio, A. Qualitative analysis of second-order models of tumor—Immune system competition. *Math. Comput. Model.* **2008**, *47*, 1339–1355. [CrossRef]
7. Cattani, C.; Ciancio, A. Hybrid two scales mathematical tools for active particles modelling complex systems with learning hiding dynamics. *Math. Models Methods Appl. Sci.* **2007**, *17*, 171–187. [CrossRef]
8. Kirschner, D. Dynamics of Co-infection with M. tuberculosis and HIV-1. *Theor. Popul. Biol.* **1999**, *55*, 94–109. [CrossRef] [PubMed]
9. Naresh, R.; Tripathi, A. Modelling and analysis of HIV-TB co-infection in a variable size population. *Math. Model. Anal.* **2005**, *10*, 275–286. [CrossRef]
10. Huo, H.F.; Chen, R.; Wang, X.Y. Modelling and stability of HIV/AIDS epidemic model with treatment. *Appl. Math. Model.* **2016**, *40*, 6550–6559. [CrossRef]
11. Bhunu, C.P.; Garira, W.; Mukandavire, Z. Modeling HIV/AIDS and tuberculosis coinfection. *Bull. Math. Biol.* **2009**, *71*, 1745–1780. [CrossRef] [PubMed]
12. Roeger, L.I.W.; Feng, Z.; Castillo-Chavez, C. Modeling TB and HIV co-infections. *Math. Biosci. Eng.* **2009**, *6*, 815–837. [PubMed]
13. Singh, R.; Ali, S.; Jain, M. Epidemic model of HIV/AIDS transmission dynamics with different latent stages based on treatment. *Am. J. Appl. Math.* **2016**, *4*, 222–234. [CrossRef]
14. Silva, C.J.; Torres, D.F. Modeling TB-HIV syndemic and treatment. *J. Appl. Math.* **2014**, *248407*. [CrossRef]
15. Available online: https://aidsinfo.nih.gov/understanding-hiv-aids/fact-sheets/26/90/hiv-and-tuberculosis--tb- (accessed on 3 January 2020).
16. Diekmann, O.; Heesterbeek, J.A.P.; Metz, J.A. On the definition and the computation of the basic reproduction ratio R0 in models for infectious diseases in heterogeneous populations. *J. Math. Biol.* **1990**, *28*, 365–382. [CrossRef] [PubMed]
17. Freedman, H.I.; Ruan, S.; Tang, M. Uniform persistence and flows near a closed positively invariant set. *J. Dyn. Differ. Equ.* **1994**, *6*, 583–600. [CrossRef]
18. Thieme, H.R. Epidemic and demographic interaction in the spread of potentially fatal diseases in growing populations. *Math. Biosci.* **1992**, *111*, 99–130. [CrossRef]
19. Feng, Z.; Castillo-Chavez, C.; Capurro, A.F. A model for tuberculosis with exogenous reinfection. *Theor. Popul. Biol.* **2000**, *57*, 235–247. [CrossRef] [PubMed]
20. Castillo-Chavez, C.; Blower, S.; van den Driessche, P.; Kirschner, D.; Yakubu, A.A. (Eds.) *Mathematical Approaches for Emerging and Reemerging Infectious Diseases: An Dntroduction*; Springer Science & Business Media: Berlin, Germany.
21. LaSalle, J.P. The stability of dynamical systems. *Soc. Ind. Appl. Math.* **1976**, *25*. [CrossRef]
22. Khan, M.A.; Islam, S.; Khan, S.A. Mathematical modeling towards the dynamical interaction of leptospirosis. *Appl. Math. Inf. Sci.* **2014**, *8*, 1049–1056. [CrossRef]

© 2020 by the authors. Licensee MDPI, Basel, Switzerland. This article is an open access article distributed under the terms and conditions of the Creative Commons Attribution (CC BY) license (http://creativecommons.org/licenses/by/4.0/).

Article

On a Harmonic Univalent Subclass of Functions Involving a Generalized Linear Operator

Abdeljabbar Talal Yousef and Zabidin Salleh *

Department of Mathematics, Faculty of Ocean Engineering Technology and Informatics, Universiti Malaysia Terengganu, Kuala Nerus 21030, Terengganu, Malaysia; abduljabaryousef@gmail.com
* Correspondence: zabidin@umt.edu.my

Received: 5 March 2020; Accepted: 21 March 2020; Published: 24 March 2020

Abstract: In this paper, a subclass of complex-valued harmonic univalent functions defined by a generalized linear operator is introduced. Some interesting results such as coefficient bounds, compactness, and other properties of this class are obtained.

Keywords: harmonic univalent functions; generalized linear operator; differential operator; Salagean operator; coefficient bounds

1. Introduction

Let H represent the continuous harmonic functions which are harmonic in the open unit disk $U = \{z : z \in \mathbb{C}, |z| < 1\}$ and let A be a subclass of H which represents the functions which are analytic in U. A harmonic function in U could be expressed as $f = h + \overline{g}$, where h and g are in A, h is the analytic part of f, g is the co-analytic part of f and $|h'(z)| > |g'(z)|$ is a necessary and sufficient condition for f to be locally univalent and sense-preserving in U (see Clunie and Sheil-Small [1]). Now we write,

$$h(z) = z + \sum_{n=2}^{\infty} a_n z^n, \quad g(z) = \sum_{n=2}^{\infty} b_n z^n. \tag{1}$$

Let SH represents the functions of the form $f = h + \overline{g}$ which are harmonic and univalent in U, which normalized by the condition $f(0) = f_z(0) - 1 = 0$. The subclass SH^0 of SH consists of all functions in SH which have the additional property $f_{\overline{z}}(0) = 0$. The class SH was investigated by Clunie and Sheil-Smallas [1]. Since then, many researchers have studied the class SH and even investigated some subclasses of it. Also, we observe that the class SH reduces to the class S of normalized analytic univalent functions in U, if the co-analytic part of f is equal to zero. For $f \in S$, the Salagean differential operator $D^n (n \in \mathbb{N}_0 = \mathbb{N} \cup \{0\})$ was defined by Salagean [2]. For $f = h + \overline{g}$ given by (1), Jahangiri et al. [3] defined the modified Salagean operator of f as

$$D^m f(z) = D^m h(z) + (-1)^m \overline{D^m g(z)},$$

where

$$D^m h(z) = z + \sum_{n=2}^{\infty} n^m a_n z^n, \quad D^m g(z) = \sum_{n=2}^{\infty} n^m b_n z^n.$$

Next, for functions $f \in A$, For $n \in \mathbb{N}_0, \beta \geq \gamma \geq 0$, Yalçın and Altınkaya [4] defined the differential operator of $I^m_{\gamma,\beta} f : SH^0 \to SH^0$. Now we define our differential operator:

$$I^0_{\delta,\mu,\lambda,\eta,\varsigma,\tau}f(z) = h(z) + \overline{g(z)}$$

$$I^1_{\delta,\mu,\lambda,\varsigma,\tau}f(z) = z + \sum_{n=2}^{\infty}\left(\frac{\mu+\lambda-(\delta-\varsigma)(\lambda-\tau)D^0 f(z)+(\delta-\varsigma)(\lambda-\tau)D^1 f(z)}{\mu+\lambda}\right) \quad (2)$$

$$= \frac{\mu+\lambda-(\delta-\varsigma)(\lambda-\tau)\left(h(z)+\overline{g(z)}\right)+(\delta-\varsigma)(\lambda-\tau)\left(zh(z)+\overline{zg'(z)}\right)}{\mu+\lambda}$$

$$I^m_{\delta,\mu,\lambda,\varsigma,\tau}f(z) = I^1_{\delta,\mu,\lambda,\varsigma,\tau}\left(I^{m-1}_{\delta,\mu,\lambda,\varsigma,\tau}f(z)\right). \quad (3)$$

If f is given by (1), then from (2) and (3), we get (see [5])

$$I^m_{\delta,\mu,\lambda,\varsigma,\tau}f(z) = z + \sum_{n=2}^{\infty}\left(\frac{\mu+\lambda+(\delta-\varsigma)(\lambda-\tau)(n-1)}{\mu+\lambda}\right)^m a_n z^n$$

$$+ (-1)^m \sum_{n=2}^{\infty}\left(\frac{\mu+\lambda+(\delta-\varsigma)(\lambda-\tau)(n-1)}{\mu+\lambda}\right)^m \overline{b_n z^n}. \quad (4)$$

The operator $I^m_{\delta,\mu,\lambda,\varsigma,\tau}f(z)$ generalizes the following differential operators:

If $f \in A$, then when we take $\mu = 1$, $\lambda = 0$, $\delta = 0$, $\tau = 1$, $\varsigma = 1$ we obtain $I^m_{0,\tau,\delta,\varsigma}f(z)$ was introduced and studied by Ramadan and Darus [6]. By taking different choices of μ, λ, δ, τ and ς we get $I^m_{1-\lambda,\tau,0,\varsigma}f(z)$ was introduced and studied by Darus and Ibrahim [7], $I^m_{\mu,\lambda,0,1,0}f(z)$ was introduced and studied by Swamy [8], $I^m_{1-\lambda,0,1,0}f(z)$ was introduced and studied by Al-Oboudi [9] and $I^m_{0,0,1,0}f(z)$ was introduced and studied by Salagean [2].

If $f \in H$, then $I^m_{\mu,\lambda,0,1,0}f(z)$ becomes the modified Salagean operator introduced by Yasar and Yalçin [10].

A function $f : U \to C$ is subordinate to the function $g : U \to C$ denoted by $f(z) \prec g(z)$, if there exists an analytic function $w : U \to U$ with $w(0) = 0$ such that

$$f(z) = g(w(z)), (z \in U).$$

Moreover, if the function g is univalent in U, then we have (see [11,12]):

$$f(z) \prec g(z) \text{ if and only if } f(0) = g(0), \ f(U) \subset g(U).$$

Denote by $SH^0(\delta,\mu,\lambda,\varsigma,\tau,m,A,B)$ the subclass of SH^0 consisting of functions of the form (1) that satisfy the condition

$$\frac{I^{m+1}_{\delta,\mu,\lambda,\varsigma,\tau}f(z)}{I^m_{\delta,\mu,\lambda,\varsigma,\tau}f(z)} \prec \frac{1+Az}{1+Bz}, -1 \leq A < B \leq 1 \quad (5)$$

where $I^m_{\delta,\mu,\lambda,\eta,\varsigma,\tau}f(z)$ is defined by (4). For relevant and recent references related to this work, we refer the reader to [13–20].

In this paper we use the same techniques that have been used earlier by Dziok [21] and Dziok et al. [22], to investigate coefficient bound, distortion bounds, and some other properties for the class $SH^0(\delta,\mu,\lambda,\varsigma,\tau,m,A,B)$.

2. Coefficient Bounds

In this section we find the coefficient bound for the class $SH^0(\delta,\mu,\lambda,\varsigma,\tau,m,A,B)$.

Theorem 1. *Let the function $f(z) = h + \overline{g}$ be defined by (1). Then $f \in SH^0(\delta,\mu,\lambda,\varsigma,\tau,m,A,B)$ if*

$$\sum_{n=2}^{\infty}(C_n|a_n| + D_n|b_n|) \leq B - A \quad (6)$$

where

$$C_n = \left(\frac{\mu+\lambda+(\delta-\varsigma)(\lambda-\tau)(n-1)}{\mu+\lambda}\right)^m \left\{\frac{(\delta-\varsigma)(\lambda-\tau)(n-1)[B+1]-(\mu+\lambda)(B-A)}{\mu+\lambda}\right\} \quad (7)$$

and

$$D_n = \left(\frac{\mu+\lambda+(\delta-\varsigma)(\lambda-\tau)(n-1)}{\mu+\lambda}\right)^m \left\{\frac{[A+B(2+(\delta-\varsigma)(\lambda-\tau)(n-1))](\mu+\lambda)}{\mu+\lambda}\right\}. \quad (8)$$

Proof. Let $a_n \neq 0$ or $b_n \neq 0$ for $n \geq 2$. Since $C_n, D_n \geq n(B-A)$ by (6), we obtain

$$|h'(z)| - |g'(z)| \geq 1 - \sum_{n=2}^{\infty} n|a_n||z|^{n-1} - \sum_{n=2}^{\infty} n|b_n||z|^{n-1}$$

$$\geq 1 - |z| \sum_{n=2}^{\infty} (n|a_n| + n|b_n|)$$

$$\geq 1 - \frac{|z|}{B-A} \sum_{n=2}^{\infty} (C_n|a_n| + D_n|b_n|)$$

$$\geq 1 - |z| > 0.$$

Therefore, f is univalent in U. To ensure the univalence condition, consider $z_1, z_2 \in U$ so that $z_1 \neq z_2$. Then

$$\left|\frac{z_1^n - z_2^n}{z_1 - z_2}\right| = \left|\sum_{m=1}^{n} z_1^{m-1} - z_2^{n-m}\right| \leq \sum_{m=1}^{n} |z_1^{m-1}||z_2^{n-m}| < n, \; n \geq 2.$$

So, we have

$$\left|\frac{f(z_1)-f(z_2)}{h(z_1)-h(z_2)}\right| \geq 1 - \left|\frac{g(z_1)-g(z_2)}{h(z_1)-h(z_2)}\right| = 1 - \left|\frac{\sum_{n=2}^{\infty} b_k(z_1^n-z_2^n)}{z_1-z_2+\sum_{n=2}^{\infty} a_n(z_1^n-z_2^n)}\right|$$

$$> 1 - \frac{\sum_{n=2}^{\infty} n|b_n|}{1-\sum_{n=2}^{\infty} n|a_n|} \geq 1 - \frac{\sum_{n=2}^{\infty} \frac{D_n}{B-A}|b_n|}{\sum_{n=2}^{\infty} \frac{C_n}{B-A}|a_n|} \geq 0,$$

which proves univalences.

On the other hand, $f \in SH^0(\delta, \mu, \lambda, \varsigma, \tau, m, A, B)$ if and only if there exists a function w; with $w(0) = 0$, and $|w(z)| < 1 (z \in U)$ such that

$$\frac{I_{\delta,\mu,\lambda,\varsigma,\tau}^{m+1} f(z)}{I_{\delta,\mu,\lambda,\varsigma,\tau}^{m} f(z)} < \frac{1+Az}{1+Bz}$$

or

$$\frac{I_{\delta,\mu,\lambda,\varsigma,\tau}^{m+1} f(z) - I_{\delta,\mu,\lambda,\varsigma,\tau}^{m} f(z)}{BI_{\delta,\mu,\lambda,\varsigma,\tau}^{m+1} f(z) - AI_{\delta,\mu,\lambda,\varsigma,\tau}^{m} f(z)} < 1, \; (z \in U). \quad (9)$$

The above inequality (9) holds, since for $|z| = r$ $(0 < r < 1)$ we obtain

$$\left| I^{m+1}_{\delta,\mu,\lambda,\varsigma,\tau} f(z) - I^m_{\delta,\mu,\lambda,\varsigma,\tau} f(z) \right| - \left| B I^{m+1}_{\delta,\mu,\lambda,\varsigma,\tau} f(z) - A I^m_{\delta,\mu,\lambda,\varsigma,\tau} f(z) \right|$$

$$= \left| \sum_{n=2}^{\infty} \left(\frac{\mu + \lambda + (\delta - \varsigma)(\lambda - \tau)(n-1)}{\mu + \lambda} \right)^m \frac{(\delta - \varsigma)(\lambda - \tau)(n-1)}{\mu + \lambda} a_n z^n \right.$$

$$\left. + (-1)^m \sum_{n=2}^{\infty} \left(\frac{\mu + \lambda + (\delta - \varsigma)(\lambda - \tau)(n-1)}{\mu + \lambda} \right)^m \frac{2(\mu + \lambda) + (\delta - \varsigma)(\lambda - \tau)(n-1)}{\mu + \lambda} \overline{b_n z^n} \right|$$

$$- \left| (B - A)z + \sum_{n=2}^{\infty} \left(\frac{\mu + \lambda + (\delta - \varsigma)(\lambda - \tau)(n-1)}{\mu + \lambda} \right)^m \left(B \frac{\mu + \lambda + (\delta - \varsigma)(\lambda - \tau)(n-1)}{\mu + \lambda} - A \right) a_n z^n \right.$$

$$\left. - (-1)^m \sum_{n=2}^{\infty} \left(\frac{\mu + \lambda + (\delta - \varsigma)(\lambda - \tau)(n-1)}{\mu + \lambda} \right)^m \left(B, \frac{2(\mu + \lambda) + \delta(-\varsigma)(\lambda - \tau)(1-n)}{\mu + \lambda} + A \right) \overline{b_n z^n} \right|$$

$$\leq \sum_{n=2}^{\infty} \left(\frac{\mu + \lambda + (\delta - \varsigma)(\lambda - \tau)(n-1)}{\mu + \lambda} \right)^m \frac{(\delta - \varsigma)(\lambda - \tau)(n-1)}{\mu + \lambda} |a_n| r^n +$$

$$\sum_{n=2}^{\infty} \left(\frac{\mu + \lambda + (\delta - \varsigma)(\lambda - \tau)(n-1)}{\mu + \lambda} \right)^m \frac{2(\mu + \lambda) + (\delta - \varsigma)(\lambda - \tau)(1-n)}{\mu + \lambda} |b_n| r^n - (B - A)r$$

$$+ \sum_{n=2}^{\infty} \left(\frac{\mu + \lambda + (\delta - \varsigma)(\lambda - \tau)(n-1)}{\mu + \lambda} \right)^m \left(B \frac{\mu + \lambda + (\delta - \varsigma)(\lambda - \tau)(n-1) + A}{\mu + \lambda} - A \right) |a_n| r^n$$

$$+ \sum_{n=2}^{\infty} \left(\frac{\mu + \lambda + (\delta - \varsigma)(\lambda - \tau)(n-1)}{\mu + \lambda} \right)^m \left(B \frac{2(\mu + \lambda) + (\delta - \varsigma)(\lambda - \tau)(n-1)}{\mu + \lambda} + A \right) |b_n| r^n$$

$$\leq r \left\{ \sum_{n=2}^{\infty} (C_n |a_n| + D_n |b_n|) r^{n-1} - (B - A) \right\} < 0.$$

Therefore, $f \in SH^0(\delta, \mu, \lambda, \varsigma, \tau, m, A, B)$, and so the proof is completed.

Next we show that the condition (6) is also necessary for the functions $f \in H$ to be in the class $SH^0_T(\delta, \mu, \lambda, \varsigma, \tau, m, A, B) = T^m \cap SH^0(\delta, \mu, \lambda, \varsigma, \tau, m, A, B)$ where T^m is the class of functions $f = h + \overline{g} \in SH^0$ so that

$$f = h + \overline{g} = z - \sum_{n=2}^{\infty} a_n z^n + (-1)^m \sum_{n=2}^{\infty} |b_n| \overline{z^n} \quad (z \in U). \tag{10}$$

□

Theorem 2. Let $f = h + \overline{g}$ be defined by (10). Then $f \in SH^0_T(\delta, \mu, \lambda, \varsigma, \tau, m, A, B)$ if and only if the condition (6) holds.

Proof. For this proof, we let the fractions $\frac{(\delta - \varsigma)(\lambda - \tau)(n-1)}{\mu + \lambda} = L$ and $\frac{2(\mu + \lambda) + (\delta - \varsigma)(\lambda - \tau)(n-1)}{\mu + \lambda} = K$. The first part "if statement" follows from Theorem 1. Conversely, we suppose that $f \in SH^0_T(\delta, \mu, \lambda, \varsigma, \tau, m, A, B)$, then by (9) we have

$$\left| \frac{\sum_{n=2}^{\infty} \left[(L)^m \frac{(\delta - \varsigma)(\lambda - \tau)(n-1)}{\mu + \lambda} |a_n| z^n + (K)^m \frac{2(\mu + \lambda) + (\delta - \varsigma)(\lambda - \tau)(n-1)}{\mu + \lambda} \overline{|b_n| z^n} \right]}{(B - A)z - \sum_{n=2}^{\infty} \left[(L)^m (BL - A) |a_n| z^n + (K)^m (BK + A) |b_n| z^n \right]} \right| < 1.$$

For $|z| = r < 1$, we obtain

$$\frac{\sum_{n=2}^{\infty}\left[(L)^m \frac{(\delta-\varsigma)(\lambda-\tau)(n-1)}{\mu+\lambda}|a_n| + (K)^m \frac{2(\mu+\lambda)+(\delta-\varsigma)(\lambda-\tau)(n-1)}{\mu+\lambda}\overline{|b_n|}\right]r^{n-1}}{(B-A) - \sum_{n=2}^{\infty}\left[(L)^m(BL-A)|a_n| + (K)^m(BK+A)|b_n|\right]r^{n-1}} < 1.$$

Thus, for C_n and D_n as defined by (7) and (8), we have

$$\sum_{n=2}^{\infty}[C_n|a_n| + D_n|b_n|]r^{n-1} < B - A (0 \leq r < 1). \tag{11}$$

Let $\{\rho_n\}$ be the sequence of partial sums of the series

$$\sum_{k=2}^{n}[C_k|a_k| + D_k|b_k|].$$

Then $\{\rho_n\}$ is a non-decreasing sequence and by (11) it is bounded above by $B - A$. Thus, it is convergent and

$$\sum_{n=2}^{\infty}[C_n|a_n| + D_n|b_n|] = \lim_{n \to +\infty} \rho_n \leq B - A.$$

This gives us the condition (6). □

3. Compactness and Convex

In this section we obtain the compactness and the convex relation for the class $SH^0(\delta, \mu, \lambda, \varsigma, \tau, m, A, B)$.

Theorem 3. *The class $SH^0_T(\delta, \mu, \lambda, \varsigma, \tau, m, A, B)$ is convex and compact subset of SH.*

Proof. Let $f_t \in SH^0_T(\delta, \mu, \lambda, \varsigma, \tau, m, A, B)$, where

$$f_t(z) = z - \sum_{n=2}^{\infty}|a_{t,n}|z^n + (-1)^m \sum_{n=2}^{\infty}|b_{t,n}|\overline{z^n} (z \in U, t \in \mathbb{N}). \tag{12}$$

Then for $0 \leq \psi \leq 1$, let $f_1, f_2 \in SH^0_T(\delta, \mu, \lambda, \varsigma, \tau, m, A, B)$ be defined by (12). Then

$$\xi(z) = \psi f_1(z) + (1-\psi)f_2(z)$$
$$= z - \sum_{n=2}^{\infty}\left(\psi|a_{1,n}| + (1-\psi)|a_{2,n}|\right)z^n + (-1)^m \sum_{n=2}^{\infty}\left(\psi|b_{1,n}| + (1-\psi)|b_{2,n}|\right)\overline{z^n}$$

and

$$\sum_{n=2}^{\infty}\left\{C_n\left(\psi|a_{1,n}| + (1-\psi)|a_{2,n}|\right) + D_n\left(\psi|b_{1,n}| + (1-\psi)|b_{2,n}|\right)\right\}$$
$$= \psi \sum_{n=2}^{\infty}\left\{C_n|a_{1,n}| + D_n|b_{1,n}|\right\} + (1-\psi)\sum_{n=2}^{\infty}\left\{C_n|a_{2,n}| + D_n|b_{2,n}|\right\}$$
$$\leq \psi(B-A) + (1-\psi)(B-A) = B - A.$$

Thus, the function $\xi = \psi f_1(z) + (1-\psi)f_2(z)$ is in the class $SH^0_T(\delta, \mu, \lambda, \varsigma, \tau, m, A, B)$. This implies that $SH^0_T(\delta, \mu, \lambda, \varsigma, \tau, m, A, B)$ is convex.

For $f_t \in SH_T^0(\delta, \mu, \lambda, \varsigma, \tau, m, A, B), t \in \mathbb{N}$ and $|z| \le r \; (0 < r < 1)$, then we have

$$|f_t(z)| \le r + \sum_{n=2}^{\infty}\{|a_{t,n}| + |b_{t,n}|\}r^n$$

$$\le r + \sum_{n=2}^{\infty}\{C_n|a_{t,n}| + D_n|b_{t,n}|\}r^n$$

$$\le r + (B - A)r^2.$$

Therefore, $SH_T^0(\delta, \mu, \lambda, \varsigma, \tau, m, A, B)$ is uniformly bounded. Let

$$f_t(z) = z - \sum_{n=2}^{\infty}|a_{t,n}|z^n + (-1)^m \sum_{n=2}^{\infty}|b_{t,n}|\overline{z^n} \; (z \in U, \; t \in \mathbb{N}).$$

also, let $f = h + \overline{g}$ where h and g are given by (1). Then by Theorem 2 we get

$$\sum_{n=2}^{\infty}\{C_n|a_n| + D_n|b_{t,n}|\} \le B - A. \tag{13}$$

If we assume $f_t \to f$, then we get that $|a_{t,n}| \to |a_n|$ and $|b_{t,n}| \to |b_n|$ as $n \to +\infty$ ($t \in \mathbb{N}$). Let $\{\rho_n\}$ be the sequence of partial sums of the series $\sum_{n=2}^{\infty}\{C_n|a_{t,n}| + D_n|b_{t,n}|\}$. Then $\{\rho_n\}$ is a non-decreasing sequence and by (13) it is bounded above by $B - A$. Thus, it is convergent and

$$\sum_{n=2}^{\infty}\{C_n|a_{t,n}| + D_n|b_{t,n}|\} = \lim_{n\to\infty}\rho_n \le B - A.$$

Therefore, $f \in SH_T^0(\delta, \mu, \lambda, \varsigma, \tau, m, A, B)$ and therefore the class $SH_T^0(\delta, \mu, \lambda, \varsigma, \tau, m, A, B)$ is closed. As a result, the class is closed, and the class $SH_T^0(\delta, \mu, \lambda, \varsigma, \tau, m, A, B)$ is also compact subset of SH, which completes the proof. □

Lemma 1 [23]. Let $f = h + \overline{g}$ be so that h and g are given by (1). Furthermore, let

$$\sum_{n=2}^{\infty}\left\{\frac{n-\alpha}{1-\alpha}|a_n| + \frac{n+\alpha}{1-\alpha}|b_n|\right\} \le 1 \; (z \in U)$$

where $0 \le \alpha < 1$. Then f is harmonic, orientation preserving, univalent in U and f is starlike of order α.

Theorem 4. Let $0 \le \alpha < 1$, C_n and D_n be defined by (7) and (8). Then

$$r_\alpha^*\left(SH_T^0(\delta, \mu, \lambda, \varsigma, \tau, n, A, B)\right) = \inf_{n\ge 2}\left[\frac{1-\alpha}{B-A}\min\left\{\frac{C_n}{n+\alpha}, \frac{D_n}{n+\alpha}\right\}\right]^{\frac{1}{n-1}}, \tag{14}$$

where r_α^* is the radius of starlikeness of order α.

Proof. Let $f \in SH_T^0(\delta, \mu, \lambda, \varsigma, \tau, m, A, B)$ be of the form (10). Then, for $|z| = r < 1$, we get

$$\left| \frac{I_{0,\eta} f(z) - (1+\alpha) f(z)}{I_{0,\eta} f(z) + (1+\alpha) f(z)} \right|$$

$$= \left| \frac{-\alpha z - \sum_{n=2}^{\infty}(n-1-\alpha)|a_n|z^n - (-1)^m \sum_{n=2}^{\infty}(n+1+\alpha)|b_n|\overline{z^n}}{(2-\alpha)z - \sum_{n=2}^{\infty}(n-1-\alpha)|a_n|z^n - (-1)^m \sum_{n=2}^{\infty}(n-1+\alpha)|b_n|\overline{z^n}} \right|$$

$$\leq \frac{\alpha - \sum_{n=2}^{\infty}\left\{(n-1-\alpha)|a_n| - (-1)^m \sum_{n=2}^{\infty}(n+1+\alpha)|b_n|\right\} r^{n-1}}{2 - \alpha - \sum_{n=2}^{\infty}\left\{(n-1-\alpha)|a_n| - (-1)^m \sum_{n=2}^{\infty}(n-1+\alpha)|b_n|\right\}}.$$

By using Lemma 1, we observe that f is starlike of order α in U_r if and only if

$$\left| \frac{I_{0,\eta} f(z) - (1+\alpha) f(z)}{I_{0,\eta} f(z) + (1+\alpha) f(z)} \right| < 1, z \in U_r$$

or

$$\sum_{n=2}^{\infty} \left\{ \frac{n-\alpha}{1-\alpha}|a_n| + \frac{n+\alpha}{1-\alpha}|b_n| \right\} r^{n-1} \leq 1. \tag{15}$$

Furthermore, by using Theorem 2, we get

$$\sum_{n=2}^{\infty} \left\{ \frac{C_n}{1-\alpha}|a_n| + \frac{D_n}{1-\alpha}|b_n| \right\} r^{n-1} \leq 1.$$

Condition (15) is true if

$$\frac{n-\alpha}{1-\alpha} r^{n-1} \leq \frac{C_n}{B-A} r^{n-1}.$$

This proves

$$\frac{n+\alpha}{1-\alpha} r^{n-1} \leq \frac{D_n}{B-A} r^{n-1} (n = 2, 3 \ldots).$$

So, the function f is starlike of order α in the disk $U_{r_\alpha^*}$ where

$$r_\alpha^* = \inf_{n \geq 2} \left[\frac{1-\alpha}{B-A} \min \left\{ \frac{C_n}{n+\alpha}, \frac{D_n}{n+\alpha} \right\} \right]^{\frac{1}{n-1}},$$

and the function

$$f_n(z) = h_n(z) + \overline{g_n(z)} = z - \frac{B-A}{C_n} z^n + (-1)^m \frac{B-A}{D_n} \overline{z^n}.$$

So, the radius r_α^* cannot be larger. Then we get (14). □

4. Extreme Points

In this section we find the extreme points for the class $SH^0(\delta, \mu, \lambda, \varsigma, \tau, m, A, B)$.

Theorem 5. *The extreme points of $SH_T^0(\delta, \mu, \lambda, \varsigma, \tau, m, A, B)$ are the functions f of the form (1) where $h = h_k$ and $g = g_k$ are of the form*

$$h_1(z) = z,$$

$$h_n(z) = z - \frac{B-A}{C_n} z^n, \tag{16}$$

$$g_n(z) = (-1)^m \frac{B-A}{D_n} \overline{z^n}, (z \in U, n \geq 2).$$

Proof. Suppose that $g_n = \psi f_1 + (1-\psi) f_2$ where $0 < \psi < 1$ and $f_1, f_2 \in SH_T^0(\delta, \mu, \lambda, \varsigma, \tau, m, A, B)$ are written in the form

$$f_t(z) = z - \sum_{n=2}^{\infty} |a_{t,n}| z^n + (-1)^m \sum_{n=2}^{\infty} |b_{t,n}| \overline{z^n} \quad (z \in U, \; t \in \{1,2\}).$$

Then, by (16), we get

$$|b_{1,n}| = |b_{2,n}| = \frac{B-A}{D_n},$$

and $a_{1,t} = a_{2,t} = 0$ for $t \in \{2,3\ldots\}$ and $b_{1,t} = b_{2,t} = 0$ for $t \in \{2,3\ldots\} \setminus \{n\}$. It follows that $g_n(z) = f_1(z) = f_2(z)$ and g_n are in the class of extreme points of the class $SH_T^0(\delta, \mu, \lambda, \varsigma, \tau, m, A, B)$. We also can ensure that the functions $h_n(z)$ are the extreme points of the class $SH_T^0(\delta, \mu, \lambda, \varsigma, \tau, m, A, B)$. Now, assume that a function f of the form (1) is in the class of the extreme points of the class $SH_T^0(\delta, \mu, \lambda, \varsigma, \tau, m, A, B)$ and f is not of the form (16). Then there exists $k \in \{2, 3\ldots\}$ such that

$$0 < |a_k| < \frac{B-A}{\left(\frac{\mu+\lambda+(\delta-\varsigma)(\lambda-\tau)(k-1)}{\mu+\lambda}\right)^m \left\{\frac{(\delta-\varsigma)(\lambda-\tau)[(k-1)(B+1)] + (\mu+\lambda)(B-A)}{\mu+\lambda}\right\}}$$

or

$$0 < |b_k| < \frac{B-A}{\left(\frac{\mu+\lambda-(\delta-\varsigma)(\lambda-\tau)(n+1)}{\mu+\lambda}\right)^m \left\{\frac{[A+B(2+(\delta-\varsigma)(\lambda-\tau)(n-1))](\mu+\lambda)}{\mu+\lambda}\right\}}.$$

If

$$0 < |a_k| < \frac{B-A}{\left(\frac{\mu+\lambda+(\delta-\varsigma)(\lambda-\tau)(n-1)}{\mu+\lambda}\right)^m \left\{\frac{(\delta-\varsigma)(\lambda-\tau)(n-1)[B+1] - (\mu+\lambda)(B-A)}{\mu+\lambda}\right\}}$$

then putting

$$\psi = \frac{|a_k| \left[\left(\frac{\mu+\lambda+(\delta-\varsigma)(\lambda-\tau)(n-1)}{\mu+\lambda}\right)^m \left\{\frac{(\delta-\varsigma)(\lambda-\tau)(n-1)[B+1] - (\mu+\lambda)(B-A)}{\mu+\lambda}\right\}\right]}{B-A}$$

and

$$\chi = \frac{f - \psi h_k}{1-\psi},$$

we have $0 < \psi < 1$, $h_k \neq \chi$. Therefore, f is not in the class of the extreme points of the class $SH_T^0(\delta, \mu, \lambda, \eta, \varsigma, \tau, m, A, B)$. Similarly, if

$$0 < |b_k| < \frac{B-A}{\left(\frac{\mu+\lambda+(\delta-\varsigma)(\lambda-\tau)(n-1)}{\mu+\lambda}\right)^m \left\{\frac{[A+B(2+(\delta-\varsigma)(\lambda-\tau)(n-1))](\mu+\lambda)}{\mu+\lambda}\right\}}$$

then putting

$$\psi = \frac{|b_k| \left(\frac{\mu+\lambda+(\delta-\varsigma)(\lambda-\tau)(n-1)}{\mu+\lambda}\right)^m \left\{\frac{[A+B(2+(\delta-\varsigma)(\lambda-\tau)(n-1))](\mu+\lambda)}{\mu+\lambda}\right\}}{B-A}$$

and

$$\chi = \frac{f - \psi g_k}{1-\psi},$$

we have $0 < \psi < 1$, $g_k \neq \chi$. It follows that f is not in the family of extreme points of the class $SH_T^0(\delta, \mu, \lambda, \varsigma, \tau, m, A, B)$ and so the proof is completed. □

Author Contributions: Conceptualization, A.T.Y. and Z.S.; methodology, A.T.Y.; software, A.T.Y.; validation, A.T.Y. and Z.S.; formal analysis, A.T.Y.; investigation, A.T.Y.; resources, A.T.Y.; data curation, A.T.Y.; writing—original draft preparation, A.T.Y.; writing—review and editing, Z.S.; visualization, Z.S.; supervision, Z.S.; project administration, Z.S.; funding acquisition, Z.S. All authors have read and agreed to the published version of the manuscript.

Funding: This research received no external funding.

Conflicts of Interest: The authors declare no conflict of interest.

References

1. Clunie, J.; Sheil-Small, T. Harmonic Univalent Functions. *Ann. Acad. Sci. Fenn. Ser. A I Math.* **1984**, *9*, 3–25. [CrossRef]
2. Salagean, G.S. Subclasses of univalent functions. In *Complex Analysis - Fifth Roumanian-Finnish Seminar*; Cazanu, C.A., Jurchescu, M., Suciu, I., Eds.; Lectures Notes in Mathematics; Springer Nature: Cham, Switzerland, 1983; Volume 1013, pp. 362–372.
3. Jahangiri, J.; Magesh, N.; Murugesan, C. Certain Subclasses of Starlike Harmonic Functions Defined by Subordination. *J. Frac. Cal. App.* **2017**, *8*, 88–100.
4. Yalçın, S.; Altınkaya, Ş. On a Subclass of Harmonic Univalent Functions involving a Linear Operator. *AIP Conference Proceedings Vol. 1926, no. 1, p. 020045, Budapest, Hungary, 15–18 August 2017*; Tosun, M., Ersoy, S., Ilarslan, K., Eds.; AIP Publishing LLC: New York, NY, USA, 2018.
5. Bayram, H.; Yalçın, S. A Subclass of Harmonic Univalent Functions Defined by a Linear Operator. *Palestine J. Math.* **2017**, *6*, 320–326.
6. Ramadan, S.F.; Darus, M. On the Fekete-Szegö Inequality for a Class of Analytic Functions Defined by using Generalized Differential Operator. *Acta Universitatis Apulensis* **2011**, *26*, 167–178.
7. Darus, M.; Ibrahim, R.W. On Subclasses for Generalized Operators of Complex Order. *Far East J. Math. Sci.* **2009**, *33*, 299–308.
8. Swamy, S.R. Inclusion Properties of Certain Subclasses of Analytic Functions. *Int. Math. Forum* **2012**, *7*, 1751–1760.
9. Al-Oboudi, F.M. On Univalent Functions Defined by a Generalized Sălăgean Operator. *Int. J. Math. Math. Sci.* **2004**, *27*, 1429–1436. [CrossRef]
10. Yasar, E.; Yalçın, S. Generalized Salagean-type Harmonic Univalent Functions. *Stud. Univ. Babes-Bolyai Math.* **2012**, *57*, 395–403.
11. Yousef, F.; Frasin, B.A.; Al-Hawary, T. Fekete-Szegö Inequality for Analytic and Bi-univalent Functions Subordinate to Chebyshev Polynomials. *Filomat* **2018**, *32*, 3229–3236. [CrossRef]
12. Yousef, F.; Alroud, S.; Illafe, M. A Comprehensive Subclass of Bi-univalent Functions Associated with Chebyshev Polynomials of the Second Kind. *Bol. Soc. Mat. Mex.* **2019**, 1–11. [CrossRef]
13. Al-Hawary, T.; Frasin, B.A.; Yousef, F. Coefficients Estimates for Certain Classes of Analytic Functions of Complex Order. *Afr. Mat.* **2018**, *29*, 1265–1271. [CrossRef]
14. Amourah, A.A.; Yousef, F. Some Properties of a Class of Analytic Functions Involving a New Generalized Differential Operator. *Bol. Soc. Paran. Mat.* **2020**, *38*, 33–42. [CrossRef]
15. Frasin, B.A.; Al-Hawary, T.; Yousef, F. Necessary and Sufficient Conditions for Hypergeometric Functions to be in a Subclass of Analytic Functions. *Afr. Mat.* **2019**, *30*, 223–230. [CrossRef]
16. Klén, R.; Manojlović, V.; Simić, S.; Vuorinen, M. Bernoulli Inequality and Hypergeometric Functions. *Proc. Amer. Math. Soc.* **2014**, *142*, 559–573. [CrossRef]
17. Malkowsky, E.; Rakočević, V. *Advanced Functional Analysis*; CRC Press, Taylor and Francis Group: Boca Raton, FL, USA, 2019.
18. Pavlović, M. *Function Classes on the Unit Disc: An Introduction, Vol. 52*; Walter de Gruyter GmbH & Co KG: Berlin, Germany, 2019.
19. Silverman, H. Univalent Functions with Negative Coefficients. *Proc. Amer. Math. Soc.* **1975**, *51*, 109–116. [CrossRef]
20. Todorčević, V. *Harmonic Quasiconformal Mappings and Hyperbolic Type Metrics*; Springer Nature AG: Cham, Switzerland, 2019.
21. Dziok, J. On Janowski Harmonic Functions. *J. App. Ana.* **2015**, *21*, 99–107. [CrossRef]

22. Dziok, J.; Jahangiri, J.; Silverman, H. Harmonic Functions with Varying Coefficients. *J. Ineq. App.* **2016**, *2016*, 139. [CrossRef]
23. Jahangiri, J.M. Harmonic Functions Starlike in the Unit Disk. *J. Math. Ana. App.* **1999**, *235*, 470–477. [CrossRef]

© 2020 by the authors. Licensee MDPI, Basel, Switzerland. This article is an open access article distributed under the terms and conditions of the Creative Commons Attribution (CC BY) license (http://creativecommons.org/licenses/by/4.0/).

Article

Coincidence Continuation Theory for Multivalued Maps with Selections in a Given Class

Donal O'Regan

School of Mathematics, Statistics and Applied Mathematics, National University of Ireland, H91 TK33 Galway, Ireland; donal.oregan@nuigalway.ie

Received: 20 March 2020; Accepted: 3 April 2020; Published: 10 April 2020

Abstract: This paper considers the topological transversality theorem for general multivalued maps which have selections in a given class of maps.

Keywords: essential maps; coincidence points; topological principles; selections

1. Introduction

To motivate this study first fix a map Φ (an important case is when Φ is the identity). Many coincidence problems between a map F and Φ (i.e., finding a (coincidence) point x with $F(x) \cap \Phi(x) \neq \emptyset$) arise naturally in applications. For a complicated map F the idea here is to try to relate it to a simpler and solvable coincidence problem between a map G and Φ (i.e., we assume we have a (coincidence) point y with $G(y) \cap \Phi(y) \neq \emptyset$) where the map G is homotopic (in an appropriate way) to F and from this we hope to deduce that there is a coincidence point between F and Φ (i.e., we hope to deduce that there is a (coincidence) point x with $F(x) \cap \Phi(x) \neq \emptyset$). To achieve this we consider general (instead of specific) classes of maps and we present the notion of homotopy for this class of maps which are coincidence free on the boundary of the set considered. In particular, in this paper, we look at multivalued maps F and G with selections in a given class of maps and with $F \cong G$ in this setting. The topological transversality theorem in this setting will state that F is Φ–essential if and only if G is Φ–essential (essential maps were introduced in [1] and extended by many authors in [2–5]). In this paper we discuss the topological transversality theorem in a very general setting using a simple and effective approach. In this paper, we consider a generalization of Φ–essential maps, namely the d–Φ–essential maps.

2. Topological Transversality Theorems

A multivalued map G from a space X to a space Y is a correspondence which associates to every $x \in X$ a subset $G(x) \subseteq Y$. In this paper let E be a completely regular topological space and U an open subset of E.

We will consider classes **A**, **B** and **D** of maps.

Definition 1. We say $F \in D(\overline{U}, E)$ (respectively $F \in B(\overline{U}, E)$) if $F : \overline{U} \to 2^E$ and $F \in \mathbf{D}(\overline{U}, E)$ (respectively $F \in \mathbf{B}(\overline{U}, E)$); here 2^E denotes the family of nonempty subsets of E and $\overline{U}$ denotes the closure of U in E.

In this paper we use bold face only to indicate the properties of our maps and usually $D = \mathbf{D}$ etc. Examples of $F \in \mathbf{D}(\overline{U}, E)$ might be that $F : \overline{U} \to K(E)$ is an upper semicontinuous compact map and F has convex values or $F : \overline{U} \to K(E)$ is an upper semicontinuous compact map and F has acyclic values; here $K(E)$ denotes the family of nonempty compact subsets of E.

Definition 2. We say $F \in A(\overline{U}, E)$ if $F : \overline{U} \to 2^E$ and $F \in \mathbf{A}(\overline{U}, E)$ and there exists a selection $\Psi \in D(\overline{U}, E)$ of F.

Remark 1. Let Z and W be subsets of Hausdorff topological vector spaces Y_1 and Y_2 and F a multifunction. We say $F \in PK(Z, W)$ if W is convex and there exists a map $S: Z \to W$ with $Z = \cup \{int\, S^{-1}(w): w \in W\}$, $co\,(S(x)) \subseteq F(x)$ for $x \in Z$ and $S(x) \neq \emptyset$ for each $x \in Z$; here $S^{-1}(w) = \{z: w \in S(z)\}$, int denotes the interior and co denotes the convex hull. Let E be a Hausdorff topological vector space (note topological vector spaces are completely regular), U an open subset of E and $\overline{U}$ paracompact. In this case we say $F \in \mathbf{A}(\overline{U}, E)$ if $F \in PK(\overline{U}, E)$ is a compact map, and we say $\Psi \in \mathbf{D}(\overline{U}, E)$ if Ψ is a single valued, continuous, compact map. Now [6] guarantees that there exists a continuous, compact selection $f: \overline{U} \to E$ of F.

In this section we fix a $\Phi \in B(\overline{U}, E)$ and now we present the notion of coincidence free on the boundary, Φ–essentiality and homotopy.

Definition 3. We say $F \in A_{\partial U}(\overline{U}, E)$ (respectively $F \in D_{\partial U}(\overline{U}, E)$) if $F \in A(\overline{U}, E)$ (respectively $F \in D(\overline{U}, E)$) with $F(x) \cap \Phi(x) = \emptyset$ for $x \in \partial U$; here ∂U denotes the boundary of U in E.

Definition 4. We say $F \in A_{\partial U}(\overline{U}, E)$ is Φ–essential in $A_{\partial U}(\overline{U}, E)$ if for any selection $\Psi \in D(\overline{U}, E)$ of F and any map $J \in D_{\partial U}(\overline{U}, E)$ with $J|_{\partial U} = \Psi|_{\partial U}$ there exists a $x \in U$ with $J(x) \cap \Phi(x) \neq \emptyset$.

Remark 2. If $F \in A_{\partial U}(\overline{U}, E)$ is Φ–essential in $A_{\partial U}(\overline{U}, E)$ and if $\Psi \in D(\overline{U}, E)$ is any selection of F then there exists an $x \in U$ with $\Psi(x) \cap \Phi(x) \neq \emptyset$ (take $J = \Psi$ in Definition 4), and $\emptyset \neq \Psi(x) \cap \Phi(x) \subseteq F(x) \cap \Phi(x)$.

Definition 5. Let E be a completely regular (respectively, normal) topological space and let $\Psi, \Lambda \in D_{\partial U}(\overline{U}, E)$. We say Ψ is homotopic to Λ in the class $D_{\partial U}(\overline{U}, E)$ and we write $\Psi \cong \Lambda$ in $D_{\partial U}(\overline{U}, E)$ if there exists a map $H: \overline{U} \times [0, 1] \to 2^E$ with $H(., \eta(.)) \in \mathbf{D}(\overline{U}, E)$ for any continuous function $\eta: \overline{U} \to [0, 1]$ with $\eta(\partial U) = 0$, $\Phi(x) \cap H_t(x) = \emptyset$ for any $x \in \partial U$ and $t \in (0, 1)$, $\{x \in \overline{U}: \Phi(x) \cap H(x, t) \neq \emptyset$ for some $t \in [0, 1]\}$ is compact (respectively, closed), $H_0 = \Psi$ and $H_1 = \Lambda$ (here $H_t(x) = H(x, t)$).

Remark 3. It is of interest to note that in our results below alternatively we could use the following definition for $\cong$ in $D_{\partial U}(\overline{U}, E)$: $\Psi \cong \Lambda$ in $D_{\partial U}(\overline{U}, E)$ if there exists a map $H: \overline{U} \times [0, 1] \to 2^E$ with $H \in \mathbf{D}(\overline{U} \times [0, 1], E)$, $\Phi(x) \cap H_t(x) = \emptyset$ for any $x \in \partial U$ and $t \in (0, 1)$ (here $H_t(x) = H(x, t)$), $\{x \in \overline{U}: \Phi(x) \cap H(x, t) \neq \emptyset$ for some $t \in [0, 1]\}$ is compact (respectively, closed), $H_0 = \Psi$ and $H_1 = \Lambda$. Note here if we use this definition then we will also assume for any map $\Theta \in \mathbf{D}(\overline{U} \times [0, 1], E)$ and any map $f \in \mathbf{C}(\overline{U}, \overline{U} \times [0, 1])$ then $\Theta \circ f \in \mathbf{D}(\overline{U}, E)$; here $\mathbf{C}$ denotes the class of single valued continuous functions.

Now we are in a position to define homotopy ($\cong$) in our class $A_{\partial U}(\overline{U}, E)$.

Definition 6. Let $F, G \in A_{\partial U}(\overline{U}, E)$. We say F is homotopic to G in the class $A_{\partial U}(\overline{U}, E)$ and we write $F \cong G$ in $A_{\partial U}(\overline{U}, E)$ if for any selection $\Psi \in D_{\partial U}(\overline{U}, E)$ (respectively, $\Lambda \in D_{\partial U}(\overline{U}, E)$) of F (respectively, of G) we have $\Psi \cong \Lambda$ in $D_{\partial U}(\overline{U}, E)$.

Next, we present a simple and crucial result that will immediately yield the topological transversality theorem in this setting.

Theorem 1. Let E be a completely regular (respectively, normal) topological space, U an open subset of E, $F \in A_{\partial U}(\overline{U}, E)$ and $G \in A_{\partial U}(\overline{U}, E)$ is Φ–essential in $A_{\partial U}(\overline{U}, E)$. Suppose also

$$\begin{cases} \text{for any selection } \Psi \in D_{\partial U}(\overline{U}, E) \text{ (respectively, } \Lambda \in D_{\partial U}(\overline{U}, E)) \\ \text{of } F \text{ (respectively, of } G) \text{ and any map } J \in D_{\partial U}(\overline{U}, E) \\ \text{with } J|_{\partial U} = \Psi|_{\partial U} \text{ we have } \Lambda \cong J \text{ in } D_{\partial U}(\overline{U}, E). \end{cases} \quad (1)$$

Then F is Φ–essential in $A_{\partial U}(\overline{U}, E)$.

Proof. Let $\Psi \in D_{\partial U}(\overline{U}, E)$ be any selection of F and consider any map $J \in D_{\partial U}(\overline{U}, E)$ with $J|_{\partial U} = \Psi|_{\partial U}$. It remains to show that there exists an $x \in U$ with $J(x) \cap \Phi(x) \neq \emptyset$. Let $\Lambda \in D_{\partial U}(\overline{U}, E)$ be any selection of G. Now (1) guarantees that there exists a map $H : \overline{U} \times [0,1] \to 2^E$ with $H(.,\eta(.)) \in D(\overline{U}, E)$ for any continuous function $\eta : \overline{U} \to [0,1]$ with $\eta(\partial U) = 0$, $\Phi(x) \cap H_t(x) = \emptyset$ for any $x \in \partial U$ and $t \in (0,1)$, $\{x \in \overline{U} : \Phi(x) \cap H(x,t) \neq \emptyset \text{ for some } t \in [0,1]\}$ is compact (respectively, closed), $H_0 = \Lambda$, and $H_1 = J$ (here $H_t(x) = H(x,t)$). Let

$$\Omega = \{x \in \overline{U} : \Phi(x) \cap H(x,t) \neq \emptyset \text{ for some } t \in [0,1]\}.$$

Now since G is Φ–essential in $A_{\partial U}(\overline{U}, E)$ then Remark 2 (note $H_0 = \Lambda$) guarantees that $\Omega \neq \emptyset$. Ω is compact (respectively, closed) if E is a completely regular (respectively, normal) topological space. Next note $\Omega \cap \partial U = \emptyset$ and now we can deduce that there exists a continuous map (called a Urysohn map) $\mu : \overline{U} \to [0,1]$ with $\mu(\partial U) = 0$ and $\mu(\Omega) = 1$. Define a map R by $R(x) = H(x, \mu(x))$ for $x \in \overline{U}$. Note $R \in D_{\partial U}(\overline{U}, E)$ with $R|_{\partial U} = H_0|_{\partial U} = \Lambda|_{\partial U}$. Now since G is Φ–essential in $A_{\partial U}(\overline{U}, E)$ then there exists $x \in U$ with $R(x) \cap \Phi(x) \neq \emptyset$ (i.e., $H_{\mu(x)}(x) \cap \Phi(x) \neq \emptyset$) and so $x \in \Omega$. As a result $\mu(x) = 1$ so $\emptyset \neq H_1(x) \cap \Phi(x) = J(x) \cap \Phi(x)$, and we are finished. □

Now assume

$$\cong \text{ in } D_{\partial U}(\overline{U}, E) \text{ is an equivalence relation} \tag{2}$$

and

$$\begin{cases} \text{if } F \in A_{\partial U}(\overline{U}, E) \text{ and if } \Psi \in D_{\partial U}(\overline{U}, E) \text{ is any} \\ \text{selection of } F \text{ and } J \in D_{\partial U}(\overline{U}, E) \text{ is any map} \\ \text{with } \Psi|_{\partial U} = J|_{\partial U} \text{ then } \Psi \cong J \text{ in } D_{\partial U}(\overline{U}, E). \end{cases} \tag{3}$$

Theorem 2. *Let E be a completely regular (respectively, normal) topological space, U an open subset of E, and assume (2) and (3) hold. Suppose F and G are two maps in $A_{\partial U}(\overline{U}, E)$ with $F \cong G$ in $A_{\partial U}(\overline{U}, E)$. Now F is Φ–essential in $A_{\partial U}(\overline{U}, E)$ if and only if G is Φ–essential in $A_{\partial U}(\overline{U}, E)$.*

Proof. Assume G is Φ–essential in $A_{\partial U}(\overline{U}, E)$. We use Theorem 1 to show F is Φ–essential in $A_{\partial U}(\overline{U}, E)$. Let $\Psi \in D_{\partial U}(\overline{U}, E)$ be any selection of F, $\Lambda \in D_{\partial U}(\overline{U}, E)$ be any selection of G and consider any map $J \in D_{\partial U}(\overline{U}, E)$ with $J|_{\partial U} = \Psi|_{\partial U}$. Now (3) guarantees that $\Psi \cong J$ in $D_{\partial U}(\overline{U}, E)$ and this together with $F \cong G$ in $A_{\partial U}(\overline{U}, E)$ (so $\Psi \cong \Lambda$ in $D_{\partial U}(\overline{U}, E)$) and (2) guarantees that $\Lambda \cong J$ in $D_{\partial U}(\overline{U}, E)$. Thus (1) holds so Theorem 1 guarantees that F is Φ–essential in $A_{\partial U}(\overline{U}, E)$. A similar argument shows if F is Φ–essential in $A_{\partial U}(\overline{U}, E)$ then G is Φ–essential in $A_{\partial U}(\overline{U}, E)$. □

Now we consider a generalization of Φ–essential maps, namely the d–Φ–essential maps (these maps were motivated from the notion of the degree of a map). Let E be a completely regular topological space and U an open subset of E. For any map $\Psi \in D(\overline{U}, E)$ let $\Psi^\star = I \times \Psi : \overline{U} \to 2^{\overline{U} \times E}$, with $I : \overline{U} \to \overline{U}$ given by $I(x) = x$, and let

$$d : \left\{ (\Psi^\star)^{-1}(B) \right\} \cup \{\emptyset\} \to K \tag{4}$$

be any map with values in the nonempty set K; here $B = \{(x, \Phi(x)) : x \in \overline{U}\}$.

Next we present the notions of d–Φ–essentiality and homotopy.

Definition 7. *Let $F \in A_{\partial U}(\overline{U}, E)$ and write $F^\star = I \times F$. We say $F^\star : \overline{U} \to 2^{\overline{U} \times E}$ is d–Φ–essential if for any selection $\Psi \in D(\overline{U}, E)$ of F and any map $J \in D_{\partial U}(\overline{U}, E)$ with $J|_{\partial U} = \Psi|_{\partial U}$ we have that $d\left((\Psi^\star)^{-1}(B)\right) = d\left((J^\star)^{-1}(B)\right) \neq d(\emptyset)$; here $\Psi^\star = I \times \Psi$ and $J^\star = I \times J$.*

Remark 4. *If $F^\star$ is d–Φ–essential then for any selection $\Psi \in D(\overline{U}, E)$ of F (with $\Psi^\star = I \times \Psi$) we have*

$$\emptyset \neq (\Psi^\star)^{-1}(B) = \{x \in \overline{U} : (x, \Psi(x)) \cap (x, \Phi(x)) \neq \emptyset\},$$

so there exists a $x \in U$ with $(x, \Psi(x)) \cap (x, \Phi(x)) \neq \emptyset$ (i.e., $\Phi(x) \cap \Psi(x) \neq \emptyset$ so in particular $\Phi(x) \cap F(x) \neq \emptyset$).

Now we define homotopy in this setting for our class $D_{\partial U}(\overline{U}, E)$.

Definition 8. *Let E be a completely regular (respectively, normal) topological space and let $\Psi, \Lambda \in D_{\partial U}(\overline{U}, E)$. We say Ψ is homotopic to Λ in the class $D_{\partial U}(\overline{U}, E)$ and we write $\Psi \cong \Lambda$ in $D_{\partial U}(\overline{U}, E)$ if there exists a map $H : \overline{U} \times [0,1] \to 2^E$ with $H(.,\eta(.)) \in \mathbf{D}(\overline{U}, E)$ for any continuous function $\eta : \overline{U} \to [0,1]$ with $\eta(\partial U) = 0$, $\Phi(x) \cap H_t(x) = \emptyset$ for any $x \in \partial U$ and $t \in (0,1)$, $\{x \in \overline{U} : (x, \Phi(x)) \cap (x, H(x,t)) \neq \emptyset \text{ for some } t \in [0,1]\}$ is compact (respectively, closed), $H_0 = \Psi$ and $H_1 = \Lambda$ (here $H_t(x) = H(x,t)$).*

Remark 5. *There is an analogue Remark 3 in this situation.*

Definition 9. *Let $F, G \in A_{\partial U}(\overline{U}, E)$. We say $F \cong G$ in $A_{\partial U}(\overline{U}, E)$ if for any selection $\Psi \in D_{\partial U}(\overline{U}, E)$ (respectively, $\Lambda \in D_{\partial U}(\overline{U}, E)$) of F (respectively, of G) we have $\Psi \cong \Lambda$ in $D_{\partial U}(\overline{U}, E)$ (Definition 8).*

Theorem 3. *Let E be a completely regular (respectively, normal) topological space, U an open subset of E, $B = \{(x, \Phi(x)) : x \in \overline{U}\}$, d is defined in (4), $F \in A_{\partial U}(\overline{U}, E)$, $G \in A_{\partial U}(\overline{U}, E)$ with $F^\star = I \times F$ and $G^\star = I \times G$. Suppose $G^\star$ is d–Φ–essential and*

$$\begin{cases} \text{for any selection } \Psi \in D_{\partial U}(\overline{U}, E) \text{ (respectively, } \Lambda \in D_{\partial U}(\overline{U}, E)) \\ \text{of } F \text{ (respectively, of } G) \text{ and any map } J \in D_{\partial U}(\overline{U}, E) \text{ with} \\ J|_{\partial U} = \Psi|_{\partial U} \text{ we have } \Lambda \cong J \text{ in } D_{\partial U}(\overline{U}, E) \text{ (Definition 8) and} \\ d\left((\Psi^\star)^{-1}(B)\right) = d\left((\Lambda^\star)^{-1}(B)\right); \text{ here } \Psi^\star = I \times \Psi \text{ and } \Lambda^\star = I \times \Lambda. \end{cases} \quad (5)$$

Then $F^\star$ is d–Φ–essential.

Proof. Let $\Psi \in D_{\partial U}(\overline{U}, E)$ be any selection of F and consider any map $J \in D_{\partial U}(\overline{U}, E)$ with $J|_{\partial U} = \Psi|_{\partial U}$. It remains to show $d\left((\Psi^\star)^{-1}(B)\right) = d\left((J^\star)^{-1}(B)\right) \neq d(\emptyset)$; here $\Psi^\star = I \times \Psi$ and $J^\star = I \times J$. Let $\Lambda \in D_{\partial U}(\overline{U}, E)$ be any selection of G and let $\Lambda^\star = I \times \Lambda$. Now (5) guarantees that there exists a map $H : \overline{U} \times [0,1] \to 2^E$ with $H(.,\eta(.)) \in \mathbf{D}(\overline{U}, E)$ for any continuous function $\eta : \overline{U} \to [0,1]$ with $\eta(\partial U) = 0$, $\Phi(x) \cap H_t(x) = \emptyset$ for any $x \in \partial U$ and $t \in (0,1)$, $\{x \in \overline{U} : (x, \Phi(x)) \cap (x, H(x,t)) \neq \emptyset \text{ for some } t \in [0,1]\}$ is compact (respectively, closed), $H_0 = \Lambda$ and $H_1 = J$ (here $H_t(x) = H(x,t)$) and $d\left((\Psi^\star)^{-1}(B)\right) = d\left((\Lambda^\star)^{-1}(B)\right)$. Let

$$\Omega = \{x \in \overline{U} : (x, \Phi(x)) \cap (x, H(x,t)) \neq \emptyset \text{ for some } t \in [0,1]\}.$$

Now $\Omega \neq \emptyset$ since $G^\star$ is d–Φ–essential (and $H_0 = \Lambda$). Ω is compact (respectively, closed) if E is a completely regular (respectively, normal) topological space. Next note $\Omega \cap \partial U = \emptyset$ and so there exists a Urysohn map $\mu : \overline{U} \to [0,1]$ with $\mu(\partial U) = 0$ and $\mu(\Omega) = 1$. Define a map R by $R(x) = H(x, \mu(x))$ for $x \in \overline{U}$ and write $R^\star = I \times R$. Note $R \in D_{\partial U}(\overline{U}, E)$ with $R|_{\partial U} = H_0|_{\partial U} = \Lambda|_{\partial U}$. Since $G^\star$ is d–Φ–essential then

$$d\left((\Lambda^\star)^{-1}(B)\right) = d\left((R^\star)^{-1}(B)\right) \neq d(\emptyset). \quad (6)$$

Now since $\mu(\Omega) = 1$ we have

$$\begin{aligned}(R^\star)^{-1}(B) &= \{x \in \overline{U} : (x, \Phi(x)) \cap (x, H(x, \mu(x))) \neq \emptyset\} \\ &= \{x \in \overline{U} : (x, \Phi(x)) \cap (x, H(x,1)) \neq \emptyset\} = (J^\star)^{-1}(B),\end{aligned}$$

so from (6) we have $d\left((\Lambda^\star)^{-1}(B)\right) = d\left((J^\star)^{-1}(B)\right) \neq d(\emptyset)$. Now combine with the above and we have $d\left((\Psi^\star)^{-1}(B)\right) = d\left((J^\star)^{-1}(B)\right) \neq d(\emptyset)$. □

Now assume

$$\cong \text{ in } D_{\partial U}(\overline{U}, E) \text{ (Definition 8) is an equivalence relation} \tag{7}$$

and

$$\begin{cases} \text{if } F \in A_{\partial U}(\overline{U}, E) \text{ and if } \Psi \in D_{\partial U}(\overline{U}, E) \text{ is any selection} \\ \text{of } F \text{ and } J \in D_{\partial U}(\overline{U}, E) \text{ is any map with } \Psi|_{\partial U} = J|_{\partial U} \\ \text{then } \Psi \cong J \text{ in } D_{\partial U}(\overline{U}, E) \text{ (Definition 8).} \end{cases} \tag{8}$$

Now we establish the topological transversality theorem in this setting.

Theorem 4. *Let E be a completely regular (respectively, normal) topological space, U an open subset of E, $B = \{(x, \Phi(x)) : x \in \overline{U}\}$, d is defined in (4), and assume (7) and (8) hold. Suppose F and G are two maps in $A_{\partial U}(\overline{U}, E)$ with $F^\star = I \times F$, $G^\star = I \times G$ and $F \cong G$ in $A_{\partial U}(\overline{U}, E)$ (Definition 9). Then $F^\star$ is d–Φ–essential if and only if $G^\star$ is d–Φ–essential.*

Proof. Assume $G^\star$ is d–Φ–essential. Let $\Psi \in D_{\partial U}(\overline{U}, E)$ be any selection of F, $\Lambda \in D_{\partial U}(\overline{U}, E)$ be any selection of G and consider any map $J \in D_{\partial U}(\overline{U}, E)$ with $J|_{\partial U} = \Psi|_{\partial U}$. If we show (5) then $F^\star$ is d–Φ–essential from Theorem 3. Now (8) guarantees that $\Psi \cong J$ in $D_{\partial U}(\overline{U}, E)$ (Definition 8) and this together with $F \cong G$ in $A_{\partial U}(\overline{U}, E)$ (Definition 9) (so $\Psi \cong \Lambda$ in $D_{\partial U}(\overline{U}, E)$ (Definition 8)) guarantees that $\Lambda \cong J$ in $D_{\partial U}(\overline{U}, E)$ (Definition 8). To complete (5) it remains to show $d\left((\Psi^\star)^{-1}(B)\right) = d\left((\Lambda^\star)^{-1}(B)\right)$; here $\Psi^\star = I \times \Psi$ and $\Lambda^\star = I \times \Lambda$. Note $G \cong F$ in $A_{\partial U}(\overline{U}, E)$ (Definition 9) so let $H : \overline{U} \times [0,1] \to 2^E$ with $H(.,\eta(.)) \in \mathbf{D}(\overline{U}, E)$ for any continuous function $\eta : \overline{U} \to [0,1]$ with $\eta(\partial U) = 0$, $\Phi(x) \cap H_t(x) = \emptyset$ for any $x \in \partial U$ and $t \in (0,1)$, $\{x \in \overline{U} : (x, \Phi(x)) \cap (x, H(x,t)) \neq \emptyset \text{ for some } t \in [0,1]\}$ is compact (respectively, closed), $H_0 = \Lambda$ and $H_1 = \Psi$ (here $H_t(x) = H(x,t)$). Let

$$\Omega = \{x \in \overline{U} : (x, \Phi(x)) \cap (x, H(x,t)) \neq \emptyset \text{ for some } t \in [0,1]\}.$$

Now $\Omega \neq \emptyset$ and there exists a Urysohn map $\mu : \overline{U} \to [0,1]$ with $\mu(\partial U) = 0$ and $\mu(\Omega) = 1$. Define the map R by $R(x) = H(x, \mu(x))$ and write $R^\star = I \times R$. Now $R \in D_{\partial U}(\overline{U}, E)$ with $R|_{\partial U} = \Lambda|_{\partial U}$ so since $G^\star$ is d–Φ–essential then $d\left((\Lambda^\star)^{-1}(B)\right) = d\left((R^\star)^{-1}(B)\right) \neq d(\emptyset)$. Now since $\mu(\Omega) = 1$ we have (see the argument in Theorem 3) $(R^\star)^{-1}(B) = (\Psi^\star)^{-1}(B)$ and as a result we have $d\left((\Psi^\star)^{-1}(B)\right) = d\left((\Lambda^\star)^{-1}(B)\right)$. □

Remark 6. *It is also easy to extend the above ideas to other natural situations [3,4]. Let E be a (Hausdorff) topological vector space (so automatically completely regular), Y a topological vector space, and U an open subset of E. Let $L : \text{dom } L \subseteq E \to Y$ be a linear (not necessarily continuous) single valued map; here $\text{dom } L$ is a vector subspace of E. Finally $T : E \to Y$ will be a linear, continuous single valued map with $L + T : \text{dom } L \to Y$ an isomorphism (i.e., a linear homeomorphism); for convenience we say $T \in H_L(E, Y)$. We say $F \in A(U, Y; L, T)$ if $(L + T)^{-1}(F + T) \in A(\overline{U}, E)$ and we could discuss Φ–essential and d–Φ–essential in this situation.*

Finally, we consider the above in the weak topology situation. Let X be a Hausdorff locally convex topological vector space and U a weakly open subset of C where C is a closed convex subset of X. We will consider classes **A**, **B** and **D** of maps.

Definition 10. *We say $F \in WD(\overline{U^w}, C)$ (respectively $F \in WB(\overline{U^w}, C)$) if $F : \overline{U^w} \to 2^C$ and $F \in \mathbf{D}(\overline{U^w}, C)$ (respectively $F \in \mathbf{B}(\overline{U^w}, C)$); here $\overline{U^w}$ denotes the weak boundary of U in C.*

Definition 11. *We say $F \in WA(\overline{U^w}, C)$ if $F : \overline{U^w} \to 2^C$ and $F \in \mathbf{A}(\overline{U^w}, C)$ and there exists a selection $\Psi \in WD(\overline{U^w}, C)$ of F.*

Now we fix a $\Phi \in WB(\overline{U^w}, C)$ and present the notion of coincidence free on the boundary, Φ–essentiality and homotopy in this setting.

Definition 12. *We say $F \in WA_{\partial U}(\overline{U^w}, C)$ (respectively $F \in WD_{\partial U}(\overline{U^w}, C)$) if $F \in WA(\overline{U^w}, C)$ (respectively $F \in WD(\overline{U^w}, C)$) with $F(x) \cap \Phi(x) = \emptyset$ for $x \in \partial U$; here ∂U denotes the weak boundary of U in C.*

Definition 13. *We say $F \in WA_{\partial U}(\overline{U^w}, C)$ is Φ–essential in $WA_{\partial U}(\overline{U^w}, C)$ if for any selection $\Psi \in WD(\overline{U^w}, C)$ of F and any map $J \in WD_{\partial U}(\overline{U^w}, C)$ with $J|_{\partial U} = \Psi|_{\partial U}$ there exists a $x \in U$ with $J(x) \cap \Phi(x) \neq \emptyset$.*

Definition 14. *Let $\Psi, \Lambda \in WD_{\partial U}(\overline{U^w}, C)$. We say $\Psi \cong \Lambda$ in $WD_{\partial U}(\overline{U^w}, C)$ if there exists a map $H : \overline{U^w} \times [0,1] \to 2^C$ with $H(., \eta(.)) \in \mathbf{D}(\overline{U^w}, C)$ for any weakly continuous function $\eta : \overline{U^w} \to [0,1]$ with $\eta(\partial U) = 0$, $\Phi(x) \cap H_t(x) = \emptyset$ for any $x \in \partial U$ and $t \in (0,1)$, $\{x \in \overline{U^w} : \Phi(x) \cap H(x,t) \neq \emptyset \text{ for some } t \in [0,1]\}$ is weakly compact, $H_0 = \Psi$ and $H_1 = \Lambda$ (here $H_t(x) = H(x,t)$).*

Definition 15. *Let $F, G \in WA_{\partial U}(\overline{U^w}, C)$. We say $F \cong G$ in $WA_{\partial U}(\overline{U^w}, C)$ if for any selection $\Psi \in WD_{\partial U}(\overline{U^w}, C)$ (respectively, $\Lambda \in WD_{\partial U}(\overline{U^w}, C)$) of F (respectively, of G) we have $\Psi \cong \Lambda$ in $WD_{\partial U}(\overline{U^w}, C)$.*

Theorem 5. *Let X be a Hausdorff locally convex topological vector space and U a weakly open subset of C where C is a closed convex subset of X. Suppose $F \in WA_{\partial U}(\overline{U^w}, C)$ and $G \in WA_{\partial U}(\overline{U^w}, C)$ is Φ–essential in $WA_{\partial U}(\overline{U^w}, C)$ and*

$$\begin{cases} \text{for any selection } \Psi \in WD_{\partial U}(\overline{U^w}, C) \text{ (respectively, } \Lambda \in WD_{\partial U}(\overline{U^w}, C)) \\ \text{of } F \text{ (respectively, of } G) \text{ and any map } J \in WD_{\partial U}(\overline{U^w}, C) \\ \text{with } J|_{\partial U} = \Psi|_{\partial U} \text{ we have } \Lambda \cong J \text{ in } WD_{\partial U}(\overline{U^w}, C). \end{cases} \quad (9)$$

Then F is Φ–essential in $WA_{\partial U}(\overline{U^w}, C)$.

Proof. A slight modification of the argument in Theorem 1 guarantees the result; we just need to note that $X = (X, w)$, the space X endowed with the weak topology, is completely regular. □

Assume

$$\cong \text{ in } WD_{\partial U}(\overline{U^w}, C) \text{ is an equivalence relation} \quad (10)$$

and

$$\begin{cases} \text{if } F \in WA_{\partial U}(\overline{U^w}, C) \text{ and if } \Psi \in WD_{\partial U}(\overline{U^w}, C) \text{ is any} \\ \text{selection of } F \text{ and } J \in WD_{\partial U}(\overline{U^w}, C) \text{ is any map} \\ \text{with } \Psi|_{\partial U} = J|_{\partial U} \text{ then } \Psi \cong J \text{ in } WD_{\partial U}(\overline{U^w}, C). \end{cases} \quad (11)$$

A slight modification of the proof of Theorem 2 guarantees the topological transversality theorem in this setting.

Theorem 6. *Let X be a Hausdorff locally convex topological vector space and U a weakly open subset of C where C is a closed convex subset of X and assume (10) and (11) hold. Suppose F and G are two maps in $WA_{\partial U}(\overline{U^w}, C)$ with $F \cong G$ in $WA_{\partial U}(\overline{U^w}, C)$. Now F is Φ–essential in $WA_{\partial U}(\overline{U^w}, C)$ if and only if G is Φ–essential in $WA_{\partial U}(\overline{U^w}, C)$.*

Funding: This research received no external funding.

Conflicts of Interest: The author declares no conflict of interest.

References

1. Granas, A. Sur la méthode de continuité de Poincaré. *C. R. Acad. Sci. Paris* **1976**, *282*, 983–985.
2. Gabor, G.; Gorniewicz, L.; Slosarski, M. Generalized topological essentiality and coincidence points of multivalued maps. *Set-Valued Anal.* **2009**, *17*, 1–19. [CrossRef]
3. O'Regan, D. Essential maps and coincidence principles for general classes of maps. *Filomat* **2017**, *31*, 3553–3558. [CrossRef]
4. O'Regan, D. Topological transversality principles and general coincidence theory. *An. Stiint. Univ. Ovidius Constanta Ser. Mat.* **2017**, *25*, 159–170. [CrossRef]
5. Precup, R. On the topological transversality principle. *Nonlinear Anal.* **1993**, *20*, 1–9. [CrossRef]
6. Lin, L.J.; Park, S.; You, Z.T. Remarks on fixed points, maximal elements and equilibria of generalized games. *J. Math. Anal. Appl.* **1999**, *233*, 581–596. [CrossRef]

© 2020 by the author. Licensee MDPI, Basel, Switzerland. This article is an open access article distributed under the terms and conditions of the Creative Commons Attribution (CC BY) license (http://creativecommons.org/licenses/by/4.0/).

Article

Generalized Briot-Bouquet Differential Equation Based on New Differential Operator with Complex Connections

Rabha W. Ibrahim [1,2], Rafida M. Elobaid [3,*] and Suzan J. Obaiys [4]

1. Informetrics Research Group, Ton Duc Thang University, Ho Chi Minh City 758307, Vietnam; rabhaibrahim@tdtu.edu.vn
2. Faculty of Mathematics & Statistics, Ton Duc Thang University, Ho Chi Minh City 758307, Vietnam
3. Department of General Sciences, Prince Sultan University, Riyadh 12435, Saudi Arabia
4. School of Mathematical and Computer Sciences, Heriot-Watt University Malaysia, Putrajaya 62200, Malaysia; s.obaiys@hw.ac.uk
* Correspondence: robaid@psu.edu.sa

Received: 30 March 2020; Accepted: 16 April 2020; Published: 21 April 2020

Abstract: A class of Briot–Bouquet differential equations is a magnificent part of investigating the geometric behaviors of analytic functions, using the subordination and superordination concepts. In this work, we aim to formulate a new differential operator with complex connections (coefficients) in the open unit disk and generalize a class of Briot–Bouquet differential equations (BBDEs). We study and generalize new classes of analytic functions based on the new differential operator. Consequently, we define a linear operator with applications.

Keywords: differential operator; univalent function; analytic function; subordination; unit disk

MSC: 30C55; 30C45

1. Introduction

Inequalities in a complex domain play a massive role in function theory. They have been employed to introduce the geometric interpolation of analytic functions in the open unit disk. Moreover, they have been utilized to formulate generalized classes of analytic functions. Recently, Lupas [1] suggested a combination of two famous differential operators given by Ruscheweyh [2] and Sàlàgean [3] to present a set of inequalities and inclusions by using the concept of subordination.

In this study, we shall define a new differential operator of complex coefficients and study its behaviors based on the properties of the theory of geometric functions. The new operator will be formulated in generalized sub-classes of starlike functions. Subordination inequalities include the generalized operator, and some well-known functions are discussed. Sharp results are indicated in the sequel. As an application, we introduce a generalization of a class of Briot–Bouquet differential equations (BBDEs) in the complex domain. Consequently, examples are illustrated utilizing the time-space BBDEs. A comparison with recent works is shown in the sequel.

2. Differential Operators

The theory of special functions in one variable has a long and ironic past; the rising importance in special functions of several variables is moderately contemporary. Currently, there has been quick progress specifically in the area of special functions with the consideration of symmetries and harmonic analysis connected with root systems. The drive for this work comes from some generalizations of the theory of symmetric spaces, whose functions can be written as special functions depending on definite sets of parameters. A key implementation in the study of special functions with reflection symmetries is Dunkl operators, which are known as a class of differential-difference operators. In this effort, we present a Dunkl differential-difference operator of the first type in a complex domain, under a special class of analytic functions, called a class of normalized analytic functions. This class plays an important role in the field of geometric function theory. Based on this connection between the Dunkl operator and geometric function theory, we impose a major class of geometric presentations called the starlike class of analytic functions. A significant motivation to study Dunkl operators is created by their application in the analysis of quantum many-body systems of a special type. These operators describe integrated systems in one dimension and have seen considerable increased attention in mathematical physics, especially in conformal field theory (see [4,5] for recent works).

Let $\wedge$ be the class of the analytic functions taking the expansion:

$$\Upsilon(\xi) = \xi + \sum_{n=2}^{\infty} \Upsilon_n \xi^n, \quad \xi \in \cup = \{\xi : |\xi| < 1\}. \tag{1}$$

For a function $\Upsilon \in \wedge$, the Ruscheweyh formulation of the derivative is given by the following expansion formula:

$$R^m \Upsilon(\xi) = \xi + \sum_{n=2}^{\infty} C_{m+n-1}^m \Upsilon_n \xi^n,$$

where the term C_{m+n-1}^m is the combination of coefficients. Moreover, the Sàlàgean derivation expansion is defined by:

$$S^m \Upsilon(\xi) = \xi + \sum_{n=2}^{\infty} n^m \Upsilon_n \xi^n.$$

Consequently, Lupas combined the above operators to get a linear operator as follows [1]:

$$L_\alpha^m \Upsilon(\xi) = \xi + \sum_{n=2}^{\infty} [\alpha n^m + (1-\alpha) C_{m+n-1}^m] \Upsilon_n \xi^n, \quad \xi \in \cup, \alpha \in [0,1].$$

Here, we introduce a differential operator taking the following expansion:

$$\begin{aligned} D_\lambda^0 \Upsilon(\xi) &= \Upsilon(\xi) \\ D_\lambda^1 \Upsilon(\xi) &= \xi \Upsilon'(\xi) + \lambda\Big((\Upsilon(\xi) - \xi) - (\Upsilon(-\xi) + \xi)) \Big), \quad \lambda \in \mathbb{C} \\ &\vdots \\ D_\lambda^m \Upsilon(\xi) &= D_\lambda(D_\lambda^{m-1} \Upsilon(\xi)) \\ &= \xi + \sum_{n=2}^{\infty} [n + \lambda(1 + (-1)^{n+1})]^m \Upsilon_n \xi^n. \end{aligned} \tag{2}$$

For $\lambda = 0$, the operator reduces to the Sàlàgean differential operator. Moreover, the operator D_λ^m imposes a modification of the Dunkl operator of the first type [6,7], where λ is the Dunkl parameter, which indicates the balance between the differential and difference part in Equation (2). One of its applications is recognizing the harmonic and oscillation behaviors of the solution, and $\Upsilon(-\xi)$ is the reflection of the function $\Upsilon(\xi)$, which plays a significant role in the symmetry problem. Moreover, when $m = 2$, the operator reduces to the generalized Dunkl-Coulomb operator [8].

Remark 1. *We note that the original Dunkl operator admits the formula (see [9]):*

$$D \Upsilon(\xi) = \Upsilon'(\xi) + \frac{\lambda}{2} (\Upsilon(\xi) - \Upsilon(-\xi)),$$

which implies that $D \Upsilon (\xi) \notin \bigwedge$. Therefore, (2) is a modification that gives $D \Upsilon (\xi) \in \bigwedge$ (the class of normalized functions in the geometric function theory).

We proceed with discussing the behavior of the term $\lambda(1 + (-1)^{n+1})$. Obviously, when:

$$\lambda := \frac{1}{1 - e^{2i\pi}} = \lim_{n \to \infty} \frac{1}{(1 + (-1)^{n+1})},$$

we get the shifted Sàlàgean differential operator:

$$D^m \Upsilon (\xi) = \xi + \sum_{n=2}^{\infty} [n+1]^m \, \Upsilon_n \, \xi^n. \tag{3}$$

Furthermore, we have:

$$\lim_{n \to \infty} (1 + (-1)^{n+1}) = 1 + e^{2i\pi},$$

which implies that for $\lambda := \frac{1}{1+e^{2i\pi}}$, we get (3). The term $(1 + (-1)^{n+1})$ plays an important role in the oscillation problem, which was discussed in [8] (see Figure 1):

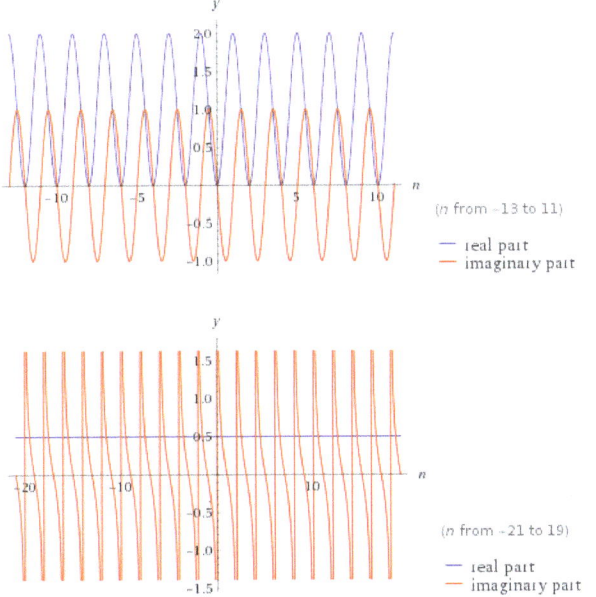

Figure 1. The first graph is $(1 + (-1)^{n+1})$, and the second is $1/(1 + (-1)^{n+1})$.

For functions Υ and λ in $\bigwedge$, we say that Υ is subordinated to λ, denoted by $\Upsilon \prec \lambda$, if there occurs a Schwarz function $\top \in \cup$ with $\top(0) = 0$ and $|\top(\xi)| < 1$, $\xi \in \cup$ so that $\Upsilon(\xi) = \lambda(\top(\xi))$ for all $\xi \in \cup$ (see [10]). Basically, $\Upsilon(\xi) \prec \lambda(\xi)$ is equivalent to $\Upsilon(0) = \lambda(0)$ and $\Upsilon(\cup) \subset \lambda(\cup)$.

3. Briot–Bouquet Differential Equation

The investigation of the complex Briot–Bouquet differential equations (BBDEs) is the study of a special class of differential equations whose consequences are designed in a complex domain (such as the open unit disk). The chief formula of BBDE is:

$$\frac{\xi(\Upsilon(\xi))'}{\Upsilon(\xi)} = \Lambda(\xi), \; \Upsilon \in \bigwedge, \; \xi \in \cup.$$

One can find different applications of these equations in dynamic and control systems (see [11–13]). The operator (2) can be used to generalize BBDE as follows:

$$\frac{\xi(D_\lambda^m \Upsilon(\xi))'}{D_\lambda^m \Upsilon(\xi)} = \Lambda(\xi), \quad \xi \in \cup, \Upsilon \in \bigwedge, \tag{4}$$

where $\Lambda(\xi)$ is univalent convex in $\cup$. Our aim is to study the upper outcome of (4) by using subordination inequalities.

Theorem 1. *Let $\Upsilon \in \bigwedge$ and $\Lambda(\xi)$ be univalent convex in $\cup$ fulfilling the subordination formula:*

$$\frac{\xi(D_\lambda^m \Upsilon(\xi))'}{D_\lambda^m \Upsilon(\xi)} \prec \Lambda(\xi). \tag{5}$$

Then, the upper bound of the solution of (5) is:

$$D_\lambda^m \Upsilon(\xi) \prec \xi \exp\left(\int_0^\xi \frac{\Lambda(\Psi(\iota)) - 1}{\iota} d\iota\right),$$

where $\Psi(\xi)$ is analytic in $\cup$, with $\Psi(0) = 0$ and $|\Psi(\xi)| < 1$. In addition, for $|\xi| = \iota$, $D_\lambda^m \Upsilon(\xi)$ achieves the inequality:

$$\exp\left(\int_0^1 \frac{\Lambda(\Psi(-\iota)) - 1}{\iota} d\iota\right) \leq \left|\frac{D_\lambda^m \Upsilon(\xi)}{\xi}\right| \leq \exp\left(\int_0^1 \frac{\Lambda(\Psi(\iota)) - 1}{\iota} d\iota\right).$$

Proof. By the definition of the subordination, Inequality (5) satisfies that there exists a Schwarz function with $\Psi(0) = 0$ and $|\Psi(\xi)| < 1$ such that:

$$\frac{\xi(D_\lambda^m \Upsilon(\xi))'}{D_\lambda^m \Upsilon(\xi)} = \Lambda(\Psi(\xi)), \quad \xi \in \cup.$$

This leads to the equation:

$$\frac{(D_\lambda^m \Upsilon(\xi))'}{D_\lambda^m \Upsilon(\xi)} - \frac{1}{\xi} = \frac{\Lambda(\Psi(\xi)) - 1}{\xi}.$$

By integrating both sides, we obtain:

$$\log D_\lambda^m \Upsilon(\xi) - \log \xi = \int_0^\xi \frac{\Lambda(\Psi(\iota)) - 1}{\iota} d\iota.$$

A computation yields:

$$\log\left(\frac{D_\lambda^m \Upsilon(\xi)}{\xi}\right) = \int_0^\xi \frac{\Lambda(\Psi(\iota)) - 1}{\iota} d\iota, \tag{6}$$

which is equivalent to the fact:

$$D_\lambda^m \Upsilon(\xi) \prec \xi \exp\left(\int_0^\xi \frac{\Lambda(\Psi(\iota)) - 1}{\iota} d\iota\right).$$

Further, the function Λ designs the disk $0 < |\xi| < \iota < 1$ on a territory, which is symmetric convex, agreeing with the real axis, that is:

$$\Lambda(-\iota|\xi|) \leq \Re(\Lambda(\Psi(\iota\xi))) \leq \Lambda(\iota|\xi|), \quad \iota \in (0,1);$$

thus, we attain the following inequalities:

$$\Lambda(-\iota) \leq \Lambda(-\iota|\xi|), \quad \Lambda(\iota|\xi|) \leq \Lambda(\iota).$$

By employing the above inequalities, we obtain the integral inequalities:

$$\int_0^1 \frac{\Lambda(\Psi(-\iota|\xi|)) - 1}{\iota} d\iota \leq \Re\left(\int_0^1 \frac{\Lambda(\Psi(\iota)) - 1}{\iota} d\iota\right) \leq \int_0^1 \frac{\Lambda(\Psi(\iota|\xi|)) - 1}{\iota} d\iota,$$

which leads to the next inequalities:

$$\int_0^1 \frac{\Lambda(\Psi(-\iota|\xi|)) - 1}{\iota} d\iota \leq \log\left|\frac{D_\lambda^m \Upsilon(\xi)}{\xi}\right| \leq \int_0^1 \frac{\Lambda(\Psi(\iota|\xi|)) - 1}{\iota} d\iota,$$

and:

$$\exp\left(\int_0^1 \frac{\Lambda(\Psi(-\iota|\xi|)) - 1}{\iota} d\iota\right) \leq \left|\frac{D_\lambda^m \Upsilon(\xi)}{\xi}\right| \leq \exp\left(\int_0^1 \frac{\Lambda(\Psi(\iota|\xi|)) - 1}{\iota} d\iota\right).$$

We conclude that:

$$\exp\left(\int_0^1 \frac{\Lambda(\Psi(-\iota)) - 1}{\iota} d\iota\right) \leq \left|\frac{D_\lambda^m \Upsilon(\xi)}{\xi}\right| \leq \exp\left(\int_0^1 \frac{\Lambda(\Psi(\iota)) - 1}{\iota} d\iota\right).$$

□

Theorem 2. *Suppose that $\Upsilon \in \bigwedge$ with non-negative connections. If $\Re(\lambda) > 0$ and Λ, in Equation (4), is univalent convex in $\cup$, then there occurs a solution fulfilling upper bound inequality:*

$$D_\lambda^m \Upsilon(\xi) \prec \xi \exp\left(\int_0^\xi \frac{\Lambda(\Psi(\iota)) - 1}{\iota} d\iota\right), \tag{7}$$

where $\Psi(\xi)$ is analytic in $\cup$, with $\Psi(0) = 0$ and $|\Psi(\xi)| < 1$.

Proof. In view of the assumptions, we attain:

$$\Re\left(\frac{\xi(D_\lambda^m \Upsilon(\xi))'}{D_\lambda^m \Upsilon(\xi)}\right) > 0$$

$$\Leftrightarrow \Re\left(\frac{\xi + \sum_{n=2}^\infty n[n + \lambda(1 + (-1)^{n+1})]^m \Upsilon_n \xi^n}{\xi + \sum_{n=2}^\infty [n + \lambda(1 + (-1)^{n+1})]^m \Upsilon_n \xi^n}\right) > 0$$

$$\Leftrightarrow \Re\left(\frac{1 + \sum_{n=2}^\infty n[n + \lambda(1 + (-1)^{n+1})]^m \Upsilon_n \xi^{n-1}}{1 + \sum_{n=2}^\infty [n + \lambda(1 + (-1)^{n+1})]^m \Upsilon_n \xi^{n-1}}\right) > 0$$

$$\Leftrightarrow \left(\frac{1 + \sum_{n=2}^\infty n[n + \lambda(1 + (-1)^{n+1})]^m \Upsilon_n}{1 + \sum_{n=2}^\infty [n + \lambda(1 + (-1)^{n+1})]^m \Upsilon_n}\right) > 0, \quad \xi \to 1^+$$

$$\Leftrightarrow \left(1 + \sum_{n=2}^\infty n[n + \lambda(1 + (-1)^{n+1})]^m \Upsilon_n\right) > 0.$$

In addition, we confirm that $(D_\lambda^m \Upsilon)(0) = 0$, which implies that:

$$\frac{\xi(D_\lambda^m \Upsilon(\xi))'}{D_\lambda^m \Upsilon(\xi)} \in \mathcal{P}.$$

Hence, according to Theorem 1, we arrive at (7). □

Numerical Examples

We deal with the following examples.

Example 1. *Suppose the parametric BB-control system (time-space equation):*

$$\frac{\xi(D_\lambda^m \Upsilon_\tau(\xi))'}{D_\lambda^m \Upsilon_\tau(\xi)} = \frac{1+\xi}{1-\xi}, \quad (8)$$

where $0 < \tau < 1, |\xi| < 1$ and:

$$\Upsilon_\tau(\xi) = \frac{\xi}{(1-\tau\xi)^2}$$

$$= \xi + 2\tau\xi^2 + 3\tau^2\xi^3 + 4\tau^3\xi^4 + 5\tau^4\xi^5 + 6\tau^5\xi^6 + O(\xi^7).$$

Our aim is to apply Theorem 2. By operating the formula of (2) for different values of $\lambda > 0$, we have:

$$D_{0.1}^1\left(\frac{\xi}{(1-\tau\xi)^2}\right) = \xi + 4\tau\xi^2 + 9.6\tau^2\xi^3 + 16\tau^3\xi^4 + 26\tau^4\xi^5 + O(\xi^6),$$

$$D_{0.5}^1\left(\frac{\xi}{(1-\tau\xi)^2}\right) = \xi + 4\tau\xi^2 + 12\tau^2\xi^3 + 16\tau^3\xi^4 + 30\tau^4\xi^5 + O(\xi^6),$$

$$D_1^1\left(\frac{\xi}{(1-\tau\xi)^2}\right) = \xi + 4\tau\xi^2 + 15\tau^2\xi^3 + 16\tau^3\xi^4 + 35\tau^4\xi^5 + O(\xi^6)$$

$$D_2^1\left(\frac{\xi}{(1-\tau\xi)^2}\right) = \xi + 4\tau\xi^2 + 21\tau^2\xi^3 + 16\tau^3\xi^4 + 45\tau^4\xi^5 + O(\xi^6).$$

Now, a computation implies that:

$$\xi \exp\left(\int_0^\xi \frac{\Lambda(\Psi(\iota))-1}{\iota} d\iota\right) = \xi \exp\left(\int_0^\xi \frac{\frac{1+\iota}{1-\iota}-1}{\iota} d\iota\right)$$
$$\approx \xi \exp(-2\log(\xi-1)), \quad \Re(\xi) < 1 \quad (9)$$
$$= \xi + 2\xi^2 + 3\xi^3 + 4\xi^4 + 5\xi^5 + O(\xi^6).$$

Comparing the connection values of $D_\lambda^1\left(\frac{\xi}{(1-\tau\xi)^2}\right)$ and (9), we conclude that $\tau \in [0.5, 1)$ implies that:

$$D_\lambda^1\left(\frac{\xi}{(1-\tau\xi)^2}\right) \prec \xi \exp\left(\int_0^\xi \frac{\frac{1+\iota}{1-\iota}-1}{\iota} d\iota\right).$$

Therefore, $D_\lambda^1\left(\frac{\xi}{(1-\tau\xi)^2}\right)$ is a solution of Equation (8).

Example 2. *In this example, we consider a wave equation taking the formula:*

$$\frac{\xi(D_\lambda^m \Upsilon_\tau(\xi))'}{D_\lambda^m \Upsilon_\tau(\xi)} = 1 + \sin(\xi), \quad (10)$$

where $0 < \tau < 1, |\xi| < 1$ and $\Upsilon_\tau(\xi) = \frac{\xi}{(1-\tau\xi)^2}$.

It is clear that:
$$\int_0^\xi (\sin(\iota)/\iota)d\iota = Si(\xi) = \xi - \xi^3/18 + \xi^5/600 + O(\xi^6),$$

where Si is the sin integral function. Consequently, we have:

$$\xi \exp\left(\int_0^\xi \frac{(1+\sin(\iota))-1}{\iota}d\iota\right) = \xi - \xi^3/18 + \xi^5/600 + O(\xi^6).$$

By compering the connection values, we indicate that $\tau \in [0, 14.7]$, and Equation (10) has an upper univalent solution for all λ satisfying:

$$D_\lambda^1\left(\frac{\xi}{(1-\tau\xi)^2}\right) \prec \xi \exp\left(Si(\xi)\right).$$

Remark 2. *Theorem 2 admits the following facts:*

- *The nonlinear model that we studied has no computational complexity cost. It is, fairly enough, not high speed because we have one variable and one parameter.*
- *It focuses on a starlike formula, which corresponds to the diffusion of the natural system of differential equations. Therefore, we reformulated the Dunkl operator to be suitable for this study.*
- *Theorem 2 gives the upper analytic solution in the open unit disk. Moreover, the upper bound solution is convex univalent; thus, all the trajectories approximate slightly the solution of Equation (7).*

4. Linear Combination Operator

This work deals with a new operator combining R^m and D_λ^m as follows:

$$\begin{aligned}J_{\alpha,\lambda}^m \Upsilon(\xi) &= (1-\alpha)R^m \Upsilon(\xi) + \alpha D_\lambda^m \Upsilon(\xi) \\ &= \xi + \sum_{n=2}^\infty [(1-\alpha)C_{m+n-1}^m + \alpha(n+\lambda(1+(-1)^{n+1}))^m] \Upsilon_n \xi^n.\end{aligned} \quad (11)$$

Remark 3.

- $m = 0 \Longrightarrow J_{\alpha,\lambda}^0 \Upsilon(\xi) = \Upsilon(\xi);$
- $\lambda = 0 \Longrightarrow J_{\alpha,0}^m \Upsilon(\xi) = L_\alpha^m \Upsilon(\xi);$
- $\alpha = 0 \Longrightarrow J_{0,\lambda}^m \Upsilon(\xi) = R^m \Upsilon(\xi);$
- $\alpha = 1 \Longrightarrow J_{1,\lambda}^m \Upsilon(\xi) = D_\lambda^m \Upsilon(\xi);$
- $\lambda = 0, \alpha = 1 \Longrightarrow J_{1,0}^m \Upsilon(\xi) = S^m \Upsilon(\xi).$

Definition 1. *Let $\alpha \geq 0, \lambda \in \mathbb{C}$, and $m \in \mathbb{N}$. A function $\Upsilon \in \Lambda$ belongs to $\mathfrak{S}_m^*(\alpha, \lambda, \sigma)$ if and only if:*

$$\frac{\xi(J_{\alpha,\lambda}^m \Upsilon(\xi))'}{J_{\alpha,\lambda}^m \Upsilon(\xi)} \prec \sigma(\xi), \quad \xi \in \cup,$$

where σ is a univalent function with a positive real part in $\cup$ satisfying $\sigma(0) = 1, \Re(\sigma'(\xi)) > 0$.

Note that the class $\mathfrak{S}_m^*(\alpha, \lambda, \sigma)$ is a generalization of some classes of analytic functions. Moreover, this class is a specialist of the Ma and Minda class [14] given as follows ($\mathfrak{S}^*(\sigma)$):

$$\frac{\xi \Upsilon'(\xi)}{\Upsilon(\xi)} \prec \sigma(\xi).$$

Moreover, when $\sigma(\xi) = 1 + \sin(\xi)$ and $m = 0$, the class:

$$\frac{\xi \Upsilon'(\xi)}{\Upsilon(\xi)} \prec 1 + \sin(\xi)$$

was studied by Cho et al. [15]. Our class is a generalization of two classes given by Khatter et al. [16] as follows:

$$\frac{\xi \Upsilon'(\xi)}{\Upsilon(\xi)} \prec \beta + (1-\beta)\sqrt{1+\xi}$$

and

$$\frac{\xi \Upsilon'(\xi)}{\Upsilon(\xi)} \prec \beta + (1-\beta)e^{\xi},$$

where $\beta = 0$ introduces the class [17]:

$$\frac{\xi \Upsilon'(\xi)}{\Upsilon(\xi)} \prec e^{\xi}.$$

Kumar et al. [18] defined the class by using Bell numbers as follows:

$$\frac{\xi \Upsilon'(\xi)}{\Upsilon(\xi)} \prec e^{e^{\xi}-1}.$$

Theorem 3. *If $\beta \in [0,1], \xi \in \mathbb{U}$, then each function of the form:*

- $\sigma(\xi) = \beta + (1-\beta)\sqrt{1+\xi}$,
- $\sigma(\xi) = \beta + (1-\beta)e^{\xi}$,
- $\sigma(\xi) = \beta + (1-\beta)(1+\sin(\xi))$,
- $\sigma(\xi) = \beta + (1-\beta)e^{e^{\xi}-1}$,

has the upper and lower bound for all $r \in (0,1), \theta \in [0, 2\pi)$ as follows:

$$\min_{|\xi|=r} \Re(\sigma(\xi)) = \sigma(-r) = \min_{|\xi|=r} |\sigma(\xi)|$$

and

$$\max_{|\xi|=r} \Re(\sigma(\xi)) = \sigma(r) = \max_{|\xi|=r} |\sigma(\xi)|.$$

Proof. The first and second type can be located in [16]. We only need to prove the third type. For $\beta = 0$, we have the function $\sigma(\xi) = 1 + \sin(\xi)$ (see [15]). It is clear that:

$$\sin(\xi) = \sin(re^{i\theta}) = \sin(r\cos(\theta))\cosh(r\sin(\theta)) + i\cos(r\cos(\theta))\sinh(r\sin(\theta))$$

therefore, we have

$$\Re(\sigma(\xi)) = 1 + \sin(r\cos(\theta))\cosh(r\sin(\theta)).$$

Consequently, by taking $r \to 0$, we obtain:

$$\min_{|\xi|=r} \Re(\sigma(\xi)) = 1 - \sin(r) = \min_{|\xi|=r} |\sigma(\xi)| = 1.$$

Moreover, we have:

$$|\sin(re^{i\theta})|^2 = \cos^2(r\cos\theta)\sinh 2(r\sin\theta) + \sin^2 2(r\cos\theta)\cosh 2(r\sin r) \leq \sinh^2(r);$$

thus, this yields:

$$\max_{|\xi|=r} \Re(\sigma(\xi)) = 1 + \sin(r) = \max_{|\xi|=r} |\sigma(\xi)| \leq 1 + \sinh^2(r).$$

Extending the above result, for $\beta > 0$, we have:

$$\min_{|\xi|=r} \Re(\sigma(\xi)) = \beta + (1-\beta)(1 - \sin(r)) = \min_{|z|=r} |\sigma(\xi)| = 1,$$

and
$$\max_{|\xi|=r} \Re(\sigma(\xi)) = \beta + (1-\beta)(1+\sin(r)) = \max_{|\xi|=r} |\sigma(\xi)| \leq \beta + (1-\beta)(1+\sinh^2(r)).$$

This is similar for the last assertion. □

The next result can be found in [10].

Lemma 1. *If $\tau > 0$ and $\sigma \in \mathfrak{H}[1,n]$, then there are constants $\wp > 0$ and $\nu > 0$ with $\nu = \nu(\wp, \tau, n)$, so that:*
$$\sigma(\xi) + \tau \xi \sigma'(\xi) \prec \left[\frac{1+\xi}{1-\xi}\right]^\nu \Rightarrow \sigma(\xi) \prec \left[\frac{1+\xi}{1-\xi}\right]^\wp.$$

Lemma 2. *Let $\varphi(\xi)$ be a convex function in $\mathbb{U}$, $h(\xi) = \varphi(\xi) + n\nu(\xi \varphi'(\xi))$ for $\nu > 0$, and n be a positive integer. If $\varrho \in \mathfrak{H}[\varphi(0), n]$, and:*
$$\varrho(\xi) + \nu \xi \varrho'(\xi) \prec h(\xi), \quad \xi \in \mathbb{U},$$

then
$$\varrho(\xi) \prec \varphi(\xi),$$

and this result is sharp.

5. Subordination Inequalities

Here, we are concerned with the class $\mathfrak{S}_m^*(\alpha, \lambda, \sigma)$ for special types of $\sigma(\xi)$ that are given in Theorem 3.

Theorem 4. *The class $\mathfrak{S}_m^*(\alpha, \lambda, \sigma)$ achieves the following inclusion:*
$$\mathfrak{S}_m^*(\alpha, \lambda, \sigma) \subset \mathfrak{S}_m^*(\alpha, \lambda, \gamma) \subset \mathfrak{S}_m^*(\alpha, \lambda),$$

where σ is one of the types in Theorem 3 and:
$$\mathfrak{S}_m^*(\alpha, \lambda, \gamma) := \{Y \in \bigwedge \Re\left(\frac{\xi(J_{\alpha,\lambda}^m Y(\xi)))'}{J_{\alpha,\lambda}^m Y(\xi)}\right) > \gamma\};$$

$$\mathfrak{S}_m^*(\alpha, \lambda) := \{Y \in \bigwedge \Re\left(\frac{\xi(J_{\alpha,\lambda}^m Y(\xi))'}{J_{\alpha,\lambda}^m Y(\xi)}\right) > 0\}.$$

Proof. Let $Y \in \mathfrak{S}_m^*(\alpha, \lambda, \sigma)$, and let $\sigma(\xi) = \beta + (1-\beta)\sqrt{1+\xi}$, then we have the inequality:
$$\frac{\xi(J_{\alpha,\lambda}^m Y(\xi))'}{J_{\alpha,\lambda}^m Y(\xi)} \prec \beta + (1-\beta)\sqrt{1+\xi}, \quad \xi \in \mathbb{U}.$$

In view of Theorem 3, we obtain:
$$\min_{|\xi|=1^-} \Re(\beta + (1-\beta)\sqrt{1+\xi}) < \Re\left(\frac{\xi(J_{\alpha,\lambda}^m Y(\xi))'}{J_{\alpha,\lambda}^m Y(\xi)}\right) < \max_{|\xi|=1^+} \Re(\beta + (1-\beta)\sqrt{1+\xi}),$$

which yields:
$$\beta < \Re\left(\frac{\xi(J_{\alpha,\lambda}^m Y(\xi))'}{J_{\alpha,\lambda}^m Y(\xi)}\right) < \beta + (1-\beta)\sqrt{2}.$$

Hence, we have:
$$\Re\left(\frac{\xi(J_{\alpha,\lambda}^m Y(\xi))'}{J_{\alpha,\kappa}^m Y(\xi)}\right) > \beta := \gamma \geq 0,$$

and consequently, we get the requested result. Consider $\sigma(\xi) = \beta + (1-\beta)e^\xi$; we have:

$$\min_{|\xi|=1} \Re(\beta + (1-\beta)e^\xi) < \Re\left(\frac{\xi(J_{\alpha,\lambda}^m \Upsilon(\xi))'}{J_{\alpha,\lambda}^m \Upsilon(\xi)}\right) < \max_{|\xi|=1} \Re(\beta + (1-\beta)e^\xi),$$

which implies:

$$(\beta + (1-\beta)\frac{1}{e}) < \Re\left(\frac{\xi(J_{\alpha,\lambda}^m \Upsilon(\xi))'}{J_{\alpha,\lambda}^m \Upsilon(\xi)}\right) < (\beta + (1-\beta)e),$$

that is:

$$\Re\left(\frac{\xi(J_{\alpha,\lambda}^m \Upsilon(\xi))'}{J_{\alpha,\lambda}^m \Upsilon(\xi)}\right) > (\beta + (1-\beta)\frac{1}{e}) := \gamma \geq 0.$$

Similarly, by letting $\sigma(\xi) = \beta + (1-\beta)(1+\sin(\xi))$, then we have:

$$\min_{|\xi|=1} \Re(\beta + (1-\beta)(1+\sin(\xi))) < \Re\left(\frac{\xi(J_{\alpha,\lambda}^m \Upsilon(\xi))'}{J_{\alpha,\lambda}^m \Upsilon(\xi)}\right) < \max_{|\xi|=1} \Re(\beta + (1-\beta)(1+\sin(\xi))),$$

which leads to:

$$(\beta + 0.158(1-\beta)) < \Re\left(\frac{\xi(J_{\alpha,\lambda}^m \Upsilon(\xi))'}{J_{\alpha,\lambda}^m \Upsilon(\xi)}\right) < (\beta + 1.841(1-\beta)),$$

and this brings the inequality:

$$\Re\left(\frac{\xi(J_{\alpha,\lambda}^m \Upsilon(\xi))'}{J_{\alpha,\lambda}^m \Upsilon(\xi)}\right) > (\beta + 0.158(1-\beta)) := \gamma \geq 0.$$

□

Remark 4. *In Theorem 4,*

- $m = 0, \beta = 0, \sigma(\xi) = 1 + \sin\xi \Longrightarrow$ [15];
- $m = 0 \Longrightarrow$ [16];
- $m = 0, \beta = 0, \sigma(\xi) = e^\xi \Longrightarrow$ [19];
- $m = 0, \beta = 0, \sigma(\xi) = \sqrt{1+\xi} \Longrightarrow$ [19].

Theorem 5. *The class $\mathfrak{S}_m^*(\alpha, \lambda, \sigma)$ achieves the following inclusion:*

$$\mathfrak{S}_m^*(\alpha, \lambda, \sigma) \subset \mathfrak{M}_m(\alpha, \lambda, \gamma) := \{\Upsilon \in \bigwedge \Re\left(\frac{\xi(J_{\alpha,\lambda}^m \Upsilon(\xi))'}{J_{\alpha,\lambda}^m \Upsilon(\xi)}\right) < \gamma, \gamma > 1\}.$$

where σ is given in Theorem 3.

The set $\mathfrak{M}_m(\alpha, \lambda, \gamma)$ is a generalization of the set:

$$\mathfrak{M}(\gamma) := \{\Upsilon \in \bigwedge \Re\left(\frac{\xi(\Upsilon(\xi))'}{\Upsilon(\xi)}\right) < \gamma, \gamma > 1\}$$

given by Uralegaddi et al. [20].

Proof. Let $\Upsilon \in \mathfrak{S}_m^*(\alpha, \lambda, \sigma)$, where σ is given in Theorem 3. By the proof of Theorem 4, we have:

$$\Re\left(\frac{\xi(J_{\alpha,\lambda}^m \Upsilon(\xi))'}{J_{\alpha,\lambda}^m \Upsilon(\xi)}\right) < \beta + (1-\beta)\sqrt{2} := \gamma,$$

and:
$$\Re\left(\frac{\xi(J_{\alpha,\lambda}^m \Upsilon(\xi))'}{J_{\alpha,\lambda}^m \Upsilon(\xi)}\right) < \beta + (1-\beta)e := \gamma$$

$$\Re\left(\frac{\xi(J_{\alpha,\kappa}^m \Upsilon(\xi))'}{J_{\alpha,\lambda}^m \Upsilon(\xi)}\right) < (\beta + 1.841(1-\beta)) := \gamma,$$

Hence, $\Upsilon \in \mathfrak{M}_m(\alpha, \Upsilon(\xi), \gamma)$, where the value of γ is based on the function σ, which completes the proof. □

Remark 5. *In Theorem 5,*

- $m = 0, \beta = 0, \sigma(\xi) = 1 + \sin\xi \implies$ [15];
- $m = 0, \sigma(\xi) = \beta + (1-\beta)e^\xi \implies$ [16], Theorem 2.5;
- $m = 0, \sigma(\xi) = \beta + (1-\beta)(\sqrt{1+\xi}) \implies$ [16], Theorem 2.6;
- $m = 0, \beta = 0, \sigma(\xi) = (\sqrt{1+\xi}) \implies$ [16], Corollary 2.7.

Theorem 6. *If $\Upsilon \in \bigwedge$ satisfies the subordination:*

$$\left(\frac{\xi(J_{\alpha,\lambda}^m \Upsilon(\xi))'}{J_{\alpha,\lambda}^m \Upsilon(\xi)}\right)\left(2 + \frac{\xi(J_{\alpha,\lambda}^m \Upsilon(\xi))''}{(J_{\alpha,\lambda}^m \Upsilon(\xi))'} - \frac{\xi(J_{\alpha,\lambda}^m \Upsilon(\xi))'}{J_{\alpha,\Upsilon(\xi)}^m \Upsilon(\xi)}\right) \prec \left[\frac{1+\xi}{1-\xi}\right]^\tau$$

then $\Upsilon \in \mathfrak{S}_m^*(\alpha, \lambda, \sigma)$, where $\sigma(\xi) = \left[\frac{1+\xi}{1-\xi}\right]^\wp$ for $\wp > 0, \tau > 0$.

Proof. To employ Lemma 1, a calculation implies that:

$$\left(\frac{\xi(J_{\alpha,\lambda}^m \Upsilon(\xi))'}{J_{\alpha,\lambda}^m \Upsilon(\xi)}\right) + \xi\left(\frac{\xi(J_{\alpha,\lambda}^m \Upsilon(\xi))'}{J_{\alpha,\lambda}^m \Upsilon(\xi)}\right)'$$

$$= \left(\frac{\xi(J_{\alpha,\lambda}^m \Upsilon(\xi))'}{J_{\alpha,\lambda}^m \Upsilon(\xi)}\right)\left(2 + \frac{\xi(J_{\alpha,\lambda}^m \Upsilon(\xi))''}{(J_{\alpha,\lambda}^m \Upsilon(\xi))'} - \frac{\xi(J_{\alpha,\lambda}^m \Upsilon(\xi))'}{J_{\alpha,\lambda}^m \Upsilon(\xi)}\right)$$

$$\prec \left[\frac{1+\xi}{1-\xi}\right]^\tau.$$

Thus, in view of Lemma 1, we have:

$$\left(\frac{\xi(J_{\alpha,\lambda}^m \Upsilon(\xi))'}{J_{\alpha,\lambda}^m \Upsilon(\xi)}\right) \prec \left[\frac{1+\xi}{1-\xi}\right]^\wp := \sigma(\xi),$$

which implies that $\Upsilon \in \mathfrak{S}_m^*(\alpha, \lambda, \sigma)$. □

Theorem 7. *Let φ be a convex function such that $\varphi(0) = 0$, and let $\hbar$ be the function:*

$$\hbar(\xi) = \varphi(\xi) + \frac{\xi}{1-\ell}\varphi'(\xi), \quad \xi \in \mathbb{U}, \ell \in (0,1).$$

If for a function, $\Upsilon \in \bigwedge$ satisfies the subordination:

$$\left(\frac{\xi}{J_{\alpha,\lambda}^{m+1} \Upsilon(\xi)}\right)^\ell \frac{J_{\alpha,\lambda}^m \Upsilon(\xi)}{1-\ell}\left(\frac{(J_{\alpha,\lambda}^{m+1} \Upsilon(\xi))'}{J_{\alpha,\lambda}^{m+1} \Upsilon(\xi)} - \ell\frac{(J_{\alpha,\lambda}^m \Upsilon(\xi))'}{J_{\alpha,\lambda}^m \Upsilon(\xi)}\right) \prec \hbar(\xi)$$

then:

$$\left(\frac{J_{\alpha,\lambda}^{m+1} \Upsilon(\xi)}{\xi}\right)\left(\frac{\xi}{J_{\alpha,\lambda}^{m+1} \Upsilon(\xi)}\right)^\ell \prec \varphi(\xi), \quad \xi \in \mathbb{U}.$$

The outcome is sharp.

Proof. We aim to apply Lemma 2. Let:

$$\varrho(\xi) = \Big(\frac{J_{\alpha,\lambda}^{m+1} \Upsilon(\xi)}{\xi}\Big)\Big(\frac{\xi}{J_{\alpha,\lambda}^{m+1} \Upsilon(\xi)}\Big)^{\ell}.$$

A differentiation implies that:

$$\Big(\frac{\xi}{J_{\alpha,\lambda}^{m+1} \Upsilon(\xi)}\Big)^{\ell} \frac{J_{\alpha,\lambda}^{m} \Upsilon(\xi)}{1-\ell} \Big(\frac{(J_{\alpha,\lambda}^{m+1} \Upsilon(\xi))'}{J_{\alpha,\lambda}^{m+1} \Upsilon(\xi)} - \ell \frac{(J_{\alpha,\lambda}^{m}) \Upsilon(\xi))'}{J_{\alpha,\lambda}^{m} \Upsilon(\xi)}\Big) = \varrho(\xi) + \Big(\frac{1}{1-\ell}\Big)\xi \varrho'(\xi)$$

Thus, by the assumption, we have:

$$\varrho(\xi) + \Big(\frac{1}{1-\ell}\Big)\xi \varrho'(\xi) \prec \hbar(\xi) = \varphi(\xi) + \frac{\xi}{1-\ell}\varphi'(\xi), \quad \xi \in \cup.$$

Employing Lemma 2 yields $\varrho(\xi) \prec \hbar(\xi)$, which means:

$$\Big(\frac{J_{\alpha,\lambda}^{m+1} \Upsilon(\xi)}{\xi}\Big)\Big(\frac{\xi}{J_{\alpha,\lambda}^{m+1} \Upsilon(\xi)}\Big)^{\ell} \prec \varphi(\xi), \quad \xi \in \cup.$$

This result is sharp. □

Remark 6. *In Theorem 6, $\lambda = 0 \Longrightarrow$ [21] Theorem 2.14.*

6. Conclusions

This study was concerned with a class of Briot–Bouquet differential equations utilizing a new differential operator of complex connections. Some inequalities involving the subordination concept were investigated. For future work, the idea of [22] will be used to present a harmonic class of Briot–Bouquet differential equations.

Author Contributions: R.W.I., R.M.E. and S.J.O. contributed equally and significantly to writing this article. All authors read and agreed to the published version of the manuscript.

Funding: This research received no external funding.

Acknowledgments: The authors would like to express their full thanks to the respected reviewers for the deep comments, which improved our paper.

Conflicts of Interest: The authors declare no conflict of interest.

References

1. Lupas, A.A. On special differential subordinations using Salagean and Ruscheweyh operators. *Math. Inequal. Appl.* **2009**, *12*, 781–790.
2. Ruscheweyh, S. New criteria for univalent functions. *Proc. Am. Math. Soc.* **1975**, *49*, 109–115. [CrossRef]
3. Salagean, G.S. Subclasses of univalent functions. In *Complex Analysis—Fifth Romanian-Finnish Seminar*; Springer: Berlin, Germany, 1983; Volume 1013, pp. 362–372.
4. Ibrahim, R.W. Arched foot based on conformal complex neural network testing. *Math. Comput. Simul.* **2020**, *174*, 175–182. [CrossRef]
5. Ibrahim, R.W. Conformal Geometry of the Turtle Shell. *J. King Saud Univ. Sci.* **2020**, *32*, 2202–2206. [CrossRef]
6. Ibrahim, R.W.; Darus, M. Subordination inequalities of a new Salagean-difference operator. *Int. J. Math. Comput. Sci.* **2019**, *14*, 573–582.
7. Dunkl, C.F. Differential-difference operators associated with reflections groups. *Trans. Am. Math. Soc.* **1989**, *311*, 167–183. [CrossRef]

8. Genest, V.X.; Lapointe, A.; Vinet, L. The Dunkl-Coulomb problem in the plane. *Phys. Lett. A* **2015**, *379*, 923–927. [CrossRef]
9. Rösler, M. Dunkl Operators: Theory and Applications. In *Orthogonal Polynomials and Special Functions*; Koelink, E., Van Assche, W., Eds.; Springer: Berlin/Heidelberg, Germany, 2003; Volume 1817.
10. Miller, S.S.; Mocanu, P.T. *Differential Subordinations: Theory and Applications*; CRC Press: Boca Raton, FL, USA, 2000.
11. Ebrahimi, F.; Mohammadi, K.; Barouti, M.M.; Habibi, M. Wave propagation analysis of a spinning porous graphene nanoplatelet-reinforced nanoshell. *Waves Random Complex Media* **2019**, 1–27. [CrossRef]
12. Habibi, M.; Mohammadgholiha, M.; Safarpour, H. Wave propagation characteristics of the electrically GNP-reinforced nanocomposite cylindrical shell. *J. Braz. Soc. Mech. Sci. Eng.* **2019**, *41*, 221. [CrossRef]
13. Ibrahim, R.W.; Hadid, S.B.; Momani, S. Generalized Briot–Bouquet differential equation by a quantum difference operator in a complex domain. *Int. J. Dyn. Control* **2020**, 1–10. [CrossRef]
14. Ma, W.C.; Minda, D. A unified treatment of some special classes of univalent functions. In Proceedings of the Conference on Complex Analysis, Tianjin, China, 19–23 June 1992.
15. Cho, N.E.; Kumar, V.; Kumar, S.S.; Ravichandran, V. Radius problems for starlike functions associated with the sine function. *Bull. Iran. Math. Soc.* **2019**, *45*, 213. [CrossRef]
16. Khatter, K.; Ravichandran, V.; Kumar, S.S. Starlike functions associated with exponential function and the lemniscate of Bernoulli. *Rev. Real Acad. Cienc. Exactas Ser. A Mat.* **2019**, *113*, 233–253. [CrossRef]
17. Mendiratta, R.; Nagpal, S.; Ravichandran, V. On a subclass of strongly starlike functions associated with exponential function. *Bull. Malays. Math. Sci. Soc.* **2015**, *38*, 365–386. [CrossRef]
18. Kumar, V.; Cho, N.E.; Ravichandran, V.; Srivastava, H.M. Sharp coefficient bounds for starlike functions associated with the Bell numbers. *Math. Slovaca* **2019**, *69*, 1053–1064. [CrossRef]
19. Kanas, S.; Wisniowska, A. Conic domains and starlike functions. *Rev. Roum. Math. Pures Appl.* **2000**, *45*, 647–657.
20. Uralegaddi, B.A.; Ganigi, M.D.; Sarangi, S.M. Univalent functions with positive coefficients. *Tamkang J. Math.* **1994**, *25*, 225–230.
21. Lupaş, A.A. Some differential subordinations using Ruscheweyh derivative and Salaagean operator. *Adv. Differ. Equ.* **2013**, *2013*, 150. [CrossRef]
22. Yousef, A.T.; Salleh, Z. On a Harmonic Univalent Subclass of Functions Involving a Generalized Linear Operator. *Axioms* **2020**, *9*, 32. [CrossRef]

© 2020 by the authors. Licensee MDPI, Basel, Switzerland. This article is an open access article distributed under the terms and conditions of the Creative Commons Attribution (CC BY) license (http://creativecommons.org/licenses/by/4.0/).

Article

Nonlocal Inverse Problem for a Pseudohyperbolic-Pseudoelliptic Type Integro-Differential Equations

Tursun K. Yuldashev

Uzbek-Israel Joint Faculty of High Technology and Engineering Mathematics, National University of Uzbekistan, Tashkent 100174, Uzbekistan; t.yuldashev@nuu.uz or tursun.k.yuldashev@gmail.com; Tel.: +998-99-519-59-31

Received: 23 March 2020; Accepted: 24 April 2020; Published: 27 April 2020

Abstract: The questions of solvability of a nonlocal inverse boundary value problem for a mixed pseudohyperbolic-pseudoelliptic integro-differential equation with spectral parameters are considered. Using the method of the Fourier series, a system of countable systems of ordinary integro-differential equations is obtained. To determine arbitrary integration constants, a system of algebraic equations is obtained. From this system regular and irregular values of the spectral parameters were calculated. The unique solvability of the inverse boundary value problem for regular values of spectral parameters is proved. For irregular values of spectral parameters is established a criterion of existence of an infinite set of solutions of the inverse boundary value problem. The results are formulated as a theorem.

Keywords: integro-differential equation; mixed type equation; spectral parameters; integral conditions; solvability

---

1. Statement of the Inverse Problem

From the point of applications, partial differential and integro-differential equations are of great interest [1,2]. The presence of the integral term in the differential equation plays an important role [3,4]. Also important to study the spectral questions of solvability of the differential and integro-differential equations [5–10]. In References [11–13], using the results of the theory of complete generalized Jordan sets it is considered the reduction of the partial differential equations with irreversible linear operator of finite index in the main differential expression to the regular problems.

Direct and inverse boundary value problems, where the type of differential equation in the domain under consideration changes, have important applications. Direct boundary value problems for differential and integro-differential equations of mixed type were studied in the works of many authors, in particular, in References [14–24]. In References [25,26] the inverse problems for second order mixed type differential equations were considered in rectangular domain. In this paper, we study the unique classical solvability of a nonlocal inverse boundary value problem of mixed pseudohyperbolic-pseudoelliptic integro-differential equation for regular values of spectral parameters. We also study the solvability conditions of the inverse boundary value problem for irregular values of spectral parameters.

In multidimensional domain $\Omega = \{-T < t < T, 0 < x_1, x_2, ..., x_m < l\}$ a mixed integro-differential equation of the following form is considered

$$\begin{cases} U_{tt} - \sum_{i=1}^{m} [U_{ttx_ix_i} - U_{x_ix_i}] = \nu \int_0^T K_1(t,s) U(s,x) \, ds + f_1(t) g_1(x), \ t > 0, \\ U_{tt} - \sum_{i=1}^{m} [U_{ttx_ix_i} + \omega^2 U_{x_ix_i}] = \nu \int_{-T}^0 K_2(t,s) U(s,x) \, ds + f_2(t) g_2(x), \ t < 0, \end{cases} \quad (1)$$

where T and l are given positive real numbers, ω is positive spectral parameter, $x \in \mathbb{R}^m$, ν is real non-zero spectral parameter, $0 \neq K_j(t,s) = a_j(t) b_j(s)$, a_j, $b_j \in C[-T;T]$, $0 \neq f_1 \in C[0;T]$, $0 \neq f_2 \in C[-T;0]$, $g_j \in C(\Omega_l^m)$ are redefinition functions, $\Omega_l^m = [0;l]^m$, $j = 1, 2$.

Problem 1. *Find in the domain Ω a triple of unknown functions*

$$U(t,x) \in C(\overline{\Omega}) \cap C^1(\Omega') \cap C^{2,2}(\Omega) \cap C^{2+2}_{t,x}(\Omega) \cap$$

$$\cap C^{2+2+0+\ldots+0}_{t,x_1,x_2,\ldots,x_m}(\Omega) \cap C^{2+0+2+0+\ldots+0}_{t,x_1,x_2,x_3,\ldots,x_m}(\Omega) \cap \ldots \cap C^{2+0+\ldots+0+2}_{t,x_1,\ldots,x_{m-1},x_m}(\Omega),$$

$$g_i(x) \in C(\Omega_l^m), \ i = 1,2,$$

satisfying the mixed integro-differential Equation (1) *and the following nonlocal boundary conditions*

$$\int_0^T U(t,x)\, dt = \varphi_1(x), \ x \in \Omega_l^m, \tag{2}$$

$$\int_{-T}^0 U(t,x)\, dt = \varphi_2(x), \ x \in \Omega_l^m, \tag{3}$$

$$U(t,0,x_2,x_3,\ldots,x_m) = U(t,l,x_2,x_3,\ldots,x_m) =$$
$$= U(t,x_1,0,x_3,\ldots,x_m) = U(t,x_1,l,x_3,\ldots,x_m) = \ldots =$$
$$= U(t,x_1,\ldots,x_{m-1},0) = U(t,x_1,\ldots,x_{m-1},l) =$$
$$= U_{x_1 x_1}(t,0,x_2,x_3,\ldots,x_m) = U_{x_1 x_1}(t,l,x_2,x_3,\ldots,x_m) =$$
$$= U_{x_1 x_1}(t,x_1,0,x_3,\ldots,x_m) = U_{x_1 x_1}(t,x_1,l,x_3,\ldots,x_m) = \ldots =$$
$$= U_{x_1 x_1}(t,x_1,\ldots,x_{m-1},0) = U_{x_1 x_1}(t,x_1,\ldots,x_{m-1},l) = \ldots =$$
$$= U_{x_m x_m}(t,0,x_2,x_3,\ldots,x_m) = U_{x_m x_m}(t,l,x_2,x_3,\ldots,x_m) =$$
$$= U_{x_m x_m}(t,x_1,0,x_3,\ldots,x_m) = U_{x_m x_m}(t,x_1,l,x_3,\ldots,x_m) = \ldots =$$
$$= U_{x_m x_m}(t,x_1,\ldots,x_{m-1},0) = U_{x_m x_m}(t,x_1,\ldots,x_{m-1},l) = 0, \ 0 \leq t \leq T, \tag{4}$$

and additional conditions

$$U(t_i, x) = \psi_i(x), \ i = 1,2, \ x \in \Omega_l^m, \tag{5}$$

where $\varphi_i(x)$, $\psi_i(x)$ are given smooth functions, $\varphi_i(0) = \varphi_i(l) = 0$, $\psi_i(0) = \psi_i(l) = 0$, $i = 1,2$, $t_1 \in (0;T)$, $t_2 \in (-T;0)$, $\Omega' = \Omega \cup \{x_1, x_2, \ldots, x_m = 0\} \cup \{x_1, x_2, \ldots, x_m = l\}$, $\Omega = \Omega_- \cup \Omega_+$, $\Omega_- = \{-T < t < 0, \ 0 < x_1, x_2, \ldots, x_m < l\}$, $\Omega_+ = \{0 < t < T, \ 0 < x_1, x_2, \ldots, x_m < l\}$, $\overline{\Omega} = \{-T \leq t \leq T, \ 0 \leq x_1, x_2, \ldots, x_m \leq l\}$.

2. Expansion of the Solution of the Direct Problem (1)–(4) into Fourier Series. Regular Case

The solution of the integro-differential Equation (1) in domain Ω is sought in the form of a Fourier series

$$U(t,x) = \sum_{n_1,\ldots,n_m = 1}^{\infty} u_{n_1,\ldots,n_m}(t)\, \vartheta_{n_1,\ldots,n_m}(x), \tag{6}$$

where

$$u_{n_1,\ldots,n_m}(t) = \int_{\Omega_l^m} U(t,x)\, \vartheta_{n_1,\ldots,n_m}(x)\, dx, \tag{7}$$

$$\int_{\Omega_l^m} U(t,x)\,\vartheta_{n_1,\ldots,n_m}(x)\,dx = \int_0^l \ldots \int_0^l U(t,x)\,\vartheta_{n_1,\ldots,n_m}(x)\,dx_1\ldots dx_m$$

$$\vartheta_{n_1,\ldots,n_m}(x) = \left(\sqrt{\frac{2}{l}}\right)^m \sin\frac{\pi n_1}{l}x_1 \ldots \sin\frac{\pi n_m}{l}x_m,$$

$$\Omega_l^m = [0;l]^m, n_1,\ldots,n_m = 1,2,\ldots$$

Also suppose that

$$g_i(x) = \sum_{n_1,\ldots,n_m=1}^{\infty} g_{i n_1,\ldots,n_m}\,\vartheta_{n_1,\ldots,n_m}(x), \tag{8}$$

where

$$g_{i n_1,\ldots,n_m} = \int_{\Omega_l^m} g_i(x)\,\vartheta_{n_1,\ldots,n_m}(x)\,dx,\ i=1,2.$$

Substituting series (6) and (8) into Equation (1), we obtain a countable system of integro-differential equations

$$u''_{n_1,\ldots,n_m}(t) - \lambda^2_{n_1,\ldots,n_m} u_{n_1,\ldots,n_m}(t) =$$

$$= \nu \int_0^T a_1(t)\,b_1(s)\,u_{n_1,\ldots,n_m}(s)\,ds + f_1(t)\,g_{1 n_1,\ldots,n_m},\ t>0, \tag{9}$$

$$u''_{n_1,\ldots,n_m}(t) + \lambda^2_{n_1,\ldots,n_m}\omega^2 u_{n_1,\ldots,n_m}(t) =$$

$$= \nu \int_{-T}^0 a_2(t)\,b_2(s)\,u_{n_1,\ldots,n_m}(s)\,ds + f_2(t)\,g_{2 n_1,\ldots,n_m},\ t<0, \tag{10}$$

where $\lambda^2_{n_1,\ldots,n_m} = \frac{\mu^2_{n_1,\ldots,n_m}}{1+\mu^2_{n_1,\ldots,n_m}}$, $\mu_{n_1,\ldots,n_m} = \frac{\pi}{T}\sqrt{n_1^2 + \ldots + n_m^2}$.

By the aid of notations

$$\alpha_{n_1,\ldots,n_m} = \int_0^T b_1(s)\,u_{n_1,\ldots,n_m}(s)\,ds, \tag{11}$$

$$\beta_{n_1,\ldots,n_m} = \int_{-T}^0 b_2(s)\,u_{n_1,\ldots,n_m}(s)\,ds, \tag{12}$$

we rewrite the countable systems of Equations (9) and (10) as follows

$$u''_{n_1,\ldots,n_m}(t) - \lambda^2_{n_1,\ldots,n_m} u_{n_1,\ldots,n_m}(t) = \nu a_1(t)\,\alpha_{n_1,\ldots,n_m} + f_1(t)\,g_{1 n_1,\ldots,n_m},\ t>0, \tag{13}$$

$$u''_{n_1,\ldots,n_m}(t) + \lambda^2_{n_1,\ldots,n_m}\omega^2 u_{n_1,\ldots,n_m}(t) = \nu a_2(t)\,\beta_{n_1,\ldots,n_m} + f_2(t)\,g_{2 n_1,\ldots,n_m},\ t<0. \tag{14}$$

Countable systems of differential Equations (13) and (14) are solved by the method of variation of arbitrary constants:

$$u_{n_1,\ldots,n_m}(t) = A_{1 n_1,\ldots,n_m}\exp\{\lambda_{n_1,\ldots,n_m}t\} + A_{2 n_1,\ldots,n_m}\exp\{-\lambda_{n_1,\ldots,n_m}t\} +$$

$$+ \eta_{1 n_1,\ldots,n_m}(t),\ t>0, \tag{15}$$

$$u_{n_1,\ldots,n_m}(t) = B_{1 n_1,\ldots,n_m}\cos\lambda_{n_1,\ldots,n_m}\omega t + B_{2 n_1,\ldots,n_m}\sin\lambda_{n_1,\ldots,n_m}\omega t +$$

$$+ \eta_{2 n_1,\ldots,n_m}(t),\ t<0, \tag{16}$$

where $A_{in_1,\ldots,n_m}$, $B_{in_1,\ldots,n_m}$ $(i=1,2)$ are unknown constants to be uniquely determined,

$$\eta_{1n_1,\ldots,n_m}(t) = \nu\, \alpha_{n_1,\ldots,n_m} h_{1n_1,\ldots,n_m}(t) + g_{1n_1,\ldots,n_m} h_{2n_1,\ldots,n_m}(t),$$

$$\eta_{2n_1,\ldots,n_m}(t) = \nu\, \beta_{n_1,\ldots,n_m} \delta_{1n_1,\ldots,n_m}(t) + g_{2n_1,\ldots,n_m} \delta_{2n_1,\ldots,n_m}(t),$$

$$h_{1n_1,\ldots,n_m}(t) = \frac{1}{\lambda_{n_1,\ldots,n_m}} \int_0^t \sinh \lambda_{n_1,\ldots,n_m}(t-s)\, a_1(s)\, ds,$$

$$h_{2n_1,\ldots,n_m}(t) = \frac{1}{\lambda_{n_1,\ldots,n_m}} \int_0^t \sinh \lambda_{n_1,\ldots,n_m}(t-s)\, f_1(s)\, ds,$$

$$\delta_{1n_1,\ldots,n_m}(t) = \frac{1}{\lambda_{n_1,\ldots,n_m}\omega} \int_0^t \sin \lambda_{n_1,\ldots,n_m}\omega(t-s)\, a_2(s)\, ds,$$

$$\delta_{2n_1,\ldots,n_m}(t) = \frac{1}{\lambda_{n_1,\ldots,n_m}\omega} \int_0^t \sin \lambda_{n_1,\ldots,n_m}\omega(t-s)\, f_2(s)\, ds.$$

From the statement of the problem it follows that the continuous conjugation conditions are fulfilled: $U(0+0, x) = U(0-0, x)$ and $U'(0+0, x) = U'(0-0, x)$. So, taking the Formula (7) into account, we have

$$u_{n_1,\ldots,n_m}(0+0) = \int_{\Omega_l^m} U(0+0, x)\, \vartheta_{n_1,\ldots,n_m}(x)\, dx =$$

$$= \int_{\Omega_l^m} U(0-0, x)\, \vartheta_{n_1,\ldots,n_m}(x)\, dx = u_{n_1,\ldots,n_m}(0-0). \tag{17}$$

Differentiating functions (7) once with respect to t, similarly to (17) we obtain

$$u'_{n_1,\ldots,n_m}(0+0) = \int_{\Omega_l^m} U_t(0+0, x)\, \vartheta_{n_1,\ldots,n_m}(x)\, dx =$$

$$= \int_{\Omega_l^m} U_t(0-0, x)\, \vartheta_{n_1,\ldots,n_m}(x)\, dx = u'_{n_1,\ldots,n_m}(0-0). \tag{18}$$

Taking conditions (17) and (18) into account from representations (15) and (16) we obtain

$$A_{1n_1,\ldots,n_m} = \frac{1}{2}\left(B_{1n_1,\ldots,n_m} + \omega B_{2n_1,\ldots,n_m}\right), \quad A_{2n_1,\ldots,n_m} = \frac{1}{2}\left(B_{1n_1,\ldots,n_m} - \omega B_{2n_1,\ldots,n_m}\right).$$

Then the functions (15) and (16) take the forms

$$u_{n_1,\ldots,n_m}(t) = B_{1n_1,\ldots,n_m} \cosh \lambda_{n_1,\ldots,n_m} t + \omega B_{2n_1,\ldots,n_m} \sinh \lambda_n t + \eta_{1n_1,\ldots,n_m}(t),\ t>0, \tag{19}$$

$$u_{n_1,\ldots,n_m}(t) = B_{1n_1,\ldots,n_m} \cos \lambda_{n_1,\ldots,n_m}\omega t + B_{2n_1,\ldots,n_m} \sin \lambda_{n_1,\ldots,n_m}\omega t + \eta_{2n_1,\ldots,n_m}(t),\ t<0. \tag{20}$$

Taking formula (7) into account we will rewrite conditions (2) and (3) in the following forms

$$\int_0^T u_{n_1,\ldots,n_m}(t)\, dt = \int_{\Omega_l^m} \int_0^T U(t, x)\, dt\, \vartheta_{n_1,\ldots,n_m}(x)\, dx =$$

$$= \int_{\Omega_l^m} \varphi_1(x)\,\vartheta_{n_1,\ldots,n_m}(x)\,dx = \varphi_{1\,n_1,\ldots,n_m}, \qquad (21)$$

$$\int_{-T}^{0} u_{n_1,\ldots,n_m}(t)\,dt = \int_{\Omega_l^m} \int_{-T}^{0} U(t,x)\,dt\,\vartheta_{n_1,\ldots,n_m}(x)\,dx = \qquad (22)$$

$$= \int_{\Omega_l^m} \varphi_2(x)\,\vartheta_{n_1,\ldots,n_m}(x)\,dx = \varphi_{2\,n_1,\ldots,n_m}.$$

The coefficients $B_{1\,n_1,\ldots,n_m}$ and $B_{2\,n_1,\ldots,n_m}$ in (19) and (20) are unknown. To find them we use the conditions (21) and (22):

$$\int_0^T u_{n_1,\ldots,n_m}(t)\,dt =$$

$$= \int_0^T \left[B_{1\,n_1,\ldots,n_m} \cosh \lambda_{n_1,\ldots,n_m} t + \omega B_{2\,n_1,\ldots,n_m} \sinh \lambda_{n_1,\ldots,n_m} t + \eta_{1\,n_1,\ldots,n_m}(t) \right] dt =$$

$$= \frac{1}{\lambda_{n_1,\ldots,n_m}} \left[B_{1\,n_1,\ldots,n_m} \sinh \lambda_{n_1,\ldots,n_m} T + \omega B_{2\,n_1,\ldots,n_m} \left(\cosh \lambda_{n_1,\ldots,n_m} T - 1\right) \right] +$$

$$+ \xi_{1\,n_1,\ldots,n_m} = \varphi_{1\,n_1,\ldots,n_m}, \qquad (23)$$

$$\int_{-T}^{0} u_{n_1,\ldots,n_m}(t)\,dt =$$

$$= \int_{-T}^{0} \left[B_{1\,n_1,\ldots,n_m} \cos \lambda_{n_1,\ldots,n_m} \omega t + B_{2\,n_1,\ldots,n_m} \sin \lambda_{n_1,\ldots,n_m} \omega t + \eta_{2\,n_1,\ldots,n_m}(t) \right] dt =$$

$$= \frac{1}{\lambda_{n_1,\ldots,n_m}\omega} \left[B_{1\,n_1,\ldots,n_m} \sin \lambda_{n_1,\ldots,n_m} \omega T + B_{2\,n_1,\ldots,n_m} \left(\cos \lambda_{n_1,\ldots,n_m} \omega T - 1\right) \right] +$$

$$+ \xi_{2\,n_1,\ldots,n_m} = \varphi_{2\,n_1,\ldots,n_m}, \qquad (24)$$

where $\xi_{1\,n_1,\ldots,n_m} = \int_0^T \eta_{1\,n_1,\ldots,n_m}(t)\,dt,\ \xi_{2\,n_1,\ldots,n_m} = \int_{-T}^{0} \eta_{2\,n_1,\ldots,n_m}(t)\,dt.$

Relations (23) and (24) are considered as a system of algebraic equations (SAE) with respect to unknown coefficients $B_{1\,n_1,\ldots,n_m}$ and $B_{2\,n_1,\ldots,n_m}$

$$\begin{cases} B_{1\,n_1,\ldots,n_m} \sinh \lambda_{n_1,\ldots,n_m} T + \omega B_{2\,n_1,\ldots,n_m} \left(\cosh \lambda_{n_1,\ldots,n_m} T - 1\right) = \\ = \lambda_{n_1,\ldots,n_m} \varphi_{1\,n_1,\ldots,n_m} - \lambda_{n_1,\ldots,n_m} \xi_{1\,n_1,\ldots,n_m}, \\ B_{1\,n_1,\ldots,n_m} \sin \lambda_{n_1,\ldots,n_m} \omega T + B_{2\,n_1,\ldots,n_m} \left(\cos \lambda_{n_1,\ldots,n_m} \omega T - 1\right) = \\ = \lambda_{n_1,\ldots,n_m} \varphi_{2\,n_1,\ldots,n_m} \omega - \lambda_{n_1,\ldots,n_m} \omega\, \xi_{2\,n_1,\ldots,n_m}. \end{cases}$$

If we assume that

$$\sigma_{n_1,\ldots,n_m} =$$

$$= \sinh \lambda_{n_1,\ldots,n_m} T \left(\cos \lambda_{n_1,\ldots,n_m} \omega T - 1\right) - \omega \sin \lambda_{n_1,\ldots,n_m} \omega T \left(\cosh \lambda_{n_1,\ldots,n_m} T - 1\right) \neq 0, \qquad (25)$$

then SAE with respect to $B_{1\,n_1,\ldots,n_m}$ and $B_{2\,n_1,\ldots,n_m}$ is uniquely solvable. Solving this system from (19) and (20) we arrive at the following representations

$$u_{n_1,\ldots,n_m}(t,\omega) = \frac{\lambda_{n_1,\ldots,n_m}}{\sigma_{n_1,\ldots,n_m}} \left[\varphi_{1\,n_1,\ldots,n_m} M_{1\,n_1,\ldots,n_m}(t,\omega) + \varphi_{2\,n_1,\ldots,n_m} M_{2\,n_1,\ldots,n_m}(t,\omega) + \right.$$

$$+\xi_{1n_1,\ldots,n_m}M_{3n_1,\ldots,n_m}(t,\omega)+\xi_{2n_1,\ldots,n_m}M_{4n_1,\ldots,n_m}(t,\omega)\big]+\eta_{1n_1,\ldots,n_m}(t),\quad t>0,\qquad(26)$$

$$u_{n_1,\ldots,n_m}(t,\omega)=\frac{\lambda_{n_1,\ldots,n_m}}{\sigma_{n_1,\ldots,n_m}}\big[\varphi_{1n_1,\ldots,n_m}N_{1n_1,\ldots,n_m}(t,\omega)+\varphi_{2n_1,\ldots,n_m}N_{2n_1,\ldots,n_m}(t,\omega)+$$

$$+\xi_{1n_1,\ldots,n_m}N_{3n_1,\ldots,n_m}(t,\omega)+\xi_{2n_1,\ldots,n_m}N_{4n_1,\ldots,n_m}(t,\omega)\big]+\eta_{2n_1,\ldots,n_m}(t),\quad t<0,\qquad(27)$$

where

$$M_{1n_1,\ldots,n_m}(t,\omega)=(\cos\lambda_{n_1,\ldots,n_m}\omega T-1)\cosh\lambda_{n_1,\ldots,n_m}t-\sin\lambda_{n_1,\ldots,n_m}\omega T\sinh\lambda_{n_1,\ldots,n_m}t,$$

$$M_{2n_1,\ldots,n_m}(t,\omega)=\omega^2(1-\cosh\lambda_{n_1,\ldots,n_m}T)\cosh\lambda_{n_1,\ldots,n_m}t+\omega\sinh\lambda_{n_1,\ldots,n_m}T\sinh\lambda_{n_1,\ldots,n_m}t,$$

$$M_{3n_1,\ldots,n_m}(t,\omega)=(1-\cos\lambda_{n_1,\ldots,n_m}\omega T)\cosh\lambda_{n_1,\ldots,n_m}t+\sin\lambda_{n_1,\ldots,n_m}\omega T\sinh\lambda_{n_1,\ldots,n_m}t,$$

$$M_{4n_1,\ldots,n_m}(t,\omega)=\omega^2(\cosh\lambda_{n_1,\ldots,n_m}T-1)\cosh\lambda_{n_1,\ldots,n_m}t-\omega\sinh\lambda_{n_1,\ldots,n_m}T\sinh\lambda_{n_1,\ldots,n_m}t,$$

$$N_{1n_1,\ldots,n_m}(t,\omega)=(\cos\lambda_{n_1,\ldots,n_m}\omega T-1)\cos\lambda_{n_1,\ldots,n_m}\omega t-\sin\lambda_{n_1,\ldots,n_m}\omega T\sin\lambda_{n_1,\ldots,n_m}\omega t,$$

$$N_{2n_1,\ldots,n_m}(t,\omega)=\omega(1-\cosh\lambda_{n_1,\ldots,n_m}T)\cos\lambda_{n_1,\ldots,n_m}\omega t+\omega\sinh\lambda_{n_1,\ldots,n_m}T\sin\lambda_{n_1,\ldots,n_m}\omega t,$$

$$N_{3n_1,\ldots,n_m}(t,\omega)=(1-\cos\lambda_{n_1,\ldots,n_m}\omega T)\cos\lambda_{n_1,\ldots,n_m}\omega t+\sin\lambda_{n_1,\ldots,n_m}\omega T\sin\lambda_{n_1,\ldots,n_m}\omega t,$$

$$N_{4n_1,\ldots,n_m}(t,\omega)=\omega^2(\cosh\lambda_{n_1,\ldots,n_m}T-1)\cos\lambda_{n_1,\ldots,n_m}\omega t-\omega\sinh\lambda_{n_1,\ldots,n_m}T\sin\lambda_{n_1,\ldots,n_m}\omega t.$$

Taking the following presentations

$$\eta_{1n_1,\ldots,n_m}(t)=\nu\alpha_{n_1,\ldots,n_m}h_{1n_1,\ldots,n_m}(t)+g_{1n_1,\ldots,n_m}h_{2n_1,\ldots,n_m}(t),$$

$$\eta_{2n_1,\ldots,n_m}(t)=\nu\beta_{n_1,\ldots,n_m}\delta_{1n_1,\ldots,n_m}(t)+g_{2n_1,\ldots,n_m}\delta_{2n_1,\ldots,n_m}(t)$$

into account representations (26) and (27) are written in the following forms

$$u_{n_1,\ldots,n_m}(t,\omega)=$$

$$=P_{1n_1,\ldots,n_m}(t,\omega)+\nu\alpha_{n_1,\ldots,n_m}P_{2n_1,\ldots,n_m}(t,\omega)+\nu\beta_{n_1,\ldots,n_m}P_{3n_1,\ldots,n_m}(t,\omega)+$$

$$+g_{1n_1,\ldots,n_m}P_{4n_1,\ldots,n_m}(t,\omega)+g_{2n_1,\ldots,n_m}P_{5n_1,\ldots,n_m}(t,\omega),\quad t>0,\qquad(28)$$

$$u_{n_1,\ldots,n_m}(t,\omega)=$$

$$=Q_{1n_1,\ldots,n_m}(t,\omega)+\nu\alpha_{n_1,\ldots,n_m}Q_{2n_1,\ldots,n_m}(t,\omega)+\nu\beta_{n_1,\ldots,n_m}Q_{3n_1,\ldots,n_m}(t,\omega)+$$

$$+g_{1n_1,\ldots,n_m}Q_{4n_1,\ldots,n_m}(t,\omega)+g_{2n_1,\ldots,n_m}Q_{5n_1,\ldots,n_m}(t,\omega),\quad t<0,\qquad(29)$$

where

$$P_{1n_1,\ldots,n_m}(t,\omega)=\frac{\lambda_{n_1,\ldots,n_m}}{\sigma_{n_1,\ldots,n_m}}\big[\varphi_{1n_1,\ldots,n_m}M_{1n_1,\ldots,n_m}(t,\omega)+\varphi_{2n_1,\ldots,n_m}M_{2n_1,\ldots,n_m}(t,\omega)\big],$$

$$P_{2n_1,\ldots,n_m}(t,\omega)=\frac{\lambda_{n_1,\ldots,n_m}}{\sigma_{n_1,\ldots,n_m}}M_{3n_1,\ldots,n_m}(t,\omega)\int_0^T h_{1n_1,\ldots,n_m}(t)\,dt+h_{1n_1,\ldots,n_m}(t),$$

$$P_{3n_1,\ldots,n_m}(t,\omega)=\frac{\lambda_{n_1,\ldots,n_m}}{\sigma_{n_1,\ldots,n_m}}M_{4n_1,\ldots,n_m}(t,\omega)\int_{-T}^0 \delta_{1n_1,\ldots,n_m}(t)\,dt,$$

$$P_{4n_1,\ldots,n_m}(t,\omega)=\frac{\lambda_{n_1,\ldots,n_m}}{\sigma_{n_1,\ldots,n_m}}M_{3n_1,\ldots,n_m}(t,\omega)\int_0^T h_{2n_1,\ldots,n_m}(t)\,dt+h_{2n_1,\ldots,n_m}(t),$$

$$P_{5\,n_1,\ldots,n_m}(t,\omega) = \frac{\lambda_{n_1,\ldots,n_m}}{\sigma_{n_1,\ldots,n_m}} M_{4\,n_1,\ldots,n_m}(t,\omega) \int_{-T}^{0} \delta_{2\,n_1,\ldots,n_m}(t)\,dt,$$

$$Q_{1\,n_1,\ldots,n_m}(t,\omega) = \frac{\lambda_{n_1,\ldots,n_m}}{\sigma_{n_1,\ldots,n_m}} \left[\varphi_{1\,n_1,\ldots,n_m} N_{1\,n_1,\ldots,n_m}(t,\omega) + \varphi_{2\,n_1,\ldots,n_m} N_{2\,n_1,\ldots,n_m}(t,\omega) \right],$$

$$Q_{2\,n_1,\ldots,n_m}(t,\omega) = \frac{\lambda_{n_1,\ldots,n_m}}{\sigma_{n_1,\ldots,n_m}} N_{3\,n_1,\ldots,n_m}(t,\omega) \int_{0}^{T} h_{1\,n_1,\ldots,n_m}(t)\,dt,$$

$$Q_{3\,n_1,\ldots,n_m}(t,\omega) = \frac{\lambda_{n_1,\ldots,n_m}}{\sigma_{n_1,\ldots,n_m}} N_{4\,n_1,\ldots,n_m}(t,\omega) \int_{-T}^{0} \delta_{1\,n_1,\ldots,n_m}(t)\,dt + \delta_{1\,n_1,\ldots,n_m}(t),$$

$$Q_{4\,n_1,\ldots,n_m}(t,\omega) = \frac{\lambda_{n_1,\ldots,n_m}}{\sigma_{n_1,\ldots,n_m}} N_{3\,n_1,\ldots,n_m}(t,\omega) \int_{0}^{T} h_{2\,n_1,\ldots,n_m}(t)\,dt,$$

$$Q_{5\,n_1,\ldots,n_m}(t,\omega) = \frac{\lambda_{n_1,\ldots,n_m}}{\sigma_{n_1,\ldots,n_m}} N_{4\,n_1,\ldots,n_m}(t,\omega) \int_{-T}^{0} \delta_{2\,n_1,\ldots,n_m}(t)\,dt + \delta_{2\,n_1,\ldots,n_m}(t).$$

We substitute (28) and (29) into (11) and (12), respectively. Then we obtain a countable system of two algebraic equations (CSTAE)

$$\begin{cases} \alpha_{n_1,\ldots,n_m}(1 - \nu E_{n_1,\ldots,n_m}) - \beta_{n_1,\ldots,n_m} \nu F_{n_1,\ldots,n_m}(\omega) = \Phi_{n_1,\ldots,n_m}(\omega), \\ -\alpha_{n_1,\ldots,n_m} \nu H_{n_1,\ldots,n_m}(\omega) + \beta_{n_1,\ldots,n_m}(1 - \nu G_{n_1,\ldots,n_m}) = \Psi_{n_1,\ldots,n_m}(\omega), \end{cases} \quad (30)$$

where

$$E_{n_1,\ldots,n_m} = \int_{0}^{T} b_1(t) P_{2\,n_1,\ldots,n_m}(t)\,dt,\ F_{n_1,\ldots,n_m}(\omega) = \int_{0}^{T} b_1(t) P_{3\,n_1,\ldots,n_m}(t,\omega)\,dt,$$

$$H_{n_1,\ldots,n_m}(\omega) = \int_{-T}^{0} b_2(t) Q_{2\,n_1,\ldots,n_m}(t,\omega)\,dt,\ G_{n_1,\ldots,n_m} = \int_{-T}^{0} b_2(t) Q_{3\,n_1,\ldots,n_m}(t)\,dt,$$

$$\Phi_{n_1,\ldots,n_m}(\omega) = \varphi_{1\,n_1,\ldots,n_m} P_{01\,n_1,\ldots,n_m} +$$
$$+ \varphi_{2\,n_1,\ldots,n_m} P_{02\,n_1,\ldots,n_m} + g_{1\,n_1,\ldots,n_m} P_{03\,n_1,\ldots,n_m} + g_{2\,n_1,\ldots,n_m} P_{04\,n_1,\ldots,n_m}, \quad (31)$$

$$\Psi_{n_1,\ldots,n_m}(\omega) = \varphi_{1\,n_1,\ldots,n_m} Q_{01\,n_1,\ldots,n_m} +$$
$$+ \varphi_{2\,n_1,\ldots,n_m} Q_{02\,n_1,\ldots,n_m} + g_{1\,n_1,\ldots,n_m} Q_{03\,n_1,\ldots,n_m} + g_{2\,n_1,\ldots,n_m} Q_{04\,n_1,\ldots,n_m}, \quad (32)$$

$$P_{0i\,n_1,\ldots,n_m} = \frac{\lambda_{n_1,\ldots,n_m}}{\sigma_{n_1,\ldots,n_m}} \int_{-T}^{0} b_2(t) M_{i\,n_1,\ldots,n_m}(t,\omega)\,dt,\ i = 1, 2,$$

$$P_{0j\,n_1,\ldots,n_m} = \frac{\lambda_{n_1,\ldots,n_m}}{\sigma_{n_1,\ldots,n_m}} \int_{-T}^{0} b_2(t) P_{1+j,\,n_1,\ldots,n_m}(t,\omega)\,dt,\ j = 3, 4,$$

$$Q_{0i\,n_1,\ldots,n_m} = \frac{\lambda_{n_1,\ldots,n_m}}{\sigma_{n_1,\ldots,n_m}} \int_{-T}^{0} b_2(t) N_{i\,n_1,\ldots,n_m}(t,\omega)\,dt,\ i = 1, 2,$$

$$Q_{0jn_1,\ldots,n_m} = \frac{\lambda_{n_1,\ldots,n_m}}{\sigma_{n_1,\ldots,n_m}} \int_{-T}^{0} b_2(t)\, Q_{1+j,n_1,\ldots,n_m}(t,\omega)\, dt, \quad j = 3, 4.$$

For the unique solvability of CSTAE (30) the following condition is required

$$\Delta_{n_1,\ldots,n_m}(\nu) = \begin{vmatrix} 1 - \nu\, E_{n_1,\ldots,n_m} & -\nu\, F_{n_1,\ldots,n_m}(\omega) \\ -\nu\, H_{n_1,\ldots,n_m}(\omega) & 1 - \nu\, G_{n_1,\ldots,n_m} \end{vmatrix} =$$

$$= (E_{n_1,\ldots,n_m} G_{n_1,\ldots,n_m} - H_{n_1,\ldots,n_m}(\omega) F_{n_1,\ldots,n_m}(\omega))\, \nu^2 -$$

$$- (E_{n_1,\ldots,n_m} + G_{n_1,\ldots,n_m})\, \nu + 1 \neq 0. \tag{33}$$

A quadratic equation has no real roots, if its discriminant is negative. Therefore, from condition (33) we arrive at the following condition

$$(E_{n_1,\ldots,n_m} - G_{n_1,\ldots,n_m})^2 + 4 H_{n_1,\ldots,n_m}(\omega) F_{n_1,\ldots,n_m}(\omega) < 0. \tag{34}$$

Let condition (34) be fulfilled. Then we solve the CSTAE (30):

$$\alpha_{n_1,\ldots,n_m} = \frac{\Phi_{n_1,\ldots,n_m}(\omega) + \nu\, (\Psi_{n_1,\ldots,n_m}(\omega) F_{n_1,\ldots,n_m}(\omega) - \Phi_{n_1,\ldots,n_m}(\omega) G_{n_1,\ldots,n_m})}{\Delta_{n_1,\ldots,n_m}(\nu)},$$

$$\beta_{n_1,\ldots,n_m} = \frac{\Psi_{n_1,\ldots,n_m}(\omega) + \nu\, (\Phi_{n_1,\ldots,n_m}(\omega) H_{n_1,\ldots,n_m}(\omega) - \Psi_{n_1,\ldots,n_m}(\omega) E_{n_1,\ldots,n_m})}{\Delta_{n_1,\ldots,n_m}(\nu)}.$$

Substituting these solutions into (28) and (29), we obtain

$$u_{n_1,\ldots,n_m}(t,\omega,\nu) = P_{1\,n_1,\ldots,n_m}(t,\omega) +$$

$$+ \frac{\nu}{\Delta_{n_1,\ldots,n_m}(\nu)} [\Phi_{n_1,\ldots,n_m}(\omega)(1 - \nu G_{n_1,\ldots,n_m}) + \nu \Psi_{n_1,\ldots,n_m}(\omega) F_{n_1,\ldots,n_m}(\omega)] P_{2\,n_1,\ldots,n_m}(t,\omega) +$$

$$+ \frac{\nu}{\Delta_{n_1,\ldots,n_m}(\nu)} [\nu \Phi_{n_1,\ldots,n_m}(\omega) H_{n_1,\ldots,n_m}(\omega) + \Psi_{n_1,\ldots,n_m}(\omega)(1 - \nu E_{n_1,\ldots,n_m})] P_{3\,n_1,\ldots,n_m}(t,\omega) +$$

$$+ g_{1\,n_1,\ldots,n_m} P_{4\,n_1,\ldots,n_m}(t,\omega) + g_{2\,n_1,\ldots,n_m} P_{5\,n_1,\ldots,n_m}(t,\omega), \quad t > 0, \tag{35}$$

$$u_{n_1,\ldots,n_m}(t,\omega,\nu) = Q_{1\,n_1,\ldots,n_m}(t,\omega) +$$

$$+ \frac{\nu}{\Delta_{n_1,\ldots,n_m}(\nu)} [\Phi_{n_1,\ldots,n_m}(\omega)(1 - \nu G_{n_1,\ldots,n_m}) + \nu \Psi_{n_1,\ldots,n_m}(\omega) F_{n_1,\ldots,n_m}(\omega)] Q_{2\,n_1,\ldots,n_m}(t,\omega) +$$

$$+ \frac{\nu}{\Delta_{n_1,\ldots,n_m}(\nu)} [\nu \Phi_{n_1,\ldots,n_m}(\omega) H_{n_1,\ldots,n_m}(\omega) + \Psi_{n_1,\ldots,n_m}(\omega)(1 - \nu E_{n_1,\ldots,n_m})] Q_{3\,n_1,\ldots,n_m}(t,\omega) +$$

$$+ g_{1\,n_1,\ldots,n_m} Q_{4\,n_1,\ldots,n_m}(t,\omega) + g_{2\,n_1,\ldots,n_m} Q_{5\,n_1,\ldots,n_m}(t,\omega), \quad t < 0, \tag{36}$$

Taking (31), (32) and the following relations

$$P_{1\,n_1,\ldots,n_m}(t,\omega) = \frac{\lambda_{n_1,\ldots,n_m}}{\sigma_{n_1,\ldots,n_m}} \left[\varphi_{1\,n_1,\ldots,n_m} M_{1\,n_1,\ldots,n_m}(t,\omega) + \varphi_{2\,n_1,\ldots,n_m} M_{2\,n_1,\ldots,n_m}(t,\omega) \right],$$

$$Q_{1\,n_1,\ldots,n_m}(t,\omega) = \frac{\lambda_{n_1,\ldots,n_m}}{\sigma_{n_1,\ldots,n_m}} \left[\varphi_{1\,n_1,\ldots,n_m} N_{1\,n_1,\ldots,n_m}(t,\omega) + \varphi_{2\,n_1,\ldots,n_m} N_{2\,n_1,\ldots,n_m}(t,\omega) \right]$$

into account the representations (35) and (36) we rewrite in the following views

$$u_{n_1,\ldots,n_m}(t,\omega,\nu) = \varphi_{1\,n_1,\ldots,n_m} V_{1\,n_1,\ldots,n_m}(t,\omega,\nu) + \varphi_{2\,n_1,\ldots,n_m} V_{2\,n_1,\ldots,n_m}(t,\omega,\nu) +$$

$$+ g_{1\,n_1,\ldots,n_m} V_{3\,n_1,\ldots,n_m}(t,\omega,\nu) + g_{2\,n_1,\ldots,n_m} V_{4\,n_1,\ldots,n_m}(t,\omega,\nu), \quad t > 0, \tag{37}$$

$$u_{n_1,\ldots,n_m}(t,\omega,\nu) = \varphi_{1\,n_1,\ldots,n_m} W_{1\,n_1,\ldots,n_m}(t,\omega,\nu) + \varphi_{2\,n_1,\ldots,n_m} W_{2\,n_1,\ldots,n_m}(t,\omega,\nu) +$$
$$+ g_{1\,n_1,\ldots,n_m} W_{3\,n_1,\ldots,n_m}(t,\omega,\nu) + g_{2\,n_1,\ldots,n_m} W_{4\,n_1,\ldots,n_m}(t,\omega,\nu),\ t<0, \tag{38}$$

where

$$V_{i\,n_1,\ldots,n_m}(t,\omega,\nu) = \frac{\lambda_{n_1,\ldots,n_m}}{\sigma_{n_1,\ldots,n_m}} M_{i\,n_1,\ldots,n_m}(t,\omega) +$$
$$+ P_{0i\,n_1,\ldots,n_m} V_{01\,n_1,\ldots,n_m}(t,\omega,\nu) + Q_{0i\,n_1,\ldots,n_m} V_{02\,n_1,\ldots,n_m}(t,\omega,\nu),\ i=1,2,$$
$$V_{j\,n_1,\ldots,n_m}(t,\omega,\nu) = P_{(i+1)\,n_1,\ldots,n_m}(t,\omega) +$$
$$+ P_{0j\,n_1,\ldots,n_m} V_{01\,n_1,\ldots,n_m}(t,\omega,\nu) + Q_{0j\,n_1,\ldots,n_m} V_{02\,n_1,\ldots,n_m}(t,\omega,\nu),\ j=3,4,$$
$$W_{i\,n_1,\ldots,n_m}(t,\omega,\nu) = \frac{\lambda}{\sigma_{n_1,\ldots,n_m}} N_{i\,n_1,\ldots,n_m}(t,\omega) +$$
$$+ P_{0i\,n_1,\ldots,n_m} W_{01\,n_1,\ldots,n_m}(t,\omega,\nu) + Q_{0i\,n_1,\ldots,n_m} W_{02\,n_1,\ldots,n_m}(t,\omega,\nu),\ i=1,2,$$
$$W_{j\,n_1,\ldots,n_m}(t,\omega,\nu) = Q_{(i+1)\,n_1,\ldots,n_m}(t,\omega) +$$
$$+ P_{0j\,n_1,\ldots,n_m} W_{01\,n_1,\ldots,n_m}(t,\omega,\nu) + Q_{0j\,n_1,\ldots,n_m} W_{02\,n_1,\ldots,n_m}(t,\omega,\nu),\ j=3,4,$$
$$V_{01\,n_1,\ldots,n_m}(t,\omega,\nu) +$$
$$= \frac{\nu}{\Delta_{n_1,\ldots,n_m}(\nu)} \left[(1-\nu G_{n_1,\ldots,n_m}) P_{2\,n_1,\ldots,n_m}(t,\omega) + \nu H_{n_1,\ldots,n_m}(\omega) P_{3\,n_1,\ldots,n_m}(t,\omega) \right],$$
$$V_{02\,n_1,\ldots,n_m}(t,\omega,\nu) = \nu P_{2\,n_1,\ldots,n_m}(t,\omega) + (1-\nu E_{n_1,\ldots,n_m}) P_{3\,n_1,\ldots,n_m}(t,\omega),$$
$$W_{01\,n_1,\ldots,n_m}(t,\omega,\nu) =$$
$$= \frac{\nu}{\Delta_{n_1,\ldots,n_m}(\nu)} \left[(1-\nu G_{n_1,\ldots,n_m}) Q_{2\,n_1,\ldots,n_m}(t,\omega) + \nu H_{n_1,\ldots,n_m}(\omega) Q_{3\,n_1,\ldots,n_m}(t,\omega) \right],$$
$$W_{02\,n_1,\ldots,n_m}(t,\omega,\nu) =$$
$$= \frac{\nu}{\Delta_{n_1,\ldots,n_m}(\nu)} \left[\nu F_{n_1,\ldots,n_m}(\omega) Q_{2\,n_1,\ldots,n_m}(t,\omega) + (1-\nu E_{n_1,\ldots,n_m}) Q_{3\,n_1,\ldots,n_m}(t,\omega) \right].$$

Now we substitute representations (37) and (38) into the Fourier series (6) and obtain the following formal solution of the direct problem (1)–(4)

$$U(t,x,\omega,\nu) =$$
$$= \sum_{n_1,\ldots,n_m=1}^{\infty} \vartheta_{n_1,\ldots,n_m}(x) \left[\varphi_{1\,n_1,\ldots,n_m} V_{1\,n_1,\ldots,n_m}(t,\omega,\nu) + \varphi_{2\,n_1,\ldots,n_m} V_{2\,n_1,\ldots,n_m}(t,\omega,\nu) + \right.$$
$$\left. + g_{1\,n_1,\ldots,n_m} V_{3\,n_1,\ldots,n_m}(t,\omega,\nu) + g_{2\,n_1,\ldots,n_m} V_{4\,n_1,\ldots,n_m}(t,\omega,\nu) \right],\ t>0, \tag{39}$$
$$U(t,x,\omega,\nu) =$$
$$= \sum_{n_1,\ldots,n_m=1}^{\infty} \vartheta_{n_1,\ldots,n_m}(x) \left[\varphi_{1\,n_1,\ldots,n_m} W_{1\,n_1,\ldots,n_m}(t,\omega,\nu) + \varphi_{2\,n_1,\ldots,n_m} W_{2\,n_1,\ldots,n_m}(t,\omega,\nu) + \right.$$
$$\left. + g_{1\,n_1,\ldots,n_m} W_{3\,n_1,\ldots,n_m}(t,\omega,\nu) + g_{2\,n_1,\ldots,n_m} W_{4\,n_1,\ldots,n_m}(t,\omega,\nu) \right],\ t<0. \tag{40}$$

3. Inverse Problem (1)–(5). The Regular Case of the Spectral Parameter ω

We use the additional conditions (5) and from the Fourier series (39) and (40) we obtain that

$$\psi_1(x) = U(t_1,x,\omega,\nu) = \sum_{n_1,\ldots,n_m=1}^{\infty} \vartheta_{n_1,\ldots,n_m}(x) \times$$

$$\times \left[\varphi_{1n_1,\ldots,n_m} V_{1n_1,\ldots,n_m}(t_1,\omega,\nu) + \varphi_{2n_1,\ldots,n_m} V_{2n_1,\ldots,n_m}(t_1,\omega,\nu) + \right.$$
$$\left. + g_{1n_1,\ldots,n_m} V_{3n_1,\ldots,n_m}(t_1,\omega,\nu) + g_{2n_1,\ldots,n_m} V_{4n_1,\ldots,n_m}(t_1,\omega,\nu) \right], \quad 0 < t_1 < T, \tag{41}$$

$$\psi_2(x) = U(t_2, x, \omega, \nu) = \sum_{n_1,\ldots,n_m=1}^{\infty} \vartheta_{n_1,\ldots,n_m}(x) \times$$
$$\times \left[\varphi_{1n_1,\ldots,n_m} W_{1n_1,\ldots,n_m}(t_2,\omega,\nu) + \varphi_{2n_1,\ldots,n_m} W_{2n_1,\ldots,n_m}(t_2,\omega,\nu) + \right.$$
$$\left. + g_{1n_1,\ldots,n_m} V_{3n_1,\ldots,n_m}(t_2,\omega,\nu) + g_{2n_1,\ldots,n_m} V_{4n_1,\ldots,n_m}(t_2,\omega,\nu) \right], \quad -T < t_2 < 0. \tag{42}$$

Assume that the functions $\psi_i(x)$ are expanded in Fourier series

$$\psi_i(x) = \sum_{n_1,\ldots,n_m=1}^{\infty} \psi_{in_1,\ldots,n_m} \vartheta_{n_1,\ldots,n_m}(x), \tag{43}$$

where $\psi_{in_1,\ldots,n_m} = \int_{\Omega_l^m} \psi_i(x) \vartheta_{n_1,\ldots,n_m}(x) dx, \quad i = 1, 2, \quad n_1, \ldots, n_m = 1, 2, \ldots$

Then, taking into account (43), from (41) and (42) we obtain

$$\psi_{1n_1,\ldots,n_m} = \varphi_{1n_1,\ldots,n_m} V_{1n_1,\ldots,n_m}(t_1,\omega,\nu) + \varphi_{2n_1,\ldots,n_m} V_{2n_1,\ldots,n_m}(t_1,\omega,\nu) +$$
$$+ g_{1n_1,\ldots,n_m} V_{3n_1,\ldots,n_m}(t_1,\omega,\nu) + g_{2n_1,\ldots,n_m} V_{4n_1,\ldots,n_m}(t_1,\omega,\nu), \quad 0 < t_1 < T,$$

$$\psi_{2n_1,\ldots,n_m} = \varphi_{1n_1,\ldots,n_m} W_{1n_1,\ldots,n_m}(t_2,\omega,\nu) + \varphi_{2n_1,\ldots,n_m} W_{2n_1,\ldots,n_m}(t_2,\omega,\nu) +$$
$$+ g_{1n_1,\ldots,n_m} V_{3n_1,\ldots,n_m}(t_2,\omega,\nu) + g_{2n_1,\ldots,n_m} V_{4n_1,\ldots,n_m}(t_2,\omega,\nu), \quad -T < t_2 < 0.$$

Hence we find a system of two algebraic equations for finding the coefficients of the redefinition functions $g_{1n_1,\ldots,n_m}$ and $g_{2n_1,\ldots,n_m}$

$$\begin{cases} g_{1n_1,\ldots,n_m} V_{3n_1,\ldots,n_m}(t_1,\omega,\nu) + g_{2n_1,\ldots,n_m} V_{4n_1,\ldots,n_m}(t_1,\omega,\nu) = \\ = \psi_{1n_1,\ldots,n_m} - \varphi_{1n_1,\ldots,n_m} V_{1n_1,\ldots,n_m}(t_1,\omega,\nu) - \varphi_{2n_1,\ldots,n_m} V_{2n_1,\ldots,n_m}(t_1,\omega,\nu), \\ g_{1n_1,\ldots,n_m} W_{3n_1,\ldots,n_m}(t_2,\omega,\nu) + g_{2n_1,\ldots,n_m} W_{4n_1,\ldots,n_m}(t_2,\omega,\nu) = \\ = \psi_{2n_1,\ldots,n_m} - \varphi_{1n_1,\ldots,n_m} W_{1n_1,\ldots,n_m}(t_2,\omega,\nu) - \varphi_{2n_1,\ldots,n_m} W_{2n_1,\ldots,n_m}(t_2,\omega,\nu). \end{cases}$$

Solving this system of algebraic equations, we obtain

$$g_{1n_1,\ldots,n_m}(\omega,\nu) =$$
$$= \frac{1}{r_{01n_1,\ldots,n_m}} \left[\psi_{1n_1,\ldots,n_m} W_{4n_1,\ldots,n_m}(t_2,\omega,\nu) - \psi_{2n_1,\ldots,n_m} V_{4n_1,\ldots,n_m}(t_1,\omega,\nu) + \right.$$
$$\left. + \varphi_{1n_1,\ldots,n_m} r_{11n_1,\ldots,n_m} + \varphi_{2n_1,\ldots,n_m} r_{12n_1,\ldots,n_m} \right], \tag{44}$$

$$g_{2n_1,\ldots,n_m}(\omega,\nu) =$$
$$= \frac{1}{r_{01n_1,\ldots,n_m}} \left[-\psi_{1n_1,\ldots,n_m} W_{3n_1,\ldots,n_m}(t_2,\omega,\nu) + \psi_{2n_1,\ldots,n_m} V_{3n_1,\ldots,n_m}(t_1,\omega,\nu) + \right.$$
$$\left. + \varphi_{1n_1,\ldots,n_m} r_{21n_1,\ldots,n_m} + \varphi_{2n_1,\ldots,n_m} r_{22n_1,\ldots,n_m} \right], \tag{45}$$

where $r_{01n_1,\ldots,n_m} =$
$$= V_{3n_1,\ldots,n_m}(t_1,\omega,\nu) W_{4n_1,\ldots,n_m}(t_2,\omega,\nu) - V_{4n_1,\ldots,n_m}(t_1,\omega,\nu) W_{3n_1,\ldots,n_m}(t_2,\omega,\nu) \neq 0,$$

$$r_{11n_1,\ldots,n_m} = -V_{1n_1,\ldots,n_m}(t_1,\omega,\nu) W_{4n_1,\ldots,n_m}(t_2,\omega,\nu) + V_{4n_1,\ldots,n_m}(t_1,\omega,\nu) W_{1n_1,\ldots,n_m}(t_2,\omega,\nu),$$

$$r_{12n_1,\ldots,n_m} = -V_{2n_1,\ldots,n_m}(t_1,\omega,\nu) W_{4n_1,\ldots,n_m}(t_2,\omega,\nu) + V_{4n_1,\ldots,n_m}(t_1,\omega,\nu) W_{2n_1,\ldots,n_m}(t_2,\omega,\nu),$$

$$r_{21\,n_1,\ldots,n_m} = -V_{3\,n_1,\ldots,n_m}(t_1,\omega,\nu)\,W_{1\,n_1,\ldots,n_m}(t_2,\omega,\nu) + V_{1\,n_1,\ldots,n_m}(t_1,\omega,\nu)\,W_{3\,n_1,\ldots,n_m}(t_2,\omega,\nu),$$

$$r_{22\,n_1,\ldots,n_m} = -V_{3\,n_1,\ldots,n_m}(t_1,\omega,\nu)\,W_{2\,n_1,\ldots,n_m}(t_2,\omega,\nu) + V_{2\,n_1,\ldots,n_m}(t_1,\omega,\nu)\,W_{3\,n_1,\ldots,n_m}(t_2,\omega,\nu).$$

Substituting representations (44) and (45) into the Fourier series (8), we obtain

$$g_1(x,\omega,\nu) = \frac{1}{r_{01\,n_1,\ldots,n_m}} \sum_{n_1,\ldots,n_m=1}^{\infty} \vartheta_{n_1,\ldots,n_m}(x)\,\big[\psi_{1\,n_1,\ldots,n_m} W_{4\,n_1,\ldots,n_m}(t_2,\omega,\nu) -$$

$$-\psi_{2\,n_1,\ldots,n_m} V_{4\,n_1,\ldots,n_m}(t_1,\omega,\nu) + \varphi_{1\,n_1,\ldots,n_m} r_{11\,n_1,\ldots,n_m} + \varphi_{2\,n_1,\ldots,n_m} r_{12\,n_1,\ldots,n_m}\big], \quad (46)$$

$$g_2(x,\omega,\nu) = \frac{1}{r_{01\,n_1,\ldots,n_m}} \sum_{n_1,\ldots,n_m=1}^{\infty} \vartheta_{n_1,\ldots,n_m}(x)\,\big[-\psi_{1\,n_1,\ldots,n_m} W_{3\,n_1,\ldots,n_m}(t_2,\omega,\nu) +$$

$$+\psi_{2\,n_1,\ldots,n_m} V_{3\,n_1,\ldots,n_m}(t_1,\omega,\nu) + \varphi_{1\,n_1,\ldots,n_m} r_{21\,n_1,\ldots,n_m} + \varphi_{2\,n_1,\ldots,n_m} r_{22\,n_1,\ldots,n_m}\big]. \quad (47)$$

Now we substitute representations (44) and (45) into the main series (39) and (40):

$$U(t,x,\omega,\nu) =$$

$$= \sum_{n_1,\ldots,n_m=1}^{\infty} \vartheta_{n_1,\ldots,n_m}(x)\,\big[\varphi_{1\,n_1,\ldots,n_m} D_{11\,n_1,\ldots,n_m}(t,\omega,\nu) + \varphi_{2\,n_1,\ldots,n_m} D_{12\,n_1,\ldots,n_m}(t,\omega,\nu) +$$

$$+\psi_{1\,n_1,\ldots,n_m} D_{13\,n_1,\ldots,n_m}(t,\omega,\nu) + \psi_{2\,n_1,\ldots,n_m} D_{14\,n_1,\ldots,n_m}(t,\omega,\nu)\big], \quad t > 0, \quad (48)$$

$$U(t,x,\omega,\nu) =$$

$$= \sum_{n_1,\ldots,n_m=1}^{\infty} \vartheta_{n_1,\ldots,n_m}(x)\,\big[\varphi_{1\,n_1,\ldots,n_m} D_{21\,n_1,\ldots,n_m}(t,\omega,\nu) + \varphi_{2\,n_1,\ldots,n_m} D_{22\,n_1,\ldots,n_m}(t,\omega,\nu) +$$

$$+\psi_{1\,n_1,\ldots,n_m} D_{23\,n_1,\ldots,n_m}(t,\omega,\nu) + \psi_{2\,n_1,\ldots,n_m} D_{24\,n_1,\ldots,n_m}(t,\omega,\nu)\big], \quad t < 0, \quad (49)$$

where

$$D_{1i\,n_1,\ldots,n_m}(t,\omega,\nu) = V_{i\,n_1,\ldots,n_m}(t,\omega,\nu) +$$

$$+ \frac{r_{1i\,n_1,\ldots,n_m}}{r_{01\,n_1,\ldots,n_m}} V_{3\,n_1,\ldots,n_m}(t,\omega,\nu) + \frac{r_{2i\,n_1,\ldots,n_m}}{r_{01\,n_1,\ldots,n_m}} V_{4\,n_1,\ldots,n_m}(t,\omega,\nu), \quad i = 1, 2,$$

$$D_{13\,n_1,\ldots,n_m}(t,\omega,\nu) = \frac{1}{r_{01\,n_1,\ldots,n_m}} \times$$

$$\times \big[V_{3\,n_1,\ldots,n_m}(t,\omega,\nu)\,W_{4\,n_1,\ldots,n_m}(t_2,\omega,\nu) - V_{4\,n_1,\ldots,n_m}(t,\omega,\nu)\,W_{3\,n_1,\ldots,n_m}(t_2,\omega,\nu)\big],$$

$$D_{14\,n_1,\ldots,n_m}(t,\omega,\nu) = \frac{1}{r_{01\,n_1,\ldots,n_m}} \times$$

$$\times \big[-V_{3\,n_1,\ldots,n_m}(t,\omega,\nu)\,V_{4\,n_1,\ldots,n_m}(t_1,\omega,\nu) + V_{4\,n_1,\ldots,n_m}(t,\omega,\nu)\,V_{3\,n_1,\ldots,n_m}(t_1,\omega,\nu)\big],$$

$$D_{2i\,n_1,\ldots,n_m}(t,\omega,\nu) = W_{i\,n_1,\ldots,n_m}(t,\omega,\nu) +$$

$$+ \frac{r_{1i\,n_1,\ldots,n_m}}{r_{01\,n_1,\ldots,n_m}} W_{3\,n_1,\ldots,n_m}(t,\omega,\nu) + \frac{r_{2i\,n_1,\ldots,n_m}}{r_{01\,n_1,\ldots,n_m}} W_{4\,n_1,\ldots,n_m}(t,\omega,\nu), \quad i = 1, 2,$$

$$D_{23\,n_1,\ldots,n_m}(t,\omega,\nu) = \frac{1}{r_{01\,n_1,\ldots,n_m}} \times$$

$$\times \big[W_{4\,n_1,\ldots,n_m}(t,\omega,\nu)\,W_{3\,n_1,\ldots,n_m}(t_2,\omega,\nu) - W_{3\,n_1,\ldots,n_m}(t,\omega,\nu)\,W_{4\,n_1,\ldots,n_m}(t_2,\omega,\nu)\big],$$

$$D_{24\,n_1,\ldots,n_m}(t,\omega,\nu) = \frac{1}{r_{01\,n_1,\ldots,n_m}} \times$$

$$\times \big[-W_{3\,n_1,\ldots,n_m}(t,\omega,\nu)\,V_{4\,n_1,\ldots,n_m}(t_1,\omega,\nu) + W_{4\,n_1,\ldots,n_m}(t,\omega,\nu)\,V_{3\,n_1,\ldots,n_m}(t_1,\omega,\nu)\big].$$

4. Convergence of Series (46)–(49)

We show that under certain conditions with respect to the functions $\varphi_i(x)$ and $\psi_i(x)$ ($i = 1, 2$) the series (46)–(49) converge absolutely and uniformly in the domain $\overline{\Omega}$. Indeed, according to the statement of the problem the functions $D_{ijn_1,\ldots,n_m}(t, \omega, v)$ ($i = 1, 2$; $j = \overline{1, 4}$) uniformly bounded on the segment $[-T; T]$. So $|D_{ijn_1,\ldots,n_m}(t, \omega, v)| < \infty$ for all $i = 1, 2$, $j = \overline{1, 4}$. Since $0 < \lambda_{n_1,\ldots,n_m} < 1$, then for any positive integers $n_1, \ldots, n_m$ there exist finite constant numbers C_{0i} ($i = 1, 2$), that there take place the following estimates

$$\max_{n_1,\ldots,n_m \in N} \left\{ \max_{t \in [0;T]} |D_{1jn_1,\ldots,n_m}(t, \omega, v)|; \max_{t \in [-T;0]} |D_{2jn_1,\ldots,n_m}(t, \omega, v)| \right\} \leq C_{01},$$

$$\max_{n_1,\ldots,n_m \in N} \left\{ \max_{t \in [0;T]} |D''_{1jn_1,\ldots,n_m}(t, \omega, v)|; \max_{t \in [-T;0]} |D''_{2jn_1,\ldots,n_m}(t, \omega, v)| \right\} \leq C_{02}, \quad (50)$$

$j = \overline{1, 4}$.

Condition A. We suppose that the functions $\varphi_i, \psi_i \in C^2[0; l]^m$, $i = 1, 2$ on the domain $[0; l]^m$ have piecewise continuous third order derivatives. Then by integrating in parts the following integrals three times with respect to the variable x_1

$$\varphi_{in_1,\ldots,n_m} = \int_{\Omega_l^m} \varphi_i(x) \vartheta_{n_1,\ldots,n_m}(x) dx, \quad \psi_{in_1,\ldots,n_m} = \int_{\Omega_l^m} \psi_i(x) \vartheta_{n_1,\ldots,n_m}(x) dx, \quad i = 1, 2$$

we derive that

$$\varphi_{in_1,\ldots,n_m} = -\left(\frac{l}{\pi}\right)^3 \frac{\varphi'''_{in_1,\ldots,n_m}}{n_1^3}, \quad \psi_{in_1,\ldots,n_m} = -\left(\frac{l}{\pi}\right)^3 \frac{\psi'''_{in_1,\ldots,n_m}}{n_1^3}, \quad (51)$$

where

$$\varphi'''_{in_1,\ldots,n_m} = \int_{\Omega_l^m} \frac{\partial^3 \varphi_i(x)}{\partial x_1^3} \vartheta_{n_1,\ldots,n_m}(x) dx, \quad \psi'''_{in_1,\ldots,n_m} = \int_{\Omega_l^m} \frac{\partial^3 \psi_i(x)}{\partial x_1^3} \vartheta_{n_1,\ldots,n_m}(x) dx. \quad (52)$$

By integrating in parts the integrals (52) three times with respect to the variable x_2 we obtain that

$$\varphi'''_{in_1,\ldots,n_m} = -\left(\frac{l}{\pi}\right)^3 \frac{\varphi^{(6)}_{in_1,\ldots,n_m}}{n_2^3}, \quad \psi'''_{in_1,\ldots,n_m} = -\left(\frac{l}{\pi}\right)^3 \frac{\psi^{(6)}_{in_1,\ldots,n_m}}{n_2^3}, \quad (53)$$

where

$$\varphi^{(6)}_{in_1,\ldots,n_m} = \int_{\Omega_l^m} \frac{\partial^6 \varphi_i(x)}{\partial x_1^3 \partial x_2^3} \vartheta_{n_1,\ldots,n_m}(x) dx, \quad \psi^{(6)}_{in_1,\ldots,n_m} = \int_{\Omega_l^m} \frac{\partial^6 \psi_i(x)}{\partial x_1^3 \partial x_2^3} \vartheta_{n_1,\ldots,n_m}(x) dx.$$

Continuing this process, by induction we obtain

$$\varphi^{(3m-3)}_{in_1,\ldots,n_m} = -\left(\frac{l}{\pi}\right)^3 \frac{\varphi^{(3m)}_{in_1,\ldots,n_m}}{n_m^3}, \quad \psi^{(3m-3)}_{in_1,\ldots,n_m} = -\left(\frac{l}{\pi}\right)^3 \frac{\psi^{(3m)}_{in_1,\ldots,n_m}}{n_m^3}, \quad (54)$$

where

$$\varphi^{(3m)}_{in_1,\ldots,n_m} = \int_{\Omega_l^m} \frac{\partial^{3m} \varphi_i(x)}{\partial x_1^3 \partial x_2^3 \ldots \partial x_m^3} \vartheta_{n_1,\ldots,n_m}(x) dx, \quad \psi^{(3m)}_{in_1,\ldots,n_m} = \int_{\Omega_l^m} \frac{\partial^{3m} \psi_i(x)}{\partial x_1^3 \partial x_2^3 \ldots \partial x_m^3} \vartheta_{n_1,\ldots,n_m}(x) dx.$$

Here the Bessel inequalities are true

$$\sum_{n_1,\ldots,n_m=1}^{\infty} \left[\varphi_{in_1,\ldots,n_m}^{(3m)}\right]^2 \leq \left(\frac{2}{l}\right)^m \int_{\Omega_l^m} \left[\frac{\partial^{3m}\varphi_i(x)}{\partial x_1^3 \partial x_2^3 \ldots \partial x_m^3}\right]^2 dx, \quad (55)$$

$$\sum_{n_1,\ldots,n_m=1}^{\infty} \left[\psi_{in_1,\ldots,n_m}^{(3m)}\right]^2 \leq \left(\frac{2}{l}\right)^m \int_{\Omega_l^m} \left[\frac{\partial^{3m}\psi_i(x)}{\partial x_1^3 \partial x_2^3 \ldots \partial x_m^3}\right]^2 dx. \quad (56)$$

From (51), (53) and (54) implies that

$$\varphi_{in_1,\ldots,n_m} = \left(\frac{l}{\pi}\right)^{3m} \frac{\varphi_{in_1,\ldots,n_m}^{(3m)}}{n_1^3 \ldots n_m^3}, \quad \psi_{in_1,\ldots,n_m} = \left(\frac{l}{\pi}\right)^{3m} \frac{\psi_{in_1,\ldots,n_m}^{(3m)}}{n_1^3 \ldots n_m^3}, \quad i=1,2. \quad (57)$$

Taking formulas (50), (55)–(57) into account and applying the Cauchy-Schwarz inequality and Bessel inequality, for series (48) and (49) we obtain

$$|U(t,x,\omega,\nu)| \leq \sum_{n_1,\ldots,n_m=1}^{\infty} |u_{n_1,\ldots,n_m}(t,\omega,\nu)| \cdot |\vartheta_{n_1,\ldots,n_m}(x)| \leq$$

$$\leq \left(\sqrt{\frac{2}{l}}\right)^m C_{01} \sum_{n_1,\ldots,n_m=1}^{\infty} \left[|\varphi_{1n_1,\ldots,n_m}| + |\varphi_{2n_1,\ldots,n_m}| + |\psi_{1n_1,\ldots,n_m}| + |\psi_{2n_1,\ldots,n_m}|\right] \leq$$

$$\leq \gamma_1 \left[\sum_{n_1,\ldots,n_m=1}^{\infty} \frac{1}{n_1^3 \ldots n_m^3} \left|\varphi_{1n_1,\ldots,n_m}^{(3m)}\right| + \sum_{n_1,\ldots,n_m=1}^{\infty} \frac{1}{n_1^3 \ldots n_m^3} \left|\varphi_{2n_1,\ldots,n_m}^{(3m)}\right| + \right.$$

$$\left. + \sum_{n_1,\ldots,n_m=1}^{\infty} \frac{1}{n_1^3 \ldots n_m^3} \left|\psi_{1n_1,\ldots,n_m}^{(3m)}\right| + \sum_{n_1,\ldots,n_m=1}^{\infty} \frac{1}{n_1^3 \ldots n_m^3} \left|\psi_{2n_1,\ldots,n_m}^{(3m)}\right|\right] \leq$$

$$\leq \left(\sqrt{\frac{2}{l}}\right)^m \gamma_1 \sqrt{\sum_{n_1,\ldots,n_m=1}^{\infty} \frac{1}{n_1^6 \ldots n_m^6}} \left[\sqrt{\int_{\Omega_l^m} \left[\frac{\partial^{3m}\varphi_1(x)}{\partial x_1^3 \partial x_2^3 \ldots \partial x_m^3}\right]^2 dx} + \right.$$

$$+ \sqrt{\int_{\Omega_l^m} \left[\frac{\partial^{3m}\varphi_2(x)}{\partial x_1^3 \partial x_2^3 \ldots \partial x_m^3}\right]^2 dx} + \sqrt{\int_{\Omega_l^m} \left[\frac{\partial^{3m}\psi_1(x)}{\partial x_1^3 \partial x_2^3 \ldots \partial x_m^3}\right]^2 dx} +$$

$$\left. + \sqrt{\int_{\Omega_l^m} \left[\frac{\partial^{3m}\psi_2(x)}{\partial x_1^3 \partial x_2^3 \ldots \partial x_m^3}\right]^2 dx}\right] < \infty, \quad (58)$$

where $\gamma_1 = \left(\sqrt{\frac{2}{l}}\right)^m C_{01} \left(\frac{l}{\pi}\right)^{3m}$.

It follows from estimate (58) that the series (48) and (49) converge absolutely and uniformly in the domain $\bar{\Omega}$ under conditions (25) and (33).

From the convergence of series (48) and (49), in particular, it follows that the series (46) and (47) converge absolutely and uniformly in the domain Ω_l^m.

5. Possibility of Term Differentiation of the Series (48) and (49)

Functions (48) and (49) formally differentiate the required number of times

$$U_{tt}(t,x,\omega,\nu) =$$

$$= \sum_{n_1,\ldots,n_m=1}^{\infty} \vartheta_{n_1,\ldots,n_m}(x) \left[\varphi_{1\,n_1,\ldots,n_m} D''_{11\,n_1,\ldots,n_m}(t,\omega,\nu) + \varphi_{2\,n_1,\ldots,n_m} D''_{12\,n_1,\ldots,n_m}(t,\omega,\nu) + \right.$$
$$\left. + \psi_{1\,n_1,\ldots,n_m} D''_{13\,n_1,\ldots,n_m}(t,\omega,\nu) + \psi_{2\,n_1,\ldots,n_m} D''_{14\,n_1,\ldots,n_m}(t,\omega,\nu) \right], \quad t > 0, \tag{59}$$

$$U_{tt}(t,x,\omega,\nu) =$$

$$= \sum_{n_1,\ldots,n_m=1}^{\infty} \vartheta_{n_1,\ldots,n_m}(x) \left[\varphi_{1\,n_1,\ldots,n_m} D''_{21\,n_1,\ldots,n_m}(t,\omega,\nu) + \varphi_{2\,n_1,\ldots,n_m} D''_{22\,n_1,\ldots,n_m}(t,\omega,\nu) + \right.$$
$$\left. + \psi_{1\,n_1,\ldots,n_m} D''_{23\,n_1,\ldots,n_m}(t,\omega,\nu) + \psi_{2\,n_1,\ldots,n_m} D''_{24\,n_1,\ldots,n_m}(t,\omega,\nu) \right], \quad t < 0, \tag{60}$$

$$U_{x_1 x_1}(t,x,\omega,\nu) = - \sum_{n_1,\ldots,n_m=1}^{\infty} \left(\frac{\pi n_1}{l} \right)^2 \vartheta_{n_1,\ldots,n_m}(x) \times$$
$$\times \left[\varphi_{1\,n_1,\ldots,n_m} D_{11\,n_1,\ldots,n_m}(t,\omega,\nu) + \varphi_{2\,n_1,\ldots,n_m} D_{12\,n_1,\ldots,n_m}(t,\omega,\nu) + \right.$$
$$\left. + \psi_{1\,n_1,\ldots,n_m} D_{13\,n_1,\ldots,n_m}(t,\omega,\nu) + \psi_{2\,n_1,\ldots,n_m} D_{14\,n_1,\ldots,n_m}(t,\omega,\nu) \right], \quad t > 0, \tag{61}$$

$$U_{x_1 x_1}(t,x,\omega,\nu) = - \sum_{n_1,\ldots,n_m=1}^{\infty} \left(\frac{\pi n_1}{l} \right)^2 \vartheta_{n_1,\ldots,n_m}(x) \times$$
$$\times \left[\varphi_{1\,n_1,\ldots,n_m} D_{21\,n_1,\ldots,n_m}(t,\omega,\nu) + \varphi_{2\,n_1,\ldots,n_m} D_{22\,n_1,\ldots,n_m}(t,\omega,\nu) + \right.$$
$$\left. + \psi_{1\,n_1,\ldots,n_m} D_{23\,n_1,\ldots,n_m}(t,\omega,\nu) + \psi_{2\,n_1,\ldots,n_m} D_{24\,n_1,\ldots,n_m}(t,\omega,\nu) \right], \quad t < 0, \tag{62}$$

$$U_{x_2 x_2}(t,x,\omega,\nu) = - \sum_{n_1,\ldots,n_m=1}^{\infty} \left(\frac{\pi n_2}{l} \right)^2 \vartheta_{n_1,\ldots,n_m}(x) \times$$
$$\times \left[\varphi_{1\,n_1,\ldots,n_m} D_{11\,n_1,\ldots,n_m}(t,\omega,\nu) + \varphi_{2\,n_1,\ldots,n_m} D_{12\,n_1,\ldots,n_m}(t,\omega,\nu) + \right.$$
$$\left. + \psi_{1\,n_1,\ldots,n_m} D_{13\,n_1,\ldots,n_m}(t,\omega,\nu) + \psi_{2\,n_1,\ldots,n_m} D_{14\,n_1,\ldots,n_m}(t,\omega,\nu) \right], \quad t > 0, \tag{63}$$

$$U_{x_2 x_2}(t,x,\omega,\nu) = - \sum_{n_1,\ldots,n_m=1}^{\infty} \left(\frac{\pi n_2}{l} \right)^2 \vartheta_{n_1,\ldots,n_m}(x) \times$$
$$\times \left[\varphi_{1\,n_1,\ldots,n_m} D_{21\,n_1,\ldots,n_m}(t,\omega,\nu) + \varphi_{2\,n_1,\ldots,n_m} D_{22\,n_1,\ldots,n_m}(t,\omega,\nu) + \right.$$
$$\left. + \psi_{1\,n_1,\ldots,n_m} D_{23\,n_1,\ldots,n_m}(t,\omega,\nu) + \psi_{2\,n_1,\ldots,n_m} D_{24\,n_1,\ldots,n_m}(t,\omega,\nu) \right], \quad t < 0. \tag{64}$$

The expansions of the following functions into Fourier series are defined in the domain Ω_l^m in a similar way

$$U_{x_3 x_3}(t,x,\omega,\nu), \ldots, U_{x_m x_m}(t,x,\omega,\nu), U_{tt x_1 x_1}(t,x,\omega,\nu), U_{tt x_2 x_2}(t,x,\omega,\nu), \ldots, U_{tt x_m x_m}(t,x,\omega,\nu).$$

The convergence of series (59) and (60) is proved similarly to the proof of the convergence of series (48) and (49). Let us show the convergence of series (61)–(64). Taking into account Formulas (50), (55)–(57) and applying the Cauchy-Schwarz inequality and Bessel inequality, we obtain

$$|U_{x_1 x_1}(t,x,\omega,\nu)| \leq \sum_{n_1,\ldots,n_m=1}^{\infty} \left(\frac{\pi n_1}{l} \right)^2 |u_{n_1,\ldots,n_m}(t,\omega,\nu)| \cdot |\vartheta_{n_1,\ldots,n_m}(x)| \leq$$

$$\leq \left(\sqrt{\frac{2}{l}} \right)^m \left(\frac{\pi}{l} \right)^2 C_{01} \sum_{n_1,\ldots,n_m=1}^{\infty} n_1^2 \left[|\varphi_{1\,n_1,\ldots,n_m}| + |\varphi_{2\,n_1,\ldots,n_m}| + |\psi_{1\,n_1,\ldots,n_m}| + |\psi_{2\,n_1,\ldots,n_m}| \right] \leq$$

$$\leq \gamma_2 \left[\sum_{n_1,\ldots,n_m=1}^{\infty} \frac{1}{n_1 n_2^3 \cdots n_m^3} \left| \varphi_{1\,n_1,\ldots,n_m}^{(3m)} \right| + \sum_{n_1,\ldots,n_m=1}^{\infty} \frac{1}{n_1 n_2^3 \cdots n_m^3} \left| \varphi_{2\,n_1,\ldots,n_m}^{(3m)} \right| + \right.$$

$$+ \sum_{n_1,\ldots,n_m=1}^{\infty} \frac{1}{n_1 n_2^3 \ldots n_m^3} \left| \psi_{1n_1,\ldots,n_m}^{(3m)} \right| + \sum_{n_1,\ldots,n_m=1}^{\infty} \frac{1}{n_1 n_2^3 \ldots n_m^3} \left| \psi_{2n_1,\ldots,n_m}^{(3m)} \right| \right] \leq$$

$$\leq \left(\sqrt{\frac{2}{l}}\right)^m \gamma_2 \sqrt{\sum_{n_1,\ldots,n_m=1}^{\infty} \frac{1}{n_1^2 n_2^6 \ldots n_m^6}} \left[\sqrt{\int_{\Omega_l^m} \left[\frac{\partial^{3m} \varphi_1(x)}{\partial x_1^3 \partial x_2^3 \ldots \partial x_m^3}\right]^2 dx} + \right.$$

$$+ \sqrt{\int_{\Omega_l^m} \left[\frac{\partial^{3m} \varphi_2(x)}{\partial x_1^3 \partial x_2^3 \ldots \partial x_m^3}\right]^2 dx} + \sqrt{\int_{\Omega_l^m} \left[\frac{\partial^{3m} \psi_1(x)}{\partial x_1^3 \partial x_2^3 \ldots \partial x_m^3}\right]^2 dx} +$$

$$\left. + \sqrt{\int_{\Omega_l^m} \left[\frac{\partial^{3m} \psi_2(x)}{\partial x_1^3 \partial x_2^3 \ldots \partial x_m^3}\right]^2 dx} \right] < \infty,$$

where $\gamma_2 = \left(\sqrt{\frac{2}{l}}\right)^m C_{01} \left(\frac{l}{\pi}\right)^{3m-2}$;

$$|U_{x_2 x_2}(t,x,\omega,\nu)| \leq \sum_{n_1,\ldots,n_m=1}^{\infty} \left(\frac{\pi n_2}{l}\right)^2 |u_{n_1,\ldots,n_m}(t,\omega,\nu)| \cdot |\vartheta_{n_1,\ldots,n_m}(x)| \leq$$

$$\leq \left(\sqrt{\frac{2}{l}}\right)^m \left(\frac{\pi}{l}\right)^2 C_{01} \sum_{n_1,\ldots,n_m=1}^{\infty} n_2^2 \left[|\varphi_{1n_1,\ldots,n_m}| + |\varphi_{2n_1,\ldots,n_m}| + |\psi_{1n_1,\ldots,n_m}| + |\psi_{2n_1,\ldots,n_m}| \right] \leq$$

$$\leq \gamma_2 \left[\sum_{n_1,\ldots,n_m=1}^{\infty} \frac{1}{n_1^3 n_2 n_3^3 \ldots n_m^3} \left| \varphi_{1n_1,\ldots,n_m}^{(3m)} \right| + \sum_{n_1,\ldots,n_m=1}^{\infty} \frac{1}{n_1^3 n_2 n_3^3 \ldots n_m^3} \left| \varphi_{2n_1,\ldots,n_m}^{(3m)} \right| + \right.$$

$$\left. + \sum_{n_1,\ldots,n_m=1}^{\infty} \frac{1}{n_1^3 n_2 n_3^3 \ldots n_m^3} \left| \psi_{1n_1,\ldots,n_m}^{(3m)} \right| + \sum_{n_1,\ldots,n_m=1}^{\infty} \frac{1}{n_1^3 n_2 n_3^3 \ldots n_m^3} \left| \psi_{2n_1,\ldots,n_m}^{(3m)} \right| \right] \leq$$

$$\leq \left(\sqrt{\frac{2}{l}}\right)^m \gamma_2 \sqrt{\sum_{n_1,\ldots,n_m=1}^{\infty} \frac{1}{n_1^6 n_2^2 n_3^6 \ldots n_m^6}} \left[\sqrt{\int_{\Omega_l^m} \left[\frac{\partial^{3m} \varphi_1(x)}{\partial x_1^3 \partial x_2^3 \ldots \partial x_m^3}\right]^2 dx} + \right.$$

$$+ \sqrt{\int_{\Omega_l^m} \left[\frac{\partial^{3m} \varphi_2(x)}{\partial x_1^3 \partial x_2^3 \ldots \partial x_m^3}\right]^2 dx} + \sqrt{\int_{\Omega_l^m} \left[\frac{\partial^{3m} \psi_1(x)}{\partial x_1^3 \partial x_2^3 \ldots \partial x_m^3}\right]^2 dx} +$$

$$\left. + \sqrt{\int_{\Omega_l^m} \left[\frac{\partial^{3m} \psi_2(x)}{\partial x_1^3 \partial x_2^3 \ldots \partial x_m^3}\right]^2 dx} \right] < \infty,$$

The convergence of Fourier series for functions $U_{x_3 x_3}(t,x,\omega,\nu), \ldots, U_{x_m x_m}(t,x,\omega,\nu)$, $U_{ttx_1 x_1}(t,x,\omega,\nu), U_{ttx_2 x_2}(t,x,\omega,\nu), \ldots, U_{ttx_m x_m}(t,x,\omega,\nu)$ is proved in a similar way in the domain Ω_l^m.

Therefore, the functions $U(t,x,\omega,\nu)$, $g_1(x,\omega,\nu)$ and $g_2(x,\omega,\nu)$ defined by series (46)–(49) satisfy the conditions of the given problem.

To establish the uniqueness of the function $U(t,x,\omega,\nu)$ we show that, under the zero integral conditions $\int_0^T U(t,x,\omega,\nu)\,dt = 0$, $\int_{-T}^0 U(t,x,\omega,\nu)\,dt = 0$, $0 \leq x \leq l$ the inverse boundary value problem (1)–(5) has only a trivial solution. We suppose that $\varphi_i(x) \equiv 0$, $\psi_i(x) \equiv 0$. Then $\varphi_{in_1,\ldots,n_m} = 0$, $\psi_{in_1,\ldots,n_m} = 0$ and from formulas (48) and (49) in the domain Ω_l^m implies that

$$\int_{\Omega_l^m} U(t, x, \omega, \nu) \vartheta_{n_1,\ldots,n_m}(x)\, dx = 0.$$

Hence, by virtue of completeness of systems of the eigenfunctions $\left\{ \sqrt{\frac{2}{l}} \sin \frac{\pi n_1}{l} x_1 \right\}$, $\left\{ \sqrt{\frac{2}{l}} \sin \frac{\pi n_2}{l} x_2 \right\}, \ldots, \left\{ \sqrt{\frac{2}{l}} \sin \frac{\pi n_m}{l} x_m \right\}$ in the space $L_2(\Omega_l^m)$ we deduce that $U(t, x, \omega, \nu) \equiv 0$ for all $x \in \Omega_l^m$ and $t \in [-T; T]$.

Therefore, under conditions (25) and (33), the inverse problem has a unique pair of solutions in the domain Ω_l^m.

6. Calculation of Values of Spectral Parameters

Let condition (25) be violated, that is, we suppose that

$$\sigma_{n_1,\ldots,n_m} = \sinh \lambda_{n_1,\ldots,n_m} T \left(\cos \lambda_{n_1,\ldots,n_m} \omega T - 1 \right) -$$
$$- \omega \sin \lambda_{n_1,\ldots,n_m} \omega T \left(\cosh \lambda_{n_1,\ldots,n_m} T - 1 \right) = 0 \quad (65)$$

for some values of ω, where $\lambda_{n_1,\ldots,n_m}^2 = \frac{\mu_{n_1,\ldots,n_m}^2}{1 + \mu_{n_1,\ldots,n_m}^2}$, $\mu_{n_1,\ldots,n_m} = \frac{\pi}{l} \sqrt{n_1^2 + \ldots + n_m^2}$.

From equality (65) with respect to the spectral parameter ω we arrive at the quadratic trigonometric equation

$$(a_{n_1,\ldots,n_m} + 1) \tan^2 \frac{y_{n_1,\ldots,n_m}}{2} + 2 b_{n_1,\ldots,n_m} \omega \tan \frac{y_{n_1,\ldots,n_m}}{2} + (a_{n_1,\ldots,n_m} - 1) = 0,$$

where

$$y_{n_1,\ldots,n_m} = \lambda_{n_1,\ldots,n_m} \omega T, \quad a_{n_1,\ldots,n_m} = \sinh \lambda_{n_1,\ldots,n_m} T,$$
$$b_{n_1,\ldots,n_m} = \coth \lambda_{n_1,\ldots,n_m} T - \sinh^{-1} \lambda_{n_1,\ldots,n_m} T.$$

The set of positive solutions of this equation with respect to the spectral parameter ω for some $k_1, \ldots, k_m$ is denoted by $\Im_1$. We call the numbers $\omega \in \Im_1$ as irregular, since because the condition (25) is violated for them. The set $\Lambda_1 = (0; \infty) \setminus \Im_1$ is called the set of regular values of the spectral parameter ω, for which condition (25) is fulfilled. If condition (34) is violated, then the kernels of the mixed integro-differential Equation (1) have at most two values of ν_1 and ν_2. We call these real nonzero numbers as an irregular kernel numbers of the mixed integro-differential Equation (1) and denote their set $\{\nu_1, \nu_2\}$ by $\Im_2$. We take away the values ν_1 and ν_2 of the spectral parameter ν from the set of nonzero real numbers $(-\infty; 0) \cup (0; \infty)$. The resulting set $\Lambda_3 = (-\infty; 0) \cup (0; \infty) \setminus \{\nu_1, \nu_2\}$ is called the set of regular values of the parameter ν. For all values of $\nu \in \Lambda_2$ condition (33) is satisfied.

We use the following notations for sets

$$\aleph_1 = \{(\omega, \nu) \mid \omega \in \Lambda_1; \nu \in \Lambda_3\}; \ \aleph_2 = \{(\omega, \nu) \mid \omega \in \Im_1; \nu \in (-\infty; 0) \cup (0; \infty)\},$$

$$\aleph_3 = \{(\omega, \nu) \mid \omega \in \Lambda_1; \nu \in \Im_2\}.$$

For $(\omega, \nu) \in \aleph_1$ formulas (46)–(49) hold. This is the case when all values of the spectral parameters ω and ν are regular. Therefore, in this case, the unique solution of the inverse boundary value problem (1)–(5) in the domain Ω_l^m is represented in the form of series (46)–(49).

7. Expansion of the Solution of the Direct Problem (1)–(4) in a Fourier Series. Irregular Case of a Spectral Parameter ω

For some $k_1, \ldots, k_m$ and $(\nu, \omega) \in \aleph_2$, where $\aleph_2 = \{(\omega, \nu) \mid \omega \in \Im_1; \nu \in (-\infty; 0) \cup (0; \infty)\}$ we first find a formal solution of the direct problem (1)–(4). In this case, instead of (28) and (29), we have the representations

$$u_{k_1,\ldots,k_m}(t, \omega) = C_{1k_1,\ldots,k_m} \cosh \lambda_{k_1,\ldots,k_m} t + \omega C_{2k_1,\ldots,k_m} \sinh \lambda_{k_1,\ldots,k_m} t +$$
$$+ \nu \alpha_{k_1,\ldots,k_m} h_{1k_1,\ldots,k_m}(t) + g_{1k_1,\ldots,k_m} h_{2k_1,\ldots,k_m}(t), \quad t > 0, \tag{66}$$

$$u_{k_1,\ldots,k_m}(t, \omega) = C_{1k_1,\ldots,k_m} \cos \lambda_{k_1,\ldots,k_m} \omega t + C_{2k_1,\ldots,k_m} \sin \lambda_{k_1,\ldots,k_m} \omega t +$$
$$+ \nu \beta_{k_1,\ldots,k_m} \delta_{1k_1,\ldots,k_m}(t) + g_{2k_1,\ldots,k_m} \delta_{2k_1,\ldots,k_m}(t), \quad t < 0, \tag{67}$$

where $C_{ik_1,\ldots,k_m}$ ($i = 1, 2$) are arbitrary constants.

Substituting (66) into (11) and (67) into (12), we obtain

$$\tau_{ik_1,\ldots,k_m} \alpha_{k_1,\ldots,k_m} = C_{1k_1,\ldots,k_m} \chi_{i1k_1,\ldots,k_m} + C_{2k_1,\ldots,k_m} \chi_{i2k_1,\ldots,k_m} + g_{ik_1,\ldots,k_m} \chi_{i4k_1,\ldots,k_m}, \tag{68}$$

where

$$\tau_{ik_1,\ldots,k_m} = 1 - \nu \chi_{i3k_1,\ldots,k_m} \neq 0, \quad i = 1, 2, \tag{69}$$

$$\chi_{11k_1,\ldots,k_m} = \int_0^T b_1(s) \cosh \lambda_{k_1,\ldots,k_m} s \, ds, \quad \chi_{12k_1,\ldots,k_m} = \omega \int_0^T b_1(s) \sinh \lambda_{k_1,\ldots,k_m} s \, ds,$$

$$\chi_{13k_1,\ldots,k_m} = \int_0^T b_1(s) h_{1k_1,\ldots,k_m}(s) \, ds, \quad \chi_{14k_1,\ldots,k_m} = \int_0^T b_1(s) h_{2k_1,\ldots,k_m}(s) \, ds,$$

$$\chi_{21k_1,\ldots,k_m} = \int_{-T}^0 b_2(s) \cos \lambda_{k_1,\ldots,k_m} \omega s \, ds, \quad \chi_{22k_1,\ldots,k_m} = \int_{-T}^0 b_2(s) \sin \lambda_{k_1,\ldots,k_m} \omega s \, ds,$$

$$\chi_{23k_1,\ldots,k_m} = \int_{-T}^0 b_2(s) \delta_{1k_1,\ldots,k_m}(s) \, ds, \quad \chi_{24k_1,\ldots,k_m} = \int_{-T}^0 b_2(s) \delta_{2k_1,\ldots,k_m}(s) \, ds.$$

We show that condition (69) is always fulfilled, that is,

$$1 - \nu \chi_{13k_1,\ldots,k_m} \neq 0, \quad 1 - \nu \chi_{23k_1,\ldots,k_m} \neq 0.$$

First, we suppose that simultaneously take place

$$1 - \nu \chi_{13k_1,\ldots,k_m} = 0, \quad 1 - \nu \chi_{23k_1,\ldots,k_m} = 0. \tag{70}$$

Then we come to the conclusion that $\chi_{13k_1,\ldots,k_m} = \nu^{-1}$, $\chi_{23k_1,\ldots,k_m} = \nu^{-1}$, that is, $\chi_{13k_1,\ldots,k_m} = \chi_{23k_1,\ldots,k_m}$. It cannot be, since because $\chi_{13k_1,\ldots,k_m}$ and $\chi_{13k_1,\ldots,k_m}$ are different quantities. Therefore, (70) does not hold.

Now suppose that

$$1 - \nu \chi_{13k_1,\ldots,k_m} = 0, \quad 1 - \nu \chi_{23k_1,\ldots,k_m} \neq 0. \tag{71}$$

Then we have consider the quadratic equation

$$(1 - \nu \chi_{13k_1,\ldots,k_m})(1 - \nu \chi_{23k_1,\ldots,k_m}) =$$
$$= \nu^2 \chi_{13k_1,\ldots,k_m} \chi_{23k_1,\ldots,k_m} - \nu (\chi_{13k_1,\ldots,k_m} + \chi_{23k_1,\ldots,k_m}) + 1 = 0.$$

Solving this equation we derive the roots: $\nu_1 = \frac{1}{\chi_{13 k_1,\ldots,k_m}}$, $\nu_2 = \frac{1}{\chi_{23 k_1,\ldots,k_m}}$. But, by our assumption: $1 - \nu \chi_{23 k_1,\ldots,k_m} \neq 0$. We came to a contradiction. Therefore, (71) does not hold. Similarly, it can be shown that there is no

$$1 - \nu \chi_{13 k_1,\ldots,k_m} \neq 0, \quad 1 - \nu \chi_{23 k_1,\ldots,k_m} = 0.$$

Therefore, the condition (69) is always fulfilled. Then from (68) we find that

$$\alpha_{k_1,\ldots,k_m} = C_{1 k_1,\ldots,k_m} \tilde{\chi}_{11 k_1,\ldots,k_m} + C_{2 k_1,\ldots,k_m} \tilde{\chi}_{12 k_1,\ldots,k_m} + g_{1 k_1,\ldots,k_m} \tilde{\chi}_{13 k_1,\ldots,k_m}, \tag{72}$$

$$\beta_{k_1,\ldots,k_m} = C_{1 k_1,\ldots,k_m} \tilde{\chi}_{21 k_1,\ldots,k_m} + C_{2 k_1,\ldots,k_m} \tilde{\chi}_{22 k_1,\ldots,k_m} + g_{2 k_1,\ldots,k_m} \tilde{\chi}_{23 k_1,\ldots,k_m}, \tag{73}$$

where

$$\tilde{\chi}_{ij k_1,\ldots,k_m} = \frac{\chi_{ij k_1,\ldots,k_m}}{\tau_{i k_1,\ldots,k_m}}, \quad \tilde{\chi}_{i3 k_1,\ldots,k_m} = \frac{\chi_{i4 k_1,\ldots,k_m}}{\tau_{i k_1,\ldots,k_m}}, \quad i = 1, 2, \ j = 1, 2.$$

Substitution of values (72) into (66) and (73) into (67) gives us the following representations

$$u_{k_1,\ldots,k_m}(t, \omega, \nu) = C_{1 k_1,\ldots,k_m} \gamma_{11 k_1,\ldots,k_m}(t, \omega, \nu) + C_{2 k_1,\ldots,k_m} \gamma_{12 k_1,\ldots,k_m}(t, \omega, \nu) +$$

$$+ g_{1 k_1,\ldots,k_m}(\omega, \nu) \gamma_{13 k_1,\ldots,k_m}(t, \omega, \nu), \quad t > 0, \tag{74}$$

$$u_{k_1,\ldots,k_m}(t, \omega, \nu) = C_{1 k_1,\ldots,k_m} \gamma_{21 k_1,\ldots,k_m}(t, \omega, \nu) + C_{2 k_1,\ldots,k_m} \gamma_{22 k_1,\ldots,k_m}(t, \omega, \nu) +$$

$$+ g_{2 k_1,\ldots,k_m}(\omega, \nu) \gamma_{23 k_1,\ldots,k_m}(t, \omega, \nu), \quad t < 0, \tag{75}$$

where

$$\gamma_{11 k_1,\ldots,k_m}(t, \omega, \nu) = \cosh \lambda_{k_1,\ldots,k_m} t + \nu h_{1 k_1,\ldots,k_m}(t) \tilde{\chi}_{11 k_1,\ldots,k_m},$$

$$\gamma_{12 k_1,\ldots,k_m}(t, \omega, \nu) = \omega \sinh \lambda_{k_1,\ldots,k_m} t + \nu h_{1 k_1,\ldots,k_m}(t) \tilde{\chi}_{12 k_1,\ldots,k_m},$$

$$\gamma_{13 k_1,\ldots,k_m}(t, \omega, \nu) = h_{2 k_1,\ldots,k_m}(t) + \nu h_{1 k_1,\ldots,k_m}(t) \tilde{\chi}_{13 k_1,\ldots,k_m},$$

$$\gamma_{21 k_1,\ldots,k_m}(t, \omega, \nu) = \cos \lambda_{k_1,\ldots,k_m} \omega t + \nu \delta_{1 k_1,\ldots,k_m}(t) \tilde{\chi}_{21 k_1,\ldots,k_m},$$

$$\gamma_{21 k_1,\ldots,k_m}(t, \omega, \nu) = \sin \lambda_{k_1,\ldots,k_m} \omega t + \nu \delta_{1 k_1,\ldots,k_m}(t) \tilde{\chi}_{22 k_1,\ldots,k_m},$$

$$\gamma_{23 k_1,\ldots,k_m}(t, \omega, \nu) = \delta_{2 k_1,\ldots,k_m}(t) + \nu \delta_{1 k_1,\ldots,k_m}(t) \tilde{\chi}_{23 k_1,\ldots,k_m}.$$

Then from (74) and (75) yields that the solution of the direct problem (1)–(4) in the domain Ω_l^m for $(\nu, \omega) \in \aleph_2$ can be represented as the following Fourier series

$$U(t, x, \omega, \nu) = \sum_{k_1,\ldots,k_m=1}^{\infty} \vartheta_{k_1,\ldots,k_m}(x) \left[C_{1 k_1,\ldots,k_m} \gamma_{11 k_1,\ldots,k_m}(t, \omega, \nu) + \right.$$

$$+ C_{2 k_1,\ldots,k_m} \gamma_{12 k_1,\ldots,k_m}(t, \omega, \nu) + g_{1 k_1,\ldots,k_m}(\omega, \nu) \gamma_{13 k_1,\ldots,k_m}(t, \omega, \nu) \right], \quad t > 0, \tag{76}$$

$$U(t, x, \omega, \nu) = \sum_{k_1,\ldots,k_m=1}^{\infty} \vartheta_{k_1,\ldots,k_m}(x) \left[C_{1 k_1,\ldots,k_m} \gamma_{21 k_1,\ldots,k_m}(t, \omega, \nu) + \right.$$

$$+ C_{2 k_1,\ldots,k_m} \gamma_{22 k_1,\ldots,k_m}(t, \omega, \nu) + g_{2 k_1,\ldots,k_m}(\omega, \nu) \gamma_{23 k_1,\ldots,k_m}(t, \omega, \nu) \right], \quad t < 0, \tag{77}$$

where $C_{i k_1,\ldots,k_m}$ ($i = 1, 2$) are arbitrary constants.

8. Inverse Problem (1)–(5). Irregular Case of a Spectral Parameter ω

We apply the additional conditions (5) and from the Fourier series (76) and (77) we obtain that

$$\psi_i(x, \omega, \nu) = U(t_i, x, \omega, \nu) = \sum_{k_1,\ldots,k_m=1}^{\infty} \vartheta_{k_1,\ldots,k_m}(x) \left[C_{1 k_1,\ldots,k_m} \gamma_{i1 k_1,\ldots,k_m}(t_i, \omega, \nu) + \right.$$

$$+ C_{2k_1,\ldots,k_m} \gamma_{i2k_1,\ldots,k_m}(t_i, \omega, \nu) + g_{ik_1,\ldots,k_m}(\omega, \nu) \gamma_{i3k_1,\ldots,k_m}(t_i, \omega, \nu)], \ i = 1, 2. \quad (78)$$

Taking the expansions (43) and $\gamma_{i3k_1,\ldots,k_m}(t, \omega, \nu) \neq 0$, $i = 1, 2$ into account from the relations (78) we derive

$$g_{ik_1,\ldots,k_m}(\omega, \nu) = \psi_{1k_1,\ldots,k_m}(\omega, \nu) \cdot \tilde{\gamma}_{i3k_1,\ldots,k_m}(t_i, \omega, \nu) -$$
$$- C_{1k_1,\ldots,k_m} \cdot \tilde{\gamma}_{i1k_1,\ldots,k_m}(t_i, \omega, \nu) - C_{2k_1,\ldots,k_m} \cdot \tilde{\gamma}_{i2k_1,\ldots,k_m}(t_i, \omega, \nu), \ i = 1, 2, \quad (79)$$

where

$$\tilde{\gamma}_{i3k_1,\ldots,k_m}(t_i, \omega, \nu) = (\gamma_{i3k_1,\ldots,k_m}(t_i, \omega, \nu))^{-1}, \ i = 1, 2,$$

$$\tilde{\gamma}_{ijk_1,\ldots,k_m}(t_i, \omega, \nu) = \frac{\gamma_{ijk_1,\ldots,k_m}(t_i, \omega, \nu)}{\gamma_{i3k_1,\ldots,k_m}(t_i, \omega, \nu)}, \ i, j = 1, 2.$$

Substituting representations (79) into the Fourier series (8), we obtain

$$g_i(x, \omega, \nu) = \sum_{k_1,\ldots,k_m=1}^{\infty} \vartheta_{k_1,\ldots,k_m}(x) \left[\psi_{1k_1,\ldots,k_m}(\omega, \nu) \cdot \tilde{\gamma}_{i3k_1,\ldots,k_m}(t_i, \omega, \nu) -\right.$$
$$\left. -C_{1k_1,\ldots,k_m} \cdot \tilde{\gamma}_{i1k_1,\ldots,k_m}(t_i, \omega, \nu) - C_{2k_1,\ldots,k_m} \cdot \tilde{\gamma}_{i2k_1,\ldots,k_m}(t_i, \omega, \nu)\right], \ i = 1, 2. \quad (80)$$

Substitution of the representations (79) into the series (76) and (77) gives

$$U(t, x, \nu) = \sum_{k_1,\ldots,k_m=1}^{\infty} \vartheta_{k_1,\ldots,k_m}(x) \left[\psi_{1k_1,\ldots,k_m} Z_{11k_1,\ldots,k_m}(t, \omega, \nu) +\right.$$
$$\left. + C_{1k_1,\ldots,k_m} Z_{12k_1,\ldots,k_m}(t, \omega, \nu) + C_{2k_1,\ldots,k_m} Z_{13k_1,\ldots,k_m}(t, \omega, \nu)\right], \ t > 0, \quad (81)$$

$$U(t, x, \nu) = \sum_{k_1,\ldots,k_m=1}^{\infty} \vartheta_{k_1,\ldots,k_m}(x) \left[\psi_{2k_1,\ldots,k_m} Z_{21k_1,\ldots,k_m}(t, \omega, \nu) +\right.$$
$$\left. + C_{1k_1,\ldots,k_m} Z_{22k_1,\ldots,k_m}(t, \omega, \nu) + C_{2k_1,\ldots,k_m} Z_{23k_1,\ldots,k_m}(t, \omega, \nu)\right], \ t < 0, \quad (82)$$

where

$$Z_{i1k_1,\ldots,k_m}(t, \omega, \nu) = \gamma_{i3k_1,\ldots,k_m}(t, \omega, \nu) \cdot \tilde{\gamma}_{i3k_1,\ldots,k_m}(t_i, \omega, \nu),$$
$$Z_{i2k_1,\ldots,k_m}(t, \omega, \nu) = \gamma_{i1k_1,\ldots,k_m}(t, \omega, \nu) - \gamma_{i3k_1,\ldots,k_m}(t, \omega, \nu) \cdot \tilde{\gamma}_{i1k_1,\ldots,k_m}(t_i, \omega, \nu),$$
$$Z_{i3k_1,\ldots,k_m}(t, \omega, \nu) = \gamma_{i2k_1,\ldots,k_m}(t, \omega, \nu) - \gamma_{i3k_1,\ldots,k_m}(t, \omega, \nu) \cdot \tilde{\gamma}_{i2k_1,\ldots,k_m}(t_i, \omega, \nu), \ i = 1, 2.$$

By virtue of the fact that $Z_{ijk_1,\ldots,k_m}(t, \omega, \nu)$ ($i = 1, 2$; $j = 1, 2, 3$) are uniformly bounded functions and the conditions **A** are satisfied for the functions $\psi_i(x)$, the arbitrary constants $C_{ik_1,\ldots,k_m}$ can be chosen such that the series (80)–(82) converge absolutely and uniformly. The proof of this statement is carried out in exactly the same way as in the case of regular values of spectral parameters.

9. Statement of the Theorem. Conclusions

The questions of solvability of a nonlocal inverse boundary value problem for a mixed pseudohyperbolic-pseudoelliptic integro-differential Equation (1) with spectral parameters ω and ν are considered. Using the method of the Fourier series in the form (6), a system of countable systems of ordinary integro-differential Equations (9) and (10) is obtained. To determine arbitrary integration constants, a system of algebraic equations is obtained. From this system, regular and irregular values of the spectral parameter ω were calculated (condition (25)). From the condition (34) we calculate regular and irregular values of the spectral parameter ν. The following theorem is proved.

Theorem 1. *Let conditions* **A** *be fulfilled. Then for values* $(\nu, \omega) \in \aleph_1$ *the inverse problem* (1)–(5) *is uniquely solvable in the domain* Ω_l^m *and this solution is represented in the form of series* (46)–(49)*. And for values* $(\nu, \omega) \in \aleph_2$ *the inverse problem* (1)–(5) *in the domain* Ω_l^m *has an infinite number of solutions. These solution is represented in the form of series* (80)–(82)*. Moreover, a necessary conditions for the existence of solutions of the problem are:* $\varphi_1(x) \equiv 0$, $\varphi_2(x) \equiv 0$.

In the case of all possible values $(\nu, \omega) \in \aleph_3$, where $\aleph_3 = \{(\omega, \nu) \mid \omega \in \Lambda_1; \nu \in \Im_2\}$, the questions of solvability of the inverse problem (1)–(5) are studied in a similar way.

Funding: This research received no external funding.

Conflicts of Interest: The author declares no conflicts of interest.

References

1. Benney, D.J.; Luke, J.C. Interactions of permanent waves of finite amplitude. *J. Math. Phys.* **1964**, *43*, 309–313. [CrossRef]
2. Cavalcanti, M.M.; Domingos Cavalcanti, V.N.; Ferreira, J. Existence and uniform decay for a nonlinear viscoelastic equation with strong damping. *Math. Methods Appl. Sci.* **2001**, *24*, 1043–1053. [CrossRef]
3. Yuldashev, T.K. Nonlocal boundary value problem for a nonlinear Fredholm integro-differential equation with degenerate kernel. *Differ. Equ.* **2018**, *54*, 1646–1653. [CrossRef]
4. Yuldashev, T.K. On the solvability of a boundary value problem for the ordinary Fredholm integrodifferential equation with a degenerate kernel. *Comput. Math. Math. Phys.* **2019**, *59*, 241–252. [CrossRef]
5. Il'in, V.A.; Moiseev, E.I. An upper bound taken on the diagonal for the spectral function of the multidimensional Schrödinger operator with a potential satisfying the Kato condition. *Diff. Equ.* **1998**, *34*, 358–368.
6. Kapustin, N.Y.; Moiseev, E.I. A spectral problem for the Laplace operator in the square with a spectral parameter in the boundary condition. *Differ. Equ.* **1998**, *34*, 663–668.
7. Kapustin, N.Y.; Moiseev, E.I. Convergence of spectral expansions for functions of the Hölder class for two problems with a spectral parameter in the boundary condition. *Differ. Equ.* **2000**, *36*, 1182–1188. [CrossRef]
8. Kapustin, N.Y. Moiseev, E.I. A Remark on the Convergence Problem for Spectral Expansions Corresponding to a Classical Problem with Spectral Parameter in the Boundary Condition. *Differ. Equ.* **2001**, *37*, 1677–1683. [CrossRef]
9. Yuldashev, T.K. On inverse boundary value problem for a Fredholm integro-differential equation with degenerate kernel and spectral parameter. *Lobachevskii J. Math.* **2019**, *40*, 230–239. [CrossRef]
10. Yuldashev, T.K. Spectral features of the solving of a Fredholm homogeneous integro-differential equation with integral conditions and reflecting deviation. *Lobachevskii J. Math.* **2019**, *40*, 2116–2123. [CrossRef]
11. Sidorov, D.N.; Sidorov, N.A. Solution of irregular systems using skeleton decomposition of linear operators. *VEstnik Yuzhno Ural. State Univ. Seria Mat. Model. Progr.* **2017**, *10*, 63–73.
12. Sidorov, N.A. Classic solutions of boundary value problems for partial differential equations with operator of finite index in the main part of equation. *Bull. Irkutsk. State Univ. Ser. Math.* **2019**, *27*, 55–70. [CrossRef]
13. Sidorov, N.A.; Sidorov, D.N. Solving the Hammerstein integral equation in the irregular case by successive approximations. *Sib. Math. J.* **2010**, *51*, 325–329. [CrossRef]
14. Apakov, Y.P. A three-dimensional analog of the Tricomi problem for a parabolic-hyperbolic equation. *J. Appl. Industr. Math.* **2012**, *6*, 12–21. [CrossRef]
15. Dzhuraev, T.D.; Apakov, Y.P. The Gellerstedt problem for a parabolic-hyperbolic equation in a three-dimensional space. *Differ. Equ.* **1990**, *26*, 322–330.
16. Islomov, B. Analogues of the Tricomi problem for an equation of mixed parabolic-hyperbolic type with two lines and different order of degeneracy. *Differ. Equ.* **1991**, *27*, 713–719.
17. Polosin, A.A. Gellerstedt Type Directional Derivative Problem for an Equation of the Mixed Type with a Spectral Parameter. *Differ. Equ.* **2019**, *55*, 1373–1383. [CrossRef]
18. Ruziev, M.K. On the Boundary-Value Problem for a Class of Equations of Mixed Type in an Unbounded Domain. *Math. Notes* **2012**, *92*, 70–78. [CrossRef]

19. Soldatov, A.P. On Dirichlet-type problems for the Lavrent'ev–Bitsadze equation. *Proc. Steklov. Inst. Math.* **2012**, *278*, 233–240. [CrossRef]
20. Salakhitdinov, M.S.; Islomov, B.I. A nonlocal boundary-value problem with conormal derivative for a mixed-type equation with two inner degeneration lines and various orders of degeneracy. *Russ. Math. (Iz. VUZ)* **2011**, *55*, 42–49. [CrossRef]
21. Salakhitdinov, M.S.; Urinov, A.K. Eigenvalue problems for a mixed-type equation with two singular coefficients. *Siberian Math. J.* **2007**, *48*, 707–717. [CrossRef]
22. Urinov, A.K.; Nishonova, S.T. A Problem with Integral Conditions for an Elliptic-Parabolic Equation. *Math. Notes* **2017**, *102*, 68–80. [CrossRef]
23. Yuldashev, T.K. Nonlocal problem for a mixed type differential equation in rectangular domain. *Proc. Yerevan State Univ. Phys. Math. Sci.* **2016**, *3*, 70–78.
24. Yuldashev, T.K. On an integro-differential equation of pseudoparabolic-pseudohyperbolic type with degenerate kernels. *Proc. Yerevan State Univ. Phys. Math. Sci.* **2018**, *52*, 19–26.
25. Sabitov, K.B.; Safin, E.M. The inverse problem for a mixed-type parabolic-hyperbolic equation in a rectangular domain. *Russ. Math. (Iz. VUZ)* **2010**, *54*, 48–54. [CrossRef]
26. Sabitov, K.B.; Martem'yanova, N.V. A nonlocal inverse problem for a mixed-type equation. *Russ. Math. (Iz. VUZ)* **2011**, *55*, 61–74. [CrossRef]

© 2020 by the author. Licensee MDPI, Basel, Switzerland. This article is an open access article distributed under the terms and conditions of the Creative Commons Attribution (CC BY) license (http://creativecommons.org/licenses/by/4.0/).

Article

Lyapunov Type Theorems for Exponential Stability of Linear Skew-Product Three-Parameter Semiflows with Discrete Time

Davor Dragičević [1],* and Ciprian Preda [2,3]

[1] Department of Mathematics, University of Rijeka, Radmile Matejčić 2, 51000 Rijeka, Croatia
[2] Department of Mathematics, Faculty of Mathematics and Computer Science, West University of Timişoara, V. Pârvan Blvd. No. 4, 300223 Timişoara, Romania; ciprian.preda@e-uvt.ro
[3] Institute for Economic Forecasting, Romanian Academy, 050711 Bucharest, Romania
* Correspondence: ddragicevic@math.uniri.hr

Received: 27 March 2020; Accepted: 23 April 2020; Published: 27 April 2020

Abstract: For linear skew-product three-parameter semiflows with discrete time acting on an arbitrary Hilbert space, we obtain a complete characterization of exponential stability in terms of the existence of appropriate Lyapunov functions. As a nontrivial application of our work, we prove that the notion of an exponential stability persists under sufficiently small linear perturbations.

Keywords: exponential stability; linear skew-product semiflows; Lyapunov functions

1. Introduction

The main objective of this paper is to obtain a complete characterization of exponential stability for linear skew-product semiflows with discrete time acting on an arbitrary Hilbert space in terms of the existence of appropriate Lyapunov functions. We then use this characterization to prove that the notion of an exponential stability persists under sufficiently small linear perturbations.

We stress that the use of Lyapunov functions in the study of the stability of trajectories in the theories of differential equations and dynamical systems has a long history that goes back to the landmark work of Lyapunov [1]. For some early contributions to the theory, we refer to books by LaSalle and Lefschetz [2], Hahn [3] and Bhatia and Szegö [4]. For the first contributions dealing with infinite-dimensional dynamics, we refer to the work of Daleckij and Krein [5].

In the context of nonautonomous dynamics, the relationship between exponential dichotomies and the existence of appropriate Lyapunov functions was first considered by Maizel [6]. His results were further developed by Coppel [7,8] as well as Muldowney [9]. We note that these results considered only the case of continuous time. To the best of our knowledge, the first contributions in the case of discrete time are due to Papaschinopoulos [10]. In the recent years, there has been a renewed interest in this topic. More precisely, various characterizations of nonuniform exponential behaviour in terms of Lyapunov functions were obtained (see [11–13]). In addition, the authors have obtained first results in the context of infinite-dimensional dynamics [14] (see also [15]) which lead to further developments [16–18]. Finally, for some related results in the context of ergodic theory, we refer to [19] and references therein.

The purpose of this paper is to show that techniques we developed in our previous work [14] can be used to obtain Lyapunov-type characterization of exponential stability for a very general type of nonautonomous dynamics. More precisely, we consider the so-called linear skew-product three-parameter semiflows. This notion was introduced by Megan and Stoica [20] and includes various previously studied notions as a particular case (see Examples 1 and 2).

Finally, we would like to mention that Lyapunov type characterizations of exponential stability are certainly not the only tool used to study stability of nonautonomous dynamics. Indeed, there is a vast literature devoted to the so-called Perron type characterizations of exponential stability (see [21–26] and references therein) as well as to Datko-Pazy-Rolewicz techniques (see [27–33]). For some other approaches to the study of exponential stability for nonautonomous systems, we refer to [34,35].

The paper is organized as follows. In Section 2 we introduce all relevant notions and recall auxiliarly results which will be used in the paper. In Section 3 we state and prove the main results of our paper. Finally, in Section 4 we apply the main result to the study of the robustness property of exponential stability for linear skew-product three-parameter semiflows.

2. Preliminaries

Let (Θ, d) be a metric space and let X be a Hilbert space over $\mathbb{C}$. By $B(X)$ we will denote the space of all bounded operators on X.

Definition 1. *A map $\sigma \colon \Theta \times \mathbb{Z} \times \mathbb{Z} \to \Theta$ is said to be a continuous three-parameter flow (with discrete time) if:*

1. $\sigma(\theta, n, n) = \theta$ *for each* $\theta \in \Theta$ *and* $n \in \mathbb{Z}$;
2. $\sigma(\sigma(\theta, m, n), k, m) = \sigma(\theta, k, n)$ *for every* $\theta \in \Theta$ *and* $n, m, k \in \mathbb{Z}$;
3. $\sigma(\cdot, m, n)$ *is a continuous map for each* $(m, n) \in \mathbb{Z} \times \mathbb{Z}$.

Set $\Delta = \{(m, n) \in \mathbb{Z} \times \mathbb{Z} : m \geq n\}$.

Definition 2. *Let σ be a continuous three-parameter flow. A map $\Phi \colon \Theta \times \Delta \to B(X)$ is said to be a linear skew-product three-parameter semiflow (with discrete time) over σ if:*

1. $\Phi(\theta, n, n) = \mathrm{Id}$ *for* $\theta \in \Theta$ *and* $n \in \mathbb{Z}$;
2. $\Phi(\sigma(\theta, m, n), k, m)\Phi(\theta, m, n) = \Phi(\theta, k, n)$ *for* $\theta \in \Theta$ *and* $(m, n), (k, m) \in \Delta$;
3. $\theta \mapsto \Phi(\theta, m, n)x$ *is continuous for each* $x \in X$ *and* $(m, n) \in \Delta$.

Let us give some examples.

Example 1. *Assume that Θ is a singleton, i.e., that $\Theta = \{p\}$ and let $\sigma(p, m, n) = p$ for $m, n \in \mathbb{Z}$. Furthermore, let $(A_n)_{n \in \mathbb{Z}}$ be a sequence in $B(X)$. For $(m, n) \in \Delta$, set*

$$\Phi(p, m, n) = \begin{cases} A_{m-1} \cdots A_n & \text{for } m > n; \\ \mathrm{Id} & \text{for } m = n. \end{cases}$$

Then, one can easily verify that Φ is a linear skew-product three-parameter semiflow over σ.

Example 2. *Let Θ be an arbitrary Banach space and $\rho \colon \Theta \to \Theta$ a homeomorphism. We define $\sigma \colon \Theta \times \mathbb{Z} \times \mathbb{Z} \to \Theta$ by*

$$\sigma(\theta, m, n) = \rho^{m-n}(\theta), \quad \text{for } \theta \in \Theta \text{ and } m, n \in \mathbb{Z}.$$

One can easily verify that σ is a continuous three-parameter flow. Let $\mathcal{A} \colon \Theta \times \mathbb{N}_0 \to B(X)$ be a linear cocycle over ρ, i.e., $\mathcal{A}$ satisfies the following conditions:

- $\mathcal{A}(\theta, 0) = \mathrm{Id}$ *for* $\theta \in \Theta$;
- $\mathcal{A}(\theta, m+n) = \mathcal{A}(\rho^m(\theta), n)\mathcal{A}(\theta, m)$ *for* $\theta \in \Theta$ *and* $n, m \in \mathbb{N}_0$;
- $\theta \mapsto \mathcal{A}(\theta, 1)x$ *is continuous for each* $x \in X$.

For $\theta \in \Theta$ and $(m, n) \in \Delta$, set

$$\Phi(\theta, m, n) = \mathcal{A}(\theta, m - n).$$

Then, it is easy to show that Φ is a linear skew-product three-parameter semiflow over σ.

Example 3. Let σ be a continuous three-parameter flow on a metric space Θ. Furthermore, take a map $A\colon \Theta \to B(X)$ such that $\theta \mapsto A(\theta)x$ is continuous for each $x \in X$. For $(\theta, n) \in \Theta \times \mathbb{Z}$ and $x \in X$, let us consider a Cauchy problem

$$y_{m+1} = A(\sigma(\theta, m, n))y_m \quad m \geq n, \quad y_n = x.$$

Let $\Phi(\theta, m, n)x$ denote the value of the solution of this problem at time m. Then, Φ is a linear skew-product three-parameter semiflow over σ. We observe that

$$\Phi(\theta, m, n) = A(\sigma(\theta, m-1, n)) \cdots A(\sigma(\theta, n+1, n))A(\theta),$$

for $\theta \in \Theta$ and $(m, n) \in \Delta$.

We now introduce the notion of exponential stability.

Definition 3. For a linear skew-product three-parameter semiflow Φ we say that it is exponentially stable if there exist $D, \lambda > 0$ such that

$$\|\Phi(\theta, m, n)\| \leq D e^{-\lambda(m-n)}, \quad \text{for } \theta \in \Theta \text{ and } (m, n) \in \Delta. \tag{1}$$

We also introduce some additional notation that will be used throughout this paper. More precisely, for a linear skew-product three-parameter semiflow Φ over σ, we introduce a map $\tilde{\Phi}\colon \Theta \times \mathbb{Z} \to B(X)$ by

$$\tilde{\Phi}(\theta, n) = \Phi(\theta, n+1, n), \quad \text{for } (\theta, n) \in \Theta \times \mathbb{Z}.$$

Furthermore, we define $\tilde{\sigma}\colon \Theta \times \mathbb{Z} \to \Theta \times \mathbb{Z}$ by

$$\tilde{\sigma}(\theta, n) = (\sigma(\theta, n+1, n), n+1) \quad \text{for } (\theta, n) \in \Theta \times \mathbb{Z}.$$

Clearly, $\tilde{\sigma}$ is invertible and in fact,

$$\tilde{\sigma}^m(\theta, n) = (\sigma(\theta, n+m, n), n+m), \quad \text{for } (\theta, n) \in \Theta \times \mathbb{Z} \text{ and } m \in \mathbb{Z}.$$

Some Auxiliary Results

We also recall some useful results established by Daleckij and Krein [5].

Lemma 1. Assume that $\mathcal{H}$ is a Hilbert space and that T is a bounded operator on $\mathcal{H}$. Furthermore, suppose that the spectrum of T does not cover the whole unit circle S^1. Then every self-adjoint operator bounded operator W on $\mathcal{H}$ with the property that there exists $\delta > 0$ such that

$$T^* W T - W \leq -\delta \mathrm{Id} \tag{2}$$

is invertible.

We will also use the following result (also taken from [5]).

Lemma 2. Assume that $\mathcal{H}$ is a Hilbert space and that T is a bounded operator on $\mathcal{H}$. Furthermore, assume that there exists an invertible, self-adjoint and bounded linear operator W on $\mathcal{H}$ such that (2) holds for some $\delta > 0$. Then, the spectrum of T does not intersect S^1 and there exist $\delta' > 0$ satisfying

$$T W^{-1} T^* - W^{-1} \leq -\delta' \mathrm{Id}.$$

Moreover, if $W \geq 0$ (that is, $\langle Wx, x \rangle \geq 0$ for $x \in \mathcal{H}$) then the spectrum of T is contained in $\{z \in \mathbb{C} : |z| < 1\}$.

3. Main Results

The following is our first main result.

Theorem 1. *Assume that $\Phi \colon \Theta \times \Delta \to B(X)$ is an exponentially stable linear skew-product three-parameter semiflow over a continuous three-parameter flow σ. Then, there exists a family $S_{(\theta,n)}$, $(\theta,n) \in \Theta \times \mathbb{Z}$ of bounded, self-adjoint and invertible operators on X and $K, \delta > 0$ such that for $(\theta,n) \in \Theta \times \mathbb{Z}$:*

1. $S_{(\theta,n)} \geq 0$;
2.
$$\|S_{(\theta,n)}\| \leq K \quad \text{and} \quad \|S^{-1}_{(\theta,n)}\| \leq K; \tag{3}$$

3.
$$\Phi(\theta,n)^* S_{\bar{\sigma}(\theta,n)} \Phi(\theta,n) - S_{(\theta,n)} \leq -\delta \mathrm{Id}; \tag{4}$$

4.
$$\Phi(\theta,n) S^{-1}_{(\theta,n)} \Phi(\theta,n)^* - S^{-1}_{\bar{\sigma}(\theta,n)} \leq -\delta \mathrm{Id}; \tag{5}$$

Proof. For $(\theta,n) \in \Theta \times \mathbb{Z}$, set

$$S_{(\theta,n)} := \sum_{k=n}^{+\infty} \Phi(\theta,k,n)^* \Phi(\theta,k,n).$$

It follows from (1) that

$$\langle S_{(\theta,n)} x, x \rangle = \sum_{k=n}^{+\infty} \|\Phi(\theta,k,n)x\|^2$$
$$\leq \sum_{k=n}^{+\infty} D^2 e^{-2\lambda(k-n)} \|x\|^2$$
$$= K\|x\|^2,$$

where $K = \frac{D^2}{1-e^{-2\lambda}} > 0$. Obviously, $S_{(\theta,n)}$ is self-adjoint, $S_{(\theta,n)} \geq 0$ and therefore

$$\|S_{(\theta,n)}\| = \sup_{\|x\|=1} \langle S_{(\theta,n)} x, x \rangle \leq K, \quad \text{for } (\theta,n) \in \Theta \times \mathbb{Z}.$$

Hence, the first inequality (3) holds. Furthermore, we have that

$$\Phi(\theta,n)^* S_{\bar{\sigma}(\theta,n)} \Phi(\theta,n) - S_{(\theta,n)}$$
$$= \Phi(\theta,n)^* \sum_{k=n+1}^{+\infty} \Phi(\sigma(\theta,n+1,n),k,n+1)^* \Phi(\sigma(\theta,n+1,n),k,n+1) \Phi(\theta,n)$$
$$- \sum_{k=n}^{+\infty} \Phi(\theta,k,n)^* \Phi(\theta,k,n)$$
$$= \sum_{k=n+1}^{+\infty} \Phi(\theta,k,n)^* \Phi(\theta,k,n) - \sum_{k=n}^{+\infty} \Phi(\theta,k,n)^* \Phi(\theta,k,n)$$
$$= -\Phi(\theta,n,n)$$
$$= -\mathrm{Id},$$

which implies that (4) holds with $\delta = 1$.

Set now

$$l^2 := \left\{ \mathbf{x} = (x_n)_{n \in \mathbb{Z}} \subset X : \sum_{n=-\infty}^{+\infty} \|x_n\|^2 < +\infty \right\}.$$

Clearly, l^2 is a Hilbert space with respect to the scalar product

$$\langle \mathbf{x}, \mathbf{y} \rangle := \sum_{n=-\infty}^{+\infty} \langle x_n, y_n \rangle,$$

for $\mathbf{x} = (x_n)_{n \in \mathbb{Z}}$ and $\mathbf{y} = (y_n)_{n \in \mathbb{Z}}$ in l^2. For $(\theta, n) \in \Theta \times \mathbb{Z}$, we define $\mathbb{A}_{(\theta,n)} : l^2 \to l^2$ by

$$\begin{aligned}(\mathbb{A}_{(\theta,n)} \mathbf{x})_m &= \Phi(\tilde{\sigma}^{m-1}(\theta,n)) x_{m-1} \\ &= \Phi(\sigma(\theta, n+m-1,n), n+m-1) x_{m-1} \\ &= \Phi(\sigma(\theta, n+m-1,n), n+m, n+m-1) x_{m-1},\end{aligned}$$

for $m \in \mathbb{Z}$ and $\mathbf{x} = (x_n)_{n \in \mathbb{Z}} \in l^2$. It follows from (1) that

$$\sum_{m=-\infty}^{+\infty} \|(\mathbb{A}_{(\theta,n)} \mathbf{x})_m\|^2 = \sum_{m=-\infty}^{+\infty} \|\Phi(\sigma(\theta, n+m-1,n), n+m-1) x_{m-1}\|^2$$

$$\leq D^2 \sum_{m=-\infty}^{+\infty} \|x_{m-1}\|^2,$$

for every $\mathbf{x} = (x_n)_{n \in \mathbb{Z}} \in l^2$. Hence, $\mathbb{A}_{(\theta,n)}$ is well-defined and bounded linear operator for each $(\theta, n) \in \Theta \times \mathbb{Z}$.

We need the following auxiliary results.

Lemma 3. *We have that*

$$(\mathbb{A}^*_{(\theta,n)} \mathbf{x})_m = \Phi(\tilde{\sigma}^m(\theta, n))^* x_{m+1}, \quad \text{for } (\theta, n) \in \Theta \times \mathbb{Z} \text{ and } m \in \mathbb{Z}.$$

Proof of the Lemma. Take $(\theta, n) \in \Theta \times \mathbb{Z}$, we define $\mathbb{B}_{(\theta,n)} : l^2 \to l^2$ by

$$(\mathbb{B}_{(\theta,n)} \mathbf{x})_m = \Phi(\tilde{\sigma}^m(\theta, n))^* x_{m+1}, \quad \text{for } (\theta, n) \in \Theta \times \mathbb{Z} \text{ and } m \in \mathbb{Z}.$$

Obviously, $\mathbb{B}_{(\theta,n)}$ is a well-defined and bounded linear operator. For $\mathbf{x} = (x_n)_{n \in \mathbb{Z}}$ and $\mathbf{y} = (y_n)_{n \in \mathbb{Z}}$ in l^2, we have that

$$\begin{aligned}\langle \mathbb{A}_{(\theta,n)} \mathbf{x}, \mathbf{y} \rangle &= \sum_{m=-\infty}^{+\infty} \langle (\mathbb{A}_{(\theta,n)} \mathbf{x})_m, y_m \rangle \\ &= \sum_{m=-\infty}^{+\infty} \langle \Phi(\tilde{\sigma}^{m-1}(\theta,n)) x_{m-1}, y_m \rangle \\ &= \sum_{m=-\infty}^{+\infty} \langle x_{m-1}, \Phi(\tilde{\sigma}^{m-1}(\theta,n))^* y_m \rangle \\ &= \sum_{m=-\infty}^{+\infty} \langle x_{m-1}, (\mathbb{B}_{(\theta,n)} \mathbf{y})_{m-1} \rangle \\ &= \langle \mathbf{x}, \mathbb{B}_{(\theta,n)} \mathbf{y} \rangle,\end{aligned}$$

which readily implies the desired conclusion. □

Lemma 4. *There exists $t \in (0,1)$ such that spectrum of $\mathbb{A}_{(\theta,n)}$ is contained in $\{z \in \mathbb{C} : |z| \leq t\}$, for each $(\theta, n) \in \Theta \times \mathbb{Z}$.*

Proof of the Lemma. Fix $(\theta, n) \in \Theta \times \mathbb{Z}$. Then, for each $k \in \mathbb{N}$ and $\mathbf{x} = (x_m)_{m \in \mathbb{Z}} \in l^2$ we have that

$$(\mathbb{A}^k_{(\theta,n)} \mathbf{x})_m = \Phi(\sigma(\theta, n+m-k, n), n+m, n+m-k) x_{m-k}$$

and consequently (1) implies that

$$\|(\mathbb{A}_{(\theta,n)}^k \mathbf{x})_m\| \leq De^{-\lambda k}\|x_{m-k}\|,$$

for each $m \in \mathbb{Z}$. This readily yields that $\|(\mathbb{A}_{(\theta,n)}^k)\| \leq De^{-\lambda k}$. Since $k \in \mathbb{N}$ was arbitrary we conclude that the statement of the lemma holds with $t = e^{-\lambda} < 1$. □

For $(\theta, n) \in \Theta \times \mathbb{Z}$ we define $\mathbb{W}_{(\theta,n)} : l^2 \to l^2$ by

$$(\mathbb{W}_{(\theta,n)}\mathbf{x})_m = S_{\bar{\sigma}^m(\theta,n)} x_m, \quad \text{for } (\theta, n) \in \Theta \times \mathbb{Z} \text{ and } m \in \mathbb{Z}.$$

It follows easily from the already proved first inequality in (3) that $\mathbb{W}_{(\theta,n)}$ is a well-defined and bounded linear operator on l^2. Moreover, it is easy to show that $\mathbb{W}_{(\theta,n)}$ is self-adjoint.

On the other hand, observe that for $(\theta, n) \in \Theta \times \mathbb{Z}$ and $\mathbf{x} = (x_n)_{n \in \mathbb{Z}} \in l^2$, we have that

$$(\mathbb{A}_{(\theta,n)}^* \mathbb{W}_{(\theta,n)} \mathbb{A}_{(\theta,n)} \mathbf{x})_m = \Phi(\bar{\sigma}^m(\theta,n))^* S_{\bar{\sigma}^{m+1}(\theta,n)} \Phi(\bar{\sigma}^m(\theta,n)) x_m,$$

for each $m \in \mathbb{Z}$. Hence, the already proved inequality (4) (we recall that it holds with $\delta = 1$) implies that

$$\mathbb{A}_{(\theta,n)}^* \mathbb{W}_{(\theta,n)} \mathbb{A}_{(\theta,n)} - \mathbb{W}_{(\theta,n)} \leq -\mathrm{Id} \quad \text{on } l^2, \qquad (6)$$

for each $(\theta, n) \in \Theta \times \mathbb{Z}$. Hence, Lemmas 1 and 4 imply that $\mathbb{W}_{(\theta,n)}$ is invertible for every $(\theta, n) \in \Theta \times \mathbb{Z}$.

Lemma 5. *We have that*

$$\sup_{(\theta,n) \in \Theta \times \mathbb{Z}} \|\mathbb{W}_{(\theta,n)}^{-1}\| < +\infty.$$

Proof of the Lemma. For $(\theta, n) \in \Theta \times \mathbb{Z}$, set

$$\mathbb{H}_{(\theta,n)} := -\mathbb{A}_{(\theta,n)}^* \mathbb{W}_{(\theta,n)} \mathbb{A}_{(\theta,n)} + \mathbb{W}_{(\theta,n)}.$$

Then, $\mathbb{H}_{(\theta,n)} \geq \mathrm{Id}$. It is easy to verify that

$$(\mathbb{A}_{(\theta,n)}^* - \mathrm{Id})\mathbb{W}_{(\theta,n)}(\mathbb{A}_{(\theta,n)} + \mathrm{Id}) + (\mathbb{A}_{(\theta,n)}^* + \mathrm{Id})\mathbb{W}_{(\theta,n)}(\mathbb{A}_{(\theta,n)} - \mathrm{Id}) = -2\mathbb{H}_{(\theta,n)}$$

By multiplying this identity on the right by $(\mathbb{A}_{(\theta,n)} - \mathrm{Id})^{-1}$ and on the left by $(\mathbb{A}_{(\theta,n)}^* - \mathrm{Id})^{-1}$, we obtain that

$$\mathbb{W}_{(\theta,n)}(\mathbb{A}_{(\theta,n)} + \mathrm{Id})(\mathbb{A}_{(\theta,n)} - \mathrm{Id})^{-1} + (\mathbb{A}_{(\theta,n)}^* - \mathrm{Id})^{-1}(\mathbb{A}_{(\theta,n)}^* + \mathrm{Id})\mathbb{W}_{(\theta,n)}$$
$$= -2(\mathbb{A}_{(\theta,n)}^* - \mathrm{Id})^{-1}\mathbb{H}_{(\theta,n)}(\mathbb{A}_{(\theta,n)} - \mathrm{Id})^{-1}.$$

Therefore,

$$\langle (\mathbb{A}_{(\theta,n)}^* - \mathrm{Id})^{-1}\mathbb{H}_{(\theta,n)}(\mathbb{A}_{(\theta,n)} - \mathrm{Id})^{-1}\mathbf{x}, \mathbf{x}\rangle$$
$$\leq \frac{1}{2}\|\mathbb{W}_{(\theta,n)}\mathbf{x}\| \cdot \|\mathbb{A}_{(\theta,n)} + \mathrm{Id}\| \cdot \|(\mathbb{A}_{(\theta,n)} - \mathrm{Id})^{-1}\| \cdot \|\mathbf{x}\|,$$

for every $\mathbf{x} \in l^2$. On the other hand,

$$2\langle (\mathbb{A}_{(\theta,n)}^* - \mathrm{Id})^{-1}\mathbb{H}_{(\theta,n)}(\mathbb{A}_{(\theta,n)} - \mathrm{Id})^{-1}\mathbf{x}, \mathbf{x}\rangle$$
$$= 2\langle \mathbb{H}_{(\theta,n)}(\mathbb{A}_{(\theta,n)} - \mathrm{Id})^{-1}\mathbf{x}, (\mathbb{A}_{(\theta,n)} - \mathrm{Id})^{-1}\mathbf{x}\rangle$$
$$\geq 2\|(\mathbb{A}_{(\theta,n)} - \mathrm{Id})^{-1}\mathbf{x}\|^2$$
$$\geq 2\frac{\|\mathbf{x}\|^2}{\|\mathrm{Id} - \mathbb{A}_{(\theta,n)}\|^2}.$$

Combining the last two estimates, we obtain that

$$2\frac{\|\mathbf{x}\|^2}{\|\text{Id} - \mathbb{A}_{(\theta,n)}\|^2} \leq \|\mathbb{W}_{(\theta,n)}\mathbf{x}\| \cdot \|\mathbb{A}_{(\theta,n)} + \text{Id}\| \cdot \|(\mathbb{A}_{(\theta,n)} - \text{Id})^{-1}\| \cdot \|\mathbf{x}\|,$$

and thus

$$\|\mathbf{x}\| \leq \frac{1}{2}\|\mathbb{W}_{(\theta,n)}\mathbf{x}\| \cdot \|\mathbb{A}_{(\theta,n)} + \text{Id}\| \cdot \|(\mathbb{A}_{(\theta,n)} - \text{Id})^{-1}\| \cdot \|\text{Id} - \mathbb{A}_{(\theta,n)}\|^2,$$

for $\mathbf{x} \in l^2$. It follows from Lemma 4 that

$$\sup_{(\theta,n)\in\Theta\times\mathbb{Z}} \|(\mathbb{A}_{(\theta,n)} - \text{Id})^{-1}\| < +\infty.$$

Hence, there exist $R > 0$ such that

$$\|\mathbf{x}\| \leq R\|\mathbb{W}_{(\theta,n)}\mathbf{x}\| \quad \text{for } \mathbf{x} \in l^2 \text{ and } (\theta, n) \in \Theta \in \mathbb{Z}.$$

Hence,

$$\sup_{(\theta,n)\in\Theta\times\mathbb{Z}} \|\mathbb{W}_{(\theta,n)}^{-1}\| \leq R < +\infty,$$

and the proof of the lemma is completed. □

Lemma 6. *For each* $(\theta, n) \in \Theta \times \mathbb{Z}$, $S_{(\theta,n)}$ *is invertible. Furthermore, the second inequality in* (3) *holds.*

Proof of the Lemma. Observe that $S_{(\theta,n)} \geq \text{Id}$ and thus $S_{(\theta,n)}$ is injective. Take $v \in X$ and consider $\mathbf{y} = (y_m)_{m\in\mathbb{Z}} \in l^2$ given by $y_0 = v$ and $y_m = 0$ for $m \neq 0$. Since $\mathbb{W}_{(\theta,n)}$ is invertible, there exists $\mathbf{x} - (x_m)_{m\in\mathbb{Z}} \subset l^2$ such that $\mathbb{W}_{(\theta,n)}\mathbf{x} = \mathbf{y}$. Hence,

$$v = y_0 = (\mathbb{W}_{(\theta,n)}\mathbf{x})_0 = S_{(\theta,n)}x_0.$$

Hence, $S_{(\theta,n)}$ is also surjective and thus it is invertible. Moreover,

$$\|S_{(\theta,n)}^{-1}v\| = \|x_0\| \leq \|\mathbf{x}\| = \|\mathbb{W}_{(\theta,n)}^{-1}\mathbf{y}\| \leq \|\mathbb{W}_{(\theta,n)}^{-1}\| \cdot \|\mathbf{y}\| = \|\mathbb{W}_{(\theta,n)}^{-1}\| \cdot \|v\|.$$

Therefore, $\|S_{(\theta,n)}^{-1}\| \leq \|\mathbb{W}_{(\theta,n)}^{-1}\|$ for all $(\theta, n) \in \Theta \times \mathbb{Z}$. Now the second inequality in (3) follows directly from the previous lemma. □

It remains to establish (5). Using the same notation as in the proof of Lemma 5 we have

$$- 2\mathbb{W}_{(\theta,n)}^{-1}(\mathbb{A}_{(\theta,n)}^* - \text{Id})^{-1}\mathbb{H}_{(\theta,n)}(\mathbb{A}_{(\theta,n)} - \text{Id})^{-1}\mathbb{W}_{(\theta,n)}^{-1}$$
$$= (\mathbb{A}_{(\theta,n)} + \text{Id})(\mathbb{A}_{(\theta,n)} - \text{Id})^{-1}\mathbb{W}_{(\theta,n)}^{-1}$$
$$+ \mathbb{W}_{(\theta,n)}^{-1}(\mathbb{A}_{(\theta,n)}^* - \text{Id})^{-1}(\mathbb{A}_{(\theta,n)}^* + \text{Id}).$$

Moreover, multiplying this equality on the left by $\mathbb{A}_{(\theta,n)} - \text{Id}$ and on the right by $\mathbb{A}_{(\theta,n)}^* - \text{Id}$ yields that

$$- 2(\mathbb{A}_{(\theta,n)} - \text{Id})\mathbb{W}_{(\theta,n)}^{-1}(\mathbb{A}_{(\theta,n)}^* - \text{Id})^{-1}\mathbb{H}_{(\theta,n)}(\mathbb{A}_{(\theta,n)} - \text{Id})^{-1}\mathbb{W}_{(\theta,n)}^{-1}(\mathbb{A}_{(\theta,n)}^* - \text{Id})$$
$$= (\mathbb{A}_{(\theta,n)} + \text{Id})\mathbb{W}_{(\theta,n)}^{-1}(\mathbb{A}_{(\theta,n)}^* - \text{Id}) + (\mathbb{A}_{(\theta,n)} - \text{Id})\mathbb{W}_{(\theta,n)}^{-1}(\mathbb{A}_{(\theta,n)}^* + \text{Id})$$
$$= 2\mathbb{A}_{(\theta,n)}\mathbb{W}_{(\theta,n)}^{-1}\mathbb{A}_{(\theta,n)}^* - 2\mathbb{W}_{(\theta,n)}^{-1}.$$

Hence,
$$-(\mathbb{A}_{(\theta,n)} - \mathrm{Id})\mathbb{W}_{(\theta,n)}^{-1}(\mathbb{A}_{(\theta,n)}^* - \mathrm{Id})^{-1}\mathbb{H}_{(\theta,n)}(\mathbb{A}_{(\theta,n)} - \mathrm{Id})^{-1}\mathbb{W}_{(\theta,n)}^{-1}(\mathbb{A}_{(\theta,n)}^* - \mathrm{Id})$$
$$= \mathbb{A}_{(\theta,n)}\mathbb{W}_{(\theta,n)}^{-1}\mathbb{A}_{(\theta,n)}^* - \mathbb{W}_{(\theta,n)}^{-1}.$$

Observe that for each $\mathbf{x} \in l^2$, we have that
$$\langle (\mathbb{A}_{(\theta,n)} - \mathrm{Id})\mathbb{W}_{(\theta,n)}^{-1}(\mathbb{A}_{(\theta,n)}^* - \mathrm{Id})^{-1}\mathbb{H}_{(\theta,n)}(\mathbb{A}_{(\theta,n)} - \mathrm{Id})^{-1}\mathbb{W}_{(\theta,n)}^{-1}(\mathbb{A}_{(\theta,n)}^* - \mathrm{Id})\mathbf{x}, \mathbf{x}\rangle$$
$$\geq \|(\mathbb{A}_{(\theta,n)} - \mathrm{Id})^{-1}\mathbb{W}_{(\theta,n)}^{-1}(\mathbb{A}_{(\theta,n)}^* - \mathrm{Id})\mathbf{x}\|^2$$
$$\geq \frac{\|\mathbf{x}\|^2}{\|\mathbb{A}_{(\theta,n)} - \mathrm{Id}\| \cdot \|\mathbb{W}_{(\theta,n)}\| \cdot \|(\mathbb{A}_{(\theta,n)}^* - \mathrm{Id})^{-1}\|}.$$

Since there exists $L > 0$ such that
$$\|\mathbb{A}_{(\theta,n)} - \mathrm{Id}\| \cdot \|\mathbb{W}_{(\theta,n)}\| \cdot \|(\mathbb{A}_{(\theta,n)}^* - \mathrm{Id})^{-1}\| \leq L,$$
for $(\theta,n) \in \Theta \times \mathbb{Z}$, we conclude that
$$\langle \mathbb{A}_{(\theta,n)} \mathbb{W}_{(\theta,n)}^{-1} \mathbb{A}_{(\theta,n)}^* \mathbf{x}, \mathbf{x}\rangle - \langle \mathbb{W}_{(\theta,n)}^{-1}\mathbf{x}, \mathbf{x}\rangle \leq -\frac{1}{L}\langle \mathbf{x}, \mathbf{x}\rangle, \tag{7}$$
for every $\mathbf{x} \in l^2$. By applying (7) for $\mathbf{x} = (x_m)_{m\in\mathbb{Z}} \in l^2$ given by $x_m = 0$ for $m \neq 1$ and $x_1 = v$, where $v \in X$ is arbitrary, we conclude that (5) holds with $\delta = \frac{1}{L} > 0$. $\square$

We now establish the converse of Theorem 1.

Theorem 2. *Assume that $\Phi \colon \Theta \times \Delta \to B(X)$ is an linear skew-product three-parameter semiflow over a continuous three-parameter flow σ such that*
$$\sup_{(\theta,n)\in\Theta\times\mathbb{Z}} \|\Phi(\theta, n+1, n)\| < +\infty. \tag{8}$$

Furthermore, suppose that there exists a family $S_{(\theta,n)}$, $(\theta,n) \in \Theta \times \mathbb{Z}$ of bounded, self-adjoint and invertible operators on X and $K, \delta > 0$ such that $S_{(\theta,n)} \geq 0$ and that (3)–(5) hold for each $(\theta,n) \in \Theta \times \mathbb{Z}$. Then, Φ is exponentially stable.

Proof. For $(\theta,n) \in \Theta \times \mathbb{Z}$, let $\mathbb{A}_{(\theta,n)}$ and $\mathbb{W}_{(\theta,n)}$ are as in the proof of Theorem 1. Please note that (8) implies that $\mathbb{A}_{(\theta,n)}$ is a well-defined and bounded linear operator. Furthermore, observe that (4) and (5) imply that
$$\mathbb{A}_{(\theta,n)}^* \mathbb{W}_{(\theta,n)} \mathbb{A}_{(\theta,n)} - \mathbb{W}_{(\theta,n)} \leq -\delta \mathrm{Id} \quad \text{on } l^2,$$
and
$$\mathbb{A}_{(\theta,n)} \mathbb{W}_{(\theta,n)}^{-1} \mathbb{A}_{(\theta,n)}^* - \mathbb{W}_{(\theta,n)}^{-1} \leq -\delta \mathrm{Id} \quad \text{on } l^2.$$

Since $S_{(\theta,n)} \geq 0$ on X for $(\theta,n) \in \Theta \times \mathbb{Z}$, we have that $\mathbb{W}_{(\theta,n)} \geq 0$ and l^2 for each $(\theta,n) \in \Theta \times \mathbb{Z}$. Consequently, Lemma 2 implies that the spectrum of $\mathbb{A}_{(\theta,n)}$ is contained in $\{z \in \mathbb{C} : |z| < 1\}$, for every $(\theta,n) \in \Theta \times \mathbb{Z}$.

Lemma 7. *We have that*
$$\sup_{(\theta,n)\in\Theta\times\mathbb{Z}} \|(\mathrm{Id} - \mathbb{A}_{(\theta,n)})^{-1}\| < +\infty.$$

Proof of the Lemma. By repeating the arguments in the first part of the proof of Lemma 5 that

$$\delta\|(\mathbb{A}_{(\theta,n)} - \mathrm{Id})^{-1}\mathbf{x}\|^2 \leq \|\mathbb{W}_{(\theta,n)}\| \cdot \|\mathbb{A}_{(\theta,n)} + \mathrm{Id}\| \cdot \|(\mathrm{Id} - \mathbb{A}_{(\theta,n)})^{-1}\mathbf{x}\| \cdot \|\mathbf{x}\|,$$

for $(\theta, n) \in \Theta \times \mathbb{Z}$ and $\mathbf{x} \in l^2$. On the other hand, (3) and (8) imply that

$$\sup_{(\theta,n)\in\Theta\times\mathbb{Z}} (\|\mathbb{W}_{(\theta,n)}\| \cdot \|\mathbb{A}_{(\theta,n)} + \mathrm{Id}\|) < +\infty.$$

The conclusion of the lemma now readily follows. □

Take now $(\theta, n) \in \Theta \times \mathbb{Z}$, $v \in X$ and consider a sequence $\mathbf{y} = (y_m)_{m \in \mathbb{Z}}$ by

$$y_m = \begin{cases} v & \text{if } m = 0, \\ 0 & \text{if } m \neq 0. \end{cases}$$

Set $\mathbf{x} = (\mathrm{Id} - \mathbb{A}_{(\theta,n)})^{-1}\mathbf{y} \in l^2$. It is easy to verify that

$$x_m = \begin{cases} 0 & \text{if } m < 0, \\ \Phi(\theta, n+m, n)v & \text{if } m \geq 0. \end{cases}$$

Then, Lemma 7 implies that there exist $C > 0$ such that

$$\left(\sum_{k \geq n} \|\Phi(\theta, k, n)v\|^2\right)^{1/2} \leq C\|v\|, \quad \text{for } (\theta, n) \in \Theta \times \mathbb{Z} \text{ and } v \in X. \tag{9}$$

In particular, (9) implies that

$$\|\Phi(\theta, k, n)v\| \leq C\|v\|, \quad \text{for } (\theta, n) \in \Theta \times \mathbb{Z}, k \geq n \text{ and } v \in X. \tag{10}$$

Take now $\theta \in \Theta$, $v \in X$ and $m \geq n$. Then, for each $n \leq k \leq m$ we have that

$$\|\Phi(\theta, m, n)v\|^2 = \|\Phi(\sigma(\theta, k, n), m, k)\Phi(\theta, k, n)v\|^2 \leq C^2 \|\Phi(\theta, k, n)v\|^2.$$

Summing over k and using (9), we obtain that

$$(m - n + 1)\|\Phi(\theta, m, n)v\|^2 \leq C^2 \sum_{k \geq n} \|\Phi(\theta, k, n)v\|^2 \leq C^4 \|v\|^2.$$

Thus,

$$\|\Phi(\theta, m, n)\| \leq \frac{C^2}{\sqrt{m - n + 1}}.$$

Consequently, there exist $N_0 \in \mathbb{N}$ such that

$$\|\Phi(\theta, m, n)\| \leq e^{-1}, \quad \text{for } \theta \in \Theta \text{ and } m, n \in \mathbb{Z} \text{ such that } m - n \geq N_0. \tag{11}$$

Now, (10) and (11) easily imply that Φ is exponentially stable. □

4. Applications

In this section, we use Theorems 1 and 2 to prove that the notion of exponential stability persists under sufficiently small linear perturbations.

Theorem 3. *Assume that* $\Phi, \Psi \colon \Theta \times \Delta \to B(X)$ *are two linear skew-product three-parameter semiflows over a continuous three-parameter flow* σ. *Furthermore, suppose that:*

1. Φ is exponentially stable;
2. there exists $c > 0$ such that

$$\sup_{(\theta,n)\in\Theta\times\mathbb{Z}} \|\Phi(\theta, n+1, n) - \Psi(\theta, n+1, n)\| \leq c. \tag{12}$$

Then, if c is sufficiently small, Ψ is also exponentially stable.

Proof. We first observe that since Φ is exponentially stable, (12) implies that

$$\sup_{(\theta,n)\in\Theta\times\mathbb{Z}} \|\Psi(\theta, n+1, n)\| < +\infty.$$

Let $S_{(\theta,n)}$, $(\theta, n) \in \Theta \times \mathbb{Z}$, $K, \delta > 0$ be given by Theorem 1. For each $(\theta, n) \in \Theta \times \mathbb{Z}$ and $v \in X$, (4) implies that

$$\langle \Psi(\theta, n)^* S_{\tilde{\sigma}(\theta,n)} \Psi(\theta, n)v, v\rangle - \langle S_{(\theta,n)} v, v\rangle$$
$$= \langle (\Psi(\theta, n) - \Phi(\theta, n))^* S_{\tilde{\sigma}(\theta,n)} (\Psi(\theta, n) - \Phi(\theta, n))v, v\rangle$$
$$+ \langle (\Psi(\theta, n) - \Phi(\theta, n))^* S_{\tilde{\sigma}(\theta,n)} \Phi(\theta, n)v, v\rangle$$
$$+ \langle \Phi(\theta, n)^* S_{\tilde{\sigma}(\theta,n)} (\Psi(\theta, n) - \Phi(\theta, n))v, v\rangle$$
$$+ \langle \Phi(\theta, n)^* S_{\tilde{\sigma}(\theta,n)} \Phi(\theta, n)v, v\rangle - \langle S_{(\theta,n)} v, v\rangle.$$

It follows from (1), (3), (4) and (12) that

$$\langle \Psi(\theta, n)^* S_{\tilde{\sigma}(\theta,n)} \Psi(\theta, n)v, v\rangle - \langle S_{(\theta,n)} v, v\rangle$$
$$\leq -\delta\langle v, v\rangle + c^2 K\langle v, v\rangle + 2DcK\langle v, v\rangle$$
$$= -(\delta - c^2 K - 2DcK)\langle v, v\rangle,$$

for $v \in X$. We conclude that

$$\Psi(\theta, n)^* S_{\tilde{\sigma}(\theta,n)} \Psi(\theta, n) - S_{(\theta,n)} \leq -\tilde{\delta}\mathrm{Id},$$

where $\tilde{\delta} = \delta - c^2 K - 2DcK$. Observe that $\tilde{\delta} > 0$ if c is sufficiently small. Similarly, one can prove that there exists $\tilde{\delta}' > 0$ such that

$$\Psi(\theta, n) S_{(\theta,n)}^{-1} \Psi(\theta, n)^* - S_{\tilde{\sigma}(\theta,n)}^{-1} \leq -\tilde{\delta}'\mathrm{Id},$$

for every $(\theta, n) \in \Theta \times \mathbb{Z}$. Putting all this together, Theorem 2 implies that Ψ is exponentially stable and the proof of the theorem is completed. □

5. Conclusions

In this paper, we obtained a complete Lyapunov-type characterization of exponential stability for linear skew-product three-parameter semiflows with discrete time. More precisely, we proved that exponential stability can be described in terms of the existence of appropriate quadratic Lyapunov functions. We then applied these results and prove that the notion of exponential stability persists under sufficiently small linear perturbations.

Author Contributions: D.D. and C.P. contributed equally in the preparation of this paper. All authors have read and agreed to the published version of the manuscript.

Funding: This research was funded by the Croatian Science Foundation under the project IP-2019-04-1239 and by the University of Rijeka under the projects uniri-prirod-18-9 and uniri-prprirod-19-16

Acknowledgments: We would like to thank the referees for their comments.

Conflicts of Interest: The authors declare no conflict of interest.

References

1. Lyapunov, A. *The General Problem of the Stability of Motion*; Taylor & Francis: London, UK; Washington, DC, USA, 1992.
2. LaSalle, J.; Lefschetz, S. Stability by Liapunov's Direct Method, with Applications. In *Mathematics in Science and Engineering*; Academic Press: New York, NY, USA; London, UK, 1961; Volume 4.
3. Hahn, W. Stability of Motion. In *Grundlehren der Mathematischen Wissenschaften*; Springer: Berlin/Heidelberg, Germany, 1967; Volume 138.
4. Bhatia, N.; Szegö, G. Stability Theory of Dynamical Systems. In *Grundlehren der Mathematischen Wissenschaften*; Springer: Berlin/Heidelberg, Germany, 1970; Volume 161.
5. Daleckij, J.; Krein, M. Stability of Differential Equations in Banach Spaces. *Amer. Math. Soc.* **1974**. [CrossRef]
6. Maizel, A.D. On stability of solutions of systems of differential equations. *TRudi Uralskogo Politekh. Inst. Math.* **1954**, *51*, 20–50.
7. Coppel, W. Dichotomies in Stability Theory. In *Lecture Notes in Mathematics*; Springer: Berlin/Heidelberg, Germany, 1978; Volume 629.
8. Coppel, W. Dichotomies and Lyapunov functions. *J. Differ. Equ.* **1984**, *52*, 58–65. [CrossRef]
9. Muldowney, J.S. Dichotomies and asymptotic behaviour for linear differential systems. *Trans. Amer. Math. Soc.* **1984**, *283*, 465–484. [CrossRef]
10. Papaschinopoulos, G. Dichotomies in terms of Lyapunov functions for linear difference equations. *J. Math. Anal. Appl.* **1990**, *152*, 524–535. [CrossRef]
11. Barreira, L.; Dragičević, D.; Valls, C. Lyapunov functions for strong exponential dichotomies. *J. Math. Anal. Appl.* **2013**, *399*, 116–132. [CrossRef]
12. Barreira, L.; Dragičević, D.; Valls, C. Lyapunov functions for strong exponential contractions. *J. Differ. Equ.* **2013**, *255*, 449–468. [CrossRef]
13. Barreira, L.; Valls, C. Lyapunov sequences for exponential dichotomies. *J. Differ. Equ.* **2009**, *246*, 183–215. [CrossRef]
14. Dragičević, D.; Preda, C. Lyapunov theorems for exponential dichotomies in Hilbert spaces. *Intern. J. Math.* **2016**, *27*, 1650033 [CrossRef]
15. Barreira, L.; Dragičević, D.; Valls, C. Admissibility and hyperbolicity. In *Springer Briefs in Mathematics*; Springer: Berlin/Heidelberg, Germany, 2018.
16. Barreira, L.; Dragičević, D.; Valls, C. Lyapunov type characterization of hyperbolic behavior. *J. Differ. Equ.* **2017**, *263*, 3147–3173. [CrossRef]
17. Barreira, L.; Dragičević, D.; Valls, C. Nonuniform exponential dichotomies and Lyapunov functions. *Regul. Chaotic Dyn.* **2017**, *22*, 197–209. [CrossRef]
18. Chen, G.; Jendrej, J. Lyapunov-type characterisation of exponential dichotomies with applications to the heat and Klein-Gordon equations. *Trans. Amer. Math. Soc.* **2019**, *372*, 7461–7496. [CrossRef]
19. Barreira, L.; Dragičević, D.; Valls, C. Lyapunov functions and cone families. *J. Stat. Phys.* **2012**, *148*, 137–163. [CrossRef]
20. Megan, M.; Stoica, C. Exponential instability of skew-evolution semiflows in Banach spaces. *Stud. Univ. Babes Bolyai Math.* **2008**, *53*, 17–24.
21. Dragičević, D.; Sasu, A.L.; Sasu, B. On the asymptotic behavior of discrete dynamical systems—An ergodic theory approach. *J. Differ. Equ.* **2020**, *268*, 4786–4829. [CrossRef]
22. Lupa, N.; Popescu, L.H. Admissible Banach function spaces for linear dynamics with nonuniform behavior on the half-line. *Semigroup Forum* **2019**, *98*, 184–208. [CrossRef]
23. Megan, M.; Sasu, A.L.; Sasu, B. On uniform exponential stability of linear skew-product semiflows in Banach spaces. *Bull. Belg. Math. Soc. Simon. Stevin.* **2002**, *9*, 143–154. [CrossRef]
24. Preda, P.; Pogan, A.; Preda, C. Schäffer spaces and uniform exponential stability of linear skew-product semiflows. *J. Differ. Equ.* **2005**, *212*, 191–207. [CrossRef]
25. Sasu, B.; Sasu, A.L. Stability and stabilizability for linear systems of difference equations. *J. Differ. Equ. Appl.* **2004**, *10*, 1085–1105. [CrossRef]

26. Sasu, B. Stability of difference equations and applications to robustness problems. *Adv. Differ. Equ.* **2010**, *2010*, 869608. [CrossRef]
27. Backes, L.; Dragičević, D. A Rolewicz-type characterization of nonuniform behaviour. *Appl. Anal.* **2019**, in press. [CrossRef]
28. Dragičević, D. Datko-Pazy conditions for nonuniform exponential stability. *J. Differ. Equ. Appl.* **2018**, *24*, 344–357. [CrossRef]
29. Lupa, N.; Popescu, L.H. Admissible Banach function spaces and nonuniform stabilities. *arXiv* **2016**, arXiv:1607.01427.
30. Preda, C.; Preda, P.; Craciunescu, A. A version of a theorem of R. Datko for nonuniform exponential contractions. *J. Math Anal Appl.* **2012**, *385*, 572–581. [CrossRef]
31. Sasu, A.L.; Megan, M.; Sasu, B. On Rolewicz-Zabczyk techniques in the stability theory of dynamical systems. *Fixed Point Theory* **2012**, *13*, 205–236.
32. Sasu, B. Generalizations of a theorem of Rolewicz. *Appl Anal.* **2005**, *84*, 1165–1172. [CrossRef]
33. Sasu, B. Integral conditions for exponential dichotomy: A nonlinear approach. *Bull. Sci. Math.* **2010**, *134*, 235–246. [CrossRef]
34. Dragičević, D. A note on the nonuniform exponential stability and dichotomy for nonautonomous difference equations. *Linear Algebra Appl.* **2018**, *552*, 105–126. [CrossRef]
35. Dragičević, D. On the exponential stability and hyperbolicity of linear cocycles. *Linear Multilinear Algebra* **2019**, in press.

© 2020 by the authors. Licensee MDPI, Basel, Switzerland. This article is an open access article distributed under the terms and conditions of the Creative Commons Attribution (CC BY) license (http://creativecommons.org/licenses/by/4.0/).

Article

Nonlocal Fractional Boundary Value Problems Involving Mixed Right and Left Fractional Derivatives and Integrals

Ahmed Alsaedi [1,†], Abrar Broom [1,†], Sotiris K. Ntouyas [1,2,*,†] and Bashir Ahmad [1,†]

1. Nonlinear Analysis and Applied Mathematics (NAAM)-Research Group, Department of Mathematics, Faculty of Science, King Abdulaziz University, P.O. Box 80203, Jeddah 21589, Saudi Arabia; aalsaedi@hotmail.com (A.A.); abrarbroom1992@gmail.com (A.B.); bashirahmad_qau@yahoo.com (B.A.)
2. Department of Mathematics, University of Ioannina, 451 10 Ioannina, Greece
* Correspondence: sntouyas@uoi.gr
† These authors contributed equally to this work.

Received: 30 March 2020; Accepted: 29 April 2020; Published: 1 May 2020

Abstract: In this paper, we study the existence of solutions for nonlocal single and multi-valued boundary value problems involving right-Caputo and left-Riemann–Liouville fractional derivatives of different orders and right-left Riemann–Liouville fractional integrals. The existence of solutions for the single-valued case relies on Sadovskii's fixed point theorem. The first existence results for the multi-valued case are proved by applying Bohnenblust-Karlin's fixed point theorem, while the second one is based on Martelli's fixed point theorem. We also demonstrate the applications of the obtained results.

Keywords: fractional differential equations; fractional differential inclusions; existence; fixed point theorems

1. Introduction

Fractional calculus has emerged as an interesting and fruitful subject in view of wide applications of its tools in modeling complex dynamical systems. Mathematical models based on fractional-order operators provide insight into the past history of the underlying phenomena. Examples include constitutive equations (fractional law) in the viscoelastic materials [1], Caputo power law in transport processes [2], dynamic memory describing the economic processes, see [3,4].

Widespread applications of fractional differential equations motivated many researchers to develop the theoretical aspects of the topic. During the last few decades, one can witness the remarkable development on initial and boundary value problems of fractional differential equations and inclusions. Much of the literature on such problems include Caputo, Riemann–Liouville, Hadamard type fractional derivatives, and different kinds of classical and non-classical boundary conditions. For some recent works on fractional order boundary value problems, for example, see the articles [5–12] and the references cited therein. Fractional differential equations involving left and right fractional derivatives also received considerable attention, for instance, see [13–16]. These derivatives appear in the study of Euler-Lagrange equations [17], steady heat-transfer in fractal media [18], electromagnetic waves phenomena in a variety of dielectric media with susceptibility [19], etc.

Multivalued (inclusions) problems are found to be of great utility in studying dynamical systems and stochastic processes, for example, see [20,21]. In the text [22], one can find the details on stochastic processes, queueing networks, optimization and their application in finance, control, climate control, etc. Monotone differential inclusions were applied to study the nonlinear dynamics of wheeled vehicles in [23]. In [24], a fractional differential inclusion with oscillatory potential was studied. In [25], the authors investigated the mild solutions to the time fractional Navier-Stokes delay differential inclusions. Other applications include polynomial control systems [20], synchronization of piecewise continuous systems of fractional order [21], oscillation and nonoscillation of impulsive fractional differential inclusions [26], etc. For some recent existence and controllability results on fractional differential inclusions, we refer the reader to articles [27–33] and the references cited therein.

Recently, in [34], the authors studied existence and uniqueness of solutions for a new kind of boundary value problem involving right-Caputo and left-Riemann–Liouville fractional derivatives of different orders and right-left Riemann–Liouville fractional integrals, subject to nonlocal boundary conditions of the form

$$\begin{cases} {}^C D_{1-}^\alpha {}^{RL} D_{0+}^\beta y(t) + \lambda I_{1-}^p I_{0+}^q h(t,y(t)) = f(t,y(t)), \ t \in J := [0,1], \\ y(0) = y(\xi) = 0, \quad y(1) = \delta y(\mu), \ 0 < \xi < \mu < 1, \end{cases} \quad (1)$$

where ${}^C D_{1-}^\alpha$ and ${}^{RL} D_{0+}^\beta$ denote the right Caputo fractional derivative of order $\alpha \in (1,2]$ and the left Riemann–Liouville fractional derivative of order $\beta \in (0,1]$, I_{1-}^p and I_{0+}^q denote the right and left Riemann–Liouville fractional integrals of orders $p, q > 0$ respectively, $f, h : [0,1] \times \mathbb{R} \to \mathbb{R}$ are given continuous functions and $\delta, \lambda \in \mathbb{R}$.

Here we emphasize that the importance of nonlocal conditions can be understood in the sense that such conditions are used to model the peculiarities occurring inside the domain of physical and chemical processes as the classical initial and boundary conditions fail to cater this situation. The present problem is motivated by useful applications of nonlocal boundary data in petroleum exploitation, thermodynamics, elasticity, and wave propagation, etc., for instance, see [35,36] and the details therein.

The existence results for the problem (1) were derived by applying a fixed point theorem due to Krasnoselski and Leray–Schauder nonlinear alternative, while the uniqueness result was established via Banach contraction mapping principle.

The objective of the present work is to enrich the results on this new class of problems. We firstly prove another existence result for the problem (1) with the aid of Sadovskii's fixed point theorem. Afterwards, we initiate the study of the multi-valued analogue of the problem (1) by considering the following inclusions problem:

$$\begin{cases} {}^C D_{1-}^\alpha {}^{RL} D_{0+}^\beta y(t) \in F(t,y(t)) - \lambda I_{1-}^p I_{0+}^q H(t,y(t)), \ t \in [0,1], \\ y(0) = y(\xi) = 0, \quad y(1) = \delta y(\mu), \ 0 < \xi < \mu < 1, \end{cases} \quad (2)$$

where $F, H : [0,1] \times \mathbb{R} \to \mathcal{P}(\mathbb{R})$ are compact multivalued maps, $\mathcal{P}(\mathbb{R})$ is the family of all nonempty subsets of $\mathbb{R}$, and the other quantities are the same as defined in problem (1). Existence results for the problem (2) are established via fixed point theorems due to Bohnenblust-Karlin [37] and Martelli [38].

The rest of the paper is arranged as follows. In Section 2 we recall some preliminary concepts and a known lemma [34]. In Section 3 we prove an existence result for the problem (1) by applying Sadovskii's fixed point theorem. Section 4 presents the existence results for the problem (2). Applications and examples are discussed in Section 5.

2. Preliminaries

Let us collect some important definitions on fractional calculus.

Definition 1. *[39] The left and right Riemann–Liouville fractional integrals of order $\delta > 0$ for $g \in L_1[a,b]$, existing almost everywhere on $[a,b]$, are respectively defined by*

$$I_{a+}^\delta g(t) = \int_a^t \frac{(t-s)^{\delta-1}}{\Gamma(\delta)} g(s) ds \quad \text{and} \quad I_{b-}^\delta g(t) = \int_t^b \frac{(s-t)^{\delta-1}}{\Gamma(\delta)} g(s) ds.$$

In addition, according to the classical theorem of Vallee-Poussin and the Young convolution theorem, $I_{a+}^{\delta}g, I_{b-}^{\delta}g \in L_1[a,b], \delta > 0$.

Definition 2. *[39] For $g \in AC^n[a,b]$, the left Riemann–Liouville and the right Caputo fractional derivatives of order $\delta \in (n-1, n], n \in \mathbb{N}$, existing almost everywhere on $[a, b]$, are respectively defined by*

$$^{RL}D_{a+}^{\delta}g(t) = \frac{d^n}{dt^n}\int_a^t \frac{(t-s)^{n-\delta-1}}{\Gamma(n-\delta)}g(s)ds \text{ and } {}^{C}D_{b-}^{\delta}g(t) = (-1)^n \int_t^b \frac{(s-t)^{n-\delta-1}}{\Gamma(n-\delta)}g^{(n)}(s)ds.$$

The following known lemma [34] plays a key role in proving the main results.

Lemma 1. *Let $H, F \in C[0,1] \cap L(0,1)$ and $y \in C([0,1], \mathbb{R})$. Then the linear problem*

$$\begin{cases} {}^{C}D_{1-}^{\alpha}\,{}^{RL}D_{0+}^{\beta}y(t) + \lambda I_{1-}^p I_{0+}^q H(t) = F(t), & t \in J := [0,1], \\ y(0) = y(\xi) = 0, \quad y(1) = \delta y(\mu), \end{cases} \tag{3}$$

is equivalent to the fractional integral equation:

$$\begin{aligned}
y(t) &= \int_0^t \frac{(t-s)^{\beta-1}}{\Gamma(\beta)}\left[I_{1-}^{\alpha}F(s) - \lambda I_{1-}^{\alpha+p}I_{0+}^q H(s)\right]ds \\
&+ a_1(t)\left\{\delta \int_0^{\mu} \frac{(\mu-s)^{\beta-1}}{\Gamma(\beta)}\left[I_{1-}^{\alpha}F(s) - \lambda I_{1-}^{\alpha+p}I_{0+}^q H(s)\right]ds \right. \\
&\left. - \int_0^1 \frac{(1-s)^{\beta-1}}{\Gamma(\beta)}\left[I_{1-}^{\alpha}F(s) - \lambda I_{1-}^{\alpha+p}I_{0+}^q H(s)\right]ds\right\} \\
&+ a_2(t)\int_0^{\xi} \frac{(\xi-s)^{\beta-1}}{\Gamma(\beta)}\left[I_{1-}^{\alpha}F(s) - \lambda I_{1-}^{\alpha+p}I_{0+}^q H(s)\right]ds,
\end{aligned} \tag{4}$$

where

$$a_1(t) = \frac{1}{\Lambda}\left[\xi^{\beta+1}t^{\beta} - \xi^{\beta}t^{\beta+1}\right], \quad a_2(t) = \frac{1}{\Lambda}\left[t^{\beta}(1 - \delta\mu^{\beta+1}) - t^{\beta+1}(1 - \delta\mu^{\beta})\right], \tag{5}$$

and it is assumed that

$$\Lambda = \xi^{\beta+1}(1 - \delta\mu^{\beta}) - \xi^{\beta}(1 - \delta\mu^{\beta+1}) \neq 0. \tag{6}$$

3. Existence Result for the Single-Valued Problem (1) via Sadovskii's Fixed Point Theorem

Our existence result for the problem (1) is based on Sadovskii's fixed point theorem. Before proceeding further, let us recall some related auxiliary material. In the sequel, we use the norm $\|\cdot\| = \sup_{t \in [0,1]} |\cdot|$.

Definition 3. *Let M be a bounded set in metric space (X, d). The Kuratowski measure of noncompactness, $\alpha(M)$, is defined as*
$\inf\{\epsilon : M \text{ covered by a finitely many sets such that the diameter of each set} \leq \epsilon\}.$

Definition 4. *[40] Let $\Phi : \mathcal{D}(\Phi) \subseteq X \to X$ be a bounded and continuous operator on Banach space X. Then Φ is called a condensing map if $\alpha(\Phi(B)) < \alpha(B)$ for all bounded sets $B \subset \mathcal{D}(\Phi)$, where α denotes the Kuratowski measure of noncompactness.*

Lemma 2. *[41, Example 11.7] The map $K + C$ is a k-set contraction with $0 \leq k < 1$, and thus also condensing, if*

(i) $K, C : \mathcal{D} \subseteq X \to X$ are operators on the Banach space X;
(ii) K is k-contractive, that is, for all $x, y \in \mathcal{D}$ and a fixed $k \in [0, 1)$,

$$\|Kx - Ky\| \leq k\|x - y\|;$$

(iii) C is compact.

Lemma 3. *[42] Let B be a convex, bounded and closed subset of a Banach space X and $\Phi : B \to B$ be a condensing map. Then Φ has a fixed point.*

In the sequel, we set

$$\Lambda_1 = \frac{\Delta_1}{\Gamma(\alpha+1)},\ \Lambda_2 = \frac{|\lambda|\Delta_1}{\Gamma(\alpha+p+1)\Gamma(q+1)},\ \Lambda_3 = \frac{\Delta_2}{\Gamma(\alpha)},\ \Lambda_4 = \frac{|\lambda|\Delta_2}{\Gamma(\alpha+p)\Gamma(q)}, \qquad (7)$$

where

$$\Delta_1 = \frac{1}{\Gamma(\beta+1)}\Big[1+\bar{a}_1(|\delta|\mu^\beta+1)+\bar{a}_2\zeta^\beta\Big],\ \Delta_2 = \frac{1}{\Gamma(\beta+1)}\Big[1+\bar{a}_1(|\delta|+1)+\bar{a}_2\Big],$$
$$\bar{a}_1 = \max_{t\in[0,1]} |a_1(t)|,\ \bar{a}_2 = \max_{t\in[0,1]} |a_2(t)|.$$

Theorem 1. *Assume that:*

(B_1) *There exist $L > 0$ such that $|f(t,x) - f(t,y)| \leq L|x-y|$, $\forall t \in [0,1]$, $x,y \in \mathbb{R}$;*
(B_2) *$|f(t,y)| \leq \sigma(t)$ and $|h(t,y)| \leq \rho(t)$, where $\sigma, \rho \in C([0,1], \mathbb{R}^+)$.*

Then the problem (1) has at least one solution on $[0,1]$ if

$$Q := L\Lambda_1 < 1.$$

where Λ_1 is given by (7).

Proof. Let $B_r = \{x \in C([0,1], \mathbb{R}) : \|x\| \leq r\}$ be a closed bounded and convex subset of $C([0,1], \mathbb{R})$, where r is a fixed constant. In view of Lemma 1, we introduce an operator $\mathcal{G} : C([0,1], \mathbb{R}) \to C([0,1], \mathbb{R})$ associated with the problem (1) as follows:

$$\begin{aligned}
\mathcal{G}y(t) &= \int_0^t \frac{(t-s)^{\beta-1}}{\Gamma(\beta)} \Big[I_{1-}^\alpha f(s,y(s)) - \lambda I_{1-}^{\alpha+p} I_{0+}^q h(s,y(s)) \Big] ds \\
&+ a_1(t) \Bigg[\delta \int_0^\mu \frac{(\mu-s)^{\beta-1}}{\Gamma(\beta)} \Big[I_{1-}^\alpha f(s,y(s)) - \lambda I_{1-}^{\alpha+p} I_{0+}^q h(s,y(s)) \Big] ds \\
&- \int_0^1 \frac{(1-s)^{\beta-1}}{\Gamma(\beta)} \Big[I_{1-}^\alpha f(s,y(s)) - \lambda I_{1-}^{\alpha+p} I_{0+}^q h(s,y(s)) \Big] ds \Bigg] \\
&+ a_2(t) \int_0^\zeta \frac{(\zeta-s)^{\beta-1}}{\Gamma(\beta)} \Big[I_{1-}^\alpha f(s,y(s)) - \lambda I_{1-}^{\alpha+p} I_{0+}^q h(s,y(s)) \Big] ds.
\end{aligned}$$

Let us split the operator $\mathcal{G} : C([0,1], \mathbb{R}) \to C([0,1], \mathbb{R})$ on B_r as $\mathcal{G} = \mathcal{G}_1 + \mathcal{G}_2$, where

$$\begin{aligned}
\mathcal{G}_1 y(t) &= \int_0^t \frac{(t-s)^{\beta-1}}{\Gamma(\beta)} I_{1-}^\alpha f(s,y(s)) ds + a_1(t) \Bigg[\delta \int_0^\mu \frac{(\mu-s)^{\beta-1}}{\Gamma(\beta)} I_{1-}^\alpha f(s,y(s)) ds \\
&- \int_0^1 \frac{(1-s)^{\beta-1}}{\Gamma(\beta)} I_{1-}^\alpha f(s,y(s)) ds \Bigg] + a_2(t) \int_0^\zeta \frac{(\zeta-s)^{\beta-1}}{\Gamma(\beta)} I_{1-}^\alpha f(s,y(s)) ds, \\
\mathcal{G}_2 y(t) &= -\lambda \int_0^t \frac{(t-s)^{\beta-1}}{\Gamma(\beta)} I_{1-}^{\alpha+p} I_{0+}^q h(s,y(s)) ds - \lambda a_1(t) \Bigg[\delta \int_0^\mu \frac{(\mu-s)^{\beta-1}}{\Gamma(\beta)} I_{1-}^{\alpha+p} I_{0+}^q h(s,y(s)) ds \\
&- \int_0^1 \frac{(1-s)^{\beta-1}}{\Gamma(\beta)} I_{1-}^{\alpha+p} I_{0+}^q h(s,y(s)) ds \Bigg] - \lambda a_2(t) \int_0^\zeta \frac{(\zeta-s)^{\beta-1}}{\Gamma(\beta)} I_{1-}^{\alpha+p} I_{0+}^q h(s,y(s)) ds.
\end{aligned}$$

We shall show that the operators $\mathcal{G}_1$ and $\mathcal{G}_2$ satisfy all the conditions of Lemma 3. The proof will be given in several steps.

Step 1. $\mathcal{G}B_r \subset B_r$.

Let us select $r \geq \|\sigma\|\Lambda_1 + \|\rho\|\Lambda_2$, where Λ_1, Λ_2 are given by (7). For any $y \in B_r$, we have

$$\|\mathcal{G}y\|$$
$$\leq \sup_{t \in [0,1]} \left\{ \int_0^t \frac{(t-s)^{\beta-1}}{\Gamma(\beta)} \left[I_{1-}^{\alpha} |f(s,y(s))| + |\lambda| I_{1-}^{\alpha+p} I_{0+}^{q} |h(s,y(s))| \right] ds \right.$$
$$+ |a_1(t)| \left\{ |\delta| \int_0^{\mu} \frac{(\mu-s)^{\beta-1}}{\Gamma(\beta)} \left[I_{1-}^{\alpha} |f(s,y(s))| + |\lambda| I_{1-}^{\alpha+p} I_{0+}^{q} |h(s,y(s))| \right] ds \right.$$
$$+ \int_0^1 \frac{(1-s)^{\beta-1}}{\Gamma(\beta)} \left[I_{1-}^{\alpha} |f(s,y(s))| + |\lambda| I_{1-}^{\alpha+p} I_{0+}^{q} |h(s,y(s))| \right] ds \right\}$$
$$+ |a_2(t)| \int_0^{\xi} \frac{(\xi-s)^{\beta-1}}{\Gamma(\beta)} \left[I_{1-}^{\alpha} |f(s,y(s))| + |\lambda| I_{1-}^{\alpha+p} I_{0+}^{q} |h(s,y(s))| \right] ds \right\}$$
$$\leq \|\sigma\| \sup_{t \in [0,1]} \left\{ \int_0^t \frac{(t-s)^{\beta-1}}{\Gamma(\beta)} I_{1-}^{\alpha}(1) ds + |a_1(t)| \left[|\delta| \int_0^{\mu} \frac{(\mu-s)^{\beta-1}}{\Gamma(\beta)} I_{1-}^{\alpha}(1) ds \right. \right.$$
$$+ \int_0^1 \frac{(1-s)^{\beta-1}}{\Gamma(\beta)} I_{1-}^{\alpha}(1) ds \right] + |a_2(t)| \int_0^{\xi} \frac{(\xi-s)^{\beta-1}}{\Gamma(\beta)} I_{1-}^{\alpha}(1) ds \right\}$$
$$+ \|\rho\||\lambda| \sup_{t \in [0,1]} \left\{ \int_0^t \frac{(t-s)^{\beta-1}}{\Gamma(\beta)} I_{1-}^{\alpha+p} I_{0+}^{q}(1) ds + |a_1(t)| \left[|\delta| \int_0^{\mu} \frac{(\mu-s)^{\beta-1}}{\Gamma(\beta)} I_{1-}^{\alpha+p} I_{0+}^{q}(1) ds \right. \right.$$
$$+ \int_0^1 \frac{(1-s)^{\beta-1}}{\Gamma(\beta)} I_{1-}^{\alpha+p} I_{0+}^{q}(1) ds \right] + |a_2(t)| \int_0^{\xi} \frac{(\xi-s)^{\beta-1}}{\Gamma(\beta)} I_{1-}^{\alpha+p} I_{0+}^{q}(1) ds \right\}$$
$$\leq \left\{ \frac{\|\sigma\|}{\Gamma(\alpha+1)} + \frac{\|\rho\||\lambda|}{\Gamma(\alpha+p+1)\Gamma(q+1)} \right\} \Delta_1$$
$$= \|\sigma\|\Lambda_1 + \|\rho\|\Lambda_2 < r,$$

which implies that $\mathcal{G}B_r \subset B_r$.

Step 2. $\mathcal{G}_2$ is compact.

Observe that the operator $\mathcal{G}_2$ is uniformly bounded in view of Step 1. Let $t_1, t_2 \in J$ with $t_1 < t_2$ and $y \in B_r$. Then we have

$$|\mathcal{G}_2 y(t_2) - \mathcal{G}_2 y(t_1)| \leq |\lambda| \left| \int_0^{t_1} \frac{(t_2-s)^{\beta-1} - (t_1-s)^{\beta-1}}{\Gamma(\beta)} I_{1-}^{\alpha+p} I_{0+}^{q} |h(s,y(s))| ds \right|$$
$$+ |\lambda| \left| \int_{t_1}^{t_2} \frac{(t_2-s)^{\beta-1}}{\Gamma(\beta)} I_{1-}^{\alpha+p} I_{0+}^{q} |h(s,y(s))| ds \right|$$
$$+ |\lambda||a_1(t_2) - a_1(t_1)| \left\{ |\delta| \int_0^{\mu} \frac{(\mu-s)^{\beta-1}}{\Gamma(\beta)} I_{1-}^{\alpha+p} I_{0+}^{q} |h(s,y(s))| ds \right.$$
$$+ \int_0^1 \frac{(1-s)^{\beta-1}}{\Gamma(\beta)} I_{1-}^{\alpha+p} I_{0+}^{q} |h(s,y(s))| ds \right\}$$
$$+ |\lambda||a_2(t_2) - a_2(t_1)| \left\{ \int_0^{\xi} \frac{(\xi-s)^{\beta-1}}{\Gamma(\beta)} I_{1-}^{\alpha+p} I_{0+}^{q} |h(s,y(s))| ds \right\}$$
$$\leq \frac{|\lambda|\|\rho\|}{\Gamma(\beta+1)\Gamma(\alpha+p+1)\Gamma(q+1)} \left\{ 2(t_2-t_1)^{\beta} + |t_2^{\beta} - t_1^{\beta}| \right.$$

$$+\frac{(|\delta|\mu^\beta+1)}{|\Lambda|}\left(\xi^{\beta+1}|t_2^\beta-t_1^\beta|+\xi^\beta|t_2^{\beta+1}-t_1^{\beta+1}|\right)$$

$$+\frac{\xi^\beta}{|\Lambda|}\left(|1-\delta\mu^{\beta+1}||t_2^\beta-t_1^\beta|+|1-\delta\mu^\beta||t_2^{\beta+1}-t_1^{\beta+1}|\right)\bigg\},$$

which tends to zero independent of y as $t_2 \to t_1$. This shows that $\mathcal{G}_2$ is equicontinuous. It is clear from the foregoing arguments that the operator $\mathcal{G}_2$ is relatively compact on B_r. Hence, by the Arzelá-Ascoli theorem, $\mathcal{G}_2$ is compact on B_r.

Step 3. $\mathcal{G}_1$ is Q-contractive.

Using ($\mathbf{B_1}$) and ($\mathbf{B_2}$), it is easy to show that

$$\begin{aligned}\|\mathcal{G}_1 y - \mathcal{G}_1 x\| &\leq \sup_{t\in[0,1]}\bigg\{\int_0^t \frac{(t-s)^{\beta-1}}{\Gamma(\beta)}I_{1-}^\alpha|f(s,y(s))-f(s,x(s))|ds \\ &\quad +|a_1(t)|\bigg[\delta\int_0^\mu \frac{(\mu-s)^{\beta-1}}{\Gamma(\beta)}I_{1-}^\alpha|f(s,y(s))-f(s,x(s))|ds \\ &\quad +\int_0^1 \frac{(1-s)^{\beta-1}}{\Gamma(\beta)}I_{1-}^\alpha|f(s,y(s))-f(s,x(s))|ds\bigg] \\ &\quad +|a_2(t)|\int_0^\xi \frac{(\xi-s)^{\beta-1}}{\Gamma(\beta)}I_{1-}^\alpha|f(s,y(s))-f(s,x(s))|ds\bigg\} \\ &\leq \frac{L}{\Gamma(\beta+1)\Gamma(\alpha+1)}\left[1+\bar{a}_1(|\delta|\mu^\beta+1)+\bar{a}_2\xi^\beta\right]\|y-x\| \\ &= L\Lambda_1\|y-x\|,\end{aligned}$$

which is Q-contractive, since $Q := L\Lambda_1 < 1$.

Step 4. $\mathcal{G}$ is condensing. Since $\mathcal{G}_1$ is continuous, Q-contraction and $\mathcal{G}_2$ is compact, therefore, by Lemma 2, $\mathcal{G}: B_r \to B_r$ with $\mathcal{G} = \mathcal{G}_1 + \mathcal{G}_2$ is a condensing map on B_r.

From the above four steps, we conclude by Lemma 3 that the map $\mathcal{G}$ has a fixed point which, in turn, implies that the problem (1) has a solution on $[0,1]$. □

4. Existence Results for the Multi-Vaued Problem (2)

For a normed space $(X, \|\cdot\|)$, we have $\mathcal{P}_{cl}(X) = \{Y \in \mathcal{P}(X) : Y \text{ is closed}\}$, $\mathcal{P}_b(X) = \{Y \in \mathcal{P}(X) : Y \text{ is bounded}\}$, $\mathcal{P}_{cp}(X) = \{Y \in \mathcal{P}(X) : Y \text{ is compact}\}$, $\mathcal{P}_{cp,c}(X) = \{Y \in \mathcal{P}(X) : Y \text{ is compact and convex}\}$, $\mathcal{P}_{b,cl,c}(\mathbb{R}) = \{Y \in \mathcal{P}(X) : Y \text{ is bounded, closed and convex}\}$. We also define the *sets of selections* of the multi-valued maps F and H as

$$S_{F,y} := \{f \in L^1([0,1], \mathbb{R}) : f(t) \in F(t,y)\},$$

$$\widehat{S}_{H,y} := \{h \in L^1([0,1], \mathbb{R}) : h(t) \in H(t,y)\}.$$

By Lemma 1, we define a solution of the boundary value problem (2) as follows (see also [43,44]).

Definition 5. *A function $y \in C([0,1], \mathbb{R})$ is a solution of the boundary value problem (2) if $y(0) = y(\xi) = 0, y(1) = \delta y(\mu)$, and there exist functions $f \in S_{F,y}, h \in \widehat{S}_{H,y}$ a.e. on $[0,1]$ and*

$$\begin{aligned}y(t) &= \int_0^t \frac{(t-s)^{\beta-1}}{\Gamma(\beta)}\left[I_{1-}^\alpha f(s) - \lambda I_{1-}^{\alpha+p}I_{0+}^q h(s)\right]ds \\ &\quad +a_1(t)\bigg\{\delta\int_0^\mu \frac{(\mu-s)^{\beta-1}}{\Gamma(\beta)}\left[I_{1-}^\alpha f(s) - \lambda I_{1-}^{\alpha+p}I_{0+}^q h(s)\right]ds\end{aligned}$$

$$-\int_0^1 \frac{(1-s)^{\beta-1}}{\Gamma(\beta)}\left[I_{1-}^\alpha f(s) - \lambda I_{1-}^{\alpha+p} I_{0+}^q h(s)\right]ds\bigg\}$$
$$+a_2(t)\int_0^\xi \frac{(\xi-s)^{\beta-1}}{\Gamma(\beta)}\left[I_{1-}^\alpha f(s) - \lambda I_{1-}^{\alpha+p} I_{0+}^q h(s)\right]ds.$$

Now we provide the lemmas which will be used in the main existence results in this section.

Lemma 4. (Bohnenblust-Karlin) ([37]) *Let D be a nonempty, bounded, closed, and convex subset of X. Let $\Phi : D \to \mathcal{P}(\mathbb{R})$ be upper semi-continuous with closed, convex values such that $\Phi(D) \subset D$ and $\overline{\Phi(D)}$ is compact. Then Φ has a fixed point.*

Lemma 5. ([45]) *Let X be a separable Banach space. Let $F : J \times X \to \mathcal{P}_{cp,c}(X)$ be measurable with respect to t for each $y \in X$ and upper semi-continuous with respect to y for almost all $t \in J$ and $S_{F,y} \neq \emptyset$, for any $y \in C(J, X)$, and let Θ be a linear continuous mapping from $L^1(J, X)$ to $C(J, X)$. Then the operator*

$$\Theta \circ S_F : C(J, X) \to \mathcal{P}_{cp,c}(C(J, X)), \quad y \mapsto (\Theta \circ S_F)(y) = \Theta(S_{F,y})$$

is a closed graph operator in $C(J, X) \times C(J, X)$.

In the first result, we study the existence of the solution for the multi-valued problem (2) by applying Bohnenblust–Karlin fixed point theorem.

Theorem 2. *Suppose that:*

(M_1) *$F, H : [0,1] \times \mathbb{R} \to \mathcal{P}_{b,c,cp}(\mathbb{R})$; $(t, y) \to f(t, y)$ and $(t, y) \to h(t, y)$ be measurable with respect to t for each $y \in \mathbb{R}$, upper semi-continuous with respect to y for almost everywhere $t \in [0,1]$, and for each fixed $y \subset \mathbb{R}$, the sets $S_{F,y}$ and $\widehat{S}_{H,y}$ are nonempty for almost everywhere $t \in [0,1]$.*

(M_2) *For each $\rho > 0$, there exist functions $\phi_\rho, \psi_\rho \in L^1([0,1], \mathbb{R}_+)$ such that*

$$\|F(t,y)\| = \sup\{|f| : f(t) \in F(t,y)\} \leq \phi_\rho(t),$$

$$\|H(t,y)\| = \sup\{|h| : h(t) \in H(t,y)\} \leq \psi_\rho(t),$$

for each $(t, y) \in [0,1] \times \mathbb{R}$ with $\|y\| \leq \rho$, and

$$\liminf_{\rho \to +\infty} \frac{1}{\rho}\int_0^1 \phi_\rho(t)dt = \zeta_1 < \infty, \quad \liminf_{\rho \to +\infty} \frac{1}{\rho}\int_0^1 \psi_\rho(t)dt = \zeta_2 < \infty. \tag{8}$$

Then the boundary value problem (2) has at least one solution on $[0,1]$ provided that

$$\zeta_1 \Lambda_3 + \zeta_2 \Lambda_4 < 1, \tag{9}$$

where ζ_1, ζ_2 are defined by (8), and Λ_3, Λ_4 are given by (7).

Proof. To transform the problem (2) into a fixed point problem, we define a multi-valued map $\mathcal{U} : C([0,1], \mathbb{R}) \to \mathcal{P}(C([0,1], \mathbb{R}))$ as

$$\mathcal{U}(y) = \bigg\{g \in C([0,1], \mathbb{R}) : \ g(t) = \int_0^t \frac{(t-s)^{\beta-1}}{\Gamma(\beta)}\left[I_{1-}^\alpha f(s) - \lambda I_{1-}^{\alpha+p} I_{0+}^q h(s)\right]ds$$
$$+a_1(t)\bigg[\delta \int_0^\mu \frac{(\mu-s)^{\beta-1}}{\Gamma(\beta)}\left[I_{1-}^\alpha f(s) - \lambda I_{1-}^{\alpha+p} I_{0+}^q h(s)\right]ds$$
$$-\int_0^1 \frac{(1-s)^{\beta-1}}{\Gamma(\beta)}\left[I_{1-}^\alpha f(s) - \lambda I_{1-}^{\alpha+p} I_{0+}^q h(s)\right]ds\bigg]$$

$$+a_2(t)\int_0^\xi \frac{(\xi-s)^{\beta-1}}{\Gamma(\beta)}\Big[I_{1-}^\alpha f(s) - \lambda I_{1-}^{\alpha+p} I_{0+}^q h(s)\Big]ds\bigg\},$$

for $f \in S_{F,y}, h \in \widehat{S}_{H,y}$.

Now we prove that the operator $\mathcal{U}$ satisfies the hypothesis of Lemma 4 and thus it will have a fixed point which corresponds to a solution of problem (2). Here we show that $\mathcal{U}$ is a compact and upper semi-continuous multi-valued map with convex closed values. This will be established in a sequence of steps.

Step 1: $\mathcal{U}(y)$ *is convex for each* $y \in C([0,1],\mathbb{R})$. For that, let $g_1, g_2 \in \mathcal{U}(y)$. Then there exist $f_1, f_2 \in S_{F,y}$, $h_1, h_2 \in \widehat{S}_{H,y}$ such that, for each $t \in [0,1]$, we get

$$\begin{aligned}
g_i(t) &= \int_0^t \frac{(t-s)^{\beta-1}}{\Gamma(\beta)}\Big[I_{1-}^\alpha f_i(s) - \lambda I_{1-}^{\alpha+p} I_{0+}^q h_i(s)\Big]ds \\
&\quad + a_1(t)\bigg\{\delta \int_0^\mu \frac{(\mu-s)^{\beta-1}}{\Gamma(\beta)}\Big[I_{1-}^\alpha f_i(s) - \lambda I_{1-}^{\alpha+p} I_{0+}^q h_i(s)\Big]ds \\
&\quad - \int_0^1 \frac{(1-s)^{\beta-1}}{\Gamma(\beta)}\Big[I_{1-}^\alpha f_i(s) - \lambda I_{1-}^{\alpha+p} I_{0+}^q h_i(s)\Big]ds\bigg\} \\
&\quad + a_2(t)\int_0^\xi \frac{(\xi-s)^{\beta-1}}{\Gamma(\beta)}\Big[I_{1-}^\alpha f_i(s) - \lambda I_{1-}^{\alpha+p} I_{0+}^q h_i(s)\Big]ds, \; i=1,2.
\end{aligned}$$

For each $t \in [0,1]$ and $0 \le \nu \le 1$, we can find that

$$\begin{aligned}
&\Big[\nu g_1 + (1-\nu)g_2\Big](t) \\
&= \int_0^t \frac{(t-s)^{\beta-1}}{\Gamma(\beta)}\Big[I_{1-}^\alpha[\nu f_1(s) + (1-\nu)f_2(s)] - \lambda I_{1-}^{\alpha+p} I_{0+}^q[\nu h_1(s) + (1-\nu)h_2(s)]\Big]ds \\
&\quad + a_1(t)\bigg\{\delta \int_0^\mu \frac{(\mu-s)^{\beta-1}}{\Gamma(\beta)}\Big[I_{1-}^\alpha[\nu f_1(s) + (1-\nu)f_2(s)] - \lambda I_{1-}^{\alpha+p} I_{0+}^q[\nu h_1(s) + (1-\nu)h_2(s)]\Big]ds \\
&\quad - \int_0^1 \frac{(1-s)^{\beta-1}}{\Gamma(\beta)}\Big[I_{1-}^\alpha[\nu f_1(s) + (1-\nu)f_2(s)] - \lambda I_{1-}^{\alpha+p} I_{0+}^q[\nu h_1(s) + (1-\nu)h_2(s)]\Big]ds\bigg\} \\
&\quad + a_2(t)\int_0^\xi \frac{(\xi-s)^{\beta-1}}{\Gamma(\beta)}\Big[I_{1-}^\alpha[\nu f_1(s) + (1-\nu)f_2(s)] - \lambda I_{1-}^{\alpha+p} I_{0+}^q[\nu h_1(s) + (1-\nu)h_2(s)]\Big]ds.
\end{aligned}$$

Since $S_{F,y}, \widehat{S}_{H,y}$ are convex valued (F, H have convex values), it follows that $\nu g_1 + (1-\nu)g_2 \in \mathcal{U}(y)$.

Step 2: $\mathcal{U}(y)$ *maps bounded sets (balls) into bounded sets in* $C([0,1],\mathbb{R})$. Let us define $\mathcal{B}_\rho = \{y \in C([0,1],\mathbb{R}) : \|y\| \le \rho\}$ as a bounded closed convex set in $C([0,1],\mathbb{R})$ for each positive constant ρ. We shall prove that there exists a positive number $\bar{\rho}$ such that $\mathcal{U}(\mathcal{B}_{\bar{\rho}}) \subseteq \mathcal{B}_{\bar{\rho}}$. If it is not true, then we can find a function $y_\rho \in \mathcal{B}_\rho, g_\rho \in \mathcal{U}(y_\rho)$ with $\|\mathcal{U}(y_\rho)\| > \rho$, such that

$$\begin{aligned}
g_\rho(t) &= \int_0^t \frac{(t-s)^{\beta-1}}{\Gamma(\beta)}\Big[I_{1-}^\alpha f_\rho(s) - \lambda I_{1-}^{\alpha+p} I_{0+}^q h_\rho(s)\Big]ds \\
&\quad + a_1(t)\bigg\{\delta \int_0^\mu \frac{(\mu-s)^{\beta-1}}{\Gamma(\beta)}\Big[I_{1-}^\alpha f_\rho(s) - \lambda I_{1-}^{\alpha+p} I_{0+}^q h_\rho(s)\Big]ds \\
&\quad - \int_0^1 \frac{(1-s)^{\beta-1}}{\Gamma(\beta)}\Big[I_{1-}^\alpha f_\rho(s) - \lambda I_{1-}^{\alpha+p} I_{0+}^q h_\rho(s)\Big]ds\bigg\} \\
&\quad + a_2(t)\int_0^\xi \frac{(\xi-s)^{\beta-1}}{\Gamma(\beta)}\Big[I_{1-}^\alpha f_\rho(s) - \lambda I_{1-}^{\alpha+p} I_{0+}^q h_\rho(s)\Big]ds,
\end{aligned}$$

for some $f_\rho \in S_{F,y_\rho}, h_\rho \in \widehat{S}_{H,y_\rho}$.

According to condition ($\mathbf{M_2}$), we obtain

$$\begin{aligned}
\rho &< \|\mathcal{U}(y_\rho)\| \\
&\leq \int_0^t \frac{|t-s|^{\beta-1}}{\Gamma(\beta)}\left[I_{1-}^\alpha \phi_\rho(s) + |\lambda| I_{1-}^{\alpha+p} I_{0+}^q \psi_\rho(s)\right]ds \\
&\quad + |a_1(t)|\Bigg\{|\delta|\int_0^\mu \frac{|\mu-s|^{\beta-1}}{\Gamma(\beta)}\left[I_{1-}^\alpha \phi_\rho(s) + |\lambda| I_{1-}^{\alpha+p} I_{0+}^q \psi_\rho(s)\right]ds \\
&\quad + \int_0^1 \frac{|1-s|^{\beta-1}}{\Gamma(\beta)}\left[I_{1-}^\alpha \phi_\rho(s) + |\lambda| I_{1-}^{\alpha+p} I_{0+}^q \psi_\rho(s)\right]ds\Bigg\} \\
&\quad + |a_2(t)|\int_0^\xi \frac{|\xi-s|^{\beta-1}}{\Gamma(\beta)}\left[I_{1-}^\alpha \phi_\rho(s) + |\lambda| I_{1-}^{\alpha+p} I_{0+}^q \psi_\rho(s)\right]ds \\
&\leq \frac{1+\bar{a}_1(|\delta|+1)+\bar{a}_2}{\Gamma(\beta+1)\Gamma(\alpha)}\int_0^1 \phi_\rho(t)dt + \frac{|\lambda|(1+\bar{a}_1(|\delta|+1)+\bar{a}_2)}{\Gamma(\beta+1)\Gamma(\alpha+p)\Gamma(q)}\int_0^1 \psi_\rho(t)dt \\
&\leq \Lambda_3 \int_0^1 \phi_\rho(t)dt + \Lambda_4 \int_0^1 \psi_\rho(t)dt, \quad\quad\quad\quad\quad\quad\quad\quad\quad\quad\quad\quad (10)
\end{aligned}$$

where Λ_3, Λ_4 are given by (7). In (10), we have used the following estimates ($\alpha \in (1,2], \beta \in (0,1], p > 0, q > 1$):

$$\begin{aligned}
\int_0^t \frac{(t-s)^{\beta-1}}{\Gamma(\beta)} I_{1-}^\alpha \phi_\rho(s) ds &= \int_0^t \frac{(t-s)^{\beta-1}}{\Gamma(\beta)} \int_s^1 \frac{(u-s)^{\alpha-1}}{\Gamma(\alpha)} \phi_\rho(u) du\, ds \\
&\leq \int_0^t \frac{(t-s)^{\beta-1}}{\Gamma(\beta)} \frac{(1-s)^{\alpha-1}}{\Gamma(\alpha)} ds \int_0^1 \phi_\rho(u) du \\
&\leq \int_0^t \frac{(t-s)^{\beta-1}}{\Gamma(\beta)} \frac{1}{\Gamma(\alpha)} ds \int_0^1 \phi_\rho(u) du \\
&\leq \frac{1}{\Gamma(\beta+1)\Gamma(\alpha)} \int_0^1 \phi_\rho(u) du
\end{aligned}$$

$$\begin{aligned}
\int_0^t \frac{(t-s)^{\beta-1}}{\Gamma(\beta)} I_{1-}^{\alpha+p} I_{0+}^q \psi_\rho(s) ds &= \int_0^t \frac{(t-s)^{\beta-1}}{\Gamma(\beta)} \int_s^1 \frac{(u-s)^{\alpha+p-1}}{\Gamma(\alpha+p)} \int_0^u \frac{(u-r)^{q-1}}{\Gamma(q)} \psi_\rho(r) dr\, du\, ds \\
&\leq \int_0^t \frac{(t-s)^{\beta-1}}{\Gamma(\beta)} \int_s^1 \frac{(u-s)^{\alpha+p-1}}{\Gamma(\alpha+p)} \frac{u^{q-1}}{\Gamma(q)} du\, ds \int_0^1 \psi_\rho(r) dr \\
&\leq \int_0^t \frac{(t-s)^{\beta-1}}{\Gamma(\beta)} \int_s^1 \frac{1}{\Gamma(\alpha+p)} \frac{1}{\Gamma(q)} du\, ds \int_0^1 \psi_\rho(r) dr \\
&\leq \int_0^t \frac{(t-s)^{\beta-1}}{\Gamma(\beta)} \frac{1-s}{\Gamma(\alpha+p)\Gamma(q)} ds \int_0^1 \psi_\rho(r) dr \\
&\leq \frac{1}{\Gamma(\beta+1)\Gamma(\alpha+p)\Gamma(q)} \int_0^1 \psi_\rho(t) dt.
\end{aligned}$$

Dividing both sides of (10) by ρ and then taking the lower limit as $\rho \to \infty$, we find by (8) that $\zeta_1 \Lambda_3 + \zeta_2 \Lambda_4 > 1$, which is a contradiction to the assumption (9). Hence there exists a positive number $\bar{\rho}$ such that $\mathcal{U}(\mathcal{B}_{\bar{\rho}}) \subseteq \mathcal{B}_{\bar{\rho}}$.

Step 3: $\mathcal{U}(y)$ maps bounded sets into equicontinuous sets of $C([0,1], \mathbb{R})$. For that, let $0 \leq t_1 \leq t_2 \leq 1$, $y \in \mathcal{B}_{\bar{\rho}}$, and $g \in \mathcal{U}(y)$. Then there exist $f \in S_{F,y}, h \in \widehat{S}_{H,y}$ such that, for each $t \in [0,1]$, we find that

$$\begin{aligned}
g(t) &= \int_0^t \frac{(t-s)^{\beta-1}}{\Gamma(\beta)}\left[I_{1-}^\alpha f(s) - \lambda I_{1-}^{\alpha+p} I_{0+}^q h(s)\right] ds \\
&\quad + a_1(t)\Bigg\{\delta \int_0^\mu \frac{(\mu-s)^{\beta-1}}{\Gamma(\beta)}\left[I_{1-}^\alpha f(s) - \lambda I_{1-}^{\alpha+p} I_{0+}^q h(s)\right] ds
\end{aligned}$$

$$-\int_0^1 \frac{(1-s)^{\beta-1}}{\Gamma(\beta)}\left[I_{1-}^\alpha f(s) - \lambda I_{1-}^{\alpha+p} I_{0+}^q h(s)\right]ds\bigg\}$$
$$+a_2(t)\int_0^\xi \frac{(\xi-s)^{\beta-1}}{\Gamma(\beta)}\left[I_{1-}^\alpha f(s) - \lambda I_{1-}^{\alpha+p} I_{0+}^q h(s)\right]ds,$$

and that

$$|g(t_2) - g(t_1)|$$
$$= \int_0^{t_1} \frac{|(t_2-s)^{\beta-1} - (t_1-s)^{\beta-1}|}{\Gamma(\beta)}\left[I_{1-}^\alpha|f(s)| + |\lambda|I_{1-}^{\alpha+p} I_{0+}^q|h(s)|\right]ds$$
$$+\int_{t_1}^{t_2} \frac{|t_2-s|^{\beta-1}}{\Gamma(\beta)}\left[I_{1-}^\alpha|f(s)| + |\lambda|I_{1-}^{\alpha+p} I_{0+}^q|h(s)|\right]ds$$
$$+|a_1(t_2) - a_1(t_1)|\bigg\{|\delta|\int_0^\mu \frac{|\mu-s|^{\beta-1}}{\Gamma(\beta)}\left[I_{1-}^\alpha|f(s)| + |\lambda|I_{1-}^{\alpha+p} I_{0+}^q|h(s)|\right]ds$$
$$+\int_0^1 \frac{|1-s|^{\beta-1}}{\Gamma(\beta)}\left[I_{1-}^\alpha|f(s)| + |\lambda|I_{1-}^{\alpha+p} I_{0+}^q|h(s)|\right]ds\bigg\}$$
$$+|a_2(t_2) - a_2(t_1)|\int_0^\xi \frac{|\xi-s|^{\beta-1}}{\Gamma(\beta)}\left[I_{1-}^\alpha|f(s)| + |\lambda|I_{1-}^{\alpha+p} I_{0+}^q|h(s)|\right]ds$$
$$\leq \left[|t_2^\beta - t_1^\beta| + 2(t_2-t_1)^\beta + \frac{(|\delta|+1)}{|\Lambda|}\left(\xi^{\beta+1}|t_2^\beta - t_1^\beta| + \xi^\beta|t_1^{\beta+1} - t_2^{\beta+1}|\right)\right.$$
$$\left.+\frac{1}{|\Lambda|}\left(|1 - \delta\mu^{\beta+1}||t_2^\beta - t_1^\beta| + |1 - \delta\mu^\beta||t_1^{\beta+1} - t_2^{\beta+1}|\right)\right]$$
$$\times\left\{\frac{1}{\Gamma(\beta+1)\Gamma(\alpha)}\int_0^1 \phi_\rho(s)ds + \frac{|\lambda|}{\Gamma(\beta+1)\Gamma(\alpha+p)\Gamma(q)}\int_0^1 \psi_\rho(s)ds\right\}.$$

Clearly, the right-hand side of the above inequality tends to zero as $t_2 \to t_1$ independently of $y \in \mathcal{B}_\rho$. Hence $\mathcal{U}$ is equi-continuous. As $\mathcal{U}$ satisfies the above three steps, it follows by the Ascoli-Arzelá theorem that $\mathcal{U}$ is a compact multi-valued map.

Step 4: $\mathcal{U}$ *has a closed graph.* Let $y_n \to y_*$, $g_n \in \mathcal{U}(y_n)$ and $g_n \to g_*$. Then we need to show that $g_* \in \mathcal{U}(y_*)$. Associated with $g_n \in \mathcal{U}(y_n)$, we can find $f_n \in S_{F,y_n}, h_n \in \widehat{S}_{H,y_n}$ such that, for each $t \in [0,1]$, we have

$$g_n(t) = \int_0^t \frac{(t-s)^{\beta-1}}{\Gamma(\beta)}\left[I_{1-}^\alpha f_n(s) - \lambda I_{1-}^{\alpha+p} I_{0+}^q h_n(s)\right]ds$$
$$+a_1(t)\bigg\{\delta\int_0^\mu \frac{(\mu-s)^{\beta-1}}{\Gamma(\beta)}\left[I_{1-}^\alpha f_n(s) - \lambda I_{1-}^{\alpha+p} I_{0+}^q h_n(s)\right]ds$$
$$-\int_0^1 \frac{(1-s)^{\beta-1}}{\Gamma(\beta)}\left[I_{1-}^\alpha f_n(s) - \lambda I_{1-}^{\alpha+p} I_{0+}^q h_n(s)\right]ds\bigg\}$$
$$+a_2(t)\int_0^\xi \frac{(\xi-s)^{\beta-1}}{\Gamma(\beta)}\left[I_{1-}^\alpha f_n(s) - \lambda I_{1-}^{\alpha+p} I_{0+}^q h_n(s)\right]ds.$$

Thus it suffices to show that there exist $f_* \in S_{F,y_*}, h_* \in \widehat{S}_{H,y_*}$ such that for each $t \in [0,1]$,

$$g_*(t) = \int_0^t \frac{(t-s)^{\beta-1}}{\Gamma(\beta)}\left[I_{1-}^\alpha f_*(s) - \lambda I_{1-}^{\alpha+p} I_{0+}^q h_*(s)\right]ds$$
$$+a_1(t)\bigg\{\delta\int_0^\mu \frac{(\mu-s)^{\beta-1}}{\Gamma(\beta)}\left[I_{1-}^\alpha f_*(s) - \lambda I_{1-}^{\alpha+p} I_{0+}^q h_*(s)\right]ds$$
$$-\int_0^1 \frac{(1-s)^{\beta-1}}{\Gamma(\beta)}\left[I_{1-}^\alpha f_*(s) - \lambda I_{1-}^{\alpha+p} I_{0+}^q h_*(s)\right]ds\bigg\}$$

$$+a_2(t)\int_0^\xi \frac{(\xi-s)^{\beta-1}}{\Gamma(\beta)}\left[I_{1-}^\alpha f_*(s) - \lambda I_{1-}^{\alpha+p} I_{0+}^q h_*(s)\right]ds.$$

Let us consider the continuous linear operator $\Theta : L^1([0,1],\mathbb{R}) \to C([0,1])$ so that

$$\begin{aligned}(\Theta(f,h))(t) &= \int_0^t \frac{(t-s)^{\beta-1}}{\Gamma(\beta)}\left[I_{1-}^\alpha f(s) - \lambda I_{1-}^{\alpha+p} I_{0+}^q h(s)\right]ds \\ &+ a_1(t)\left\{\delta \int_0^\mu \frac{(\mu-s)^{\beta-1}}{\Gamma(\beta)}\left[I_{1-}^\alpha f(s) - \lambda I_{1-}^{\alpha+p} I_{0+}^q h(s)\right]ds \right. \\ &\left. - \int_0^1 \frac{(1-s)^{\beta-1}}{\Gamma(\beta)}\left[I_{1-}^\alpha f(s) - \lambda I_{1-}^{\alpha+p} I_{0+}^q h(s)\right]ds\right\} \\ &+ a_2(t)\int_0^\xi \frac{(\xi-s)^{\beta-1}}{\Gamma(\beta)}\left[I_{1-}^\alpha f(s) - \lambda I_{1-}^{\alpha+p} I_{0+}^q h(s)\right]ds.\end{aligned}$$

Observe that

$$\begin{aligned}&\|g_n(t) - g_*(t)\| \\ &= \left\|\int_0^t \frac{(t-s)^{\beta-1}}{\Gamma(\beta)}\left[I_{1-}^\alpha (f_n(s)-f_*(s)) - \lambda I_{1-}^{\alpha+p} I_{0+}^q (h_n(s)-h_*(s))\right]ds \right. \\ &+ a_1(t)\left\{\delta \int_0^\mu \frac{(\mu-s)^{\beta-1}}{\Gamma(\beta)}\left[I_{1-}^\alpha (f_n(s)-f_*(s)) - \lambda I_{1-}^{\alpha+p} I_{0+}^q (h_n(s)-h_*(s))\right]ds \right. \\ &\left. - \int_0^1 \frac{(1-s)^{\beta-1}}{\Gamma(\beta)}\left[I_{1-}^\alpha (f_n(s)-f_*(s)) - \lambda I_{1-}^{\alpha+p} I_{0+}^q (h_n(s)-h_*(s))\right]ds\right\} \\ &\left. + a_2(t)\int_0^\xi \frac{(\xi-s)^{\beta-1}}{\Gamma(\beta)}\left[I_{1-}^\alpha (f_n(s)-f_*(s)) - \lambda I_{1-}^{\alpha+p} I_{0+}^q (h_n(s)-h_*(s))\right]ds\right\| \to 0 \text{ as } n \to \infty.\end{aligned}$$

Thus, it follows by Lemma 5 that $\Theta \circ S_B$ is a closed graph operator where $S_B = S_F \cup \widehat{S}_H$. Moreover, we have $g_n(t) \in \Theta(S_{B,y_n})$. Since $y_n \to y_*$, $g_n \to g_*$, therefore, Lemma 5 yields

$$\begin{aligned}g_*(t) &= \int_0^t \frac{(t-s)^{\beta-1}}{\Gamma(\beta)}\left[I_{1-}^\alpha f_*(s) - \lambda I_{1-}^{\alpha+p} I_{0+}^q h_*(s)\right]ds \\ &+ a_1(t)\left\{\delta \int_0^\mu \frac{(\mu-s)^{\beta-1}}{\Gamma(\beta)}\left[I_{1-}^\alpha f_*(s) - \lambda I_{1-}^{\alpha+p} I_{0+}^q h_*(s)\right]ds \right. \\ &\left. - \int_0^1 \frac{(1-s)^{\beta-1}}{\Gamma(\beta)}\left[I_{1-}^\alpha f_*(s) - \lambda I_{1-}^{\alpha+p} I_{0+}^q h_*(s)\right]ds\right\} \\ &+ a_2(t)\int_0^\xi \frac{(\xi-s)^{\beta-1}}{\Gamma(\beta)}\left[I_{1-}^\alpha f_*(s) - \lambda I_{1-}^{\alpha+p} I_{0+}^q h_*(s)\right]ds,\end{aligned}$$

for some $f_* \in S_{F,y_*}, h_* \in \widehat{S}_{H,y_*}$.

Hence, we conclude that $\mathcal{U}$ is a compact and upper semi-continuous multi-valued map with convex closed values. Thus, the hypothesis of Lemma 4 holds true, and therefore its conclusion implies that the operator $\mathcal{U}$ has a fixed point y, which corresponds to a solution of problem (2). This completes the proof. □

Next, we give an existence result based upon the following form of fixed point theorem due to Martelli [38], which is applicable to completely continuous operators.

Lemma 6. *Let X a Banach space, and $T : X \to \mathcal{P}_{b,cl,c}(X)$ be a completely continuous multi-valued map. If the set $\mathcal{E} = \{x \in X : \kappa x \in T(x), \kappa > 1\}$ is bounded, then T has a fixed point.*

Theorem 3. *Assume that the following hypotheses hold:*

(M$_3$) $F, H : [0,1] \times \mathbb{R} \to \mathcal{P}_{b,cl,c}(\mathbb{R})$ *are L^1-Carathéodory multi-valued maps; that is, (i) $t \longmapsto F(t,y), t \longmapsto H(t,y)$, are measurable for each $y \in \mathbb{R}$; (ii) $y \longmapsto F(t,y), y \longmapsto H(t,y)$ are upper semicontinuous for almost all $t \in [0,1]$; (iii) for each $r > 0$, there exist $\phi_r, \psi_r \in L^1([0,1], \mathbb{R}^+)$ such that $\|F(t,y)\| = \sup\{|v| : v \in F(t,y)\} \leq \phi_r(t)$, $\|H(t,y)\| = \sup\{|v| : v \in F(t,y)\} \leq \psi_r(t)$, for all $y \in \mathbb{R}$ with $\|y\| \leq r$ and for almost every $t \in [0,1]$.*

(M$_4$) *There exist functions $z, u \in L^1([0,1], \mathbb{R}^+)$ such that*

$$\|F(t,y)\| \leq z(t), \quad \|H(t,y)\| \leq u(t), \text{ for a.e. } t \in [0,1] \text{ and each } y \in \mathbb{R}.$$

Then the problem (2) has at least one solution on $[0,1]$.

Proof. Consider $\mathcal{U}$ defined in the proof of Theorem 2. As in Theorem 2, we can show that $\mathcal{U}$ is convex and completely continuous. It remains to show that the set

$$\mathcal{E} = \{y \in C([0,1], \mathbb{R}) : \kappa y \in \mathcal{U}(y), \kappa > 1\}$$

is bounded. Let $y \in \mathcal{E}$, then $\kappa y \in \mathcal{U}(y)$ for some $\kappa > 1$ and there exist functions $f \in S_{F,y}, h \in \widehat{S}_{H,y}$ such that

$$\begin{aligned}
y(t) &= \int_0^t \frac{(t-s)^{\beta-1}}{\Gamma(\beta)} \left[I_{1-}^\alpha f(s) - \lambda I_{1-}^{\alpha+p} I_{0+}^q h(s) \right] ds \\
&\quad + a_1(t) \left\{ \delta \int_0^\mu \frac{(\mu-s)^{\beta-1}}{\Gamma(\beta)} \left[I_{1-}^\alpha f(s) - \lambda I_{1-}^{\alpha+p} I_{0+}^q h(s) \right] ds \right. \\
&\quad \left. - \int_0^1 \frac{(1-s)^{\beta-1}}{\Gamma(\beta)} \left[I_{1-}^\alpha f(s) - \lambda I_{1-}^{\alpha+p} I_{0+}^q h(s) \right] ds \right\} \\
&\quad + a_2(t) \int_0^\xi \frac{(\xi-s)^{\beta-1}}{\Gamma(\beta)} \left[I_{1-}^\alpha f(s) - \lambda I_{1-}^{\alpha+p} I_{0+}^q h(s) \right] ds.
\end{aligned}$$

For each $t \in [0,1]$, we have

$$\begin{aligned}
|y(t)| &\leq \int_0^t \frac{|t-s|^{\beta-1}}{\Gamma(\beta)} \left[I_{1-}^\alpha z(s) + |\lambda| I_{1-}^{\alpha+p} I_{0+}^q u(s) \right] ds \\
&\quad + |a_1(t)| \left\{ |\delta| \int_0^\mu \frac{|\mu-s|^{\beta-1}}{\Gamma(\beta)} \left[I_{1-}^\alpha z(s) + |\lambda| I_{1-}^{\alpha+p} I_{0+}^q u(s) \right] ds \right. \\
&\quad \left. + \int_0^1 \frac{|1-s|^{\beta-1}}{\Gamma(\beta)} \left[I_{1-}^\alpha z(s) + |\lambda| I_{1-}^{\alpha+p} I_{0+}^q u(s) \right] ds \right\} \\
&\quad + |a_2(t)| \int_0^\xi \frac{|\xi-s|^{\beta-1}}{\Gamma(\beta)} \left[I_{1-}^\alpha z(s) + |\lambda| I_{1-}^{\alpha+p} I_{0+}^q u(s) \right] ds \\
&\leq \frac{1 + \bar{a}_1(|\delta|+1) + \bar{a}_2}{\Gamma(\beta+1)\Gamma(\alpha)} \|z\|_{L^1} + \frac{|\lambda|(1 + \bar{a}_1(|\delta|+1) + \bar{a}_2)}{\Gamma(\beta+1)\Gamma(\alpha+p)\Gamma(q)} \|u\|_{L^1} \\
&\leq \Lambda_3 \|z\|_{L^1} + \Lambda_4 \|u\|_{L^1},
\end{aligned}$$

Taking the supremum over $t \in J$, we get

$$\|y\| \leq \Lambda_3 \|z\|_{L^1} + \Lambda_4 \|u\|_{L^1} < \infty.$$

Hence the set $\mathcal{E}$ is bounded. As a consequence of Lemma 6 we deduce that $\mathcal{U}$ has at least one fixed point which implies that the problem (2) has a solution on $[0,1]$. □

5. Applications

We consider four different cases for $F(t,y)$ and $H(t,y)$ (in (2)) to demonstrate applications of theorem (2): (**a**) F and H have sub-linear growth in their second variable. (**b**) F and H have linear growth in their second variable. (**c**) F has sub-linear growth in its second variable and H has linear growth. (**d**) F has linear growth in its second variable and H has sub-linear growth.

Case (**a**). For each $(t,y) \in [0,1] \times \mathbb{R}$, there exist functions $\sigma_i(t), \vartheta_i(t) \in L^1([0,1], \mathbb{R}^+), i = 1, 2, \gamma \in [0,1)$ such that $\|F(t,y)\| \le \sigma_1(t)|y|^\gamma + \vartheta_1(t)$ and $\|H(t,y)\| \le \sigma_2(t)|y|^\gamma + \vartheta_2(t)$ which correspond in this case to $\phi_\rho(t) = \sigma_1(t)\rho^\gamma + \vartheta_1(t)$ and $\psi_\rho(t) = \sigma_2(t)\rho^\gamma + \vartheta_2(t)$ and the condition (9) will take the form $0 \cdot \Lambda_3 + 0 \cdot \Lambda_4 < 1$, that is, $\zeta_1 = \zeta_2 = 0$.

Case (**b**). F and H will satisfy the assumptions $\|F(t,y)\| \le \sigma_1(t)|y| + \vartheta_1(t)$ and $\|H(t,y)\| \le \sigma_2(t)|y| + \vartheta_2(t)$, which, in view of (**M$_2$**), implies that $\phi_\rho(t) = \sigma_1(t)\rho + \vartheta_1(t)$ and $\psi_\rho(t) = \sigma_2(t)\rho + \vartheta_2(t)$, and the condition (9) becomes $\|\sigma_1\|_{L^1} \cdot \Lambda_3 + \|\sigma_2\|_{L^1} \cdot \Lambda_4 < 1$.

Similarly, one can verify the cases (**c**) and (**d**). Thus, the boundary value problem (2) has at least one solution on $[0,1]$ for all the cases (**a**)–(**d**).

Let us consider the following inclusions problem:

$$\begin{cases} {}^C D_{1^-}^{5/4} \, {}^{RL} D_{0^+}^{3/4} y(t) \in F(t, y(t)) - 2I_{1^-}^{3/2} I_{0^+}^{5/2} H(t, y(t)), & t \in [0,1], \\ y(0) = y(1/3) = 0, \quad y(1) = \frac{1}{4} y(2/3), \end{cases} \quad (11)$$

where $\alpha = 5/4$, $\beta = 3/4$, $\lambda = 2$, $p = 3/2$, $q = 5/2$, $\xi = 1/3$, $\mu = 2/3$, $\delta = 1/4$. It is easy to find that

$$\bar{a}_1 = \max_{t \in [0,1]} |a_1(t)| = |a_1(t)|_{t=1} \approx 1.101592729739686,$$

$$\bar{a}_2 = \max_{t \in [0,1]} |a_2(t)| = |a_2(t)|_{t=t_{a_2}} \approx 1.055901462873258,$$

where

$$t_{a_2} = \frac{\beta(1 - \delta \mu^{\beta+1})}{(1 - \delta \mu^\beta)(\beta + 1)} \approx 0.460880265746053 < 1.$$

Using the above given data, we find that $\Lambda_3 \approx 4.120918689155884$, $\Lambda_4 \approx 3.494023466997676$, where Λ_3, Λ_4 are given by (7).

(**a**). We consider $\|F(t,y)\| \le \sigma_1(t)|y|^{1/3} + \vartheta_1(t)$ and $\|H(t,y)\| \le \sigma_2(t)|y|^{1/3} + \vartheta_2(t)$ with $\sigma_i(t), \vartheta_i(t) \in L^1([0,1], \mathbb{R}^+), i = 1, 2, \gamma \in [0,1)$. In this case, F and H in (11) satisfy all the assumptions of Theorem 2 with $0 \cdot \Lambda_3 + 0 \cdot \Lambda_4 < 1$, which implies that the boundary value problem (11) has at least one solution on $[0,1]$.

(**b**). As a second example, let F and H be such that $\|F(t,y)\| \le \frac{1}{4(1+t)^2}|y| + 2e^t$ and $\|H(t,y)\| \le \frac{2}{(4+t)^2}|y| + e^{-t}$. In this case, the condition (9) will take the form $\frac{1}{8} \cdot \Lambda_3 + \frac{1}{10} \cdot \Lambda_4 \approx 0.864517182844253 < 1$. Thus, by the conclusion of Theorem 2, there exists at least one solution for the problem (11) on $[0,1]$.

In a similar manner, one can verify that the problem (2) has at least one solution on $[0,1]$ when we choose the cases: (**c**) $\|F(t,y)\| \le \sigma_1(t)|y|^{1/3} + \vartheta_1(t)$, $\|H(t,y)\| \le \frac{2}{(4+t)^2}|y| + e^{-t}$, and (**d**) $\|F(t,y)\| \le \frac{1}{4(1+t)^2}|y| + 2e^t$, $\|H(t,y)\| \le \sigma_2(t)|y|^{1/3} + \vartheta_2(t)$.

6. Conclusions

In this paper, we have discussed the existence of solutions for a new class of boundary value problems involving right-Caputo and left-Riemann–Liouville fractional derivatives of different orders and right-left Riemann–Liouville fractional integrals with nonlocal boundary conditions. The existence result for the single-valued case of the given problem is proven via Sadovski's fixed point theorem, while the existence results for the multi-valued case of the problem at hand are derived by means of

Bohnenblust-Karlin and Martelli fixed point theorems. Applications for the obtained results are also presented. By taking $\delta = 0$ in the results of this paper, we obtain the ones for a problem associated with three-point nonlocal boundary conditions: $y(0) = 0, y(\xi) = 0, y(1) = 0$ $(0 < \xi < 1)$ as a special case.

Author Contributions: Conceptualization, S.K.N. and B.A.; formal analysis, A.A., A.B., S.K.N. and B.A.; funding acquisition, A.A.; methodology, A.A., A.B., S.K.N. and B.A. All authors have read and agreed to the published version of the manuscript.

Funding: This research received no external funding.

Conflicts of Interest: The authors declare no conflict of interest.

References

1. Paola, M.D.; Pinnola, F.P.; Zingales, M. Fractional differential equations and related exact mechanical models. *Comput. Math. Appl.* **2013**, *66*, 608–620. [CrossRef]
2. Ahmed, N.; Vieru, D.; Fetecau, C.; Shah, N.A. Convective flows of generalized time-nonlocal nanofluids through a vertical rectangular channel. *Phys. Fluids* **2018**, *30*, 052002. [CrossRef]
3. Tarasov, V.E. *Fractional Dynamics: Applications of Fractional Calculus to Dynamics of Particles, Fields and Media*; Springer: New York, NY, USA, 2010.
4. Tarasova, V.V.; Tarasov, V.E. Logistic map with memory from economic model. *Chaos Solitons Fractals* **2017**, *95*, 84–91. [CrossRef]
5. Henderson, J.; Luca, R.; Tudorache, A. On a system of fractional differential equations with coupled integral boundary conditions. *Fract. Calc. Appl. Anal.* **2015**, *18*, 361–386. [CrossRef]
6. Peng, L.; Zhou, Y. Bifurcation from interval and positive solutions of the three-point boundary value problem for fractional differential equations. *Appl. Math. Comput.* **2015**, *257*, 458–466. [CrossRef]
7. Ahmad, B.; Alsaedi, A.; Ntouyas, S.K.; Tariboon, J. *Hadamard-Type Fractional Differential Equations, Inclusions and Inequalities*; Springer: Cham, Switzerland, 2017.
8. Ahmad, B.; Ntouyas, S.K. Nonlocal initial value problems for Hadamard-type fractional differential equations and inclusions. *Rocky Mt. J. Math.* **2018**, *48*, 1043–1068. [CrossRef]
9. Cui, Y.; Ma, W.; Sun, Q.; Su, X. New uniqueness results for boundary value problem of fractional differential equation. *Nonlinear Anal. Model. Control* **2018**, *23*, 31–39.
10. Ahmad, B.; Alghamdi, N.; Alsaedi, A.; Ntouyas, S.K. A system of coupled multi-term fractional differential equations with three-point coupled boundary conditions. *Fract. Calc. Appl. Anal.* **2019**, *22*, 601–618. [CrossRef]
11. Alsaedi, A.; Ahmad, B.; Alghanmi, M. Extremal solutions for generalized Caputo fractional differential equations with Steiltjes-type fractional integro-initial conditions. *Appl. Math. Lett.* **2019**, *91*, 113120. [CrossRef]
12. Liang, S.; Wang, L.; Yin, G. Fractional differential equation approach for convex optimization with convergence rate analysis. *Optim. Lett.* **2020**, *14*, 145–155. [CrossRef]
13. Zhang, L.; Ahmad, B.; Wang, G. The existence of an extremal solution to a nonlinear system with the right-handed Riemann-Liouville fractional derivative. *Appl. Math. Lett.* **2014**, *31*, 1–6. [CrossRef]
14. Khaldi, R.; Guezane-Lakoud, A. Higher order fractional boundary value problems for mixed type derivatives. *J. Nonlinear Funct. Anal.* **2017**. [CrossRef]
15. Lakoud, A.G.; Khaldi, R.; Kilicman, A. Existence of solutions for a mixed fractional boundary value problem. *Adv. Differ. Equ.* **2017**, *2017*, 164. [CrossRef]
16. Ahmad, B.; Ntouyas, S.K.; Alsaedi, A. Existence theory for nonlocal boundary value problems involving mixed fractional derivatives. *Nonlinear Anal. Model. Control* **2019**, *24*, 937–957. [CrossRef]
17. Atanackovic, T.M.; Stankovic, B. On a differential equation with left and right fractional derivatives. *Fract. Calc. Appl. Anal.* **2007**, *10*, 139–150. [CrossRef]
18. Yang, A.M.; Han, Y.; Zhang, Y.Z.; Wang, L.T.; Zhang, D.; Yang, X.J. On nonlocal fractional Volterra integro-differential equations in fractional steady heat transfer. *Therm. Sci.* **2016**, *20*, S789–S793. [CrossRef]
19. Tarasov, V.E. Fractional integro-differential equations for electromagnetic waves in dielectric media. *Theor. Math. Phys.* **2009**, *158*, 355–359. [CrossRef]

20. Korda, M.; Henrion, D.; Jones, C.N. Convex computation of the maximum controlled invariant set for polynomial control systems. *SIAM J. Control Optim.* **2014**, *52*, 2944–2969. [CrossRef]
21. Danca, M.F. Synchronization of piecewise continuous systems of fractional order. *Nonlinear Dyn.* **2014**, *78*, 2065–2084. [CrossRef]
22. Kisielewicz, M. *Stochastic Differential Inclusions and Applications*; Springer: New York, NY, USA, 2013.
23. Bastien, J. Study of a driven and braked wheel using maximal monotone differential inclusions: Applications to the nonlinear dynamics of wheeled vehicles. *Arch. Appl. Mech.* **2014**, *84*, 851–880. [CrossRef]
24. Yue, Y.; Tian, Y.; Bai, Z. Infinitely many nonnegative solutions for a fractional differential inclusion with oscillatory potential. *Appl. Math. Lett.* **2019**, *88*, 64–72. [CrossRef]
25. Wang, Y.; Liang, T. Mild solutions to the time fractional Navier-Stokes delay differential inclusions. *Discrete Contin. Dyn. Syst. Ser. B* **2019**, *24*, 3713–3740. [CrossRef]
26. Benchohra, M.; Hamani, S.; Zhou, Y. Oscillation and nonoscillation for Caputo-Hadamard impulsive fractional differential inclusions. *Adv. Differ. Equ.* **2019**, *2019*, 74. [CrossRef]
27. Kamenskii, M.; Obukhovskii, V.; Petrosyan, G.; Yao, J.-C. On semilinear fractional order differential inclusions in Banach spaces. *Fixed Point Theory* **2017**, *18*, 269–291. [CrossRef]
28. Cheng, Y.; Agarwal, R.P.; O'Regan, D. Existence and controllability for nonlinear fractional differential inclusions with nonlocal boundary conditions and time-varying delay. *Fract. Calc. Appl. Anal.* **2018**, *21*, 960–980. [CrossRef]
29. Abbas, S.; Benchohra, M.; Graef, J.R. Coupled systems of Hilfer fractional differential inclusions in Banach spaces. *Commun. Pure Appl. Anal.* **2018**, *17*, 2479–2493. [CrossRef]
30. Ntouyas, S.K.; Alsaedi, A.; Ahmed, B. Existence theorems for mixed Riemann-Liouville and Caputo fractional differential equations and inclusions with nonlocal fractional integro-differential boundary conditions. *Fractal Fract.* **2019**, *3*, 21. [CrossRef]
31. Vijayakumar, V. Approximate controllability results for non-densely defined fractional neutral differential inclusions with Hille-Yosida operators. *Internat. J. Control* **2019**, *92*, 2210–2222. [CrossRef]
32. Ahmad, B.; Ntouyas, S.K.; Alsaedi, A. Coupled systems of fractional differential inclusions with coupled boundary conditions. *Electron. J. Differ. Equ.* **2019**, *2019*, 1–21.
33. Ahmad, B.; Ntouyas, S.K.; Tariboon, J. On inclusion problems involving Caputo and Hadamard fractional derivatives. *Acta Math. Univ. Comenian. (N.S.)* **2020**, *89*, 169–183.
34. Ahmad, B.; Broom, A.; Alsaedi, A.; Ntouyas, S.K. Nonlinear integro-differential equations involving mixed right and left fractional derivatives and integrals with nonlocal boundary data. *Mathematics* **2020**, *8*, 336. [CrossRef]
35. Li, T. A class of nonlocal boundary value problems for partial differential equations and its applications in numerical analysis. *J. Comput. Appl. Math.* **1989**, *28*, 49–62.
36. Ahmad, B.; Nieto, J.J. Existence of solutions for nonlocal boundary value problems of higher-order nonlinear fractional differential equations. *Abstr. Appl. Anal.* **2009**, *2009*, 494720. [CrossRef]
37. Bohnenblust, H.F.; Karlin, S. On a theorem of Ville. In *Contributions to the Theory of Games*; Princeton University Press: Princeton, NJ, USA, 1950; Volume I, pp. 155–160.
38. Martelli, M. A Rothe's theorem for non compact acyclic-valued maps. *Boll. Un. Mat. Ital.* **1975**, *4*, 70–76.
39. Kilbas, A.A.; Srivastava, H.M.; Trujillo, J.J. *Theory and Applications of Fractional Differential Equations*; North-Holland Mathematics Studies, 204; Elsevier Science B.V.: Amsterdam, The Netherlands, 2006.
40. Granas, A.; Dugundji, J. *Fixed Point Theory*; Springer-Verlag: New York, NY, USA, 2005.
41. Zeidler, E. *Nonlinear Functional Analysis and Its Application: Fixed Point-Theorems*; Springer-Verlag: New York, NY, USA, 1986; Volume 1.
42. Sadovskii, B.N. On a fixed point principle. *Funct. Anal. Appl.* **1967**, *1*, 74–76. [CrossRef]
43. Deimling, K. *Multivalued Differential Equations*; De Gruyter: Berlin, Germany, 1992.

44. O'Regan, D.; Precup, R. Fixed point theorems for set-valued maps and existence principles for integral inclusions. *J. Math. Anal. Appl.* **2000**, *245*, 594–612. [CrossRef]
45. Lasota, A.; Opial, Z. An application of the Kakutani-Ky Fan theorem in the theory of ordinary differential equations. *Bull. Acad. Polon. Sci. Ser. Sci. Math. Astronom. Phys.* **1965**, *13*, 781–786.

© 2020 by the authors. Licensee MDPI, Basel, Switzerland. This article is an open access article distributed under the terms and conditions of the Creative Commons Attribution (CC BY) license (http://creativecommons.org/licenses/by/4.0/).

Article

Generalized Nabla Differentiability and Integrability for Fuzzy Functions on Time Scales

R. Leelavathi [1], G. Suresh Kumar [1], Ravi P. Agarwal [2,*], Chao Wang [3] and M.S.N. Murty [4,†]

1. Department of Mathematics, Koneru Lakshmaiah Education Foundation, Vaddeswaram 522502, Guntur, Andhra Pradesh, India; leelaammu36@gmail.com (R.L.); drgsk006@kluniversity.in (G.S.K.)
2. Department of Mathematics, Texas A&M University-Kingsville, Kingsville, TX 78363-8202, USA
3. Department of Mathematics, Yunnan University, Kunming 650091, China; chaowang@ynu.edu.cn
4. Sainivas, D.No. 21-47, Opp. State Bank of India, Bank Street, Nuzvid 521201, Krishna, Andhra Pradesh, India; drmsn2002@gmail.com
* Correspondence: ravi.agarwal@tamuk.edu
† Professor of Mathematics (Retd.).

Received: 9 April 2020; Accepted: 2 June 2020; Published: 8 June 2020

Abstract: This paper mainly deals with introducing and studying the properties of generalized nabla differentiability for fuzzy functions on time scales via Hukuhara difference. Further, we obtain embedding results on $\mathbb{E}_n$ for generalized nabla differentiable fuzzy functions. Finally, we prove a fundamental theorem of a nabla integral calculus for fuzzy functions on time scales under generalized nabla differentiability. The obtained results are illustrated with suitable examples.

Keywords: fuzzy functions time scales; Hukuhara difference; generalized nabla Hukuhara derivative; fuzzy nabla integral

1. Introduction

The theory of dynamic equations on time scales is a genuinely new subject and the research related to this area is developing rapidly. Time scale theory has been developed to unify continuous and discrete structures, and it allows solutions for both differential and difference equations at a time and extends those results to dynamic equations. Basic results in time scales and dynamic equations on time scales are found in [1–6]. In [7], the author illustrated an example where delta derivative needs more assumptions than nabla derivative. Some recent studies in economics [8], production, inventory models [9], adaptive control [10], neural networks [11], and neural cellular networks [12] suggest nabla derivative is also preferable and it has fewer restrictions than delta derivative on time scales.

On the other hand, when we expect to investigate a real world phenomenon absolutely, it is important to think about a number of unsure factors too. To specify these vague or imprecise notions, Zadeh [13] established fuzzy set theory. The theory of fuzzy differential equations (FDEs) and its applications was developed and studied by Kaleva [14], Lakshmikantham and Mohapatra [15]. The concept based on Hukuhara differentiability has a shortcoming that the solution to a FDEs exists only for increasing length of support. To overcome this shortcoming, Bede and Gal [16] studied generalized Hukuhara differentiability for fuzzy functions. In light of this preferred advantage, many authors [17–19] tend their enthusiasm to the generalized Hukuhara differentiability for fuzzy set valued functions.

The calculus of fuzzy functions on time scales was studied by Fard and Bidgoli [20]. Vasavi et al. [21–24] introduced Hukuhara delta derivative, second-type Hukuhara delta derivative, and generalized Hukuhara delta derivatives by using Hukuhara difference, and they studied fuzzy dynamic equations on time scales. Wang et al. [25] introduced and studied almost periodic fuzzy vector-valued functions on time scales. Deng et al. [26] studied fractional nabla-Hukuhara derivative on time scales. Recently, Leelavathi et al. [27] introduced and studied properties of nabla Hukuhara derivative for fuzzy functions on time scales. However, this derivative has the disadvantage that it exists only for the fuzzy functions on time scales which have a diameter with an increasing length. For the fuzzy functions with decreasing length of diameter on time scales, Leelavathi et al. [28] introduced the second-type nabla Hukuhara derivative and studied its properties. Later, they continued to study fuzzy nabla dynamic equations under the first and second-type nabla Hukuhara derivatives in [29] under generalized differentiability by using generalized Hukuhara difference in [30]. Consider a simple fuzzy function $F(s) = s \odot c, s \in \mathbb{T} \cap [-2,2]$, where $c = (1,2,3)$ is a triangular fuzzy number. Clearly, $F(s)$ has decreasing length of diameter in $\mathbb{T} \cap [-2,0]$ and increasing length of diameter in $\mathbb{T} \cap [0,2]$. Therefore, the fuzzy function $F(s)$ is neither a nabla Hukuhara differentiable (as defined in [27]) nor a second-type nabla Hukuhara differentiable (as defined in [28]) on $\mathbb{T} \cap [-2,2]$. In this context, it is required to define a nabla Hukuhara derivative for a fuzzy function which may have both increasing and decreasing length of diameter on a time scale. To address this issue, in the present work, we define a new derivative called generalized nabla derivative for fuzzy functions on time scales via Hukuhara difference and study their properties. In [31], the authors introduced a nabla integral for fuzzy functions on time scales and obtained fundamental properties. In the present work, we continue to study nabla integral for fuzzy functions on time scales and prove a fundamental theorem of nabla integral calculus for generalized nabla differentiable functions.

The rest of this paper is arranged as follows. In Section 2, we present some basic definitions, properties, and results relating to the calculus of fuzzy functions on time scales. In Section 3, we establish the nabla Hukuhara generalized derivative for fuzzy functions on time scales and obtain its fundamental properties. The results are highlighted with suitable examples. In Section 4, we prove an embedding theorem on $\mathbb{E}_n$ and obtain the results connecting to generalized nabla differentiability on time scales. Using these results, we finally prove the fundamental theorem of nabla integral calculus for fuzzy functions on time scales under generalized nabla differentiability and a numerical example is provided to verify the validity of the theorem.

2. Preliminaries

Let $\Re_k(\Re^n)$ be the family of all nonempty convex compact subsets of $\Re^n$. Define the set addition and scalar multiplication in $\Re_k(\Re^n)$ as usual. Then, by [14], $\Re_k(\Re^n)$ is a commutative semi-group under addition with cancellation laws. Further, if $\beta, \gamma \in \Re$ and $U, V \in \Re_k(\Re^n)$, then

$$\beta \odot (U \oplus V) = (\beta \odot U) \oplus (\beta \odot V), \quad \beta(\gamma \odot U) = (\beta \gamma) \odot U, \quad 1 \odot U = U, \text{ and if}$$

$$\beta, \gamma \geq 0 \text{ then} (\beta \oplus \gamma) U = \beta \odot U \oplus \gamma \odot U.$$

Let P and Q be two bounded nonempty subsets of $\Re^n$. By using the Pampeiu–Hausdorff metric, we define the distance between P and Q as follows:

$$d_H(P,Q) = \max\{\sup_{p \in P} \inf_{q \in Q} \|p - q\|, \sup_{q \in Q} \inf_{p \in P} \|p - q\|\},$$

where $\|.\|$ is the Euclidean norm in $\Re^n$. Then, $(\Re_k(\Re^n), d_H)$ becomes a separable and complete metric space [14].

Define:

$$\mathbb{E}_n = \{u : \Re^n \to [0,1] | u \text{ satisfies(a)–(d) below}\}, \text{ where}$$

(a) If there exists a $t \in \Re^n$ such that $u(t) = 1$, then u is said to be normal.
(b) u is fuzzy convex.
(c) u is upper semi-continuous.
(d) The closure of $\{t \in \Re^n / u(t) > 0\} = [u]^0$ is compact.

For $0 \leq \lambda \leq 1$, denote $[u]^\lambda = \{t \in \Re^n : u(t) \geq \lambda\}$; then, from the above conditions, we have that the λ-level set $[u]^\lambda \in \Re_k(\Re^n)$. By Zadeh's extension principle, a mapping $h : \R^n \times \R^n \to \R^n$ can br extended to $g : \mathbb{E}^n \times \mathbb{E}^n \to \mathbb{E}^n$ by

$$g(s,t)(z) = \sup_{z=g(x,y)} \min\{s(x), t(y)\}.$$

We have $[h(p,q)]^\lambda = h([p]^\lambda, [q]^\lambda)$, for all $p, q \in \mathbb{E}_n$ and h is continuous. The scalar multiplication $\odot$ and addition $\oplus$ of $p, q \in \mathbb{E}_n$ is defined as $[p \oplus q]^\lambda = [p]^\lambda + [q]^\lambda$, $[c \odot p]^\lambda = c[p]^\lambda$, where $p, q \in \mathbb{E}_n$, $c \in \Re$, $0 \leq \lambda \leq 1$.

Define $D_H : \mathbb{E}_n \times \mathbb{E}_n \to [0, \infty)$ by the equation

$$D_H(s, t) = \sup_{0 \leq \lambda \leq 1} d_H([s]^\lambda, [t]^\lambda),$$

where d_H is the Pampeiu–Hausdorff metric defined in $\Re_k(\Re^n)$. Then, $(\mathbb{E}_n, D_H)$ is a complete metric space [14]. The following theorem extends the properties of addition and scalar multiplication of fuzzy number valued functions ($\Re_F = \mathbb{E}_1$) to $\mathbb{E}_n$ [14].

The properties of addition and scalar multiplication of fuzzy number valued functions ($\Re_F = \mathbb{E}_1$) are easily extended to $\mathbb{E}_n$.

Theorem 1 ([32]).

(a) If we denote $\hat{0} = \chi_{\{0\}}$, then $\hat{0} \in \mathbb{E}_n$ is the zero element with respect to $\oplus$, i.e., $p \oplus \hat{0} = \hat{0} \oplus p = p$, $\forall s \in \mathbb{E}_n$.
(b) For any $p \in \mathbb{E}_n$ has no inverse with respect to '$\oplus$'.
(c) For any $\gamma, \beta \in \Re$ with $\gamma, \beta \geq 0$ or $\gamma, \beta \leq 0$ and $p \in \mathbb{E}_n$, $(\gamma + \beta) \odot p = (\gamma \odot p) \oplus (\beta \odot p)$.
(d) For any $\gamma \in \Re$ and $p, q \subset \mathbb{E}_n$, we have $\gamma \odot (p \oplus q) = (\gamma \odot p) \oplus (\gamma \odot q)$.
(e) For any $\gamma, \beta \in \Re$ and $p \in \mathbb{E}_n$, we have $\gamma \odot (\beta \odot p) = (\gamma \beta) \odot p$.

Definition 1 ([14]). Let $K, L \in \mathbb{E}_n$. If there exists $M \in \mathbb{E}_n$ such that $K = L \oplus M$, then we say that M is the Hukuhara difference of K and L and is denoted by $K \ominus_h L$.

For any $K, L, M, N \in \mathbb{E}_n$ and $\beta \in \Re$, the following hold:

(a) $D_H(K, L) = 0 \Leftrightarrow K = L$;
(b) $D_H(\beta \odot K, \beta \odot L) = |\beta| D_H(K, L)$;
(c) $D_H(K \oplus M, L \oplus M) = D_H(K, L)$;
(d) $D_H(K \ominus_h M, L \ominus_h M) = D_H(K, L)$;
(e) $D_H(K \oplus L, M \oplus N) \leq D_H(K, M) + D_H(L, N)$; and
(f) $D_H(K \ominus_h L, M \ominus_h N) \leq D_H(K, M) + D_H(L, N)$.

provided the Hukuhara differences exists.

A triangular fuzzy number is denoted by three points as $t = (t_1, t_2, t_3)$. This representation is denoted as membership function

$$\mu_t(x) = \begin{cases} 0, & x < t_1 \\ \dfrac{x - t_1}{t_2 - t_1}, & t_1 \leq x \leq t_2 \\ \dfrac{t_3 - x}{t_3 - t_2}, & t_2 \leq x \leq t_3 \\ 0, & x > t_3 \end{cases}$$

In addition, λ-level sets of triangular fuzzy number t is an interval defined by λ-cut operation, $t_\lambda = [(t_2 - t_1)\lambda + t_1, t_3 - (t_3 - t_2)\lambda]$, for all $\lambda \in [0, 1]$. Clearly, the triangular fuzzy number is in $\mathbb{E}_1$.

Let $T = (t_1, t_2, t_3)$, $S = (s_1, s_2, 3_3)$ be two triangular fuzzy numbers in $\mathbb{E}_1$. The addition and scalar multiplication are defined as:

$$S \oplus T = (t_1 + s_1, t_2 + s_2, t_3 + s_3),$$

$$k \odot T = \begin{cases} (kt_1, kt_2, kt_3) & if \ k > 0, \\ (kt_3, kt_2, kt_1) & if \ k < 0, \\ \hat{0} & if \ k = 0 \end{cases}$$

Remark 1. *From Theorem 1(c), we can deduce that, for any $\beta, \gamma \in \Re$ and $s \in \mathbb{E}_n$.*

(a) *If $\beta > \gamma \geq 0$, then $(\beta \odot s) \ominus_h (\gamma \odot s)$ exists and $(\beta \odot s) \ominus_h (\gamma \odot s) = (\beta - \gamma) \odot s$.*
(b) *If $\beta < \gamma \leq 0$, then $(\beta \odot s) \ominus_h (\gamma \odot s)$ exists and $(\beta \odot s) \ominus_h (\gamma \odot s) = (\beta - \gamma) \odot s$.*

Proof.

(a) Since $\beta - \gamma > 0$ and $\gamma > 0$, from Theorem 1(c), we get $(\beta - \gamma) \odot s \oplus \gamma \odot s = (\beta - \gamma + \gamma) \odot s = \beta \odot s$. Therefore, $(\beta - \gamma) \odot s \oplus \gamma \odot s = \beta \odot s$. Hence, $(\beta \odot s) \ominus_h (\gamma \odot s) = (\beta - \gamma) \odot s$.
(b) Since $\beta - \gamma < 0$ and $\gamma < 0$, from Theorem 1(c), it is easily proven that $(\beta \odot s) \ominus_h (\gamma \odot s) = (\beta - \gamma) \odot s$.
□

Now, we discuss the differentiability and integrability of fuzzy functions on $I = [a, b] \subset \Re$ (where I is a compact interval).

Definition 2 ([14]). *A mapping $\Phi : I \to \mathbb{E}_n$ is said to be strongly measurable if, for each $\lambda \in [0, 1]$, the fuzzy function $\Phi_\lambda : I \to \Re_k(\Re^n)$ defined by $\Phi_\lambda(s) = [\Phi(s)]^\lambda$ is measurable.*

Remark 2 ([14]). *A mapping $\Phi : I \to \mathbb{E}_n$ is said to be integrably bounded if there exists an integrable function h such that $\|x\| \leq h(s)$, for all $x \in \Phi_0(s)$.*

Definition 3 ([14]). *Let $\Phi : I \to \mathbb{E}_n$. The integral of Φ over I is denoted by $\int_I \Phi(s)ds$ or $\int_x^y \Phi(s)ds$,*

$$\left[\int_I \Phi(s)ds\right]^\lambda = \int_I \Phi_\lambda(s)ds$$
$$= \left\{\int_I g(s)ds \ /\phi : I \to \Re^n\right\},$$

where g is a level wise selection of measurable functions of Φ_λ for $0 < \lambda \leq 1$.

A mapping $\Phi : I \to \mathbb{E}_n$ is said to be integrable over I if Φ is integrably bounded and strongly measurable function and also $\int_I \Phi(s)ds \in \mathbb{E}_n$.

Theorem 2 ([14]). *Let $\Phi, \Psi : I \to \mathbb{E}_n$ be integrable. Then,*

(a) $\int \Phi \oplus \Psi = \int \Phi \oplus \int \Psi$;
(b) $\int \alpha \odot \Phi = \alpha \odot \int \Phi$, *where $\alpha \in \Re$;*
(c) $\int_x^y \Phi = \int_x^z \Phi \oplus \int_z^y \Phi$, *where $z \in \Re$;*
(d) $D_H(\Phi, \Psi)$ *is integrable; and*
(e) $D_H(\int \Phi, \int \Psi) \leq \int D_H(\Phi, \Psi)$.

Definition 4 ([18]). *A fuzzy function $\Phi : I \to \mathbb{E}_n$ is said to be differentiable from left at s_0 if for $\delta > 0$, there exists $P \in \mathbb{E}_n$, such that the following holds:*

(a) for $0 < \hbar < \delta$, $\Phi(s_0) \ominus_h \Phi(s_0 - \hbar)$ exist and $\lim_{\hbar \to 0^+} \frac{1}{\hbar} \odot (\Phi(s_0) \ominus_h \Phi(s_0 - \hbar)) = P$;

or

(b) for $0 < \hbar < \delta$, $\Phi(s_0 - \hbar)) \ominus_h \Phi(s_0)$ exist and $\lim_{\hbar \to 0^+} \frac{-1}{\hbar} \odot (\Phi(s_0 - \hbar)) \ominus_h \Phi(s_0) = P$.

Here, P is the derivative of Φ from left at s_0 and is denoted as $\Phi'_{-}(s_0)$.

Definition 5 ([18])**.** *A fuzzy function $\Phi : I \to \mathbb{E}_n$ is said to be differentiable from right at s_0 if, for $\delta > 0$, there exists $P \in \mathbb{E}_n$, such that the following holds:*

(a) for $0 < \hbar < \delta$, $\Phi(s_0 + \hbar) \ominus_h \Phi(s_0)$ exist and $\lim_{\hbar \to 0^+} \frac{1}{\hbar} \odot (\Phi(s_0 + \hbar) \ominus_h \Phi(s_0)) = P$;

or

(b) for $0 < \hbar < \delta$, $\Phi(s_0) \ominus_h \Phi(s_0 + \hbar)$ exist and $\lim_{\hbar \to 0^+} \frac{-1}{\hbar} \odot (\Phi(s_0) \ominus_h \Phi(s_0 + \hbar)) = P$.

Here, P is the derivative of Φ from right at s_0 and is denoted as $\Phi'_{+}(s_0)$. The limits are taken over $(\mathbb{E}_n, D_H)$.

Definition 6 ([18])**.** *If Φ is both left-differentiable and right-differentiable at s_0, then Φ is said to be differentiable at s_0 and $\Phi'_{-}(s_0) = \Phi'_{+}(s_0) = P$. Here, P is called the derivative of Φ at s_0 and we consider one-sided derivative at the end points of I.*

Remark 3 ([18])**.** *If Φ is differentiable at s_0, then there exists a $\delta > 0$, such that:*

(a) For $0 < \hbar < \delta$, $\Phi(s_0 - \hbar) \ominus_h \Phi(s_0)$ or $\Phi(s_0) \ominus_h \Phi(s_0 - \hbar)$ exists.
(b) For $0 < \hbar < \delta$, $\Phi(s_0 + \hbar) \ominus_h \Phi(s_0)$ or $\Phi(s_0) \ominus_h \Phi(s_0 + \hbar)$ exists.

3. Generalized Nabla Hukuhara Differentiability on Time Scales

This section is concerned with defining and studying the properties of ∇^g derivative for fuzzy functions on time scales. In addition, we illustrate the results with suitable examples.

Definition 7 ([21])**.** *For any given $\epsilon > 0$, there exists a $\delta > 0$, such that the fuzzy function $\Phi : \mathbb{T}^{[a,b]} \to \mathbb{E}_n$ has a unique $\mathbb{T}$-limit $P \in \mathbb{E}_n$ at $s \in \mathbb{T}^{[a,b]}$ if $D_H(\Phi(s) \ominus_h P, \hat{0}) \leq \epsilon$, for all $s \in N_{\mathbb{T}^{[a,b]}}(s, \delta)$ and it is denoted by $\mathbb{T} - \lim_{s \to s_0} \Phi(s)$.*

Here, $\mathbb{T}$-limit denotes the limit on time scale in the metric space $(\mathbb{E}_n, D_H)$.

Remark 4. *From the above definition, we have*

$$\mathbb{T} - \lim_{s \to s_0} \Phi(s) = P \in \mathbb{E}_n \iff \mathbb{T} - \lim_{s \to s_0} (\Phi(s) \ominus_h P) = \hat{0},$$

where the zero element in $\mathbb{E}_n$ is given by $\hat{0}$.

Definition 8. *A fuzzy mapping $\Phi : \mathbb{T}^{[a,b]} \to \mathbb{E}_n$ is continuous at $s_0 \in \mathbb{T}$, if $\mathbb{T} - \lim_{s \to s_0} \Phi(s) \in \mathbb{E}_n$ exists and $\mathbb{T} - \lim_{s \to s_0} \Phi(s) = \Phi(s_0)$, i.e.,*

$$\mathbb{T} - \lim_{s \to s_0} (\Phi(s) \ominus_h \Phi(s_0)) = \hat{0}.$$

Remark 5. *If $\Phi : \mathbb{T}^{[a,b]} \to \mathbb{E}_n$ is continuous at $s_0 \in \mathbb{T}^{[a,b]}$, then, for every $\epsilon > 0$, there exists a $\delta > 0$, such that*

$$D_H(\Phi(s) \ominus_h \Phi(s_0), \hat{0}) \leq \epsilon, \text{ for all } s \in N_{\mathbb{T}^{[a,b]}}.$$

Remark 6. Let $\Phi : \mathbb{T}^{[a,b]} \to \mathbb{E}_n$ and $s_0 \in \mathbb{T}^{[a,b]}$.

(a) If $\mathbb{T} - \lim_{s \to s_0^+} \Phi(s) = \Phi(s_0)$, then Φ is said to be right continuous at s_0.

(b) If $\mathbb{T} - \lim_{s \to s_0^-} \Phi(s) = \Phi(s_0)$, then Φ is said to be left continuous at s_0.

(c) If $\mathbb{T} - \lim_{s \to s_0^+} \Phi(s) = \Phi(s_0) = \mathbb{T} - \lim_{s \to s_0^-} \Phi(s)$, then Φ is continuous at s_0.

Definition 9. *A fuzzy function $\Phi : \mathbb{T}^{[a,b]} \to \mathbb{E}_n$ is said to be ∇^g left-differentiable at $s \in \mathbb{T}_k^{[a,b]}$, if there exists an element $\Phi_-^{\nabla^g}(s) \in \mathbb{E}_n$ with the property that, for any given $\epsilon > 0$, there exists a $N_{\mathbb{T}^{[a,b]}}$ of s for some $\delta > 0$ and $0 \leq \hbar \leq \delta$,*

$$D_H[\Phi(\varrho(s)) \ominus_h \Phi(s - \hbar), (\hbar - \nu(s)) \odot \Phi_-^{\nabla^g}(s)] \leq \epsilon |\hbar - \nu(s)| \tag{1}$$

or

$$D_H[\Phi(s - \hbar) \ominus_h \Phi(\varrho(s)), -(\hbar - \nu(s)) \odot \Phi_-^{\nabla^g}(s)] \leq \epsilon| - (\hbar - \nu(s))|, \tag{2}$$

for all $s - \hbar \in N_{\mathbb{T}^{[a,b]}}$, where $\nu(s) = s - \varrho(s)$, $\Phi_-^{\nabla^g}(s)$ is the generalized nabla left-derivative of Φ at s.

Definition 10. *A fuzzy function $\Phi : \mathbb{T}^{[a,b]} \to \mathbb{E}_n$ is said to be ∇^g right-differentiable at $s \in \mathbb{T}_k^{[a,b]}$, if there exists an element $\Phi_+^{\nabla^g}(s) \in \mathbb{E}_n$ with the property that, for every given $\epsilon > 0$, there exists a neighborhood $N_{\mathbb{T}^{[a,b]}}$ of s for some $\delta > 0$ and $0 \leq \hbar \leq \delta$,*

$$D_H[\Phi(s + \hbar) \ominus_h \Phi(\varrho(s)), (\hbar + \nu(s)) \odot \Phi_+^{\nabla^g}(s)] \leq \epsilon |\hbar + \nu(s)| \tag{3}$$

or

$$D_H[\Phi(\varrho(s)) \ominus_h \Phi(s + \hbar), -(\hbar + \nu(s)) \odot \Phi_+^{\nabla^g}(s)] \leq \epsilon| - (\hbar + \nu(s))|, \tag{4}$$

for all $s + \hbar \in N_{\mathbb{T}}^{[a,b]}$, where $\nu(s) = s - \varrho(s)$, $\Phi_+^{\nabla^g}(s)$ is the generalized nabla right-derivative of Φ at s.

Definition 11. *A fuzzy function $\Phi : \mathbb{T}^{[a,b]} \to \mathbb{E}_n$ is said to be ∇^g differentiable at $s \in \mathbb{T}_k^{[a,b]}$, if Φ is both right- and left-differentiable at $s \in \mathbb{T}_k^{[a,b]}$ and*

$$\Phi_+^{\nabla^g}(s) = \Phi_-^{\nabla^g}(s) = \Phi^{\nabla^g}(s).$$

Here, $\Phi_+^{\nabla^g}(s)$ or $\Phi_-^{\nabla^g}(s)$ is called ∇^g-derivative of Φ at $s \in \mathbb{T}_k^{[a,b]}$ and it is denoted by $\Phi^{\nabla^g}(s)$. Moreover, if ∇^g derivative exists at each $s \in \mathbb{T}_k^{[a,b]}$, then Φ is ∇^g differentiable on $\mathbb{T}_k^{[a,b]}$.

Theorem 3. *Let $\Phi : \mathbb{T}^{[a,b]} \to \mathbb{E}_n$ be a fuzzy function and $s \in \mathbb{T}_k^{[a,b]}$, then:*

(a) *If $\Phi : \mathbb{T}^{[a,b]} \to \mathbb{E}_n$ is ∇^g differentiable at s, then Φ is continuous at $s \in \mathbb{T}_k^{[a,b]}$.*

(b) *If s is left dense and $\Phi : \mathbb{T}^{[a,b]} \to \mathbb{E}_n$ is ∇^g differentiable at s iff the limits*

$$\lim_{\hbar \to 0^+} \frac{1}{\hbar} \odot (\Phi(s) \ominus_h \Phi(s - \hbar)) \text{ or } \lim_{\hbar \to 0^+} \frac{-1}{\hbar} \odot (\Phi(s - \hbar) \ominus_h \Phi(s))$$

and

$$\lim_{\hbar \to 0^+} \frac{1}{\hbar} \odot (\Phi(s + \hbar) \ominus_h \Phi(s)) \text{ or } \lim_{\hbar \to 0^+} \frac{-1}{\hbar} \odot (\Phi(s) \ominus_h \Phi(s + \hbar))$$

exist as a finite number and holds any one of the following:

(i) $\lim_{\hbar \to 0^+} \frac{1}{\hbar} \odot (\Phi(s) \ominus_h \Phi(s - \hbar)) = \Phi^{\nabla^g}(s) = \lim_{\hbar \to 0^+} \frac{1}{\hbar} \odot (\Phi(s + \hbar) \ominus_h \Phi(s));$

(ii) $\lim\limits_{\hbar \to 0^+} \dfrac{1}{\hbar} \odot (\Phi(s) \ominus_h \Phi(s-\hbar)) = \Phi^{\nabla^g}(s) = \lim\limits_{\hbar \to 0^+} \dfrac{-1}{\hbar} \odot (\Phi(s) \ominus_h \Phi(s+\hbar))$;

(iii) $\lim\limits_{\hbar \to 0^+} \dfrac{-1}{\hbar} \odot (\Phi(s-\hbar) \ominus_h \Phi(s)) = \Phi^{\nabla^g}(s) = \lim\limits_{\hbar \to 0^+} \dfrac{1}{\hbar} \odot (\Phi(s+\hbar) \ominus_h \Phi(s))$;

(iv) $\lim\limits_{\hbar \to 0^+} \dfrac{-1}{\hbar} \odot (\Phi(s-\hbar) \ominus_h \Phi(s)) = \Phi^{\nabla^g}(s) = \lim\limits_{\hbar \to 0^+} \dfrac{-1}{\hbar} \odot (\Phi(s) \ominus_h \Phi(s+\hbar))$.

Proof. (a) Suppose that Φ is ∇^g differentiable at s. Let $\epsilon \in (0,1)$. Choose $\epsilon^1 = \epsilon[1 + K + 2\nu(s)]^{-1}$, where $K = D_H[\Phi_-^{\nabla^g}(s), \hat{0}]$. Clearly, $\epsilon^1 \in (0,1)$. Since Φ is ∇^g left-differentiable, there exists $N_{\mathbb{T}[a,b]}$ a neighborhood of s such that, for all $\hbar \geq 0$ with $s - \hbar \in N_{\mathbb{T}[a,b]}$,

$$D_H[\Phi(\varrho(s)) \ominus_h \Phi(s-\hbar), (\hbar - \nu(s)) \odot \Phi_-^{\nabla^g}(s)] \leq \epsilon|\hbar - \nu(s)|,$$

or

$$D_H[\Phi(s-\hbar) \ominus_h \Phi(\varrho(s)), -(\hbar - \nu(s)) \odot \Phi_-^{\nabla^g}(s)] \leq \epsilon|-(\hbar - \nu(s))|.$$

For $0 \leq \hbar < \epsilon^1$ and for all $\hbar \geq 0$, to each $s - \hbar \in N_{\mathbb{T}[a,b]} \cap (s-\hbar, s+\hbar)$, we have,

$$\begin{aligned} D_H[\Phi(s), \Phi(s-\hbar)] &= D_H[\Phi(s) \ominus_h \Phi(s-\hbar), \hat{0}] \\ &= D_H[\Phi(s) \ominus_h \Phi(\varrho(s)) \oplus \Phi(\varrho(s)) \ominus_h \Phi(s-\hbar), \\ &\qquad (\hbar - \nu(s)) \odot \Phi_-^{\nabla^g}(s) \oplus \nu(s) \odot \Phi_-^{\nabla^g}(s) \\ &\qquad \oplus (-\hbar) \odot \Phi_-^{\nabla^g}(s)] \\ &\leq D_H[\Phi(\varrho(s)) \ominus_h \Phi(s-\hbar), (\hbar - \nu(s)) \odot \Phi_-^{\nabla^g}(s)] \\ &\quad + D_H[\Phi(s) \ominus_h \Phi(\varrho(s)), \nu(s) \odot \Phi_-^{\nabla^g}(s)] \\ &\quad + \hbar D_H[\Phi_-^{\nabla^g}(s), \hat{0}] \\ &\leq \epsilon^1|\hbar - \nu(s)| + \epsilon^1 \nu(s) + \hbar K \\ &= \epsilon^1 \hbar + \hbar K + 2\epsilon^1 \nu(s) \\ &< \epsilon^1(1 + K + 2\nu(s)) = \epsilon. \end{aligned}$$

Similarly, we can prove Φ is continuous at s, if ∇^g is right-differentiable at s.

(b) Suppose that Φ is ∇^g differentiable at s and s is left dense. To each $\epsilon \geq 0$, there exists a neighborhood $N_{\mathbb{T}[a,b]}$ of s such that

$$D_H\left[\Phi(\varrho(s)) \ominus_h \Phi(s-\hbar), (\hbar - \nu(s)) \odot \Phi_-^{\nabla^g}(s)\right] \leq \epsilon|\hbar - \nu(s)|$$

or

$$D_H\left[\Phi(s-\hbar) \ominus_h \Phi(\varrho(s)), (\nu(s) - \hbar) \odot \Phi_-^{\nabla^g}(s)\right] \leq \epsilon|-(\hbar - \nu(s))|,$$

and

$$D_H\left[\Phi(s+\hbar) \ominus_h \Phi(\varrho(s)), (\hbar + \nu(s)) \odot \Phi_+^{\nabla^g}(s)\right] \leq \epsilon|\hbar + \nu(s)|$$

or

$$D_H\left[\Phi(\varrho(s)) \ominus_h \Phi(s+\hbar), -(\hbar + \nu(s)) \odot \Phi_+^{\nabla^g}(s)\right] \leq \epsilon|-(\hbar + \nu(s))|,$$

for all $s - \hbar, s + \hbar \in N_{\mathbb{T}[a,b]}, 0 \leq \hbar \leq \delta$. Since s is left dense, $\varrho(s) = s, \nu(s) = 0$, we have

$$D_H\left[\frac{1}{\hbar}[\Phi(s) \ominus_h \Phi(s-\hbar)], \Phi_{-}^{\nabla^g}(s)\right] \leq \epsilon$$

or

$$D_H\left[\frac{-1}{\hbar}[\Phi(s-\hbar) \ominus_h \Phi(s)], \Phi_{-}^{\nabla^g}(s)\right] \leq \epsilon$$

and

$$D_H\left[\frac{1}{\hbar}[\Phi(s+\hbar) \ominus_h \Phi(s)], \Phi_{+}^{\nabla^g}(s)\right] \leq \epsilon$$

or

$$D_H\left[\frac{-1}{\hbar}[\Phi(s) \ominus_h \Phi(s+\hbar)], \Phi_{+}^{\nabla^g}(s)\right] \leq \epsilon,$$

for $s - \hbar, s + \hbar \in N_{\mathbb{T}|a,b|}, 0 \leq \hbar \leq \delta$. Since ϵ is arbitrary, we get any one of (i)–(iv). □

The converse proposition of Theorem 3(a) may not be true. That is a fuzzy function which is continuous may not be differentiable.

Example 1. *Let $\Phi : \mathbb{T}^{[0,4\pi]} \to \mathbb{E}_1$ be a fuzzy function defined as follows:*

$$\Phi(s) = \begin{cases} \sin(s) \odot c, & \text{if } m\pi \leq s \leq (4m+1)\frac{\pi}{4} \\ \cos(s) \odot c, & \text{if } (4m+1)\frac{\pi}{4} \leq s \leq (4m+1)\frac{\pi}{2}, \end{cases}$$

where $m = 0, 1, 2, 3$, $\mathbb{T} = P_{\frac{\pi}{2},\frac{\pi}{2}} = \bigcup_{k=0}^{\infty}\left[k\pi, k\pi + \frac{\pi}{2}\right]$ and $c = (2,4,6)$ is a triangular fuzzy number. Since

$$\mathbb{T} - \lim_{s \to \frac{\pi}{4}^-} \Phi(s) = \sin(\frac{\pi}{4}) \odot c = \frac{1}{\sqrt{2}} \odot c$$

and

$$\mathbb{T} - \lim_{s \to \frac{\pi}{4}^+} \Phi(s) = \cos(\frac{\pi}{4}) \odot c = \frac{1}{\sqrt{2}} \odot c.$$

In addition, $\mathbb{T} - \lim_{s \to \frac{\pi}{4}} \Phi(s)) = \Phi(\frac{\pi}{4}) = \frac{1}{\sqrt{2}} \odot c$. Then, from Remark 6(c), Φ is continuous at $s = \frac{\pi}{4}$ (See Figure 1). Since $s = \frac{\pi}{4}$ is dense, $\sin\frac{\pi}{4} > \sin(\frac{\pi}{4} - h) > 0$, for h sufficiently small, and, from Remark 1(a), we have

$$\Phi_{-}^{\nabla^g}(s) = \lim_{\hbar \to 0} \frac{1}{\hbar} \odot \left(\Phi(\frac{\pi}{4}) \ominus_h \Phi(\frac{\pi}{4} - \hbar)\right)$$

$$= \lim_{\hbar \to 0} \frac{1}{\hbar} \odot \left(\left(\sin\frac{\pi}{4} \odot c\right) \ominus_h \left(\sin(\frac{\pi}{4} - \hbar) \odot c\right)\right)$$

$$= \lim_{\hbar \to 0} \frac{(\sin(\frac{\pi}{4}) - \sin(\frac{\pi}{4} - \hbar))}{\hbar} \odot c$$

$$= \frac{1}{\sqrt{2}} \odot c.$$

In a similarly way,

$$\Phi_+^{\nabla^g}(s) = \lim_{\hbar \to 0} \frac{1}{-\hbar} \odot \left(\Phi(\frac{\pi}{4} \odot c) \ominus_h \Phi(\frac{\pi}{4} + \hbar) \odot c) \right)$$
$$= \lim_{\hbar \to 0} \frac{(\cos(\frac{\pi}{4}) - \cos(\frac{\pi}{4} + \hbar))}{-\hbar} \odot c$$
$$= \frac{-1}{\sqrt{2}} \odot c.$$

Therefore, $\Phi_-^{\nabla^g}(s) \neq \Phi_+^{\nabla^g}(s)$. Hence, Φ is not ∇^g differentiable at $s = \frac{\pi}{4}$.

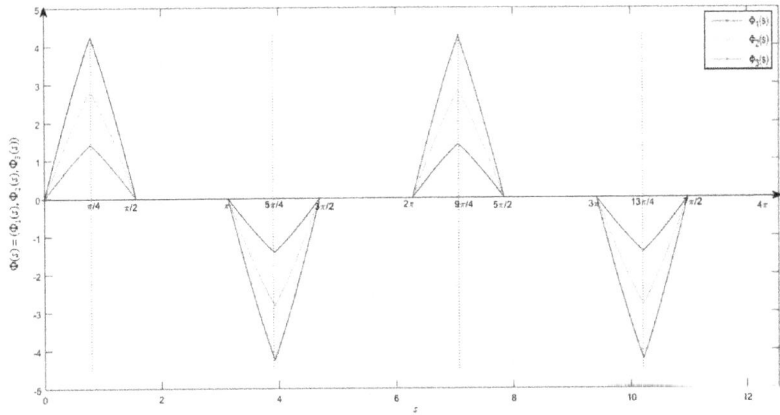

Figure 1. Graphical Representation of $\Phi(s)$ in Example 1.

Definition 11 can equivalently be written as follows:

Remark 7. *If $\Phi : \mathbb{T}^{[a,b]} \to \mathbb{E}_n$ is ∇^g differentiable at $s \in \mathbb{T}_k^{[a,b]}$ if and only if there exists an element $\Phi^{\nabla^g}(s) \in \mathbb{E}_n$, such that any one of the following holds:*

(GH1) for $0 < \hbar < \delta$, provided the Hukuhara difference $\Phi(\varrho(s)) \ominus_h \Phi(s - \hbar)$, $\Phi(s + \hbar) \ominus_h \Phi(\varrho(s))$ and the limits exist

$$\mathbb{T} - \lim_{\hbar \to 0} \frac{1}{\hbar - \nu(s)} \odot (\Phi(\varrho(s)) \ominus_h \Phi(s - \hbar))$$
$$= \mathbb{T} - \lim_{\hbar \to 0} \frac{1}{\hbar + \nu(s)} \odot (\Phi(s + \hbar) \ominus_h \Phi(\varrho(s)))$$
$$= \Phi^{\nabla^g}(s)$$

or

(GH2) for $0 < \hbar < \delta$, provided the Hukuhara difference $\Phi(s - \hbar) \ominus_h \Phi(\varrho(s))$, $\Phi(\varrho(s)) \ominus_h \Phi(s + \hbar)$ and the limits exist

$$\mathbb{T} - \lim_{\hbar \to 0} \frac{-1}{\hbar - \nu(s)} \odot (\Phi(s - \hbar) \ominus_h \Phi(\varrho(s)))$$
$$= \mathbb{T} - \lim_{\hbar \to 0} \frac{-1}{\hbar + \nu(s)} \odot (\Phi(\varrho(s)) \ominus_h \Phi(s + \hbar))$$
$$= \Phi^{\nabla^g}(s)$$

or

(GH3) *for $0 < \hbar < \delta$, provided the Hukuhara difference $\Phi(\varrho(s)) \ominus_h \Phi(s - \hbar)$, $\Phi(\varrho(s)) \ominus_h \Phi(s + \hbar)$ and the limits exist*

$$\mathbb{T} - \lim_{\hbar \to 0} \frac{1}{\hbar - \nu(s)} \odot (\Phi(\varrho(s)) \ominus_h \Phi(s - \hbar))$$
$$= \mathbb{T} - \lim_{\hbar \to 0} \frac{-1}{\hbar + \nu(s)} \odot (\Phi(\varrho(s)) \ominus_h \Phi(s + \hbar))$$
$$= \Phi^{\nabla_g}(s)$$

or

(GH4) *for $0 < \hbar < \delta$, provided the Hukuhara difference $\Phi(s - \hbar) \ominus_h \Phi(\varrho(s))$, $\Phi(s + \hbar) \ominus_h \Phi(\varrho(s))$ and the limits exist*

$$\mathbb{T} - \lim_{\hbar \to 0} \frac{-1}{\hbar - \nu(s)} \odot (\Phi(s - \hbar) \ominus_h \Phi(\varrho(s)))$$
$$= \mathbb{T} - \lim_{\hbar \to 0} \frac{1}{\hbar + \nu(s)} \odot (\Phi(s + \hbar) \ominus_h \Phi(\varrho(s)))$$
$$= \Phi^{\nabla_g}(s).$$

Thus, $\Phi^{\nabla_g} : \mathbb{T}_k^{[a,b]} \to \mathbb{E}_n$ is called the ∇_g derivative of Φ on $\mathbb{T}_k^{[a,b]}$.

Remark 8. *Let $\Phi : \mathbb{T}^{[a,b]} \to \mathbb{E}_n$ be ∇_g differentiable.*

(a) *If Φ is $(GH1)$-nabla differentiable at $s \in \mathbb{T}_k^{[a,b]}$, then there exists a $\delta > 0$, such that, for $0 \leq \lambda \leq 1$, we have*

$$diam[\Phi(s - \hbar)]^\lambda \leq diam[\Phi(\varrho(s))]^\lambda$$
$$\leq diam[\Phi(s + \hbar)]^\lambda, for \quad 0 < \hbar < \delta.$$

Thus, if Φ is $(GH1)$-nabla differentiable on $\mathbb{T}^{[a,b]}$, then $diam[\Phi(s)]^\lambda$ is non-decreasing on $\mathbb{T}^{[a,b]}$.

(b) *If Φ is $(GH2)$-nabla differentiable at $s \in \mathbb{T}_k^{[a,b]}$, then there exists a $\delta > 0$, such that, for $0 \leq \lambda \leq 1$, we have*

$$diam[\Phi(s - \hbar)]^\lambda \geq diam[\Phi(\varrho(s))]^\lambda$$
$$\geq diam[\Phi(s + \hbar)]^\lambda, \quad for \quad 0 < \hbar < \delta.$$

Thus, if Φ is $(GH2)$-nabla differentiable on $\mathbb{T}^{[a,b]}$, then $diam[\Phi(s)]^\lambda$ is non-increasing on $\mathbb{T}^{[a,b]}$.

(c) *If Φ is $(GH3)$-nabla differentiable at $s \in \mathbb{T}_k^{[a,b]}$, then there exists a $\delta > 0$, such that, for $0 \leq \lambda \leq 1$, we have*

$$diam[\Phi(s - \hbar)]^\lambda \leq diam[\Phi(\varrho(s))]^\lambda and$$
$$diam[\Phi(s + \hbar)]^\lambda \leq diam[\Phi(\varrho(s))]^\lambda, \quad for \quad 0 < \hbar < \delta.$$

Therefore, $diam[\Phi(s)]^\lambda$ is non-decreasing in the left neighborhood and non-increasing in the right neighborhood of s. Thus, monotonicity of $diam[\Phi(s)]^\lambda$ fails at s.

(d) *If Φ is $(GH4)$-nabla differentiable at $s \in \mathbb{T}_k^{[a,b]}$, then there exists a $\delta > 0$ such that, for $0 \leq \lambda \leq 1$,*

$$diam[\Phi(\varrho(s))]^\lambda \leq diam[\Phi(s - \hbar)]^\lambda and$$
$$diam[\Phi(\varrho(s))]^\lambda \leq diam[\Phi(s + \hbar)]^\lambda, \quad for \quad 0 < \hbar < \delta.$$

Therefore, $diam[\Phi(s)]^\lambda$ is non-increasing in the left neighborhood and non-decreasing in the right neighborhood of s. Thus, monotonicity of $diam[\Phi(s)]^\lambda$ fails at s.

Example 2. Let $\Phi : \mathbb{T}^{[0,3\pi]} \to \mathbb{E}_1$ be a fuzzy function defined as $\Phi(s) = \sin(s) \odot c$, where $c = (2,4,6)$ is a triangular fuzzy number. Let $\mathbb{T} = P_{\pi,\pi} = \bigcup_{k=0}^{\infty} [2k\pi, (2k+1)\pi]$.

In Figure 2, it is easily seen that $\Phi(s)$ is $(GH1)$-nabla differentiable on $\mathbb{T}^{[0,\frac{\pi}{2})\cup(2\pi,\frac{5\pi}{2})}$, $\Phi(s)$ is $(GH2)$-nabla differentiable on $\mathbb{T}^{(\frac{\pi}{2},\pi]\cup(\frac{5\pi}{2},3\pi]}$. Now, we check the ∇^g differentiability at $s = \frac{\pi}{2}$. Since $s = \frac{\pi}{2}$ is dense, $\nu(s) = 0$. In addition, $\sin(\frac{\pi}{2}) > \sin(\frac{\pi}{2} + \hbar) > 0$, and, from Remark 1(a), we have $(\sin(\frac{\pi}{2}) \odot c) \ominus_h (\sin(\frac{\pi}{2} + \hbar) \odot c) = (\sin(\frac{\pi}{2}) - \sin(\frac{\pi}{2} + \hbar)) \odot c$. Consider

$$\Phi^{\nabla^g}_+\left(\frac{\pi}{2}\right) = \lim_{\hbar \to 0^+} \frac{-1}{\hbar} \odot \left(\Phi\left(\frac{\pi}{2}\right) \ominus_h \Phi\left(\frac{\pi}{2} + \hbar\right)\right)$$
$$= \lim_{\hbar \to 0} \frac{-1}{\hbar} \odot \left(\left(\sin\frac{\pi}{2} \odot c\right) \ominus_h \left(\sin(\frac{\pi}{2} + \hbar) \odot c\right)\right)$$
$$= \lim_{\hbar \to 0} \frac{(\sin(\frac{\pi}{2}) - \sin(\frac{\pi}{2} + \hbar))}{-\hbar} \odot c$$
$$= 0 \odot c = \hat{0}.$$

In a similar way, we get $\Phi^{\nabla^g}_-(\frac{\pi}{2}) = \hat{0}$. Hence, Φ is $(GH3)$-nabla differentiable at $s = \frac{\pi}{2}$. Similarly, we can show that Φ is also $(GH3)$-nabla differentiable at $s = \frac{5\pi}{2}$.

Figure 2. Graphical Representation of $\Phi(s)$ in Example 2.

Theorem 4. *If $\Phi : \mathbb{T}^{[a,b]} \to \mathbb{E}_n$ is continuous at s and s is left scattered, then:*

(a) Φ is ∇^g differentiable at s as in $(GH1)$ or $(GH2)$ with

$$\Phi^{\nabla^g}(s) = \frac{1}{\nu(s)} \odot (\Phi(s) \ominus_h \Phi(\varrho(s)))$$
$$= \frac{-1}{\nu(s)} \odot (\Phi(\varrho(s)) \ominus_h \Phi(s))$$

and $\Phi^{\nabla^g}(s) = \hat{0}$ (or) $\Phi^{\nabla^g}(s) \in \Re^n$;

or

(b) Φ is ∇^g differentiable at s as in $(GH3)$ with $\Phi^{\nabla^g}(s) = \dfrac{-1}{\nu(s)} \odot (\Phi(\varrho(s)) \ominus_h \Phi(s))$;

or

(c) Φ is ∇^g differentiable at s as in $(GH4)$ with $\Phi^{\nabla^g}(s) = \dfrac{1}{\nu(s)} \odot (\Phi(s) \ominus_h \Phi(\varrho(s)))$.

Proof. (**a**) Suppose $s \in \mathbb{T}_k^{[a,b]}$ and Φ is continuous at left scattered point s. Then, from $(GH1)$ or $(GH2)$, we have

$$\mathbb{T} - \lim_{\hbar \to 0} \frac{1}{\hbar - \nu(s)} \odot (\Phi(\varrho(s)) \ominus_h \Phi(s-\hbar)) = \frac{-1}{\nu(s)} \odot (\Phi(\varrho(s)) \ominus_h \Phi(s)),$$

$$\mathbb{T} - \lim_{\hbar \to 0} \frac{1}{\hbar + \nu(s)} \odot (\Phi(s+\hbar) \ominus_h \Phi(\varrho(s))) = \frac{1}{\nu(s)} \odot (\Phi(s) \ominus_h \Phi(\varrho(s))).$$

Since the Hukuhara differences $(\Phi(\varrho(s)) \ominus_h \Phi(s)), (\Phi(s) \ominus \Phi(\varrho(s)))$ exists, then

$$\Phi(\varrho(s)) = \Phi(s) \oplus u(s) \quad \text{and} \quad \Phi(s) = \Phi(\varrho(s)) \oplus v(s),$$

where $u(s), v(s)$ are in $\mathbb{E}_n$. By adding the above equations, we get $u(s) \oplus v(s) = \hat{0}$. Then, $u(s) = \hat{0} = v(s)$ or $u(s), v(s)$ are in $\Re^n$ and hence the result is obvious.

(**b**) Suppose $s \in \mathbb{T}_k^{[a,b]}$ and Φ is continuous at left scattered point s. Then, from (GH3), we have

$$\mathbb{T} - \lim_{\hbar \to 0} \frac{1}{\hbar - \nu(s)} \odot (\Phi(\varrho(s)) \ominus_h \Phi(s-\hbar)) = \frac{-1}{\nu(s)} \odot (\Phi(\varrho(s)) \ominus_h \Phi(s))$$

$$\mathbb{T} - \lim_{\hbar \to 0} \frac{-1}{\hbar + \nu(s)} \odot (\Phi(\varrho(s)) \ominus_h \Phi(s+\hbar)) = \frac{-1}{\nu(s)} \odot (\Phi(\varrho(s)) \ominus_h \Phi(s))$$

Hence, $\Phi^{\nabla^g}(s) = \dfrac{-1}{\nu(s)} \odot (\Phi(\varrho(s)) \ominus_h \Phi(s))$.

(**c**) Suppose $s \in \mathbb{T}_k^{[a,b]}$ and Φ is continuous at left scattered point s. Then, from (GH4), we have

$$\mathbb{T} - \lim_{\hbar \to 0} \frac{-1}{\hbar - \nu(s)} \odot (\Phi(s-\hbar) \ominus_h \Phi(\varrho(s))) = \frac{1}{\nu(s)} \odot (\Phi(s) \ominus_h \Phi(\varrho(s))),$$

$$\mathbb{T} - \lim_{\hbar \to 0} \frac{1}{\hbar + \nu(s)} \odot (\Phi(s+\hbar) \ominus_h \Phi(\varrho(s))) = \frac{1}{\nu(s)} \odot (\Phi(s) \ominus_h \Phi(\varrho(s))).$$

Hence, $\Phi^{\nabla^g}(s) = \dfrac{1}{\nu(s)} \odot (\Phi(s) \ominus_h \Phi(\varrho(s)))$. $\square$

Remark 9. *A fuzzy function $\Phi : \mathbb{T}^{[a,b]} \to \mathbb{E}_1$ is defined as $\Phi(s) = (\phi_1(s), \phi_2(s), \phi_3(s))$, where $\phi_k : \mathbb{T}^{[a,b]} \to \mathbb{R}$, $k = 1, 2, 3$ are nabla differentiable such that $\phi_1(s) < \phi_2(s) < \phi_3(s)$, for all $s \in \mathbb{T}^{[a,b]}$.*

(a) *If Φ is ∇^g differentiable as in (GH1) at ld-point s or ∇^g differentiable as (GH4) at left scattered point s, then $\Phi^{\nabla^g}(s) = (\phi_1^\nabla, \phi_2^\nabla, \phi_3^\nabla)$, for $s \in \mathbb{T}_k^{[a,b]}$.*

(b) *If Φ is ∇^g differentiable as (GH2) at ld-point s or ∇^g differentiable as (GH3) at left scattered point s, then $\Phi^{\nabla^g}(s) = (\phi_3^\nabla, \phi_2^\nabla, \phi_1^\nabla)$, for $s \in \mathbb{T}_k^{[a,b]}$.*

Theorem 5. *Let $\Phi, \Psi : \mathbb{T}^{[a,b]} \to \mathbb{E}_n$ be ∇^g differentiable at $s \in \mathbb{T}_k^{[a,b]}$.*

(1) If Φ and Ψ are both ∇^g differentiable of same kind, then:

  (a) $(\Phi \oplus \Psi) : \mathbb{T}_k^{[a,b]} \to \mathbb{E}_n$ is also ∇^g differentiable of same kind at s with
  $$(\Phi \oplus \Psi)^{\nabla^g}(s) = \Phi^{\nabla^g}(s) \oplus \Psi^{\nabla^g}(s).$$

  (b) $(\Phi \ominus_h \Psi) : \mathbb{T}_k^{[a,b]} \to \mathbb{E}_n$ also ∇^g differentiable of same kind at s, provided $(\Phi \ominus_h \Psi)$ exists and
  $$(\Phi \ominus_h \Psi)^{\nabla^g}(s) = \Phi^{\nabla^g}(s) \ominus_h \Psi^{\nabla^g}(s).$$

(2) If Φ and Ψ are different kinds of ∇^g differentiable at s, and $(\Phi \ominus_h \Psi)$ exists for $s \in \mathbb{T}_k^{[a,b]}$, then $(\Phi \ominus_h \Psi)$ is ∇^g differentiable at s with $(\Phi \ominus_h \Psi)^{\nabla^g}(s) = \Phi^{\nabla^g}(s) \oplus (-1) \odot \Psi^{\nabla^g}(s)$.

Proof. If s is ld-point, then $\varrho(s) = s$, $\nu(s) = 0$. The proof of this theorem is similar to the proof of Lemma 4 and Theorem 4 in [17].

1(a). Suppose that Φ and Ψ are both $(GH3)$-nabla differentiable at left scattered point $s \in \mathbb{T}_k^{[a,b]}$. Then, $\Phi(\varrho(s)) \ominus_h \Phi(s)$ exists with $\Phi(\varrho(s)) = \Phi(s) \oplus u(s)$ and $\Psi(\varrho(s)) \ominus_h \Psi(s)$ exists with $\Psi(\varrho(s)) = \Psi(s) \oplus v(s)$. Now,
$$(\Phi(\varrho(s)) \ominus_h \Phi(s)) \oplus (\Psi(\varrho(s)) \ominus_h \Psi(s)) = u(s) \oplus v(s).$$
Multiplying the above equation with $\dfrac{-1}{\nu(s)}$, we get
$$\dfrac{-1}{\nu(s)} \odot ((\Phi(\varrho(s)) \oplus \Psi(\varrho(s))) \ominus_h (\Phi(s) \oplus \Psi(s)))$$
$$= \dfrac{-1}{\nu(s)} \odot (u(s) \oplus v(s)),$$
and it follows that
$$\dfrac{(\Phi \oplus \Psi)(\varrho(s)) \ominus_h (\Phi \oplus \Psi)(s)}{-\nu(s)} = \dfrac{u(s)}{-\nu(s)} \oplus \dfrac{v(s)}{-\nu(s)}.$$
Hence, $(\Phi \oplus \Psi)$ is ∇^g differentiable as in $(GH3)$ with
$$(\Phi \oplus \Psi)^{\nabla^g}(s) = \Phi^{\nabla^g}(s) \oplus \Psi^{\nabla^g}(s).$$
The case when Φ and Ψ are ∇^g differentiable as in $(GH4)$ is similar to the previous one.

1(b). Suppose Φ and Ψ are both $(GH3)$-nabla differentiable at left scattered points $s \in \mathbb{T}_k^{[a,b]}$, similar to 1(a), we have $\Phi(\varrho(s)) = \Phi(s) \oplus u(s)$ and $\Psi(\varrho(s)) = \Psi(s) \oplus v(s)$. Consider
$$(\Phi \ominus_h \Psi)(\varrho(s)) = \Phi(\varrho(s)) \ominus_h \Psi(\varrho(s))$$
$$= (\Phi(s) \oplus u(s)) \ominus_h (\Psi(s) \oplus v(s))$$
$$= (\Phi(s) \ominus_h \Psi(s)) \oplus (u(s) \ominus_h v(s)).$$
It implies that
$$(\Phi \ominus_h \Psi)(\varrho(s)) \ominus_h (\Phi \ominus_h \Psi)(s) = u(s) \ominus_h v(s).$$
Multiplying the above equation with $\dfrac{-1}{\nu(s)}$, we get the desired result. In a similar way, we can easily prove the other case.

(2). Suppose that Φ is ∇^g differentiable as in $(GH3)$ and Ψ is ∇^g differentiable as in $(GH4)$ at left scattered points $s \in \mathbb{T}_k^{[a,b]}$, then the Hukuhara difference $\Phi(\varrho(s)) \ominus_h \Phi(s)$ exists with $\Phi(\varrho(s)) =$

$\Phi(s) \oplus u(s)$ and $\Psi(s) \ominus_h \Psi(\varrho(s))$ exists with $\Psi(s) = \Psi(\varrho(s)) \oplus v(s)$. Now, by adding these equations, we get

$$\Phi(\varrho(s)) \oplus \Psi(s) = \Phi(s) \oplus u(s) \oplus \Psi(\varrho(s)) \oplus v(s).$$

Since the Hukuhara difference of $\Phi(\varrho(s)) \ominus_h \Psi(\varrho(s))$ and $\Phi(s) \ominus_h \Psi(s)$ exist, we have

$$(\Phi(\varrho(s)) \ominus_h \Psi(\varrho(s))) \ominus_h (\Phi(s) \ominus_h \Psi(s)) = u(s) \oplus v(s). \tag{5}$$

Now, by multiplying (5) with $\dfrac{1}{-\nu(s)}$, we get $\Phi \ominus \Psi$ is $(GH3)$-nabla differentiable.

In a similar way, if Φ is ∇^g differentiable as in $(GH4)$ and Ψ is ∇^g differentiable as in $(GH3)$ at left scattered points $s \in \mathbb{T}_k^{[a,b]}$, then we can easily prove that

$$(\Phi(s) \ominus_h \Psi(s)) \ominus_h (\Phi(\varrho(s)) \ominus_h \Psi(\varrho(s))) = \tilde{u}(s) + \tilde{v}(s). \tag{6}$$

Now, by multiplying (6) with $\dfrac{1}{\nu(s)}$, we get $\Phi \ominus \Psi$ is $(GH4)$-nabla differentiable. Therefore,

$$(\Phi \ominus_h \Psi)^{\nabla^g}(s) = \Phi^{\nabla^g}(s) \oplus (-1) \odot \Psi^{\nabla^g}(s).$$

□

The following example illustrates the feasibility of Theorem 5.

Example 3. Let $\Omega, \Psi : \mathbb{T}^{[0,3\pi]} \to \mathbb{E}_1$ be fuzzy functions defined as follows:

$$\Omega(s) = \begin{cases} (\frac{\pi}{2} - s) \odot c, & 0 \leq s \leq \pi \\ (s - \frac{5\pi}{2}) \odot c, & 2\pi \leq s \leq 3\pi \end{cases}$$

and

$$\Psi(s) = \begin{cases} \cos(s) \odot c, & 0 \leq s \leq \pi \\ -\cos(s) \odot c, & 2\pi \leq s \leq 3\pi \end{cases}$$

where $\mathbb{T} = P_{\pi,\pi}$, $c = (2,4,6)$ is a triangular fuzzy number.

IN Figures 3 and 4, it is easily seen that Ω and Ψ are $(GH2)$-nabla differentiable on $\mathbb{T}^{[0,\frac{\pi}{2})\cup(2\pi,\frac{5\pi}{2})}$, $(GH1)$-nabla differentiable on $\mathbb{T}^{(\frac{\pi}{2},\pi]\cup(\frac{5\pi}{2},3\pi]}$, and $(GH4)$-nabla differentiable at $s = \frac{\pi}{2}, \frac{5\pi}{2}$. Thus, $\Omega \oplus \Psi$, $\Omega \ominus_h \Psi$ are ∇^g differentiable at left scattered point $s = 2\pi$. Now, from Remark 1, we have

$$(\Omega \oplus \Psi)(s) = \begin{cases} (\frac{\pi}{2} - s + \cos(s)) \odot c, & s \in [0,\pi] \\ (s - \frac{5\pi}{2} - \cos(s)) \odot c, & s \in [2\pi, 3\pi]. \end{cases}$$

and

$$(\Omega \ominus_h \Psi)(s) = \begin{cases} (\frac{\pi}{2} - s - \cos(s)) \odot c, & \in [0,\pi] \\ (s - \frac{5\pi}{2} + \cos(s)) \odot c, & \in [2\pi, 3\pi]. \end{cases}$$

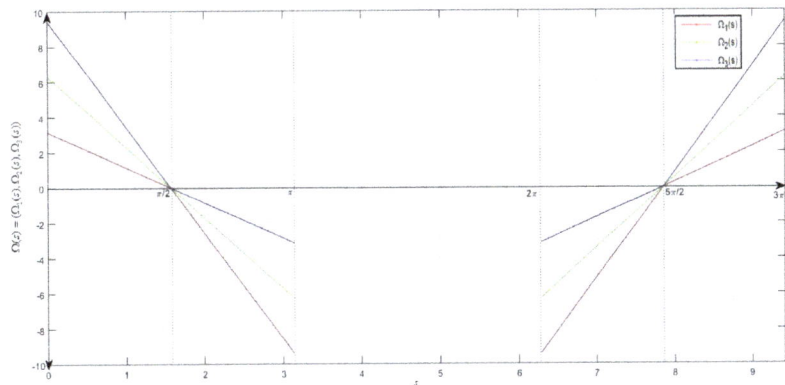

Figure 3. Graphical Representation of $\Omega(s)$ in Example 3.

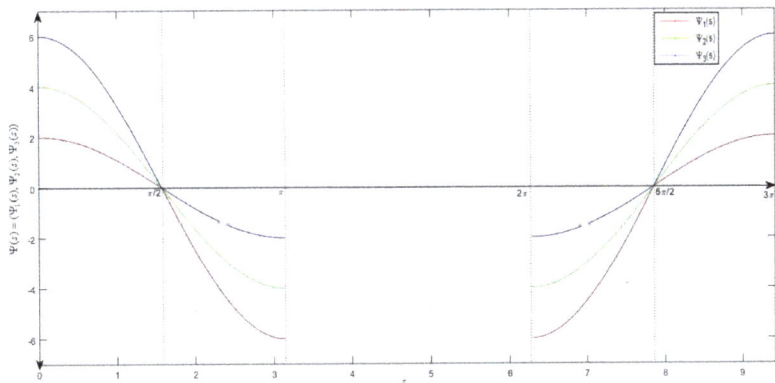

Figure 4. Graphical Representation of $\Psi(s)$ in Example 3.

In Figure 5, $(\Omega \oplus \Psi)$ is (GH2)-nabla differentiable on $\mathbb{T}^{[0,\frac{\pi}{2})\cup[2\pi,\frac{5\pi}{2})}$, (GH1)-nabla differentiable on $\mathbb{T}^{(\frac{\pi}{2},\pi]\cup(\frac{5\pi}{2},3\pi]}$. At $s = \frac{\pi}{2}$, Ω and Ψ are (GH4)-nabla differentiable with $\Omega^{\nabla g}(\frac{\pi}{2}) = (-1) \odot c$, and $\Psi^{\nabla g}(\frac{\pi}{2}) = (-1) \odot c$. Now,

$$(\Omega \oplus \Psi)^{\nabla g}_+ \left(\frac{\pi}{2}\right) = \lim_{\hbar \to 0} \frac{1}{\hbar} \left(\frac{\pi}{2} - \left(\frac{\pi}{2} + \hbar\right) + \cos\left(\frac{\pi}{2} + \hbar\right)\right) \odot c \ominus_h \left(\frac{\pi}{2} - \left(\frac{\pi}{2}\right) + \cos\left(\frac{\pi}{2}\right)\right) \odot c$$
$$= \left(\lim_{\hbar \to 0} \frac{-\hbar + \cos(\frac{\pi}{2} + \hbar)}{\hbar}\right) \odot c$$
$$= \left(-1 + (-1) \lim_{\hbar \to 0} \frac{\sin \hbar}{\hbar}\right) \odot c = -2 \odot c.$$

Similarly, we can show that $(\Omega \oplus \Psi)^{\nabla g}_-(\frac{\pi}{2}) = -2 \odot c$. Thus, $(\Omega \oplus \Psi)$ is (GH4)-nabla differentiable at $\frac{\pi}{2}$ and Theorem 5 1(a) is verified.

In Figure 6, it is easily seen that $(\Omega \ominus_h \Psi)$ is $(GH2)$-nabla differentiable on $\mathbb{T}^{[0,\frac{\pi}{2})\cup[2\pi,\frac{5\pi}{2})}$ and $(GH1)$-nabla differentiable on $\mathbb{T}^{(\frac{\pi}{2},\pi]\cup(\frac{5\pi}{2},3\pi]}$. Again, from Remark 1, we have

$$(\Omega \ominus_h \Psi)^{\nabla^g_+}(\frac{\pi}{2}) = \lim_{\hbar \to 0} \frac{1}{\hbar}\left(\frac{\pi}{2} - (\frac{\pi}{2}+\hbar) - \cos(\frac{\pi}{2}+\hbar)\right) \odot c \ominus_h \left(\frac{\pi}{2} - \frac{\pi}{2} - \cos(\frac{\pi}{2})\right) \odot c$$
$$= \left(\lim_{\hbar \to 0} \frac{-\hbar - \cos(\frac{\pi}{2}+\hbar)}{\hbar}\right) \odot c$$
$$= (-1 + \lim_{\hbar \to 0} \frac{\sin \hbar}{\hbar}) \odot c = 0 \odot c = \hat{0}.$$

Similarly, we can show that $(\Omega \ominus \Psi)^{\nabla^g_-}(\frac{\pi}{2}) = \hat{0}$. Thus, $(\Omega \ominus \Psi)$ is $(GH4)$-nabla differentiable at $\frac{\pi}{2}$ and Theorem 5 1(b) is verified.

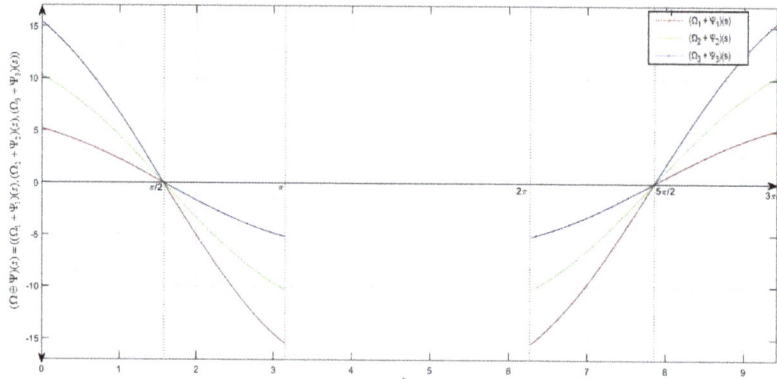

Figure 5. Graphical Representation of $(\Omega \oplus \Psi)(s)$ in Example 3.

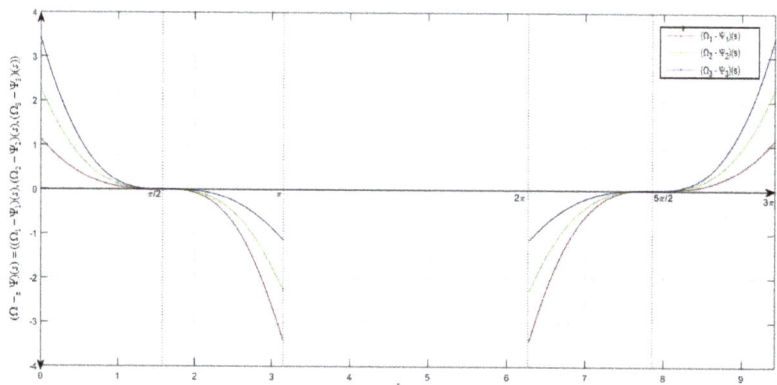

Figure 6. Graphical Representation of $(\Omega \ominus_h \Psi)(s)$ in Example 3.

Consider $\Phi(s)$ as in Example 2, Φ is (GH3)-nabla differentiability at $s = \frac{\pi}{2}$ and Ψ is (GH4)-nabla differentiability at $s = \frac{\pi}{2}$. Hence, Φ and Ψ are different kinds of ∇^g differentiable at $s = \frac{\pi}{2}$, and $(\Phi \ominus_h \Psi)$ exists at $s = \frac{\pi}{2}$. Now, from Theorem 5(2), we have

$$(\Phi \ominus_h \Psi)^{\nabla^g}_{-}(\frac{\pi}{2}) = \lim_{\hbar \to 0} \frac{1}{\hbar} \odot (\sin(\frac{\pi}{2}) - \cos(\frac{\pi}{2}) \odot c) \ominus_h ((\sin(\frac{\pi}{2} - \hbar) - \cos(\frac{\pi}{2} - \hbar) \odot c)$$
$$= \left(\lim_{\hbar \to 0} \frac{1 - \cos \hbar}{\hbar} + \lim_{\hbar \to 0} \frac{\sin \hbar}{\hbar} \right) \odot c = c.$$

Similarly, we can show that $(\Phi \ominus_h \Psi)^{\nabla^g}_{+}(\frac{\pi}{2}) = c$. Hence, Theorem 5(2) is verified.

Now, we check the ∇^g-differentiable at $s = 2\pi$. It is left scattered and $\varrho(2\pi) = \pi$, $\nu(2\pi) = \pi$. Clearly, Ω, Φ, and Ψ are (GH3)- and (GH4)-nabla differentiable at $s = 2\pi$. We get $\Omega^{\nabla^g}(2\pi) = \hat{0}$, $\Phi^{\nabla^g}(2\pi) = \hat{0}$ and $\Psi^{\nabla^g}(2\pi) = \hat{0}$. In addition, the results of Theorem 5 hold at left scattered point $s = 2\pi$.

4. Integration of Fuzzy Functions on Time Scales

In this section, we prove fundamental theorem of nabla integral calculus for fuzzy functions on time scales under generalized fuzzy nabla differentiable functions on time scales.

First, we prove an embedding theorem on $\mathbb{E}_n$ and obtain some results which are useful to prove the main theorem. To prove the these results, we make use of Definitions 1–3 and Theorem 4 in [31].

Let $\mathcal{C}[0,1]$ be the set of all functions $\mathcal{F} : [0,1] \to \Re^n$, $\mathcal{F}$ is bounded on $[0,1]$, left-continuous for each $x \in (0,1]$, right-continuous on 0, and $\mathcal{F}$ has right limit for each $x \in [0,1)$. Endowed with the norm $||\mathcal{F}||_\mathcal{C} = \sup \{|\mathcal{F}(\lambda)|_{\Re^n}; x \in [0,1]\}$, $\mathcal{C}[0,1]$ is a Banach space. It is known that the following result which embeds $\mathbb{E}_n$ into $X = \mathcal{C}[0,1] \times \mathcal{C}[0,1]$ isometrically and isomorphically.

Theorem 6. *If we define $i : \mathbb{E}_n \to X$ by $i(u) = (u_-, u_+)$, where $u_-, u_+ : [0,1] \to \Re^n$, $u_-(\lambda) = u_-^\lambda$, $u_+(\lambda) = u_+^\lambda$, then $i(\mathbb{E}_n)$ is a closed convex cone with vertex 0 in X (here X is a Banach space with the norm $||(f,g)|| = \max(||f||_\mathcal{C}, ||g||_\mathcal{C})$).*

Proof. First, we show that $X = \mathcal{C}[0,1] \times \mathcal{C}[0,1]$ is a Banach space. Consider a cauchy sequence $l_{n_0} = (f_{n_0}, g_{n_0})$ and for $\epsilon^* > 0$, there exists $N > 0$, $n_0 > N$ such that $n_0, m_0 > N$ implies $||l_{m_0} - l_{n_0}|| < \epsilon^*$, that is

$$||l_{m_0} - l_{n_0}|| = ||(f_{m_0}, g_{m_0}) - (f_{n_0}, g_{n_0})||$$
$$= ||(f_{m_0} - f_{n_0}, g_{m_0} - g_{n_0})||$$
$$= \max(||f||_\mathcal{C}, ||g||_\mathcal{C}).$$

which yields the result that $f_{n_0}(\lambda) \to f$ and $g_{n_0}(\lambda) \to g$ as $n_0 \to \infty$ where $||\mathcal{F}||_\mathcal{C} = \sup\{|\mathcal{F}(x)|; x \in [0,1]\}$, $\mathcal{C}[0,1]$ is a Banach space. Hence, $X = \mathcal{C}[0,1] \times \mathcal{C}[0,1]$ is a Banach space. To obtain i embeds $\mathbb{E}_n$ into $X = \mathcal{C}[0,1] \times \mathcal{C}[0,1]$ isometrically and isomorphically, we need to prove the following:

(a) $i(p \odot u \oplus q \odot v) = pi(u) + qi(v)$, for any $u, v \in \mathbb{E}_n$ and $p, q \geq 0$; and
(b) $D_H(u,v) = ||i(u) - i(v)||$.

Let $i(u) = (u_-, u_+)$. The λ-level set of $u \in \mathbb{E}_n$ can be written as

$$[u]^\lambda = \beta u_-^\lambda + (1-\beta) u_+^\lambda \quad \text{for all} \quad 0 \leq \beta \leq 1.$$

Now,

$$\begin{aligned}[p \odot u \oplus q \odot v]^\lambda &= p[u]^\lambda + q[v]^\lambda \\ &= p[\beta u_-^\lambda + (1-\beta)u_+^\lambda] \\ &\quad + q[\beta v_-^\lambda + (1-\beta)v_+^\lambda] \\ &= \beta\left(u_-^\lambda + v_-^\lambda\right) + (1-\beta)\left(u_+^\lambda + v_+^\lambda\right).\end{aligned}$$

Therefore,

$$\begin{aligned}i(p \odot u \oplus q \odot v) &= \left(pu_-^\lambda + qv_-^\lambda, pu_+^\lambda + qv_+^\lambda\right) \\ &= p(u_-^\lambda, u_+^\lambda) + q(v_-^\lambda, v_+^\lambda) \\ &= pi(u) + qi(v).\end{aligned}$$

Thus, (a) is proved.

Now, consider

$$\begin{aligned}\|i(u) - i(v)\| &= \|(u_-, u_+) - (v_-, v_+)\| \\ &= \|(u_- - v_-), (u_+ - v_+)\| \\ &= \max\{\|u_- - v_-\|_C, \|u_+ - v_+\|_C\} \\ &= \max\{\sup_\lambda \|u_-^\lambda - v_-^\lambda\|_{\mathbb{R}_n}, \sup_\lambda \|u_+^\lambda - v_+^\lambda\|_{\mathbb{R}_n}\} \\ &= \sup_\lambda \{\max\{\|u_-^\lambda - v_-^\lambda\|_{\mathbb{R}_n}, \|u_+^\lambda - v_+^\lambda\|_{\mathbb{R}_n}\}\}, \\ &= \sup_\lambda d_H([u]^\lambda, [v]^\lambda) \\ &= D_H(u, v).\end{aligned}$$

□

We make use the Proposition 3.1 and Remark 3.4 in [18] to prove the following results.

Theorem 7. *Suppose* $\Phi : \mathbb{T}^{[a,b]} \to \mathbb{E}_n$ *is* ∇^g *left-differentiable at* s_0*; then,* $(i \circ \Phi)(s) = i(\Phi(s))$ *is nabla-differentiable at* $s_0 \in \mathbb{T}^{[a,b]}$*. Moreover,*

(a) *If there exists a* $\delta > 0 \ni (\Phi(s_0 - \hbar) \ominus_h \Phi(\varrho(s_0)))$ *exists for* $0 < \hbar < \delta$*, then* $(i \circ \Phi)^\nabla(s_0) = -i^*(\Phi_-^{\nabla^g}(s_0))$.

(b) *If there exists a* $\delta > 0 \ni (\Phi(\varrho(s_0)) \ominus_h \Phi(s_0 - \hbar))$ *exists for* $0 < \hbar < \delta$*, then* $(i \circ \Phi)^\nabla(s_0) = i^*(\Phi_-^{\nabla^g}(s_0))$.

Proof. Let Φ be ∇^g left-differentiable at $s_0 \in \mathbb{T}^{[a,b]}$.

(a) If there exists a $\delta > 0$ such that $\Phi(s_0 - \hbar) \ominus_h \Phi(\varrho(s_0))$ exists for $0 < \hbar < \delta$, then

$$\left\| \frac{-1}{(\hbar - \nu(s_0))} [(i \circ \Phi)(s_0 - \hbar) - (i \circ \Phi)(\varrho(s_0))] - [-i^*(\Phi_-^{\nabla^g}(s_0))] \right\|$$

$$= \left\| \frac{-1}{(\hbar - \nu(s_0))} [(i \circ \Phi)(s_0 - \hbar) - (i \circ \Phi)(\varrho(s_0))] + [i^*(\Phi_-^{\nabla^g}(s_0))] \right\|$$

$$\leq \left\| \frac{1}{(\hbar - \nu(s_0))} [i(\Phi(\varrho(s_0)) - \Phi(s_0 - \hbar))] \right.$$

$$\left. + i^* \left[\frac{1}{(\hbar - \nu(s_0))} \odot [\Phi(\varrho(s_0)) \ominus_h \Phi(s_0 - \hbar)] \right] \right\|$$

$$+ \left\| -i^* \left[(\frac{1}{\hbar - \nu(s_0)}) \odot [\Phi(\varrho(s_0)) \ominus_h \Phi(s_0 - \hbar)] \right] i^*(\Phi_-^{\nabla^g}(s_0))] \right\|.$$

From Remark 3.4.1 in [18], we have

$$\left\| i^* \left[\frac{1}{(\hbar - \nu(s_0))} \odot (\Phi(\varrho(s_0)) \ominus_h \Phi(s_0 - \hbar)) \right] - i^*(\Phi_-^{\nabla^g}(s_0))] \right\|$$

$$= D_H \left[\frac{1}{(\hbar - \nu(s_0))} \odot (\Phi(\varrho(s_0)) \ominus_h \Phi(s_0 - \hbar)), \Phi_-^{\nabla^g}(s_0) \right] \to 0, \text{ as } \hbar \to 0.$$

Consider

$$i^* \left[\frac{1}{(h - \nu(s_0))} \odot (\Phi(\varrho(s_0)) \ominus_h \Phi(s_0 - \hbar)) \right] = -i \left[\frac{1}{(h - \nu(s_0))} \odot (\Phi(\varrho(s_0)) \ominus_h \Phi(s_0 - \hbar)) \right]$$

$$= \frac{-1}{(h - \nu(s_0))} [i(\Phi(\varrho(s_0)) - \Phi(s_0 - \hbar))],$$

we have

$$\left\| \frac{(i \circ \Phi)(s_0 - \hbar) - (i \circ \Phi)(\varrho(s_0))}{-(\hbar - \nu(s_0))} - [-i^*(\Phi_-^{\nabla^g}(s_0))] \right\| \to 0, \text{ as } \hbar \to 0.$$

Thus, $(i \circ \Phi)^\nabla(s_0) = -i^*(\Phi_-^{\nabla^g}(s_0))$.
Similarly, we can prove (b). □

Theorem 8. Suppose $\Phi : \mathbb{T}^{[a,b]} \to \mathbb{E}_n$ is ∇^g right-differentiable s_0; then, $(i \circ \Phi)(s) = i(\Phi(s))$ is nabla-differentiable at $s_0 \in \mathbb{T}_k^{[a,b]}$. Moreover,

(a) If there exists a $\delta > 0 \ni (\Phi(s_0 + \hbar) \ominus_h \Phi(\varrho(s_0)))$ exists for $0 < \hbar < \delta$, then

$$(i \circ \Phi)^\nabla(s_0) = i(\Phi_+^{\nabla^g}(s_0)).$$

(b) If there exists a $\delta > 0 \ni (\Phi(\varrho(s_0)) \ominus_h \Phi((s_0 + \hbar)))$ exists for $0 < \hbar < \delta$, then

$$(i \circ \Phi)^\nabla(s_0) = -i^*(\Phi_+^{\nabla^g}(s_0)).$$

Proof. The proof of this theorem is similar to that of Theorem 7. □

Theorem 9. If $\Phi : \mathbb{T}^{[a,b]} \to \mathbb{E}_n$ is ∇^g differentiable at s, then $i \circ \Phi(s)$ is nabla-differentiable and $(i \circ \Phi)^\nabla(s) \in i(\mathbb{E}_n)$. In this case, either $(i \circ \Phi)^\nabla(s) = i(\Phi^{\nabla^g}(s))$ or $(i \circ \Phi)^\nabla(s) = -i^*(\Phi^{\nabla^g}(s)), s \in \mathbb{T}_k^{[a,b]}$.

Proof. Let $\Phi : \mathbb{T}^{[a,b]} \to \mathbb{E}_n$ be ∇^g differentiable at $s \in \mathbb{T}_k^{[a,b]}$ and s is left dense; then, the proof is similar to the proof of Theorem 8 [16]. Now, for s being left scattered, we have

$$\frac{1}{\nu(s)}[i(\Phi(s)) - \Phi(\varrho(s))] = \begin{cases} \dfrac{1}{\nu(s)}[i \circ \Phi(s) - i \circ \Phi(\varrho(s))] & \text{or} \\ \dfrac{-1}{\nu(s)}[i \circ \Phi(\varrho(s)) - i \circ \Phi(s)]. \end{cases}$$

Consider

$$\left\| \frac{1}{\nu(s)}[(i \circ \Phi)(s) - (i \circ \Phi)(\varrho(s))] - i(\Phi^{\nabla^g}(s)) \right\|$$

$$= \left\| \frac{-1}{\nu(s)}[(i \circ \Phi)(\varrho(s)) - (i \circ \Phi)(s)] - i(\Phi^{\nabla^g}(s)) \right\|$$

$$= \left\| i\left(\frac{1}{\nu(s)} \odot [\Phi(s) \ominus_h \Phi(\varrho(s))]\right) - i(\Phi^{\nabla^g}(s)) \right\|$$

$$= D_H\left[\frac{1}{\nu(s)} \odot [\Phi(s) \ominus_h \Phi(\varrho(s))], \Phi^{\nabla^g}(s)\right].$$

Then, $(i \circ \Phi)^{\nabla}(s) = i(\Phi^{\nabla^g}(s))$.
Again, in the same way,

$$\left\| \frac{1}{\nu(s)}[(i \circ \Phi)(s) - (i \circ \Phi)(\varrho(s))] - [-i^*(\Phi^{\nabla^g}(s))] \right\|$$

$$= \left\| \frac{-1}{\nu(s)}[(i \circ \Phi)(\varrho(s)) - (i \circ \Phi)(s)] + [i^*(\Phi^{\nabla^g}(s))] \right\|$$

$$\leq \left\| \frac{-1}{\nu(s)}[(i \circ \Phi)(\varrho(s)) - (i \circ \Phi)(s)] \right.$$
$$\left. + i^*\left[\frac{1}{\nu(s)} \odot [\Phi(s) \ominus_h \Phi(\varrho(s))]\right] \right\|$$
$$+ \left\| i^*\left[\frac{1}{\nu(s)} \odot [\Phi(s) \ominus_h \Phi(\varrho(s))]\right] - i^*(\Phi^{\nabla^g}(s)) \right\|.$$

However,

$$\left\| i^*\left(\frac{1}{\nu(s)} \odot [\Phi(s) \ominus_h \Phi(\varrho(s))]\right) - i^*(\Phi^{\nabla^g}(s)) \right\|$$

$$= D_H\left(\frac{1}{\nu(s)} \odot (\Phi(s) \ominus_h \Phi(\varrho(s))), \Phi^{\nabla^g}(s)\right) = 0.$$

Since $i((-1) \odot \tilde{u}) = i^*(\tilde{u})$, we have

$$\left\| \frac{-1}{\nu(s)}[(i \circ \Phi)(\varrho(s)) - (i \circ \Phi)(s)] - i^*\left[\frac{\Phi(s) \ominus_h \Phi(\varrho(s))}{-\nu(s)}\right] \right\|$$

$$= \left\| \frac{-1}{\nu(s)}[(i \circ \Phi)(\varrho(s)) - (i \circ \Phi)(s)] - [-i^*(\Phi^{\nabla^g}(s))] \right\| = 0.$$

Thus, $\left\| \dfrac{(i \circ \Phi)(s) - (i \circ \Phi)(\varrho(s))}{\nu(s)} - [-i^*(\Phi^{\nabla^g}(s))] \right\| = 0$. Therefore, $(i \circ \Phi)^{\nabla}(s) = -i^*(\Phi^{\nabla^g}(s))$.
Finally, $(i \circ \Phi)^{\nabla}(s) = i(\Phi^{\nabla^g}(s)) = -i^*(\Phi^{\nabla^g}(s))$. □

From Remark 8, it is clear that, the fuzzy function $\Phi(s)$ is $(GH3)$- or $(GH4)$-nabla differentiable at discrete points. For example, if $\Phi(s)$ is ∇^g-differentiable on $\mathbb{T}^{[a,b]}$, $a < c < d < b$ and Φ is only $(GH3)$-nabla differentiable at $s = c$, $(GH4)$-nabla differentiable at $s = d$, then Φ is

$(GH1)$-nabla differentiable on $\mathbb{T}^{[a,c)\cup(d,b]}$ and $(GH2)$-nabla differentiable on $\mathbb{T}^{(c,d)}$. Therefore, if $\Phi(s)$ is ∇^g-differentiable on $\mathbb{T}^{[a,b]}$, then it is possible to partition the $\mathbb{T}^{[a,b]}$ into sub-intervals such that in each sub-interval $\Phi(s)$ is either $(GH1)$- or $(GH2)$-nabla differentiable.

Now, we prove the main theorem of this section fundamental theorem of nabla integral calculus of fuzzy functions on time scales.

Theorem 10. *Let $\Phi : \mathbb{T}^{[a,b]} \to \mathbb{E}_n$ and $a = a_0 < a_1 < a_2 < \ldots < a_k = b$ be a division of the interval $[a,b]$ such that Φ is $(GH1)$ or $(GH2)$-nabla differentiable on each of the interval $\mathbb{T}^{[a_{m-1},a_m]}$, $m = 1, 2, \ldots, k$ with same kind of differentiability on each sub-interval. Then,*

$$\int_a^b \Phi^{\nabla^g}(\tau)\nabla\tau = \sum_{m\in M}(\Phi(a_m) \ominus_h \Phi(a_{m-1})) \oplus (-1) \odot \sum_{n\in N}(\Phi(a_{n-1}) \ominus_h \Phi(a_n)),$$

where $M = \{m \in \{1,2,\ldots,k\}$ such that Φ is $(GH1)$-nabla differentiable on $\mathbb{T}^{(a_{m-1},a_m)}$ $\}$ and $N = \{n \in \{1,2,\ldots,k\}$ such that Φ is $(GH2)$-nabla differentiable on $\mathbb{T}^{(a_{n-1},a_n)}$ $\}$

Proof. Let $\Phi : \mathbb{T}^{[a,b]} \to \mathbb{E}_n$ is ∇^g differentiable on $\mathbb{T}_k^{[a,b]}$. Suppose Φ is $(GH1)$-nabla differentiable on (a_{i-1}, a_i). Then, for $m \in M$, we have

$$\int_{a_{m-1}}^{a_m} \Phi^{\nabla^g}(\tau)\nabla\tau = \Phi(a_m) \ominus_h \Phi(a_{m-1}) \text{ for all } m \in M. \tag{7}$$

Let $n \in N$; using Cauchy formula for functions with values in Banach space, we have

$$(i \circ \Phi)(a_n) = (i \circ \Phi)(a_{n-1}) + \int_{a_{n-1}}^{a_n} (i \circ \Phi)^{\nabla^g}(\tau)\nabla\tau.$$

By Theorem 9, there exists $(i \circ \Phi)^{\nabla}(s)$ and we get $(i \circ \Phi)(a_n) = (i \circ \Phi)(a_{n-1}) + \int_{a_{n-1}}^{a_n}(-i^*(\Phi^{\nabla^g})(\tau)\nabla\tau$.

Since the embedding i commutes with the integral, we obtain

$$(i \circ \Phi)(a_n) = (i \circ \Phi)(a_{n-1}) - i^*\left(\int_{a_{n-1}}^{a_n} \Phi^{\nabla^g}(\tau)\nabla\tau\right).$$

Then, it follows that

$$i^*\left(\int_{a_{n-1}}^{a_n} \Phi^{\nabla^g}(s)\nabla s\right) + (i \circ \Phi)(a_n) = (i \circ \Phi)(a_{n-1}).$$

By the definition of $i*$, we obtain

$$i\left((-1) \odot \int_{a_{n-1}}^{a_n} \Phi^{\nabla^g}(\tau)\nabla\tau\right) + i(\Phi(a_n)) = i(\Phi)(a_{n-1}).$$

By the additive property of the embedding i, we have

$$(-1) \odot \int_{a_{n-1}}^{a_n} \Phi^{\nabla^g}(\tau)\nabla\tau = \Phi(a_{n-1}) \ominus_h \Phi(a_n).$$

Finally,

$$\int_{a_{n-1}}^{a_n} \Phi^{\nabla^g}(\tau)\nabla\tau = (-1) \odot \Phi(a_{n-1}) \ominus_h \Phi(a_n), \tag{8}$$

for all $n \in N$. Adding Equations (7) and (8), we get the desired result

$$\int_a^b \Phi^{\nabla^g}(\tau)\nabla\tau = \sum_{m \in M}(\Phi(a_m) \ominus_h \Phi(a_{m-1})) \oplus (-1) \odot \sum_{n \in N}(\Phi(a_{n-1}) \ominus_h \Phi(a_n)).$$

□

Example 4. *Consider $\Phi(s)$ as in Example 2. We partition $[0, 3\pi]$ as $a_0 = 0 < a_1 = \frac{\pi}{2} < a_2 = \pi < a_3 = 2\pi < a_4 = \frac{5\pi}{2} < a_5 = 3\pi$ such that $\Phi(s)$ is (GH1)-nabla differentiable on $\mathbb{T}^{[a_{m-1}, a_m]}$, $m \in M = \{1, 3\}$ and (GH2)-nabla differentiable on $\mathbb{T}^{[a_{n-1}, a_n]}$, $n \in N = \{2, 5\}$. Thus, from Theorem 10, we have*

$$\int_a^b \Phi^{\nabla^g}(\tau)\nabla\tau = \sum_{m \in M}(\Phi(a_m) \ominus_h \Phi(a_{m-1})) \oplus (-1) \odot \sum_{n \in N}(\Phi(a_{n-1}) \ominus_h \Phi(a_n))$$

$$= (\Phi(\frac{\pi}{2}) \ominus_h \Phi(0)) \oplus (\Phi(\frac{5\pi}{2}) \ominus_h \Phi(2\pi))$$

$$\oplus (-1) \odot (\Phi(\frac{\pi}{2}) \ominus_h \Phi(\pi)) \oplus (-1)(\Phi(\frac{5\pi}{2}) \ominus_h \Phi(3\pi))$$

$$= 2 \odot c \oplus (-2) \odot c$$

$$= (4, 8, 12) \oplus (-12, -8, -4) = (-8, 0, 8).$$

5. Conclusions

This paper is concerned with investigating a new derivative called generalized nabla derivative for fuzzy functions on time scales and studies some basic properties of ∇^g derivative. In addition, we prove a fundamental theorem of nabla integral calculus for fuzzy functions on time scales under generalized differentiability on time scales. The advantage of ∇^g derivative is that it is exists even for a fuzzy function having increasing and decreasing length of diameter on a time scale. The results obtained in this paper include results of Leelavathi et al. [27], when the function having only increasing length of diameter, and the results of Leelavathi et al. [28], when the function having only decreasing length of diameter. The obtained results are illustrated with numerical examples. In the future, we propose to study fuzzy nabla dynamic equations on time scales under generalized nabla derivative and their applications.

Author Contributions: All authors contributed equally and significantly to writing this article. All authors have read and agreed to the published version of the manuscript.

Funding: This research received no external funding.

Conflicts of Interest: he authors declare no conflict of interest.

References

1. Agarwal, R.P.; Bohner, M. Basic calculus on time scales and some of its applications. *Results Math.* **1999**, *35*, 3–22. [CrossRef]
2. Bohner, M.; Peterson, A. *Dynamic Equations on Time Scales: An Introduction with Applications*; Birkhauser: Boston, MA, USA, 2001.
3. Bohner, M.; Peterson, A. *Advances in Dynamic Equations on Time Scales*; Birkhauser: Boston, MA, USA, 2003.
4. Guseinov, G.S. Integration on time scales. *J. Math. Anal. Appl.* **2003**, *285*, 107–127. [CrossRef]
5. Hilger, S. Ein Makettenkalkuls mit Anwendung auf Zentrumsmannigfaltigkeiten. Ph.D. Thesis, Universitat Wurzburg, Würzburg, Germany, 1988.
6. Hilger, S. Analysis on measure chains—A unified approach to continuous and discrete calculus. *Results Math.* **1990**, *18*, 18–56. [CrossRef]
7. Atici, F.M.; Biles, D.C. First order dynamic inclusions on time scales. *J. Math. Anal. Appl.* **2004**, *292*, 222–237. [CrossRef]

8. Atici, F.M.; Daniel, C.B.; Alex, L. An application of time scales to economics. *Math. Comput. Model.* **2006**, *43*, 718–726. [CrossRef]
9. Atici, F.M.; Usynal, F. A production-inventory model of HMMS model on time scales. *Appl. Math. Lett.* **2008**, *21*, 236–243. [CrossRef]
10. Jackson, B.J. Adaptive control in the nabla setting. *Neural Parallel Sci. Comput.* **2008**, *16*, 253–272.
11. Liu, B.; Do, Y.; Batarfi, H.A.; Alsaadi, F.E. Almost periodic solution for a neutral-type neural networks with distributed leakage delays on time scales. *Neuro Comput.* **2016**, *173*, 921–929.
12. Gao, J.; Wang, Q.R.; Zhang, L.W. Existence and stability of almost-periodic solutions for cellular neural networks with time-varying delays in leakage terms on time scales. *Appl. Math. Comput.* **2015**, *237*, 639–649. [CrossRef]
13. Zadeh, L.A. Fuzzy sets. *Inf. Control.* **1965**, *8*, 338–353. [CrossRef]
14. Kaleva, O. Fuzzy differential equations. *Fuzzy Sets Syst.* **1987**, *24*, 301–317. [CrossRef]
15. Lakshmikantham, V.; Mohapatra, R.N. *Theory of Fuzzy Differential Equations and Inclusions*; Taylor and Francis: Abingdon, UK, 2003.
16. Bede, B.; Gal, S.G. Generalizations of the differentiability of fuzzy-number-valued functions with application to fuzzy differential equations. *Fuzzy Sets Syst.* **2005**, *151*, 581–599. [CrossRef]
17. Bede, B.; Rudas, I.J.; Bencsik, A.L. First order linear fuzzy differential equations under generalized differentiability. *Inf. Sci.* **2007**, *177*, 1648–1662. [CrossRef]
18. Li, J.; Zhao, A.; Yan. J. Cauchy problem of fuzzy differential equations under generalized differentiability. *Fuzzy Sets Syst.* **2012**, *200*, 1–24. [CrossRef]
19. Stefanini, L.; Bede, B. Generalized Hukuhara differentiability of interval-valued functions and interval differential equations. *Nonlinear Anal.* **2009**, *71*, 1311–1328. [CrossRef]
20. Fard, O.S.; Bidgoli, T.A. Calculus of fuzzy functions on time scales(I). *Soft Comput.* **2015**, *19*, 293–305. [CrossRef]
21. Vasavi, C.; Suresh Kumar, G.; Murty, M.S.N. Fuzzy Hukahara delta differential and applications to fuzzy dynamic equations on time scales. *J. Uncertain Syst.* **2016**, *10*, 163–180.
22. Vasavi, C.; Suresh Kumar, G; Murty, M.S.N. Fuzzy dynamic equations on time scales under second type Hukuhara delta derivative. *Int. J. Chem. Sci.* **2016**, *14*, 49–66.
23. Vasavi, C.; Suresh Kumar, G.; Murty, M.S.N. Generalized differentiability and integrability for fuzzy set-valued functions on time scales. *Soft Comput.* **2016**, *20*, 1093–1104. [CrossRef]
24. Vasavi, C.; Suresh Kumar, G.; Murty, M.S.N. Fuzzy dynamic equations on time scales under generalized delta derivative via contractive-like mapping principles. *Indian J. Sci. Technol.* **2016**, *9*, 1–6. [CrossRef]
25. Wang, C.; Agarwal, R.P.; O'Regan, D. Calculus of fuzzy vector-valued functions and almost periodic fuzzy vector-valued functions on time scales. *Fuzzy Sets Syst.* **2019**, *375*, 1–52. [CrossRef]
26. Deng, J.; Xu, C.; Sun, L.; Cao, N.; You, X. On conformable fractional nabla-Hukuhara derivative on time scales. In Proceedings of the International Conference on Fuzzy Theory and Its Applications (iFUZZY), Yilan, Taiwan, 13–16 November 2017. [CrossRef]
27. Leelavathi, R.; Suresh Kumar, G.; Murty, M.S.N. Nabla Hukuhara differentiability for fuzzy functions on time scales. *IAENG Int. J. Appl. Math.* **2018**, *49*, 114–121.
28. Leelavathi, R.; Suresh Kumar, G.; Murty, M.S.N. Second type nabla Hukuhara differentiability for fuzzy functions on time scales. *Ital. J. Pure Appl. Math.* **2020**, *43*, 779–801.
29. Leelavathi, R.; Suresh Kumar, G.; Murty, M.S.N.; Srinivasa Rao, R.V.N. Existence-uniqueness of solutions for fuzzy nabla initial value problems on time scales. *Adv. Differ. Equ.* **2019**, *2019*, 269. [CrossRef]
30. Leelavathi, R.; Suresh Kumar, G.; Murty, M.S.N. Charaterization theorem for fuzzy functions on time scales under generalized nabla Hukuhara difference. *Int. J. Innov. Technol. Explor. Eng.* **2019**, *8*, 1704–1706.
31. Leelavathi, R.; Suresh Kumar, G.; Murty, M.S.N. Nabla Integral for Fuzzy Functions on Time Scales. *Int. J. Appl. Math.* **2018**, *31*, 669–678. [CrossRef]
32. Anastassiou, G.A.; Gal, S.G. On a fuzzy trigonometric approximation theorem of Weierstrass-type. *J. Fuzzy Math.* **2001**, *9*, 701–708.

© 2020 by the authors. Licensee MDPI, Basel, Switzerland. This article is an open access article distributed under the terms and conditions of the Creative Commons Attribution (CC BY) license (http://creativecommons.org/licenses/by/4.0/).

Article

Existence Results for Nonlocal Multi-Point and Multi-Term Fractional Order Boundary Value Problems

Bashir Ahmad [1,†], Najla Alghamdi [2,†], Ahmed Alsaedi [1,†] and Sotiris K. Ntouyas [1,3,*,†]

1. Nonlinear Analysis and Applied Mathematics (NAAM), Research Group, Department of Mathematics, Faculty of Science, King Abdulaziz University, P.O. Box 80203, Jeddah 21589, Saudi Arabia; bashirahmad_qau@yahoo.com (B.A.); aalsaedi@hotmail.com (A.A.)
2. Department of Mathematics, Faculty of Science, University of Jeddah, P.O. Box 80327, Jeddah 21589, Saudi Arabia; nmalghamdi1@uj.edu.sa
3. Department of Mathematics, University of Ioannina, 45110 Ioannina, Greece
* Correspondence: sntouyas@uoi.gr
† These authors contributed equally to this work.

Received: 25 May 2020; Accepted: 21 June 2020; Published: 24 June 2020

Abstract: In this paper, we discuss the existence and uniqueness of solutions for a new class of multi-point and integral boundary value problems of multi-term fractional differential equations by using standard fixed point theorems. We also demonstrate the application of the obtained results with the aid of examples.

Keywords: caputo fractional derivative; multi-term fractional differential equations; existence; fixed point

1. Introduction

Fractional differential equations are found to be of great utility in improving the mathematical modeling of many engineering and scientific disciplines such as physics [1] bioengineering [2], viscoelasticy [3], ecology [4], disease models [5–7], etc. For applications of differential equations containing more than one fractional order differential operators, we refer the reader to Bagley-Torvik [8], Basset equation [9] to name a few.

Fractional order boundary value problems equipped with a variety of classical and non-classical (nonlocal) boundary conditions have recently been investigated by many researchers and the literature on the topic is now much enriched, for instance, see [10–21] and the references cited therein. There has been a special focus on boundary value problems involving multi-term fractional differential equations [22–24].

The objective of the present work is to develop the existence theory for multi-term fractional differential equations equipped with nonlocal multi-point boundary conditions. Precisely, we investigate the following boundary value problem:

$$(q_2 \, {}^cD^{\sigma+2} + q_1 \, {}^cD^{\sigma+1} + q_0 \, {}^cD^{\sigma})x(t) = f(t, x(t)), \ 0 < \sigma < 1, \ 0 < t < 1, \quad (1)$$

$$x(0) = h(x), \ x(\xi) = \sum_{i=1}^{n} j_i x(\eta_i), \ x(1) = \lambda \int_0^\delta x(s)ds, \quad (2)$$

where ${}^cD^\sigma$ denote the Caputo fractional derivative of order σ, $0 < \sigma < 1$, $f : [0,1] \times \mathbb{R} \to \mathbb{R}$, $h : C([0,1], \mathbb{R}) \to \mathbb{R}$ are given continuous functions, $0 < \delta < \xi < \eta_1 < \eta_2 < \ldots < \eta_n < 1, \lambda \in \mathbb{R}$, q_0, q_1, and q_2 are real constants with $q_2 \neq 0$. One can characterize the first and second conditions in (2) as initial-nonlocal and nonlocl multi-point ones, while the last condition in (2) can be understood in the sense that the value of the unknown function x at the right-end point of the domain ($x(1)$) is proportional to the average value of x on the sub-domain $(0, \delta)$. Existence and uniqueness results are established by using the classical Banach and Krasnoselskii fixed point theorems and Leray–Schauder nonlinear alternative. Here, we emphasize that the results presented in this paper rely on the standard tools of the fixed point theory. However, their exposition to the given nonlocal problem for a multi-term (sequential) fractional differential equation produces new results which contributes to the related literature.

The rest of the paper is organized as follows: In Section 2 we recall some preliminary concepts of fractional calculus and prove a basic lemma, helping us to transform the boundary value problem (1) and (2) into a fixed point problem. The main existence and uniqueness results for the case $q_1^2 - 4q_0q_2 > 0$ are presented in details in Section 3. In Sections 4 and 5 we indicate the results for the cases $q_1^2 - 4q_0q_2 = 0$ and $q_1^2 - 4q_0q_2 < 0$ respectively. Examples illustrating the obtained results are also included.

2. Basic Results

Before presenting some auxiliary results, let us recall some preliminary concepts of fractional calculus [25,26].

Definition 1. *Let $y, y^{(m)} \in L_1[a, b]$. Then the Riemann–Liouville fractional derivative $D_a^\alpha y$ of order $\alpha \in (m-1, m]$, $m \in \mathbb{N}$, existing almost everywhere on $[a, b]$, is defined as*

$$D_a^\alpha y(t) = \frac{d^m}{dt^m} J_a^{m-\alpha} y(t) = \frac{1}{\Gamma(m-\alpha)} \frac{d^m}{dt^m} \int_a^t (t-s)^{m-1-\alpha} y(s) ds.$$

The Caputo fractional derivative $^cD_a^\alpha y$ of order $\alpha \in (m-1, m]$, $m \in \mathbb{N}$ is defined as

$$^cD_a^\alpha y(t) = D_a^\alpha \left[y(t) - y(a) - y'(a) \frac{(t-a)}{1!} - \ldots - y^{(m-1)}(a) \frac{(t-a)^{m-1}}{(m-1)!} \right].$$

Remark 1. *If $y \in AC^m[a, b]$, then the Caputo fractional derivative $^cD_a^\alpha y$ of order $\alpha \in (m-1, m]$, $m \in \mathbb{N}$, existing almost everywhere on $[a, b]$, is defined as*

$$^cD_a^\alpha y(t) = J_a^{m-\alpha} y^{(m)}(t) = \frac{1}{\Gamma(m-\alpha)} \int_a^t (t-s)^{m-1-\alpha} y^{(m)}(s) ds.$$

In the sequel, the Riemann–Liouville fractional integral I_a^α and the Caputo fractional derivative $^cD_a^\alpha$ with $a = 0$ are respectively denoted by I^α and $^cD^\alpha$.

Lemma 1. *[25] With the given notations, the following equality holds:*

$$I^\alpha(^cD^\alpha y(t)) = y(t) - c_0 - c_1 t - \ldots - c_{n-1} t^{n-1}, \quad t > 0, \ n-1 < \alpha < n, \tag{3}$$

where c_i ($i = 1, \ldots, n-1$) are arbitrary constants.

The following lemmas associated with the linear variant of problem (1) and (2) plays an important role in the sequel.

Lemma 2. *For any $\varphi \in C([0, 1], \mathbb{R})$ and $q_1^2 - 4q_0q_2 > 0$, the solution of linear multi-term fractional differential equation*

$$(q_2 \, ^cD^{\sigma+2} + q_1 \, ^cD^{\sigma+1} + q_0 \, ^cD^\sigma) x(t) = \varphi(t), \quad 0 < \sigma < 1, \ 0 < t < 1, \tag{4}$$

supplemented with the boundary conditions (2) is given by

$$\begin{aligned}
x(t) &= \frac{1}{q_2(m_2-m_1)}\Bigg\{\int_0^t\int_0^s \mathcal{A}(t)\frac{(s-u)^{\sigma-1}}{\Gamma(\sigma)}\varphi(u)du\,ds\\
&\quad + \rho_1(t)\bigg[\int_0^\zeta\int_0^s \mathcal{A}(\zeta)\frac{(s-u)^{\sigma-1}}{\Gamma(\sigma)}\varphi(u)du\,ds\\
&\quad - \sum_{i=1}^n j_i\int_0^{\eta_i}\int_0^s \mathcal{A}(\eta_i)\frac{(s-u)^{\sigma-1}}{\Gamma(\sigma)}\varphi(u)du\,ds\bigg]\\
&\quad + \rho_2(t)\bigg[\int_0^1\int_0^s \mathcal{A}(1)\frac{(s-u)^{\sigma-1}}{\Gamma(\sigma)}\varphi(u)du\,ds\\
&\quad -\lambda\int_0^\delta\int_0^s\Big(\frac{(e^{m_2(\delta-s)}-1)}{m_2}-\frac{(e^{m_1(\delta-s)}-1)}{m_1}\Big)\frac{(s-u)^{\sigma-1}}{\Gamma(\sigma)}\varphi(u)du\,ds\bigg]\Bigg\}\\
&\quad + h(x)\bigg[e^{m_2 t}+\rho_1(t)\Big(e^{m_2\zeta}-\sum_{i=1}^n j_i+\sum_{i=1}^n j_i e^{m_2\eta_i}\Big)\\
&\quad + \rho_2(t)\Big(\frac{m_2 e^{m_2}-\lambda e^{m_2\delta}-\lambda}{m_2}\Big)\bigg],
\end{aligned}\qquad (5)$$

where

$$\begin{aligned}
\mathcal{A}(\kappa) &= e^{m_2(\kappa-s)}-e^{m_1(\kappa-s)},\quad \kappa=t,1,\zeta\text{ and }\eta_i,\\
m_1 &= \frac{-q_1-\sqrt{q_1^2-4q_0 q_2}}{2q_2},\quad m_2=\frac{-q_1+\sqrt{q_1^2-4q_0 q_2}}{2q_2},\\
\rho_1(t) &= \frac{\omega_4\varrho_1(t)-\omega_3\varrho_2(t)}{\mu_1},\quad \rho_2(t)=\frac{\omega_1\varrho_2(t)-\omega_2\varrho_1(t)}{\mu_1},\\
\varrho_1(t) &= \frac{m_1(1-e^{m_2 t})-m_2(1-e^{m_1 t})}{m_1 m_2},\\
\varrho_2(t) &= q_2(m_2-m_1)(e^{m_2 t}-e^{m_1 t}),\\
\mu_1 &= \omega_1\omega_4-\omega_2\omega_3\neq 0,\\
\omega_1 &= \frac{1}{m_1 m_2}\bigg[m_2\Big(1-\sum_{i=1}^n j_i-e^{m_1\zeta}+\sum_{i=1}^n j_i e^{m_1\eta_i}\Big)\\
&\quad -m_1\Big(1-\sum_{i=1}^n j_i-e^{m_2\zeta}+\sum_{i=1}^n j_i e^{m_2\eta_i}\Big)\bigg],\\
\omega_2 &= q_2(m_2-m_1)\Big(e^{m_1\zeta}-e^{m_2\zeta}-\sum_{i=1}^n j_i e^{m_1\eta_i}+\sum_{i=1}^n j_i e^{m_2\eta_i}\Big),\\
\omega_3 &= \frac{1}{m_1 m_2}\bigg[m_2\Big(1-e^{m_1}-\lambda\delta+\lambda/m_1(e^{m_1\delta}-1)\Big)\\
&\quad -m_1\Big(1-e^{m_2}-\lambda\delta+\lambda/m_2(e^{m_2\delta}-1)\Big)\bigg],\\
\omega_4 &= q_2(m_2-m_1)\Big((e^{m_1}+\lambda/m_1(1-e^{m_1\delta}))\\
&\quad -(e^{m_2}+\lambda/m_2(1-e^{m_2\delta}))\Big).
\end{aligned}\qquad (6)$$

Proof. Applying the operator I^σ on (4) and using (3), we get

$$(q_2 D^2+q_1 D+q_0)x(t)=\int_0^t \frac{(t-s)^{\sigma-1}}{\Gamma(\sigma)}\varphi(s)ds+c_1,\qquad (7)$$

where c_1 is an arbitrary constant. By the method of variation of parameters, the solution of (7) can be written as

$$x(t) = c_1 \left[\frac{m_2(1-e^{m_1 t}) - m_1(1-e^{m_2 t})}{q_2 m_1 m_2 (m_2 - m_1)} \right] + c_2 e^{m_1 t} + c_3 e^{m_2 t}$$

$$- \frac{1}{q_2(m_2 - m_1)} \int_0^t e^{m_1(t-s)} \left(\int_0^s \frac{(s-u)^{\sigma-1}}{\Gamma(\sigma)} \varphi(u) du \right) ds$$

$$+ \frac{1}{q_2(m_2 - m_1)} \int_0^t e^{m_2(t-s)} \left(\int_0^s \frac{(s-u)^{\sigma-1}}{\Gamma(\sigma)} \varphi(u) du \right) ds, \qquad (8)$$

where m_1 and m_2 are given by (6). Using $x(0) = h(x)$ in (8), we get

$$x(t) = c_1 \left[\frac{m_2(1-e^{m_1 t}) - m_1(1-e^{m_2 t})}{q_2 m_1 m_2 (m_2 - m_1)} \right] + c_2 \left(e^{m_1 t} - e^{m_2 t} \right) + h(x) e^{m_2 t}$$

$$+ \frac{1}{q_2(m_2 - m_1)} \left[\int_0^t \left(e^{m_2(t-s)} - e^{m_1(t-s)} \right) \left(\int_0^s \frac{(s-u)^{\sigma-1}}{\Gamma(\sigma)} \varphi(u) du \right) ds \right], \qquad (9)$$

which together with the conditions $x(\xi) = \sum_{i=1}^n j_i x(\eta_i)$ and $x(1) = \lambda \int_0^\delta x(s) ds$ yields the following system of equations in the unknown constants c_1 and c_2:

$$c_1 \omega_1 + c_2 \omega_2 = V_1, \qquad (10)$$
$$c_1 \omega_3 + c_2 \omega_4 = V_2. \qquad (11)$$

where

$$V_1 = -\int_0^\xi \int_0^s A(\xi) \frac{(s-u)^{\sigma-1}}{\Gamma(\sigma)} \varphi(u) du\, ds$$

$$+ \sum_{i=1}^n j_i \int_0^{\eta_i} \int_0^s A(\eta_i) \frac{(s-u)^{\sigma-1}}{\Gamma(\sigma)} \varphi(u) du\, ds + h(x) \left(\sum_{i=1}^n j_i e^{m_2 \eta_i} - e^{m_2 \xi} \right),$$

$$V_2 = -\int_0^1 \int_0^s A(1) \frac{(s-u)^{\sigma-1}}{\Gamma(\sigma)} \varphi(u) du\, ds + h(x) \left(\frac{\lambda e^{m_2 \delta} - \lambda - m_2 e^{m_2}}{m_2} \right)$$

$$+ \lambda \int_0^\delta \int_0^s \left[\frac{(e^{m_1(\delta-s)} - 1)}{m_1} - \frac{(e^{m_2(\delta-s)} - 1)}{m_2} \right] \frac{(s-u)^{\sigma-1}}{\Gamma(\sigma)} \varphi(u) du\, ds.$$

Solving the system (10)–(11) together with the notations (6), we find that

$$c_1 = \frac{V_1 \omega_4 - V_2 \omega_2}{\mu_1}, \quad c_2 = \frac{V_2 \omega_1 - V_1 \omega_3}{\mu_1}.$$

Substituting the value of c_1 and c_2 in (9), we obtain the solution (5). The converse of the lemma follows by direct computation. This completes the proof. □

We do not provide the proofs of the following lemmas, as they are similar to that of Lemma 2.

Lemma 3. *For any $\varphi \in C([0,1], \mathbb{R})$ and $q_1^2 - 4q_0 q_2 = 0$, the solution of linear multi-term fractional differential equation*

$$(q_2 \,{}^c D^{\sigma+2} + q_1 \,{}^c D^{\sigma+1} + q_0 \,{}^c D^\sigma) x(t) = \varphi(t), \quad 0 < \sigma < 1, \ 0 < t < 1, \qquad (12)$$

supplemented with the boundary conditions (2) is given by

$$\begin{aligned}
x(t) = \frac{1}{q_2} \Bigg\{ & \int_0^t \int_0^s \mathcal{B}(t) \frac{(s-u)^{\sigma-1}}{\Gamma(\sigma)} \varphi(u) du\, ds \\
& + \chi_1(t) \Big[\int_0^\xi \int_0^s \mathcal{B}(\xi) \frac{(s-u)^{\sigma-1}}{\Gamma(\sigma)} \varphi(u) du\, ds \\
& - \sum_{i=1}^n j_i \int_0^{\eta_i} \int_0^s \mathcal{B}(\eta_i) \frac{(s-u)^{\sigma-1}}{\Gamma(\sigma)} \varphi(u) du\, ds \Big] \\
& + \chi_2(t) \Big[\int_0^1 \int_0^s \mathcal{B}(1) \frac{(s-u)^{\sigma-1}}{\Gamma(\sigma)} \varphi(u) du\, ds \\
& - \lambda \int_0^\delta \int_0^s \Big(\frac{m(\delta-s)e^{m(\delta-s)} - e^{m(\delta-s)} + 1}{m^2} \Big) \frac{(s-u)^{\sigma-1}}{\Gamma(\sigma)} \varphi(u) du\, ds \Big] \Bigg\} \\
& + h(x) \Big[e^{mt} + \chi_1(t) \Big(e^{m\xi} - \sum_{i=1}^n j_i e^{m\eta_i} \Big) + \chi_2(t) \Big(\frac{me^m - \lambda e^{m\delta} + \lambda}{m} \Big) \Big],
\end{aligned} \qquad (13)$$

where

$$\begin{aligned}
\mathcal{B}(\kappa) &= (\kappa - s)e^{m(\kappa-s)}, \quad \kappa = t, 1, \xi \text{ and } \eta_i, \\
m &= \frac{-q_1}{2q_2}, \\
\chi_1(t) &= \frac{\omega_3 v_2(t) - \omega_4 v_1(t)}{\mu_2}, \quad \chi_2(t) = \frac{\omega_2 v_1(t) - \omega_1 v_2(t)}{\mu_2}, \\
v_1(t) &= \frac{mte^{mt} - e^{mt} + 1}{m^2}, \quad v_2(t) = q_2 t e^{mt}, \\
\omega_1 &= \frac{m\xi e^{m\xi} - e^{m\xi} + 1 - \sum_{i=1}^n j_i(m\eta_i e^{m\eta_i} - e^{m\eta_i} + 1)}{m^2}, \\
\omega_2 &= q_2 \Big(\xi e^{m\xi} - \sum_{i=1}^n j_i \eta_i e^{m\eta_i} \Big), \\
\omega_3 &= \frac{m^2 e^m - me^m + m - m\lambda \delta e^{m\delta} + 2\lambda e^{m\delta} - 2\lambda - m\lambda \delta}{m^3}, \\
\omega_4 &= q_2 \Big(\frac{m^2 e^m - \lambda m \delta e^{m\delta} + \lambda e^{m\delta} - \lambda}{m^2} \Big), \\
\mu_2 &= \omega_1 \omega_4 - \omega_2 \omega_3 \neq 0.
\end{aligned} \qquad (14)$$

Lemma 4. *For any $\varphi \in C([0,1], \mathbb{R})$ and $q_1^2 - 4q_0 q_2 < 0$, the solution of linear multi-term fractional differential equation*

$$(q_2\, {}^c D^{\sigma+2} + q_1\, {}^c D^{\sigma+1} + q_0\, {}^c D^\sigma) x(t) = \varphi(t), \quad 0 < \sigma < 1, \ 0 < t < 1, \qquad (15)$$

supplemented with the boundary conditions (2) is given by

$$\begin{aligned}
x(t) = \frac{1}{q_2 b} \Bigg\{ & \int_0^t \int_0^s \mathcal{F}(t) \frac{(s-u)^{\sigma-1}}{\Gamma(\sigma)} \varphi(u) du\, ds \\
& + \tau_1(t) \Big[\int_0^\xi \int_0^s \mathcal{F}(\xi) \frac{(s-u)^{\sigma-1}}{\Gamma(\sigma)} \varphi(u) du\, ds \\
& - \sum_{i=1}^n j_i \int_0^{\eta_i} \int_0^s \mathcal{F}(\eta_i) \frac{(s-u)^{\sigma-1}}{\Gamma(\sigma)} \varphi(u) du\, ds \Big]
\end{aligned}$$

$$+\tau_2(t)\left[\int_0^1\int_0^s \mathcal{F}(1)\frac{(s-u)^{\sigma-1}}{\Gamma(\sigma)}\varphi(u)du\,ds\right.\tag{16}$$

$$-\frac{\lambda}{a^2+b^2}\int_0^\delta\int_0^s\left(b-be^{-a(\delta-s)}\cos b(\delta-s)\right.$$

$$\left.\left.-ae^{-a(\delta-s)}\sin b(\delta-s)\right)\frac{(s-u)^{\sigma-1}}{\Gamma(\sigma)}\varphi(u)du\,ds\right]\right\}$$

$$+h(x)\left[e^{-at}\cos bt+\tau_1(t)\left(e^{-a\xi}\cos b\xi-\sum_{i=1}^n j_i e^{-a\eta_i}\cos b\eta_i\right)\right.$$

$$\left.+\tau_2(t)\left(e^{-a}\cos b-\frac{\lambda}{a^2+b^2}(a-ae^{-a\delta}\cos b\delta+be^{-a\delta}\sin b\delta)\right)\right],$$

where

$$\begin{aligned}
\mathcal{F}(\kappa) &= e^{-a(\kappa-s)}\sin b(\kappa-s),\ \kappa=t,1,\xi\ \text{and}\ \eta_i,\\
m_{1,2} &= -a\pm bi,\ a=\frac{q_1}{2q_2},\ b=\frac{\sqrt{4q_0q_2-q_1^2}}{2q_2},\\
\tau_1(t) &= \frac{p_3v_2(t)-p_4v_1(t)}{\mu_3},\ \tau_2(t)=\frac{p_2v_1(t)-p_1v_2(t)}{\mu_3},\\
v_1(t) &= \frac{b-be^{-at}\cos bt-ae^{-at}\sin bt}{a^2+b^2},\ v_2(t)=q_2be^{-at}\sin bt\\
p_1 &= \frac{1}{a^2+b^2}\left[b-be^{-a\xi}\cos b\xi-ae^{-a\xi}\sin b\xi\right.\\
&\quad\left.-\sum_{i=1}^n j_i(b-be^{-a\eta_i}\cos b\eta_i-ae^{-a\eta_i}\sin b\eta_i)\right],\\
p_2 &= q_2b\left(e^{-a\xi}\sin b\xi-\sum_{i=1}^n j_i e^{-a\eta_i}\sin b\eta_i\right),\\
p_3 &= \frac{1}{a^2+b^2}\left[b-be^{-a}\cos b-ae^{-a}\sin b-b\lambda\delta\right.\\
&\quad+\frac{b\lambda}{a^2+b^2}(a-ae^{-a\delta}\cos b\delta+be^{-a\delta}\sin b\delta)\\
&\quad\left.-\frac{a\lambda}{a^2+b^2}(b-be^{-a\delta}\cos b\delta-ae^{-a\delta}\sin b\delta)\right],\\
p_4 &= q_2b\left[e^{-a}\sin b-\frac{\lambda}{a^2+b^2}(b-be^{-a\delta}\cos b\delta-ae^{-a\delta}\sin b\delta)\right],\\
\mu_3 &= p_1p_4-p_2p_3\neq 0.
\end{aligned}\tag{17}$$

3. Existence and Uniqueness Results

Denote by $\mathcal{C}=C([0,1],\mathbb{R})$ the Banach space of all continuous functions from $[0,1]$ to $\mathbb{R}$ endowed with the norm defined by $\|x\|=\sup\{|x(t)|:t\in[0,1]\}$. In relation to the problem (1) and (2) with $q_1^2-4q_0q_2>0$, we define an operator $\mathcal{J}:\mathcal{C}\to\mathcal{C}$ by Lemma 2 as

$$\begin{aligned}
(\mathcal{J}x)(t) &= \frac{1}{q_2(m_2-m_1)}\left\{\int_0^t\int_0^s \mathcal{A}(t)\frac{(s-u)^{\sigma-1}}{\Gamma(\sigma)}f(u,x(u))du\,ds\right.\\
&\quad+\rho_1(t)\left[\int_0^\xi\int_0^s \mathcal{A}(\xi)\frac{(s-u)^{\sigma-1}}{\Gamma(\sigma)}f(u,x(u))du\,ds\right.\\
&\quad\left.-\sum_{i=1}^n j_i\int_0^{\eta_i}\int_0^s \mathcal{A}(\eta_i)\frac{(s-u)^{\sigma-1}}{\Gamma(\sigma)}f(u,x(u))du\,ds\right]\\
&\quad+\rho_2(t)\left[\int_0^1\int_0^s \mathcal{A}(1)\frac{(s-u)^{\sigma-1}}{\Gamma(\sigma)}f(u,x(u))du\,ds\right.
\end{aligned}\tag{18}$$

$$-\lambda \int_0^\delta \int_0^s \left(\frac{(e^{m_2(\delta-s)}-1)}{m_2} - \frac{(e^{m_1(\delta-s)}-1)}{m_1}\right) \frac{(s-u)^{\sigma-1}}{\Gamma(\sigma)} f(u,x(u)) du\, ds \bigg]\bigg\}$$

$$+h(x)\left[e^{m_2 t} + \rho_1(t)\left(e^{m_2 \xi} - \sum_{i=1}^n j_i + \sum_{i=1}^n j_i e^{m_2 \eta_i}\right)\right.$$

$$+\rho_2(t)\left(\frac{m_2 e^{m_2} - \lambda e^{m_2 \delta} - \lambda}{m_2}\right)\bigg],$$

where $\mathcal{A}(\cdot)$, $\rho_1(t)$ and $\rho_2(t)$ are defined by (6).

Observe that the problem (1) and (2) is equivalent to the operator equation

$$x = \mathcal{J}x, \tag{19}$$

In the sequel, for the sake of computational convenience, we set

$$\begin{aligned}
\widehat{\rho}_1 &= \max_{t\in[0,1]}|\rho_1(t)|, \quad \widehat{\rho}_2 = \max_{t\in[0,1]}|\rho_2(t)|,\\
\varepsilon &= \max_{t\in[0,1]}\left|m_2(1-e^{m_1 t}) - m_1(1-e^{m_2 t})\right|,\\
\alpha &= \frac{1}{|q_2 m_1 m_2(m_2-m_1)|\Gamma(\sigma+1)}\Big\{\varepsilon + \widehat{\rho}_1\Big[\xi^\sigma |m_2(1-e^{m_1 \xi}) - m_1(1-e^{m_2 \xi})|\\
&\quad + \sum_{i=1}^n |j_i|\eta_i^\sigma |m_2(1-e^{m_1 \eta_i}) - m_1(1-e^{m_2 \eta_i})|\Big]\\
&\quad + \widehat{\rho}_2\Big[|m_2(1-e^{m_1}) - m_1(1-e^{m_2})|\\
&\quad + \frac{\delta^\sigma |\lambda|}{|m_1 m_2|}|m_2^2(m_1\delta - e^{m_1 \delta}+1) - m_1^2(m_2\delta - e^{m_2\delta}+1)|\Big]\Big\},\\
\Delta_1 &= \max_{t\in[0,1]}|e^{m_2 t}| + \widehat{\rho}_1\left(|e^{m_2 \xi}| + \sum_{i=1}^n |j_i||e^{m_2\eta_i}+1|\right) + \widehat{\rho}_2\left(\frac{|m_2 e^{m_2}| + |\lambda||e^{m_2\delta}+1|}{|m_2|}\right).
\end{aligned} \tag{20}$$

Now the platform is set to present our main results. In the first result, we use Krasnoselskii's fixed point theorem to prove the existence of solutions for the problem (1) and (2).

Theorem 1. (Krasnoselskii's fixed point theorem [27]). *Let Y be a bounded, closed, convex, and nonempty subset of a Banach space X. Let F_1 and F_2 be the operators satisfying the conditions: (i) $F_1 y_1 + F_2 y_2 \in Y$ whenever $y_1, y_2 \in Y$; (ii) F_1 is compact and continuous; (iii) F_2 is a contraction mapping. Then there exists $y \in Y$ such that $y = F_1 y + F_2 y$.*

In the forthcoming analysis, we need the following assumptions:

(G_1) $|f(t,x) - f(t,y)| \leq \ell|x-y|$, for all $t \in [0,1]$, $x,y \in \mathbb{R}$, $\ell > 0$.
(G_2) $|h(x) - h(y)| \leq L\|x-y\|$, for all $t \in [0,1]$, $x,y \in \mathcal{C}$, $L > 0$.
(G_3) $|f(t,x)| \leq \vartheta(t)$, for all $t \in [0,1]$, $x \in \mathbb{R}$ and $\vartheta \in C([0,1], \mathbb{R}^+)$.

Theorem 2. *Let $f : [0,1] \times \mathbb{R} \to \mathbb{R}$ be a continuous function satisfying the conditions (G_1) and (G_3), $h : C([0,1], \mathbb{R}) \to \mathbb{R}$ be continuous function satisfying the conditions (G_2). Then the problem (1) and (2) with $q_1^2 - 4q_0 q_2 > 0$, has at least one solution on $[0,1]$ if*

$$L\Delta_1 < 1, \tag{21}$$

where Δ_1 is given by (20).

Proof. Setting $\sup_{t\in[0,1]} |\vartheta(t)| = \|\vartheta\|$, we can fix

$$\begin{aligned}
r \geq{} & \tfrac{\|\vartheta\|}{|q_2 m_1 m_2(m_2-m_1)|\Gamma(\delta+1)}\Big\{\varepsilon + \widehat{\rho}_1\Big[\xi^\sigma |m_2(1-e^{m_1\xi}) - m_1(1-e^{m_2\xi})| \\
& + \sum_{i=1}^n |j_i|\eta_i^\sigma |m_2(1-e^{m_1\eta_i}) - m_1(1-e^{m_2\eta_i})|\Big] + \widehat{\rho}_2\Big[|m_2(1-e^{m_1}) - m_1(1-e^{m_2})| \\
& + \tfrac{\delta^\sigma|\lambda|}{|m_1 m_2|}|m_2^2(m_1\delta - e^{m_1\delta}+1) - m_1^2(m_2\delta - e^{m_2\delta}+1)|\Big]\Big\} + \Delta_1\|h\|,
\end{aligned} \qquad (22)$$

and consider $B_r = \{x \in \mathcal{C} : \|x\| \leq r\}$. Introduce the operators $\mathcal{J}_1$ and $\mathcal{J}_2$ on B_r as follows:

$$\begin{aligned}
(\mathcal{J}_1 x)(t) ={} & \tfrac{1}{q_2(m_2-m_1)}\int_0^t\int_0^s \mathcal{A}(t)\tfrac{(s-u)^{\sigma-1}}{\Gamma(\sigma)}f(u,x(u))du\,ds \\
& + \tfrac{1}{q_2(m_2-m_1)}\Big\{\rho_1(t)\Big[\int_0^\xi\int_0^s \mathcal{A}(\xi)\tfrac{(s-u)^{\sigma-1}}{\Gamma(\sigma)}f(u,x(u))du\,ds \\
& -\sum_{i=1}^n j_i \int_0^{\eta_i}\int_0^s \mathcal{A}(\eta_i)\tfrac{(s-u)^{\sigma-1}}{\Gamma(\sigma)}f(u,x(u))du\,ds\Big] \\
& +\rho_2(t)\Big[\int_0^1\int_0^s \mathcal{A}(1)\tfrac{(s-u)^{\sigma-1}}{\Gamma(\sigma)}f(u,x(u))du\,ds \\
& -\lambda\int_0^\delta\int_0^s \Big(\tfrac{(e^{m_2(\delta-s)}-1)}{m_2} - \tfrac{(e^{m_1(\delta-s)}-1)}{m_1}\Big)\tfrac{(s-u)^{\sigma-1}}{\Gamma(\sigma)}f(u,x(u))du\,ds\Big]\Big\},
\end{aligned}$$

and

$$(\mathcal{J}_2 x)(t) = h(x)\Big[e^{m_2 t} + \rho_1(t)\Big(e^{m_2\xi} - \sum_{i=1}^n j_i + \sum_{i=1}^n j_i e^{m_2\eta_i}\Big)r + \rho_2(t)\Big(\tfrac{m_2 e^{m_2} - \lambda e^{m_2\delta} - \lambda}{m_2}\Big)\Big]. \qquad (23)$$

Observe that $\mathcal{J} = \mathcal{J}_1 + \mathcal{J}_2$. For $x, y \in B_r$, we have

$$\begin{aligned}
& \|\mathcal{J}_1 x + \mathcal{J}_2 y\| \\
={} & \sup_{t\in[0,1]} |(\mathcal{J}_1 x)(t) + (\mathcal{J}_2 y)(t)| \\
\leq{} & \tfrac{1}{|q_2(m_2-m_1)|}\sup_{t\in[0,1]}\Big\{\int_0^t\int_0^s \mathcal{A}(t)\tfrac{(s-u)^{\sigma-1}}{\Gamma(\sigma)}|f(u,x(u))|du\,ds \\
& +|\rho_1(t)|\Big[\int_0^\xi\int_0^s \mathcal{A}(\xi)\tfrac{(s-u)^{\sigma-1}}{\Gamma(\sigma)}|f(u,x(u))|du\,ds \\
& +\sum_{i=1}^n |j_i|\int_0^{\eta_i}\int_0^s \mathcal{A}(\eta_i)\tfrac{(s-u)^{\sigma-1}}{\Gamma(\sigma)}f(u,x(u))du\,ds\Big] \\
& +|\rho_2(t)|\Big[\int_0^1\int_0^s \mathcal{A}(1)\tfrac{(s-u)^{\sigma-1}}{\Gamma(\sigma)}|f(u,x(u))|du\,ds \\
& +|\lambda|\int_0^\delta\int_0^s \Big(\tfrac{(e^{m_2(\delta-s)}-1)}{m_2} - \tfrac{(e^{m_1(\delta-s)}-1)}{m_1}\Big)\tfrac{(s-u)^{\sigma-1}}{\Gamma(\sigma)}|f(u,x(u))|du\,ds\Big]\Big\} \\
& +|h(y)|\Big[|e^{m_2 t}| + \rho_1(t)\Big(|e^{m_2\xi}| + \sum_{i=1}^n |j_i||e^{m_2\eta_i}+1|\Big) + \rho_2(t)\Big(\tfrac{|m_2 e^{m_2}|+|\lambda||e^{m_2\delta}+1|}{|m_2|}\Big)\Big] \\
\leq{} & \tfrac{\|\vartheta\|}{|q_2(m_2-m_1)|\Gamma(\sigma+1)}\sup_{t\in[0,1]}\Big\{t^\sigma \int_0^t \big|e^{m_2(t-s)} - e^{m_1(t-s)}\big|ds \\
& +|\rho_1(t)|\Big[\xi^\sigma \int_0^\xi \big|e^{m_2(\xi-s)} - e^{m_1(\xi-s)}\big|ds + \sum_{i=1}^n |j_i|\eta_i^\sigma \int_0^{\eta_i}\big|e^{m_2(\eta_i-s)} - e^{m_1(\eta_i-s)}\big|ds\Big]
\end{aligned}$$

$$+|\rho_2(t)|\Big[\int_0^1 \Big|e^{m_2(1-s)} - e^{m_1(1-s)}\Big|ds + |\lambda|\delta^\sigma \int_0^\delta \Big|\frac{(e^{m_2(\delta-s)}-1)}{m_2} - \frac{(e^{m_1(\delta-s)}-1)}{m_1}\Big|ds\Big]\Big\}$$

$$+\Delta_1\|h\|$$

$$\leq \frac{\|\vartheta\|}{|q_2 m_1 m_2(m_2-m_1)|\Gamma(\sigma+1)}\Big\{\varepsilon + \hat{\rho}_1[\zeta^\sigma|m_2(1-e^{m_1\zeta}) - m_1(1-e^{m_2\zeta})|$$

$$+\sum_{i=1}^n |j_i|\eta_i^\sigma|m_2(1-e^{m_1\eta_i}) - m_1(1-e^{m_2\eta_i})|] + \hat{\rho}_2[|m_2(1-e^{m_1}) - m_1(1-e^{m_2})|$$

$$+\frac{\delta^\sigma|\lambda|}{|m_1 m_2|}|m_2^2(m_1\delta - e^{m_1\delta}+1) - m_1^2(m_2\delta - e^{m_2\delta}+1)|]\Big\} + \Delta_1\|h\| \leq r,$$

where we used (22). Thus $\mathcal{J}_1 x + \mathcal{J}_2 y \in B_r$. Using the assumptions $(G_1) - (G_3)$ together with (21), we show that $\mathcal{J}_2$ is a contraction as follows:

$$\|\mathcal{J}_2 x - \mathcal{J}_2 y\|$$
$$= \sup_{t\in[0,1]} |(\mathcal{J}_2 x)(t) - (\mathcal{J}_2 y)(t)|$$
$$\leq |h(x)-h(y)|\Big[|e^{m_2 t}| + \rho_1(t)\Big(|e^{m_2\zeta}| + \sum_{i=1}^n |j_i||e^{m_2\eta_i}+1|\Big) + \rho_2(t)\Big(\frac{|m_2 e^{m_2}| + |\lambda||e^{m_2\delta}+1|}{|m_2|}\Big)\Big]$$
$$\leq L\Delta_1\|x-y\|.$$

Note that continuity of f implies that the operator $\mathcal{J}_1$ is continuous. Also, $\mathcal{J}_1$ is uniformly bounded on B_r as

$$\|\mathcal{J}_1 x\| = \sup_{t\in[0,1]} |(\mathcal{J}_1 x)(t)|$$

$$\leq \frac{\|\vartheta\|}{|q_2 m_1 m_2(m_2-m_1)|\Gamma(\sigma+1)}\Big\{\varepsilon + \hat{\rho}_1[\zeta^\sigma|m_2(1-e^{m_1\zeta}) - m_1(1-e^{m_2\zeta})|$$

$$+\sum_{i=1}^n |j_i|\eta_i^\sigma|m_2(1-e^{m_1\eta_i}) - m_1(1-e^{m_2\eta_i})|] + \hat{\rho}_2[|m_2(1-e^{m_1}) - m_1(1-e^{m_2})|$$

$$+\frac{\delta^\sigma|\lambda|}{|m_1 m_2|}|m_2^2(m_1\delta - e^{m_1\delta}+1) - m_1^2(m_2\delta - e^{m_2\delta}+1)|]\Big\}.$$

Now we prove the compactness of operator $\mathcal{J}_1$. We define $\sup_{(t,x)\in[0,1]\times B_r}|f(t,x)| = \overline{f}$. Thus, for $0 < t_1 < t_2 < 1$, we have

$$|(\mathcal{J}x)(t_2) - (\mathcal{J}x)(t_1)|$$

$$\leq \frac{1}{|q_2(m_2-m_1)|}\Big\{\Big|\int_0^{t_1}\int_0^s [\mathcal{A}(t_2) - \mathcal{A}(t_1)]\frac{(s-u)^{\sigma-1}}{\Gamma(\sigma)}f(u,x(u))du\,ds$$

$$+ \int_{t_1}^{t_2}\int_0^s \mathcal{A}(t_2)\frac{(s-u)^{\sigma-1}}{\Gamma(\sigma)}f(u,x(u))du\,ds\Big|$$

$$+|\rho_1(t_2) - \rho_1(t_1)|\Big[\int_0^\zeta \int_0^s \mathcal{A}(\zeta)\frac{(s-u)^{\sigma-1}}{\Gamma(\sigma)}|f(u,x(u))|du\,ds$$

$$+\sum_{i=1}^n |j_i|\int_0^{\eta_i}\int_0^s \mathcal{A}(\eta_i)\frac{(s-u)^{\sigma-1}}{\Gamma(\sigma)}|f(u,x(u))|du\,ds\Big]$$

$$+|\rho_2(t_2) - \rho_2(t_1)|\Big[\int_0^1\int_0^s \mathcal{A}(1)\frac{(s-u)^{\sigma-1}}{\Gamma(\sigma)}|f(u,x(u))|du\,ds$$

$$+|\lambda|\int_0^\delta \int_0^s \Big(\frac{(e^{m_2(\delta-s)}-1)}{m_2} - \frac{(e^{m_1(\delta-s)}-1)}{m_1}\Big)\frac{(s-u)^{\sigma-1}}{\Gamma(\sigma)}|f(u,x(u))|du\,ds\Big]\Big\}$$

$$
\begin{aligned}
\leq \quad & \frac{\overline{f}}{|q_2 m_1 m_2 (m_2 - m_1)| \Gamma(\sigma+1)} \Big\{ \Big(t_1^\sigma - t_2^\sigma\Big) \Big| m_1 (1 - e^{m_2(t_2 - t_1)}) - m_2 (1 - e^{m_1(t_2 - t_1)}) \Big| \\
& + t_1^\sigma \Big| m_1(e^{m_2 t_2} - e^{m_2 t_1}) - m_2(e^{m_1 t_2} - e^{m_1 t_1}) \Big| \\
& + |\rho_1(t_2) - \rho_1(t_1)| [\xi^\sigma | m_2 (1 - e^{m_1 \xi}) - m_1 (1 - e^{m_2 \xi})|] \\
& + \sum_{i=1}^{n} |j_i| \eta_i^\sigma |m_2(1 - e^{m_1 \eta_i}) - m_1(1 - e^{m_2 \eta_i})|] \\
& + |\rho_2(t_2) - \rho_2(t_1)| [|m_2(1 - e^{m_1}) - m_1(1 - e^{m_2})| \\
& + \frac{\delta^\sigma |\lambda|}{|m_1 m_2|} |m_1^2 (m_2 \delta - e^{m_2 \delta} + 1) - m_2^2 (m_1 \delta - e^{m_1 \delta} + 1)|] \Big\} \to 0, \text{ as } t_1 \to t_2,
\end{aligned}
$$

independent of x. Thus, $\mathcal{J}_1$ is relatively compact on B_r. Hence, by the Arzelá-Ascoli Theorem, $\mathcal{J}_1$ is compact on B_r. Thus all the assumption of Theorem 1 are satisfied. So, by the conclusion of Theorem 1, the problem (1) and (2) has at least one solution on $[0,1]$. The proof is completed. □

Remark 2. *In the above theorem we can interchange the roles of the operators $\mathcal{J}_1$ and $\mathcal{J}_2$ to obtain a second result by replacing (21) by the following condition:*

$$\ell \alpha < 1.$$

Now we apply Banach's contraction mapping principle to prove existence and uniqueness of solutions for the problem (1) and (2).

Theorem 3. *Assume that $f : [0,1] \times \mathbb{R} \to \mathbb{R}$ is a continuous function such that (G_1) and (G_2) are satisfied. Then there exists a unique solution for the problem (1) and (2) on $[0,1]$ if $\ell \alpha + L \Delta_1 < 1$, where α and Δ_1 are given by (20).*

Proof. Let us define $\sup_{t \in [0,1]} |f(t,0)| = M$, $\sup_{t \in [0,1]} |h(0)| = L_0$ and select $\bar{r} \geq \dfrac{\alpha M + L_0 \Delta_1}{1 - (\ell \alpha + L \Delta_1)}$ to show that $\mathcal{J} B_{\bar{r}} \subset B_{\bar{r}}$, where $B_{\bar{r}} = \{x \in \mathcal{C} : \|x\| \leq \bar{r}\}$ and $\mathcal{J}$ is defined by (18). Using the condition (G_1) and (G_2), we have

$$
\begin{aligned}
|f(t,x)| &= |f(t,x) - f(t,0) + f(t,0)| \leq |f(t,x) - f(t,0)| + |f(x,0)| \\
&\leq \ell \|x\| + M \leq \ell \bar{r} + M,
\end{aligned}
\tag{24}
$$

$$
|h(x)| = |h(x) - h(0) + h(0)| \leq |h(x) - h(0)| + |h(0)| \leq L\|x\| + L_0 \leq L\bar{r} + L_0.
\tag{25}
$$

Then, for $x \in B_{\bar{r}}$, we obtain

$$
\begin{aligned}
\|\mathcal{J}(x)\| &= \sup_{t \in [0,1]} |\mathcal{J}(x)(t)| \\
&\leq \frac{1}{|q_2(m_2 - m_1)|} \sup_{t \in [0,1]} \Big\{ \int_0^t \int_0^s \mathcal{A}(t) \frac{(s-u)^{\sigma-1}}{\Gamma(\sigma)} |f(u, x(u))| du\, ds \\
&\quad + |\rho_1(t)| \Big[\int_0^\xi \int_0^s \mathcal{A}(\xi) \frac{(s-u)^{\sigma-1}}{\Gamma(\sigma)} |f(u, x(u))| du\, ds \\
&\quad + \sum_{i=1}^n |j_i| \int_0^{\eta_i} \int_0^s \mathcal{A}(\eta_i) \frac{(s-u)^{\sigma-1}}{\Gamma(\sigma)} |f(u, x(u))| du\, ds \Big] \\
&\quad + |\rho_2(t)| \Big[\int_0^1 \int_0^s \mathcal{A}(1) \frac{(s-u)^{\sigma-1}}{\Gamma(\sigma)} |f(u, x(u))| du\, ds
\end{aligned}
$$

$$+|\lambda|\int_0^\delta \int_0^s \left(\frac{(e^{m_2(\delta-s)}-1)}{m_2} - \frac{(e^{m_1(\delta-s)}-1)}{m_1}\right)\frac{(s-u)^{\sigma-1}}{\Gamma(\sigma)}|f(u,x(u))|du\,ds\bigg]\bigg\}$$

$$+|h(x)|\bigg[|e^{m_2 t}|+\rho_1(t)(|e^{m_2\xi}|+\sum_{i=1}^n|j_i||e^{m_2\eta_i}+1|)$$

$$+\rho_2(t)\bigg(\frac{|m_2 e^{m_2}|+|\lambda||e^{m_2\delta}+1|}{|m_2|}\bigg)\bigg]$$

$$\leq \frac{(\ell\bar{r}+M)}{|q_2(m_2-m_1)|}\sup_{t\in[0,1]}\bigg\{\int_0^t |e^{m_2(t-s)}-e^{m_1(t-s)}|\frac{s^\sigma}{\Gamma(\sigma+1)}ds$$

$$+|\rho_1(t)|\bigg[\int_0^\xi |e^{m_2(\xi-s)}-e^{m_1(\xi-s)}|\frac{s^\sigma}{\Gamma(\sigma+1)}ds$$

$$+\sum_{i=1}^n|j_i|\int_0^{\eta_i}|e^{m_2(\eta_i-s)}-e^{m_1(\eta_i-s)}|\frac{s^\sigma}{\Gamma(\sigma+1)}ds\bigg]$$

$$+ |\rho_2(t)|\bigg[\int_0^1|e^{m_2(1-s)}-e^{m_1(1-s)}|\frac{s^\sigma}{\Gamma(\sigma+1)}ds$$

$$+|\lambda|\int_0^\delta \bigg|\frac{(e^{m_2(\delta-s)}-1)}{m_2}-\frac{(e^{m_1(\delta-s)}-1)}{m_1}\bigg|\frac{s^\sigma}{\Gamma(\sigma+1)}ds\bigg]\bigg\} + (L\bar{r}+L_0)\Delta_1$$

$$\leq \frac{(\ell\bar{r}+M)}{|q_2 m_1 m_2(m_2-m_1)|\Gamma(\sigma+1)}\bigg\{\varepsilon+\hat{\rho}_1[\xi^\sigma|m_2(1-e^{m_1\xi})-m_1(1-e^{m_2\xi})|$$

$$+\sum_{i=1}^n|j_i|\eta_i^\sigma|m_2(1-e^{m_1\eta_i})-m_1(1-e^{m_2\eta_i})|]+\hat{\rho}_2[|m_2(1-e^{m_1})-m_1(1-e^{m_2})|$$

$$+\frac{\delta^\sigma|\lambda|}{|m_1 m_2|}|m_2^2(m_1\delta-e^{m_1\delta}+1)-m_1^2(m_2\delta-e^{m_2\delta}+1)|]\bigg\} + (L\bar{r}+L_0)\Delta_1$$

$$= (\ell\bar{r}+M)\alpha + (L\bar{r}+L_0)\Delta_1 \leq \bar{r},$$

which clearly shows that $\mathcal{J}x \in B_{\bar{r}}$ for any $x \in B_{\bar{r}}$. Thus $\mathcal{J}B_{\bar{r}} \subset B_{\bar{r}}$. Now, for $x,y \in \mathcal{C}$ and for each $t \in [0,1]$, we have

$$\|(\mathcal{J}x)-(\mathcal{J}y)\|$$

$$\leq \frac{1}{|q_2(m_2-m_1)|}\sup_{t\in[0,1]}\bigg\{\int_0^t\int_0^s \mathcal{A}(t)\frac{(s-u)^{\sigma-1}}{\Gamma(\sigma)}|f(u,x(u))-f(u,y(u))|du\,ds$$

$$+|\rho_1(t)|\bigg[\int_0^\xi\int_0^s \mathcal{A}(\xi)\frac{(s-u)^{\sigma-1}}{\Gamma(\sigma)}|f(u,x(u))-f(u,y(u))|du\,ds$$

$$+\sum_{i=1}^n|j_i|\int_0^{\eta_i}\int_0^s \mathcal{A}(\eta_i)\frac{(s-u)^{\sigma-1}}{\Gamma(\sigma)}|f(u,x(u))-f(u,y(u))|du\,ds\bigg]$$

$$+|\rho_2(t)|\bigg[\int_0^1\int_0^s \mathcal{A}(1)\frac{(s-u)^{\sigma-1}}{\Gamma(\sigma)}|f(u,x(u))-f(u,y(u))|du\,ds$$

$$+|\lambda|\int_0^\delta\int_0^s\left(\frac{(e^{m_2(\delta-s)}-1)}{m_2}-\frac{(e^{m_1(\delta-s)}-1)}{m_1}\right)\frac{(s-u)^{\sigma-1}}{\Gamma(\sigma)}|f(u,x(u))-f(u,y(u))|du\,ds\bigg]\bigg\}$$

$$+|h(x)-h(y)|\bigg[|e^{m_2 t}|+\rho_1(t)(|e^{m_2\xi}|+\sum_{i=1}^n|j_i||e^{m_2\eta_i}+1|)$$

$$+\rho_2(t)\bigg(\frac{|m_2 e^{m_2}|+|\lambda||e^{m_2\delta}+1|}{|m_2|}\bigg)\bigg]$$

$$\leq \frac{\ell}{|q_2(m_2-m_1)|}\sup_{t\in[0,1]}\bigg\{\int_0^t|e^{m_2(t-s)}-e^{m_1(t-s)}|\frac{s^\sigma}{\Gamma(\sigma+1)}ds$$

$$+|\rho_1(t)|\Big[\int_0^\zeta \Big|e^{m_2(\zeta-s)}-e^{m_1(\zeta-s)}\Big|\frac{s^\sigma}{\Gamma(\sigma+1)}ds$$

$$+\sum_{i=1}^n |j_i|\int_0^{\eta_i}\Big|e^{m_2(\eta_i-s)}-e^{m_1(\eta_i-s)}\Big|\frac{s^\sigma}{\Gamma(\sigma+1)}ds\Big]$$

$$+|\rho_2(t)|\Big[\int_0^1 \Big|e^{m_2(1-s)}-e^{m_1(1-s)}\Big|\frac{s^\sigma}{\Gamma(\sigma+1)}ds$$

$$+|\lambda|\int_0^\delta \Big|\frac{(e^{m_2(\delta-s)}-1)}{m_2}-\frac{(e^{m_1(\delta-s)}-1)}{m_1}\Big|\frac{s^\sigma}{\Gamma(\sigma+1)}ds\Big]\Big\}\|x-y\|+L\Delta_1\|x-y\|$$

$$\leq \frac{\ell}{|q_2 m_1 m_2 (m_2-m_1)|\Gamma(\sigma+1)}\Big\{\varepsilon+\widehat{\rho}_1[\zeta^\sigma|m_2(1-e^{m_1\zeta})-m_1(1-e^{m_2\zeta})|$$

$$+\sum_{i=1}^n |j_i|\eta_i^\sigma|m_2(1-e^{m_1\eta_i})-m_1(1-e^{m_2\eta_i})|]+\widehat{\rho}_2[|m_2(1-e^{m_1})-m_1(1-e^{m_2})|$$

$$+\frac{\delta^\sigma|\lambda|}{|m_1 m_2|}|m_2^2(m_1\delta-e^{m_1\delta}+1)-m_1^2(m_2\delta-e^{m_2\delta}+1)|]\Big\}\|x-y\|+L\Delta_1\|x-y\|$$

$$= (\ell\alpha+L\Delta_1)\|x-y\|,$$

α and Δ_1 are given by (20) and depend only on the parameters involved in the problem. In view of the condition $\ell\alpha+L\Delta_1 < 1$, it follows that $\mathcal{J}$ is a contraction. Thus, by the contraction mapping principle (Banach fixed point theorem), the problem (1) and (2) has a unique solution on $[0,1]$. This completes the proof. □

The next existence result is based on Leray–Schauder nonlinear alternative.

Theorem 4. (*Nonlinear alternative for single valued maps* [28]). *Let E be a Banach space, C a closed, convex subset of E, U an open subset of C and $0 \in U$. Suppose that $F : \overline{U} \to C$ is a continuous, compact (that is, $F(\overline{U})$ is a relatively compact subset of C) map. Then either*

(i) *F has a fixed point in $\overline{U}$, or*
(ii) *there is a $u \in \partial U$ (the boundary of U in C) and $\epsilon \in (0,1)$ with $u = \epsilon F(u)$.*

We need the following assumptions:

(H_1) There exist a function $g \in C([0,1],\mathbb{R}^+)$, and a nondecreasing function $\psi : \mathbb{R}^+ \to \mathbb{R}^+$ such that $|f(t,y)| \leq g(t)\psi(\|y\|), \ \forall (t,y) \in [0,1]\times \mathbb{R}$.

(H_2) $h : C([0,1],\mathbb{R}) \to \mathbb{R}$, is continuous function with $h(0) = 0$ and there exist constant $L_1 > 0$ with $L_1 < \Delta_1^{-1}$, such that
$$|h(x)| \leq L_1\|x\| \quad \forall \ x \in \mathcal{C}.$$

(H_3) There exists a constant $K > 0$ such that
$$\frac{(1-L_1\Delta_1)K}{\|g\|\psi(K)\alpha} > 1.$$

Theorem 5. *Let $f : [0,1]\times \mathbb{R} \to \mathbb{R}$ be a continuous function. Then the problem (1) and (2) has at least one solution on $[0,1]$, if (H_1)–(H_3) are satisfied.*

Proof. Consider the operator $\mathcal{J} : \mathcal{C} \to \mathcal{C}$ defined by (18). We show that $\mathcal{J}$ maps bounded sets into bounded sets in $\mathcal{C}$. For a positive number ζ, let $\mathcal{E}_\zeta = \{x \in \mathcal{C} : \|x\| \leq \zeta\}$ be a bounded set in $\mathcal{C}$. Then we have

$$\|\mathcal{J}(x)\| = \sup_{t\in[0,1]}|\mathcal{J}(x)(t)|$$

$$\leq \frac{1}{|q_2(m_2-m_1)|} \sup_{t\in[0,1]} \left\{ \int_0^t \int_0^s \mathcal{A}(t) \frac{(s-u)^{\sigma-1}}{\Gamma(\sigma)} |f(u,x(u))| du\, ds \right.$$
$$+ |\rho_1(t)| \Big[\int_0^\zeta \int_0^s \mathcal{A}(\zeta) \frac{(s-u)^{\sigma-1}}{\Gamma(\sigma)} |f(u,x(u))| du\, ds$$
$$+ \sum_{i=1}^n |j_i| \int_0^{\eta_i} \int_0^s \mathcal{A}(\eta_i) \frac{(s-u)^{\sigma-1}}{\Gamma(\sigma)} |f(u,x(u))| du\, ds \Big]$$
$$+ |\rho_2(t)| \Big[\int_0^1 \int_0^s \mathcal{A}(1) \frac{(s-u)^{\sigma-1}}{\Gamma(\sigma)} |f(u,x(u))| du\, ds$$
$$+ |\lambda| \int_0^\delta \int_0^s \Big(\frac{(e^{m_2(\delta-s)}-1)}{m_2} - \frac{(e^{m_1(\delta-s)}-1)}{m_1} \Big) \frac{(s-u)^{\sigma-1}}{\Gamma(\sigma)} |f(u,x(u))| du\, ds \Big] \right\}$$
$$+ |h(x)| \Big[|e^{m_2 t}| + \rho_1(t)(|e^{m_2 \zeta}| + \sum_{i=1}^n |j_i||e^{m_2 \eta_i}+1|)$$
$$+ \rho_2(t) \Big(\frac{|m_2 e^{m_2}| + |\lambda||e^{m_2 \delta}+1|}{|m_2|} \Big) \Big]$$

$$\leq \frac{\|g\|\psi(\zeta)}{|q_2(m_2-m_1)|} \sup_{t\in[0,1]} \left\{ \int_0^t \Big| e^{m_2(t-s)} - e^{m_1(t-s)} \Big| \frac{s^\sigma}{\Gamma(\sigma+1)} ds \right.$$
$$+ |\rho_1(t)| \Big[\int_0^\zeta \Big| e^{m_2(\zeta-s)} - e^{m_1(\zeta-s)} \Big| \frac{s^\sigma}{\Gamma(\sigma+1)} ds$$
$$+ \sum_{i=1}^n |j_i| \int_0^{\eta_i} \Big| e^{m_2(\eta_i-s)} - e^{m_1(\eta_i-s)} \Big| \frac{s^\sigma}{\Gamma(\sigma+1)} ds \Big]$$
$$+ |\rho_2(t)| \Big[\int_0^1 \Big| e^{m_2(1-s)} - e^{m_1(1-s)} \Big| \frac{s^\sigma}{\Gamma(\sigma+1)} ds$$
$$+ |\lambda| \int_0^\delta \Big| \frac{(e^{m_2(\delta-s)}-1)}{m_2} - \frac{(e^{m_1(\delta-s)}-1)}{m_1} \Big| \frac{s^\sigma}{\Gamma(\sigma+1)} ds \Big] \right\} + L_1 \Delta_1 \zeta$$

$$\leq \frac{\|g\|\psi(\zeta)}{|q_2 m_1 m_2 (m_2-m_1)|\Gamma(\sigma+1)} \Big\{ \varepsilon + \widehat{\rho}_1 [\zeta^\sigma |m_2(1-e^{m_1\zeta}) - m_1(1-e^{m_2\zeta})|$$
$$+ \sum_{i=1}^n |j_i| \eta_i^\sigma |m_2(1-e^{m_1 \eta_i}) - m_1(1-e^{m_2 \eta_i})|] + \widehat{\rho}_2 [|m_2(1-e^{m_1}) - m_1(1-e^{m_2})|$$
$$+ \frac{\delta^\sigma |\lambda|}{|m_1 m_2|} |m_2^2(m_1\delta - e^{m_1\delta}+1) - m_1^2(m_2\delta - e^{m_2\delta}+1)|] \Big\} + L_1 \Delta_1 \zeta,$$

which yields

$$\|\mathcal{J}x\| \leq \frac{\|g\|\psi(\zeta)}{|q_2 m_1 m_2 (m_2-m_1)|\Gamma(\sigma+1)} \Big\{ \varepsilon + \widehat{\rho}_1 \zeta^\sigma |m_2(1-e^{m_1\zeta}) - m_1(1-e^{m_2\zeta})|$$
$$+ \sum_{i=1}^n |j_i| \eta_i^\sigma |m_2(1-e^{m_1 \eta_i}) - m_1(1-e^{m_2 \eta_i})|] + \widehat{\rho}_2 [|m_2(1-e^{m_1}) - m_1(1-e^{m_2})|$$
$$+ \frac{\delta^\sigma |\lambda|}{|m_1 m_2|} |m_2^2(m_1\delta - e^{m_1\delta}+1) - m_1^2(m_2\delta - e^{m_2\delta}+1)|] \Big\} + L_1 \Delta_1 \zeta.$$

Next we show that $\mathcal{J}$ maps bounded sets into equicontinuous sets of $\mathcal{C}$. Let $t_1, t_2 \subset [0,1]$ with $t_1 < t_2$ and $y \in \mathcal{E}_\zeta$, where $\mathcal{E}_\zeta$ is a bounded set of $\mathcal{C}$. Then we obtain

$$|(\mathcal{J}x)(t_2) - (\mathcal{J}x)(t_1)|$$
$$\leq \frac{1}{|q_2(m_2-m_1)|} \left\{ \Big| \int_0^{t_1} \int_0^s [\mathcal{A}(t_2) - \mathcal{A}(t_1)] \frac{(s-u)^{\sigma-1}}{\Gamma(\sigma)} f(u,x(u)) du\, ds \right.$$

$$+ \int_{t_1}^{t_2} \int_0^s \mathcal{A}(t_2) \frac{(s-u)^{\sigma-1}}{\Gamma(\sigma)} f(u,x(u)) du\, ds \Bigg|$$

$$+ |\rho_1(t_2) - \rho_1(t_1)| \Bigg[\int_0^\xi \int_0^s \mathcal{A}(\xi) \frac{(s-u)^{\sigma-1}}{\Gamma(\sigma)} |f(u,x(u))| du\, ds$$

$$+ \sum_{i=1}^n |j_i| \int_0^{\eta_i} \int_0^s \mathcal{A}(\eta_i) \frac{(s-u)^{\sigma-1}}{\Gamma(\sigma)} |f(u,x(u))| du\, ds \Bigg]$$

$$+ |\rho_2(t_2) - \rho_2(t_1)| \Bigg[\int_0^1 \int_0^s \mathcal{A}(1) \frac{(s-u)^{\sigma-1}}{\Gamma(\sigma)} |f(u,x(u))| du\, ds$$

$$+ |\lambda| \int_0^\delta \int_0^s \Big(\frac{(e^{m_2(\delta-s)} - 1)}{m_2} - \frac{(e^{m_1(\delta-s)} - 1)}{m_1} \Big) \frac{(s-u)^{\sigma-1}}{\Gamma(\sigma)} |f(u,x(u))| du\, ds \Bigg] \Bigg\}$$

$$+ |h(x)| \Bigg[|e^{m_2 t_2} - e^{m_2 t_1}| + (\rho_1(t_2) - \rho_1(t_1))(|e^{m_2 \xi}| + \sum_{i=1}^n |j_i| |e^{m_2 \eta_i} + 1|)$$

$$+ (\rho_2(t_2) - \rho_2(t_1)) \Big(\frac{|m_2 e^{m_2}| + |\lambda| |e^{m_2 \delta} + 1|}{|m_2|} \Big) \Bigg]$$

$$\leq \frac{\overline{f}}{|q_2 m_1 m_2 (m_2 - m_1)| \Gamma(\sigma + 1)} \Bigg\{ \Big((t_1^\sigma - t_2^\sigma) \big| m_1(1 - e^{m_2(t_2 - t_1)}) - m_2(1 - e^{m_1(t_2 - t_1)}) \big|$$

$$+ t_1^\sigma \big| m_1(e^{m_2 t_2} - e^{m_2 t_1}) - m_2(e^{m_1 t_2} - e^{m_1 t_1}) \big|$$

$$+ |\rho_1(t_2) - \rho_1(t_1)| |\xi^\sigma| m_2(1 - e^{m_1 \xi}) - m_1(1 - e^{m_2 \xi})|$$

$$+ \sum_{i=1}^n |j_i| \eta_i^\sigma |m_2(1 - e^{m_1 \eta_i}) - m_1(1 - e^{m_2 \eta_i})|]$$

$$+ |\rho_2(t_2) - \rho_2(t_1)| [|m_2(1 - e^{m_1}) - m_1(1 - e^{m_2})|$$

$$+ \frac{\delta^\sigma |\lambda|}{|m_1 m_2|} |m_1^2(m_2 \delta - e^{m_2 \delta} + 1) - m_2^2(m_1 \delta - e^{m_1 \delta} + 1)|] \Bigg\} + |h(x)| \Bigg[|e^{m_2 t_2} - e^{m_2 t_1}|$$

$$+ (\rho_1(t_2) - \rho_1(t_1)) \Big(|e^{m_2 \xi}| + \sum_{i=1}^n |j_i| |e^{m_2 \eta_i} + 1| \Big)$$

$$+ (\rho_2(t_2) - \rho_2(t_1)) \Big(\frac{|m_2 e^{m_2}| + |\lambda| |e^{m_2 \delta} + 1|}{|m_2|} \Big) \Bigg],$$

which tends to zero independently of $x \in \mathcal{E}_\zeta$ as $t_2 - t_1 \to 0$. As $\mathcal{J}$ satisfies the above assumptions, therefore it follows by the Arzelá-Ascoli theorem that $\mathcal{J} : \mathcal{C} \to \mathcal{C}$ is completely continuous.

The result will follow from the Leray–Schauder nonlinear alternative once it is shown that there exists $\mathcal{U} \subseteq \mathcal{C}$ with $x \neq \theta \mathcal{J} x$ for $\theta \in (0,1)$ and $x \in \partial \mathcal{U}$.

Let $x \in \mathcal{C}$ be such that $x = \theta \mathcal{J} x$ for $\theta \in [0,1]$. Then, for $t \in [0,1]$, we have

$$|x(t)| = |\theta \mathcal{J} x(t)|$$

$$\leq \frac{1}{|q_2(m_2 - m_1)|} \sup_{t \in [0,1]} \Bigg\{ \int_0^t \int_0^s \mathcal{A}(t) \frac{(s-u)^{\sigma-1}}{\Gamma(\sigma)} |f(u,x(u))| du\, ds$$

$$+ |\rho_1(t)| \Bigg[\int_0^\xi \int_0^s \mathcal{A}(\xi) \frac{(s-u)^{\sigma-1}}{\Gamma(\sigma)} |f(u,x(u))| du\, ds$$

$$+ \sum_{i=1}^n |j_i| \int_0^{\eta_i} \int_0^s \mathcal{A}(\eta_i) \frac{(s-u)^{\sigma-1}}{\Gamma(\sigma)} |f(u,x(u))| du\, ds \Bigg]$$

$$+ |\rho_2(t)| \Bigg[\int_0^1 \int_0^s \mathcal{A}(1) \frac{(s-u)^{\sigma-1}}{\Gamma(\sigma)} |f(u,x(u))| du\, ds$$

$$+ |\lambda - \int_0^\delta \int_0^s \Big(\frac{(e^{m_2(\delta-s)} - 1)}{m_2} - \frac{(e^{m_1(\delta-s)} - 1)}{m_1} \Big) \frac{(s-u)^{\sigma-1}}{\Gamma(\sigma)} |f(u,x(u))| du\, ds \Bigg\}$$

$$+|h(x)|\Big[|e^{m_2 t}| + \rho_1(t)\Big(|e^{m_2\zeta}| + \sum_{i=1}^{n} |j_i||e^{m_2\eta_i} + 1|\Big)$$

$$+\rho_2(t)\Big(\frac{|m_2 e^{m_2}| + |\lambda||e^{m_2\delta} + 1|}{|m_2|}\Big)\Big]$$

$$\leq \frac{\|g\|\psi(\|x\|)}{|q_2(m_2-m_1)|} \sup_{t\in[0,1]} \Big\{\int_0^t |e^{m_2(t-s)} - e^{m_1(t-s)}|\frac{s^\sigma}{\Gamma(\sigma+1)}ds$$

$$+|\rho_1(t)|\Big[\int_0^\zeta |e^{m_2(\zeta-s)} - e^{m_1(\zeta-s)}|\frac{s^\sigma}{\Gamma(\sigma+1)}ds$$

$$+\sum_{i=1}^n |j_i| \int_0^{\eta_i} |e^{m_2(\eta_i-s)} - e^{m_1(\eta_i-s)}|\frac{s^\sigma}{\Gamma(\sigma+1)}ds\Big]$$

$$+|\rho_2(t)|\Big[\int_0^1 |e^{m_2(1-s)} - e^{m_1(1-s)}|\frac{s^\sigma}{\Gamma(\sigma+1)}ds$$

$$+|\lambda|\int_0^\delta \Big|\frac{(e^{m_2(\delta-s)}-1)}{m_2} - \frac{(e^{m_1(\delta-s)}-1)}{m_1}\Big|\frac{s^\sigma}{\Gamma(\sigma+1)}ds\Big]\Big\} + |h(x)|\Delta_1$$

$$\leq \frac{\|g\|\psi(\|x\|)}{|q_2 m_1 m_2(m_2-m_1)|\Gamma(\sigma+1)}\Big\{\varepsilon + \widehat{\rho}_1[\zeta^\sigma|m_2(1-e^{m_1\zeta}) - m_1(1-e^{m_2\zeta})|$$

$$+\sum_{i=1}^n |j_i|\eta_i^\sigma|m_2(1-e^{m_1\eta_i}) - m_1(1-e^{m_2\eta_i})|\Big] + \widehat{\rho}_2[|m_2(1-e^{m_1}) - m_1(1-e^{m_2})|$$

$$+\frac{\delta^\sigma|\lambda|}{|m_1 m_2|}|m_1^2(m_2\delta - e^{m_2\delta}+1) - m_2^2(m_1\delta - e^{m_1\delta}+1)|]\Big\} + L_1\Delta_1\|x\|$$

$$= \|g\|\psi(\|x\|)\alpha + L_1\Delta_1\|x\|,$$

which implies that

$$\frac{(1-L_1\Delta_1)\|x\|}{\|g\|\psi(\|x\|)\alpha} \leq 1.$$

In view of (H_3), there is no solution x such that $\|x\| \neq K$. Let us set

$$U = \{x \in \mathcal{C} : \|x\| < K\}.$$

The operator $\mathcal{J} : \overline{U} \to \mathcal{C}$ is continuous and completely continuous. From the choice of U, there is no $u \in \partial U$ such that $u = \theta\mathcal{J}(u)$ for some $\theta \in (0,1)$. Consequently, by the nonlinear alternative of Leray–Schauder type [28], we deduce that $\mathcal{J}$ has a fixed point $u \in \overline{U}$ which is a solution of the problem (1) and (2). □

Example 1. *Let us consider the following boundary value problem*

$$(2\,{}^c D^{12/5} + 3\,{}^c D^{7/5} + {}^c D^{2/5})x(t) = \frac{e^{-t}}{4\sqrt{4+t^2}} \tan^{-1} x + \cos t, \quad 0 < t < 1, \tag{26}$$

subject the boundary condition

$$x(0) = \frac{1}{9}\sin x(\hat{t}), \quad x(1/5) = x(1/4) + 2x(1/3) + x(1/2), \quad x(1) = 2\int_0^{1/6} x(s)ds. \tag{27}$$

Here, $q_2 = 2$, $q_1 = 3$, $q_0 = 1$, $\sigma = 2/5$, $\zeta = 1/5$, $\eta_1 = 1/4$, $\eta_2 = 1/3$, $\eta_3 = 1/2$, $\delta = 1/6$, $j_1 = 1$, $j_2 = 2$, $j_3 = 1$, $\lambda = 2$, $\hat{t}$ is a fixed value in $[0,1]$ and

$$f(t,x) = \frac{e^{-t}}{4\sqrt{4+t^2}} \tan^{-1} x + \cos t.$$

Clearly $q_1^2 - 4q_0q_2 = 1 > 0$, and

$$|f(t,x) - f(t,y)| \leq \frac{1}{8}|x-y|,$$

$$|h(x) - h(y)| \leq \frac{1}{9}\|x-y\|.$$

where $\ell = 1/8$, $L = 1/9$. Using the given values, we found $\alpha \approx 0.095961$, $\Delta_1 \approx 6.9171$.

It is easy to check that $|f(t,x)| \leq \frac{\pi e^{-t}}{8\sqrt{4+t^2}} + \cos t = \vartheta(t)$ and $L\Delta_1 < 1$. As all the condition of Theorem 2 are satisfied the problem (26) and (27) has at least one solution on $[0,1]$. On the other hand, $\ell\alpha + L\Delta_1 < 1$ and thus there exists a unique solution for the problem (26) and (27) on $[0,1]$ by Theorem 3.

Example 2. *Consider the following fractional differential equation*

$$(2\,^c D^{12/5} + 3\,^c D^{7/5} + \,^c D^{2/5})x(t) = \frac{1}{\pi\sqrt{9+t^2}}\left(x\tan^{-1}x + \pi/2\right),\ 0 < t < 1, \tag{28}$$

subject the boundary conditions (27).

Here

$$f(t,x) = \frac{1}{\pi\sqrt{9+t^2}}(x\tan^{-1}x + \pi/2).$$

Clearly

$$|f(t,x)| \leq \frac{1}{2\sqrt{9+t^2}}\Big(\|x\| + 1\Big),$$

with $g(t) = \frac{1}{2\sqrt{9+t^2}}$, $\psi(\|x\|) = \|x\| + 1$.

Then by using the condition (H_3), we find that $K > 0.241877$ (we have used $\alpha = 0.27045$). Thus, the conclusion of Theorem 5 applies to problem (28) and (27).

4. Existence Results for Problem (1) and (2) with $q_1^2 - 4q_0q_2 = 0$

In view of Lemma 3, we can transform problem (1) and (2) into equivalent fixed point problem as follows:

$$x = \mathcal{H}x, \tag{29}$$

where the operator $\mathcal{H} : \mathcal{C} \to \mathcal{C}$ is defined by

$$
\begin{aligned}
(\mathcal{H}x)(t) &= \frac{1}{q_2}\bigg\{\int_0^t\int_0^s \mathcal{B}(t)\frac{(s-u)^{\sigma-1}}{\Gamma(\sigma)}f(u,x(u))du\,ds \\
&\quad + \chi_1(t)\bigg[\int_0^\xi\int_0^s \mathcal{B}(\xi)\frac{(s-u)^{\sigma-1}}{\Gamma(\sigma)}f(u,x(u))du\,ds \\
&\quad - \sum_{i=1}^n j_i \int_0^{\eta_i}\int_0^s \mathcal{B}(\eta_i)\frac{(s-u)^{\sigma-1}}{\Gamma(\sigma)}f(u,x(u))du\,ds\bigg] \\
&\quad + \chi_2(t)\bigg[\int_0^1\int_0^s \mathcal{B}(1)\frac{(s-u)^{\sigma-1}}{\Gamma(\sigma)}f(u,x(u))du\,ds \\
&\quad - \lambda\int_0^\delta\int_0^s \left(\frac{m(\delta-s)e^{m(\delta-s)} - e^{m(\delta-s)} + 1}{m^2}\right)\frac{(s-u)^{\sigma-1}}{\Gamma(\sigma)}f(u,x(u))du\,ds\bigg]\bigg\} \\
&\quad + h(x)\bigg[e^{mt} + \chi_1(t)\Big(e^{m\xi} - \sum_{i=1}^n j_i e^{m\eta_i}\Big) + \chi_2(t)\Big(\frac{me^m - \lambda e^{m\delta} + \lambda}{m}\Big)\bigg],
\end{aligned}
\tag{30}
$$

where $\mathcal{B}(\cdot)$, $\chi_1(t)$ and $\chi_2(t)$ are defined by (14). We set

$$\widehat{\chi}_1 = \max_{t\in[0,1]}|\chi_1(t)|,\quad \widehat{\chi}_2 = \max_{t\in[0,1]}|\chi_2(t)|,$$

$$\beta = \frac{1}{|q_2|m^2\Gamma(\sigma+1)}\left\{(1+\widehat{\chi}_2)|me^m - e^m + 1| + \widehat{\chi}_1\left[\xi^\sigma|m\xi e^{m\xi} - e^{m\xi} + 1|\right.\right.$$
$$\left.\left. + \sum_{i=1}^{n}|j_i|\eta_i^\sigma|m\eta_i e^{m\eta_i} - e^{m\eta_i} + 1|\right] + \frac{|\lambda|\delta^\sigma\widehat{\chi}_2}{|m|}\left|m\delta(e^{m\delta}+1) + 2(1-e^{m\delta})\right|\right\}, \quad (31)$$

$$\Delta_2 = \max_{t \in [0,1]} |e^{mt}| + \widehat{\chi}_1\left(|e^{m\xi}| + \sum_{i=1}^{n}|j_i||e^{m\eta_i}|\right) + \widehat{\chi}_2\left(\frac{|me^m| + |\lambda||e^{m\delta} + 1|}{|m|}\right).$$

Now we present our main results for problem (1) and (2) with $q_1^2 - 4q_0q_2 = 0$. Since the methods for proof of these results are similar to the ones obtained in Section 3, so we omit the proofs.

Theorem 6. *Let $f : [0,1] \times \mathbb{R} \to \mathbb{R}$ be a continuous function satisfying the conditions (G_1)-(G_3). Then the problem (1) and (2) with $q_1^2 - 4q_0q_2 = 0$, has at least one solution on $[0,1]$ if*

$$L\Delta_2 < 1, \quad (32)$$

where Δ_2 is given by (31).

Theorem 7. *Assume that $f : [0,1] \times \mathbb{R} \to \mathbb{R}$ is a continuous function such that (G_1) is satisfied. Then there exists a unique solution for problem (1) and (2) with $q_1^2 - 4q_0q_2 = 0$, on $[0,1]$ if $\ell\beta + L\Delta_2 < 1$, where β and Δ_2 are given by (31).*

Theorem 8. *Let $f : [0,1] \times \mathbb{R} \to \mathbb{R}$ be a continuous function. Then the problem (1) and (2) with $q_1^2 - 4q_0q_2 = 0$, has at least one solution on $[0,1]$, if (H_1), (H_2) and the following condition hold:*

(H_3') *There exists a constant $K_1 > 0$ such that*

$$\frac{(1 - L_1\Delta_2)K_1}{\|g\|\psi(K_1)\beta} > 1,$$

where β is defined by (31).

Example 3. Consider the sequential fractional differential equation

$$(2\,{}^cD^{12/5} + 4\,{}^cD^{7/5} + 2\,{}^cD^{2/5})x(t) = \frac{|x|}{(t+6)(|x|+1)} + e^{-t}, \quad 0 < t < 1, \quad (33)$$

subject the boundary conditions (27).

Here

$$f(t,x) = \frac{|x|}{(t+6)(|x|+1)} + e^{-t}.$$

Clearly $q_1^2 - 4q_0q_2 = 0$, and

$$|f(t,x) - f(t,y)| \le \frac{1}{6}|x - y|,$$

$$|h(x) - h(y)| \le \frac{1}{9}\|x - y\|.$$

where $\ell = 1/6$, $L = 1/9$. Using the given values, we find that $\beta \approx 0.29913$, $\beta_1 \approx 0.15022$ and $\Delta_2 \approx 5.135$.

It is easy to check that $|f(t,x)| \le \frac{B}{t+6} + e^{-t} = \vartheta(t)$ and $L\Delta_2 < 1$. As all the conditions of Theorem 6 are satisfied, the problem (27)–(33) has at least one solution on $[0,1]$. On the other hand, $\ell\beta + L\Delta_2 < 1$ and thus there exists a unique solution for the problem (27)–(33) on $[0,1]$ by Theorem 7.

5. Existence Results for Problem (1) and (2) with $q_1^2 - 4q_0 q_2 < 0$

In view of Lemma 4, we can transform problem (1) and (2) into equivalent fixed point problem as follows:
$$x = \mathcal{K}x, \tag{34}$$
where the operator $\mathcal{K} : \mathcal{C} \to \mathcal{C}$ is defined by

$$\begin{aligned}(\mathcal{K}x)(t) &= \frac{1}{q_2 b}\Bigg\{ \int_0^t \int_0^s \mathcal{F}(t) \frac{(s-u)^{\sigma-1}}{\Gamma(\sigma)} f(u, x(u)) du\, ds \\
&\quad + \tau_1(t) \Bigg[\int_0^{\xi} \int_0^s \mathcal{F}(\xi) \frac{(s-u)^{\sigma-1}}{\Gamma(\sigma)} y(u) du\, ds \\
&\quad - \sum_{i=1}^n j_i \int_0^{\eta_i} \int_0^s \mathcal{F}(\eta_i) \frac{(s-u)^{\sigma-1}}{\Gamma(\sigma)} f(u, x(u)) du\, ds \Bigg] \\
&\quad + \tau_2(t) \Bigg[\int_0^1 \int_0^s \mathcal{F}(1) \frac{(s-u)^{\sigma-1}}{\Gamma(\sigma)} f(u, x(u)) du\, ds \\
&\quad - \frac{\lambda}{a^2 + b^2} \int_0^{\delta} \int_0^s \Big(b - be^{-a(\delta-s)} \cos b(\delta - s) \\
&\quad - ae^{-a(\delta-s)} \sin b(\delta - s) \Big) \frac{(s-u)^{\sigma-1}}{\Gamma(\sigma)} f(u, x(u)) du\, ds \Bigg] \Bigg\} \\
&\quad + h(x) \Bigg[e^{-at} \cos bt + \tau_1(t)(e^{-a\xi} \cos b\xi - \sum_{i=1}^n j_i e^{-a\eta_i} \cos b\eta_i) \\
&\quad + \tau_2(t)(e^{-a} \cos b - \frac{\lambda}{a^2 + b^2}(a - ae^{-a\delta} \cos b\delta + be^{-a\delta} \sin b\delta)) \Bigg],\end{aligned}$$

where $\mathcal{F}(\cdot)$, $\tau_1(t)$ and $\tau_2(t)$ are defined by (17). We set

$$\begin{aligned}\widehat{\tau}_1 &= \max_{t \in [0,1]} |\tau_1(t)|, \qquad \widehat{\tau}_2 = \max_{t \in [0,1]} |\tau_2(t)| \\
\gamma &= \frac{1}{|q_2 b(a^2+b^2)|\Gamma(\sigma+1)} \Big\{ (1+\widehat{\tau}_2)\Big[|b - be^{-a}\cos b - ae^{-a}\sin b|\Big] \\
&\quad + \widehat{\tau}_1 \Big[\xi^{\sigma}|b - be^{-a\xi}\cos b\xi - ae^{-a\xi}\sin b\xi| + \sum_{i=1}^n |j_i|\eta_i^{\sigma}|b - be^{-a\eta_i}\cos b\eta_i \\
&\quad - ae^{-a\eta_i}\sin b\eta_i| \Big] + |\lambda|\delta^{\sigma}\widehat{\tau}_2\Big[|b\delta - e^{-a\delta}\sin b\delta|\Big] \Big\}, \\
\Delta_3 &= \max_{t \in [0,1]} |e^{-at}\cos bt| + \widehat{\tau}_1(|e^{-a\xi}\cos b\xi| + \sum_{i=1}^n |j_i||e^{-a\eta_i}\cos b\eta_i|) \\
&\quad + \widehat{\tau}_2\Big(|e^{-a}\cos b| + \frac{|\lambda|}{a^2+b^2}(|a - ae^{-a\delta}\cos b\delta + be^{-a\delta}\sin b\delta|)\Big).\end{aligned} \tag{35}$$

Here are the existence and uniqueness results for problem (1) and (2) with $q_1^2 - 4q_0 q_2 < 0$. As argued in the last section, we do not provide the proofs for these results.

Theorem 9. *Let $f : [0,1] \times \mathbb{R} \to \mathbb{R}$ be a continuous function satisfying the conditions (G_1)–(G_3). Then the problem (1) and (2) with $p_1^2 - 4p_0 p_2 < 0$, has at least one solution on $[0,1]$ if*

$$L\Delta_3 < 1, \tag{36}$$

where γ_1 and Δ_3 are given by (35).

Theorem 10. Assume that $f : [0,1] \times \mathbb{R} \to \mathbb{R}$ is a continuous function such that (G_1) and (G_2) are satisfied. Then there exists a unique solution for the problem (1) and (2) with $q_1^2 - 4q_0q_2 < 0$, on $[0,1]$ if $\ell\gamma + L\Delta_3 < 1$, where γ and Δ_3 are given by (35).

Theorem 11. Let $f : [0,1] \times \mathbb{R} \to \mathbb{R}$ be a continuous function. Then the problem (1) and (2) with $q_1^2 - 4q_0q_2 < 0$, has at least one solution on $[0,1]$, if $(H_1), (H_2)$ and the following condition are satisfied:

(H_3'') There exists a constant $K_2 > 0$ such that

$$\frac{(1 - L_1\Delta_3)K_2}{\|g\|\psi(K_2)\gamma} > 1,$$

where γ and Δ_3 are defined by (35).

Example 4. Consider the following boundary value problem

$$(2\,{}^cD^{12/5} + 3\,{}^cD^{7/5} + 2\,{}^cD^{2/5})x(t) = \frac{1}{(t+4)^2}\cos x + \frac{e^{-2t}}{13}, \quad 0 < t < 1, \quad (37)$$

subject the boundary condition

$$x(0) = \frac{1}{8}x(\hat{t}), \quad x(1/5) = x(1/4) + 2x(1/3) + x(1/2), \quad x(1) = 2\int_0^{1/6} x(s)ds. \quad (38)$$

Here, $\sigma = 2/5$, $\xi = 1/5$, $\eta_1 = 1/4$, $\eta_2 = 1/3$, $\eta_3 = 1/2$, $\delta = 1/6$, $j_1 = 1$, $j_2 = 2$, $j_3 = 1$, $\lambda = 2$, $\hat{t}$ is a fixed value in $[0,1]$ and

$$f(t,x) = \frac{1}{(t+4)^2}\cos x + \frac{e^{-2t}}{13}.$$

Clearly $q_1^2 - 4q_0q_2 = -7 < 0$, and

$$|f(t,x) - f(t,y)| \leq \frac{1}{16}|x-y|,$$

$$|h(x) - h(y)| \leq \frac{1}{8}\|x-y\|,$$

where $\ell = 1/16$, $L = 1/8$. Using the given values, it is found that $\gamma \approx 0.34744$, $\gamma_1 \approx 0.17937$ and $\Delta_3 \approx 1.8499$.

Obviously $|f(t,x)| \leq \frac{1}{(t+4)^2} + \frac{e^{-2t}}{13} = \vartheta(t)$ and $L\Delta_3 < 1$. As the hypothesis of Theorem 9 holds true, the problem (37) and (38) has at least one solution on $[0,1]$. Furthermore, we have $\ell\gamma + L\Delta_3 < 1$, which implies that there exists a unique solution for the problem (37) and (38) on $[0,1]$ by Theorem 10.

6. Conclusions

We have presented a detailed analysis for a multi-term fractional differential equation supplemented with nonlocal multi-point integral boundary conditions. The existence and uniqueness results are given for all three cases depending on the coefficients of the multi-term fractional differential equation: (i) $q_1^2 - 4q_0q_2 > 0$, (ii) $q_1^2 - 4q_0q_2 = 0$ and (iii) $q_1^2 - 4q_0q_2 < 0$. Existence results are proved by means of Krasnoselskii fixed point theorem and Leray–Schauder nonlinear alternative, while Banach contraction mapping principle is applied to establish the uniqueness of solutions for the given problem. The obtained results are well-illustrated with examples. Our results are new and enrich the literature on nonlocal integro-multipoint boundary problems for multi-term Caputo type fractional differential equations.

Author Contributions: Conceptualization, B.A. and S.K.N.; Formal analysis, B.A., N.A., A.A. and S.K.N.; Funding acquisition, A.A.; Methodology, B.A., N.A., A.A. and S.K.N. All authors have read and agreed to the published version of the manuscript.

Funding: This research received no external funding.

Conflicts of Interest: The authors declare no conflict of interest.

References

1. Herrmann, R. *Fractional Calculus: An Introduction for Physicists*; World Scientific: Singapore, 2011.
2. Magin, R.L. *Fractional Calculus in Bioengineering*; Begell House Publishers: Danbury, CT, USA, 2006.
3. Mainardi, F. *Fractional Calculus and Waves in Linear Viscoelasticy*; World Scientific: Singapore, 2010.
4. Javidi, M.; Ahmad, B. Dynamic analysis of time fractional order phytoplankton-toxic phytoplankton—Zooplankton system. *Ecol. Model.* **2015**, *318*, 8–18. [CrossRef]
5. Arafa, A.A.M.; Rida, S.Z.; Khalil, M. Fractional modeling dynamics of HIV and CD4+ T-cells during primary infection. *Nonlinear Biomed. Phys.* **2012**, *6*, 1–7. [CrossRef] [PubMed]
6. Ding, Y.; Wang, Z.; Ye, H. Optimal control of a fractional-order HIV-immune system with memory. *IEEE Trans. Control Syst. Technol.* **2012**, *20*, 763–769. [CrossRef]
7. Carvalho, A.; Pinto, C.M.A. A delay fractional order model for the co-infection of malaria and HIV/AIDS. *Int. J. Dyn. Control* **2017**, *5*, 168–186. [CrossRef]
8. Torvik, P.J.; Bagley, R.L. On the appearance of the fractional derivative in the behavior of real materials. *J. Appl. Mech.* **1984**, *51*, 294–298. [CrossRef]
9. Mainardi, F. Some basic problems in continuum and statistical mechanics. In *Fractals and Fractional Calculus in Continuum Mechanics*; Carpinteri, A., Mainardi, F., Eds.; Springer: Berlin, Germany, 1997; pp. 291–348.
10. Thaiprayoon, C.; Ntouyas, S.K.; Tariboon, J. On the nonlocal Katugampola fractional integral conditions for fractional Langevin equation. *Adv. Differ. Equ.* **2015**, *2015*, 374. [CrossRef]
11. Zhou, Y. *Fractional Evolution Equations and Inclusions: Analysis and Control*; Elsevier/Academic Press: London, UK, 2016.
12. Stanek, S. Periodic problem for two-term fractional differential equations. *Fract. Calc. Appl. Anal.* **2017**, *20*, 662–678. [CrossRef]
13. Ahmad, B.; Alsaedi, A.; Ntouyas, S.K.; Tariboon, J. *Hadamard-Type Fractional Differential Equations, Inclusions and Inequalities*; Springer: Cham, Switzerland, 2017.
14. Henderson, J.; Luca, R. Positive solutions for a system of coupled fractional boundary value problems. *Lith. Math. J.* **2018**, *58*, 15–32. [CrossRef]
15. Ntouyas, S.K.; Ahmad, B.; Alghanmi, M.; Alsaedi, A. Generalized fractional differential equations and inclusions equipped with nonlocal generalized fractional integral boundary conditions. *Topol. Methods Nonlinear Anal.* **2019**, *54*, 1051–1073. [CrossRef]
16. Agarwal, R.; Hristova, S.; O'Regan, D. Mittag-Leffler stability for non-instantaneous impulsive Caputo fractional differential equations with delays. *Math. Slovaca* **2019**, *69*, 583–598. [CrossRef]
17. Ahmad, B.; Ntouyas, S.K.; Alsaedi, A. Existence theory for nonlocal boundary value problems involving mixed fractional derivatives. *Nonlinear Anal. Model. Control* **2019**, *24*, 937–957. [CrossRef]
18. Iskenderoglu, G.; Kaya, D. Symmetry analysis of initial and boundary value problems for fractional differential equations in Caputo sense. *Chaos Solitons Fractals* **2020**, *134*, 109684. [CrossRef]
19. Ahmad, B.; Broom, A.; Alsaedi, A.; Ntouyas, S.K. Nonlinear integro-differential equations involving mixed right and left fractional derivatives and integrals with nonlocal boundary data. *Mathematics* **2020**, *8*, 336. [CrossRef]
20. Cen, Z.; Liu, L.-B.; Huang, J. A posteriori error estimation in maximum norm for a two-point boundary value problem with a Riemann-Liouville fractional derivative. *Appl. Math. Lett.* **2020**, *102* 106086. [CrossRef]
21. Liu, K.; Wang, J.-R.; Zhou, Y.; O'Regan, D. Hyers-Ulam stability and existence of solutions for fractional differential equations with Mittag-Leffler kernel. *Chaos Solitons Fractals* **2020**, *132*, 109534. [CrossRef]
22. Alsaedi, A.; Alghamdi, N.; Agarwal, R.P.; Ntouyas, S.K.; Ahmad, B. Multi-term fractional-order boundary-value problems with nonlocal integral boundary conditions. *Electron. J. Differ. Equ.* **2018**, *87*, 16.

23. Ahmad, B.; Alghamdi, N.; Alsaedi, A.; Agarwal, R.P.; Ntouyas, S.K. A system of coupled multi-term fractional differential equations with three-point coupled boundary conditions. *Fract. Calc. Appl. Anal.* **2019**, *22*, 601–618. [CrossRef]
24. Ruzhansky, M.; Tokmagambetov, N.; Torebek, B.T. On a non-local problem for a multi-term fractional diffusion-wave equation. *Fract. Calc. Appl. Anal.* **2020**, *23*, 324–355. [CrossRef]
25. Kilbas, A.A.; Srivastava, H.M.; Trujillo, J.J. Theory and Applications of Fractional Differential Equations; In *North-Holland Mathematics Studies*; Elsevier Science B.V.: Amsterdam, The Netherlands, 2006; Volume 2014.
26. Zhou, Y. *Basic theory of Fractional Differential Equations*; World Scientific Publishing Co. Pte. Ltd.: Hackensack, NJ, USA, 2014.
27. Krasnoselskii, M.A. Two remarks on the method of successive approximations. *Uspekhi Mat. Nauk* **1955**, *10*, 123–127.
28. Granas, A.; Dugundji, J. *Fixed Point Theory*; Springer: New York, NY, USA, 2005.

© 2020 by the authors. Licensee MDPI, Basel, Switzerland. This article is an open access article distributed under the terms and conditions of the Creative Commons Attribution (CC BY) license (http://creativecommons.org/licenses/by/4.0/).

Article

Approximate Methods for Solving Linear and Nonlinear Hypersingular Integral Equations

Ilya Boykov [1,*], Vladimir Roudnev [2] and Alla Boykova [1]

1. Department of Mathematics, The Penza State University, 40, Krasnaya Str., 440026 Penza, Russia; allaboikova@mail.ru
2. Department of Computational Physics, Saint Petersburg State University, 7/9 Universitetskaya Emb., 199034 Saint Petersburg, Russia; v.rudnev@spbu.ru
* Correspondence: boikov@pnzgu.ru

Received: 2 June 2020; Accepted: 25 June 2020 ; Published: 1 July 2020

Abstract: We propose an iterative projection method for solving linear and nonlinear hypersingular integral equations with non-Riemann integrable functions on the right-hand sides. We investigate hypersingular integral equations with second order singularities. Today,hypersingular integral equations of this type are widely used in physics and technology. The convergence of the proposed method is based on the Lyapunov stability theory of solutions of ordinary differential equation systems. The advantage of the method for linear equations is in simplicity of unique solvability verification for the approximate equations system in terms of the operator logarithmic norm. This makes it possible to estimate the norm of the inverse matrix for an approximating system. The advantage of the method for nonlinear equations is that neither the existence or reversibility of the nonlinear operator derivative is required. Examples are given illustrating the effectiveness of the proposed method.

Keywords: hypersingular integral equations; iterative projection method; Lyapunov stability theory

1. Introduction

The importance of developing analytical and numerical methods for solving hypersingular integral equations is determined by a variety of fields of mathematics and by applications that use hypersingular integral equations.

Hadamard introduced the concept of a finite part of an integral, or the hypersingular integral in modern terminology, when studying hyperbolic equations. The Riemann boundary problem leads to hypersingular integral equations in exceptional cases. The boundary integral equations method reduces the dimensions of partial differential equations; that leads to hypersingular integral equations.

Hypersingular integral equations, singular integral equations and Riemann boundary problem are widely used in aerodynamics, electrodynamics, quantum physics, antennae theory and many other fields of physics and engineering [1–5].

Analytical methods for solving singular and hypersingular integral equations are known only for certain particular types of equations [6–8]. Thus, the importance of constructing numerical solutions is clear.

Developing numerical methods for solving singular integral equations began in the middle of the last century. By now, exhaustive results have been obtained for many types of equations. A detailed account of numerical methods for solving singular integral equations as well as numerous bibliography references can be found in [9–14].

Numerical methods for solving hypersingular integral equations have been developed to a much lesser extent. Mostly numerical methods to solve hypersingular integral equations of the first kind have been developed. Numerical methods for solving hypersingular integral equations of the second kind have been much less developed. Apparently, hypersingular integral equations of the first kind are more common. Naturally, the equations of the first kind are widely used in aerodynamics (one-dimensional [15] and multi-dimensional [5,16] Prandtl equation), electrodynamics, antennae theory, etc.

The following methods are used in solving hypersingular integral equations of the first kind.

Collocations, mechanical quadratures and Galerkin methods were employed to solve equations with $p = 2$ singularity [6,17–19].

Approximate methods for solving hypersingular integral equations having singularities of order $p = 2, 3, \ldots$, and defined on closed smooth integration contours are constructed in [20].

In [21,22] spline-collocation methods for solving hypersingular and polyhypersingular integral equations of the second kind with odd and even singularities have been developed and justified. The spline-collocation methods for solving nonlinear hypersingular and polyhypersingular integral equations have been developed and justified in [23].

An iterative projection method for solving linear and nonlinear hypersingular integral equations, and polyhyperpersingular and multidimensional hypersingular equations, was proposed in [24].

In [22] the unique solvability of hypersingular integral equations with even singularities ($p = 2, 4, \ldots$) was proven. Meanwhile the convergence of approximate solution to the exact one was not justified. In [24] a unique solvability of the spline-collocation method was proven. In addition, for hypersingular integral equations with bounded right-hand sides the convergence of an approximate solution sequence to the exact solution was proven under certain additional conditions.

The iterative projection method proposed here overcomes these limitations. It was shown that if the exact equation has a solution for large enough N, where N is the dimension of an approximate system of equations, an approximate solution converges to the exact one.

Hypersingular integral equations with bounded right-hand sides are a small subset of the hypersingular integral equations family. Therefore, the problem arises of constructing and justifying approximate methods for solutions for hypersingular integral equations with non-Riemann integrable functions on the right-hand sides. This paper is devoted to those issues.

A large number of works are devoted to approximate methods for solving hypersingular integral equations of the first kind

$$\int_{-1}^{1} \frac{x(\tau)d\tau}{(\tau-t)^2} + \int_{-1}^{1} h(t,\tau)d\tau = f(t). \qquad (1)$$

To solve the Equation (1), collocation and mechanical quadrature methods [17,18], the method of orthogonal polynomials [25], the method of discrete vortices [19], the method of homotopy [26] and others are used.

In the works [27–29] computational schemes for the approximate solution of the Equation (1) are constructed and their justification is carried out under the assumption that the solution has the forms $x(t) = (1-t^2)^{\pm 1/2}\omega(t)$ or $x(t) = ((1-t)/(1+t))^{\pm 1/2}\omega(t)$, where $\omega(t)$ is a smooth function.

The hypersingular integral equations

$$\frac{1}{\pi}\int_{-1}^{1} \frac{x(\tau)d\tau}{(\tau-t)^2} = f(t), \quad -1 < t < 1, \qquad (2)$$

are widely used in aerodynamical problems and in the theory of antennae [30,31]. In the works [30,31] the Equation (2) is investigated under the assumption that the right-hand side has the form $f(t) = 1/(t-c)$ or $f(t) = \delta(t-c)$, where $\delta(t)$ is the delta-function. An analytical solution of the Equation (2) with the indicated right-hand sides is obtained under the assumption that it has the form $x(t) = \sqrt{1-t^2}\varphi(t)$.

A fairly detailed review of analytical and numerical methods for solving hypersingular integral equations is given in [32].

In this paper, we propose an approach to solving linear and nonlinear hypersingular integral equations, the right parts of which contain functions with power features.

In particular, the right-hand sides of the form

$$f(t) = g(t) \frac{1}{t-c_1} \frac{1}{t-c_2} \cdots \frac{1}{t-c_l}, l = 1, 2, \ldots, -1 < c_1 < \cdots < c_l < 1, \quad (3)$$

are considered. Here $g(t)$ is a smooth function.

Below, for simplicity of notation, we put $l = 1$ in (3).

Remark 1. *It can be shown that if in the hypersingular integral Equation (1) of the first kind the right side $f(t) \in H$, H is a Holder class, then the solution to this equation has the form $x(t) = (1-t^2)^{\pm 1/2}$ or $x(t) = ((1+t)/(1-t))^{\pm 1/2}$. For singular right-hand sides, the classes of solutions of (1) are unknown.*

Below, when constructing and justifying the computational method, we assume that the Equation (1) with a given right-hand side has a unique solution.

The proposed method has the following advantages:

(1) It allows us to extend collocations and mechanical quadratures methods to hypersingular integral equations with non-Riemann integrable right sides;
(2) For linear hypersingular integral equations, it allows one to verify the inverse operator existence and estimate its norm quite easily;
(3) The method is stable with respect to the operator and right hand side perturbations;
(4) The method does not require the existence and reversibility of the nonlinear operator derivative.

The paper is organized as follows. The continuous method for linear and nonlinear operator equations is explained in Section 2. The numerical method for solving hypersingular integral equations is presented in Section 2.

2. Continuous Method and Its Convergence Properties

The method we employ in the next section for solving hypersingular integral equations is based on the continuous method introduced in [33].

Continuous Method for Solving Operator Equations

The continuous method for solving operator equations is based on the Lyapunov theory of stability.

Let $x(t)$ be a solution of the differential equation in a Banach space B

$$\frac{dx}{dt} = F(t, x) \quad (4)$$

which is defined for all $t \geq t_0$. The solution $x(t)$ is said to be stable if (i) for each $\varepsilon > 0$ there is a corresponding $\delta = \delta(\varepsilon) > 0$ such that any solution $\tilde{x}(t)$ of (4) which satisfies the inequality $|\tilde{x}(t_0) - x(t_0)| < \delta$ exists and satisfies the inequality $|\tilde{x}(t) - x(t)| < \varepsilon$ for all $t \geq t_0$.

It is said to be asymptotically stable if in addition (ii) $|\tilde{x}(t) - x(t)| \to 0$ if $t \to \infty$ whenever $|\tilde{x}(t_0) - x(t_0)|$ is sufficiently small.

We will use the following notation:

$B(a,r) = \{z \in B : \|z - a\| \leq r\}$, $S(a,r) = \{z \in B : \|z - a\| = r\}$, $Re(K) = \Re(K) = (K + K^*)/2$, $\Lambda(K) = \lim_{h \downarrow 0}(\|I + hK\| - 1)h^{-1}$.

Here B is a Banach space, $a \in B$, K is a linear operator on B, $\Lambda(K)$ is the logarithmic norm [34] of the operator K, K^* is the conjugate operator to K and I is the identity operator.

The analytical expressions for logarithmic norms are known for operators in many spaces. We restrict ourselves to a description of the three norms.

Let $A = \{a_{ij}\}, i, j = 1, 2, \ldots, n$, be a matrix.

In the n-dimensional space R_n of vectors $x = (x_1, \ldots, x_n)$ the following norms are often used:

$$\text{octahedral- } \|x\|_1 = \sum_{i=1}^{n} |x_i|; \quad \text{cubic- } \|x\|_2 = \max_{1 \leq i \leq n} |x_i|;$$

$$\text{spherical (Euclidean)- } \|x\|_3 = (\sum_{i=1}^{n} x_i^2)^{1/2}.$$

Here are analytical expressions of the logarithmic norm of $n \times n$ matrix $A = (a_{ij})$, due to the above norms of the vectors:

octahedral logarithmic norm Λ_1

$$\Lambda_1(A) = \max_{1 \leq j \leq n} \left(a_{jj} + \sum_{i \neq j} |a_{ij}| \right);$$

cubic logarithmic norm Λ_2

$$\Lambda_2(A) = \max_{1 \leq i \leq n} \left(a_{ii} + \sum_{j \neq i} |a_{ij}| \right);$$

spherical (Euclidean) logarithmic norm Λ_3

$$\Lambda_3(A) = \lambda_{\max} \left(\frac{A + A^*}{2} \right),$$

where A^* is the conjugate matrix for A.

Note that the logarithmic norm of the same matrix can be positive in one space and negative in another.

The logarithmic norm has the some properties which are very useful for numerical mathematics.

Let A, B be $n \times n$ matrices with complex elements; and $x = (x_1, \ldots, x_n)$, $y = (y_1, \ldots, y_n)$, $\xi = (\xi_1, \ldots, \xi_n)$ and $\eta = (\eta_1, \ldots, \eta_n)$ are n-dimensional vectors with complex components. Let the systems of algebraic equations $Ax = \xi$ and $By = \eta$ be given. The norm of a vector and its subordinate operator norm of the matrix are agreed upon; the logarithmic norm $\Lambda(A)$ corresponds to the operator norm.

Theorem 1 ([35]). *If $\Lambda(A) < 0$, the matrix A is non-singular and $\|A^{-1}\| \leq 1/|\Lambda(A)|$.*

Theorem 2 ([35]). *Let $Ax = \xi$, $By = \eta$ and $\Lambda(A) < 0$, $\Lambda(B) < 0$. Then*

$$\|x - y\| \leq \frac{\|\xi - \eta\|}{|\Lambda(B)|} + \frac{\|A - B\|}{|\Lambda(A)\Lambda(B)|}.$$

Some properties of the logarithmic norm in a Banach space, which are useful in numerical mathematics, are given in [34].

Let us consider in a Banach space B, the Cauchy problem

$$\frac{dx(t)}{dt} = A(x(t)), \tag{5}$$

$$x(0) = x_0. \tag{6}$$

Let us assume that the nonlinear operator A has a Frechet derivative and $A(0) = 0$.

The sufficiently satisfying conditions of asymptotically stability for the solution of the Cauchy problem (5), (6) were obtained in [36,37].

Theorem 3. *Let the integral $\int_0^t \Lambda(A'(\varphi(\tau)))d\tau$ be non-positive (respectively, be negative and satisfy $\lim_{t\to+\infty} \frac{1}{t}\int_0^t \Lambda(A'(\varphi(\tau)))d\tau \leq -\alpha_\varphi, \alpha_\varphi > 0$) for any differentiable curve $\varphi(t)$ lying in a ball $B(0,r)$ of some radius r. Then the trivial solution of Equation (5) is stable (respectively, asymptotically stable).*

Remark 2. *Additionally, the Theorem is valid under $r = \infty$.*

Let us consider in a Banach space B a nonlinear operator equation

$$A(x) - f = 0, \qquad (7)$$

where operator A acts from B into B.

We associate Equation (7) with the Cauchy problem

$$\frac{dx(t)}{dt} = A(x(t)) - f, \qquad (8)$$

$$x(0) = x_0. \qquad (9)$$

Let x^* be a the solution of Equation (7). Let us make the change of variable $x = x^* + v$ in Equation (8). This change reduces the Cauchy problem (8), (9) to the form

$$\frac{dv(t)}{dt} = A(x^* + v(t)) - A(x^*), \qquad (10)$$

$$v(0) = x_0 - x^*. \qquad (11)$$

It is easy to see that if the trivial solution of Equation (10) is globally asymptotically stable, then $\lim_{t\to+\infty} \|v(t)\| \to 0$. So, for any initial value the solution of the Cauchy problem (8), (9) tends to x^*. It follows from the next assertions which were proven in [33].

Theorem 4. *Let Equation (7) have a solution x^*, and let inequality*

$$\lim_{t\to+\infty} \frac{1}{t}\int_0^t \Lambda(A'(g(\tau)))d\tau \leq -\alpha_g, \ \alpha_g > 0, \qquad (12)$$

be true on each differentiable curve $g(t)$ lying in the Banach space B. Then the solution of the Cauchy problem (8), (9) converges to the solution x^* of Equation (7) for any initial value.

Theorem 5. *Let Equation (7) have a solution x^*, and let the following conditions be satisfied on any differentiable curve $g(t)$ lying in the ball $B(x^*,r)$.*

1. *The inequality*

$$\int_0^t \Lambda(A'(g(\tau)))d\tau \leq 0$$

   holds for all $t(t>0)$.
2. *Inequality (12) is satisfied.*

Then the solution of the Cauchy problem (8), (9) converges to the solution x^* of Equation (7).

Remark 3. *The sufficient condition for convergence of the Cauchy problem (8), (9) solution to the solution of the operator Equation (7) is given above. It was obtained by analysing Lyapunov stability. One of the first basic* results in accretive operator theory was a relation between the solution of operator equation $Au = 0$, where A is a locally Lipschitzian and accretive operator, and the differential equation $\frac{du}{dt} = Au$ was obtained in [38].

Later, accretive operator theory and its applications for finding fixed points and constructing iterative procedures were studied by many authors. Basic results and a detailed bibliography devoted to the subject may be found in [39–42].

3. An Solution of Hypersingular Integral Equations with the Continuous Method

Let us consider the method of mechanical quadrature for solving hypersingular integral equation of the types

$$a(t)x(t) + \int_{-1}^{1} \frac{h(t,\tau)x(\tau)d\tau}{(\tau-t)^2} = f(t). \tag{13}$$

and

$$a(t)x(t) + \int_{-1}^{1} \frac{h(t,\tau,x(\tau))d\tau}{(\tau-t)^2} = f(t). \tag{14}$$

It is assumed that in the Equations (13) and (14) the right-hand sides have features of the following types

$$f(t) = \sum_{i=1}^{l} g_i(t) \frac{1}{t-c_i}, \quad f(t) = g(t) \prod_{i=1}^{l} \frac{1}{t-c_i},$$

where $-1 < c_i < 1, i = 1, 2, \ldots, l, l = 1, 2, \ldots; g(t), g_i(t), i = 1, 2, \cdots, l,$—are continuous functions.

In what follows, without loss of generality, we set $l = 1$.

Let us recall the Hadamard definition of hypersingular integrals [43].

Definition 1 ([43]). *The integral of the type*

$$\int_a^b \frac{A(x)\,dx}{(b-x)^{p+\alpha}}$$

for an integer p and $0 < \alpha < 1$, is defined as

$$\lim_{x \to b} \left[\int_a^x \frac{A(t)\,dt}{(b-t)^{p+\alpha}} + \frac{B(x)}{(b-x)^{p+\alpha-1}} \right],$$

if $A(x)$ has p derivatives in the neighborhood of the point b. Here $B(x)$ is any function that satisfies the following two conditions:

(i) *The above limit exists;*
(ii) *$B(x)$ has at least p derivatives in the neighborhood of the point $x = b$.*

It is easy to see [43], that the conditions (i) and (ii) are sufficient for the existence of the limit. Chikin in [44] introduced the following definition.

Definition 2 ([44]). *The Cauchy–Hadamard principal value of the integral*

$$\int_a^b \frac{\varphi(\tau)\,d\tau}{(\tau-c)^p}, \quad a < c < b, \tag{15}$$

is defined as

$$\int_a^b \frac{\varphi(\tau)\,d\tau}{(\tau-c)^p} = \lim_{v \to 0} \left[\int_a^{c-v} \frac{\varphi(\tau)\,d\tau}{(\tau-c)^p} + \int_{c+v}^b \frac{\varphi(\tau)\,d\tau}{(\tau-c)^p} + \frac{\xi(v)}{v^{p-1}} \right],$$

where $\xi(v)$ is a function constructed so that the limit exists.

3.1. An Approximate Solution of Linear Hypersingular Integral Equations with Second Order Singularity

Consider a one-dimensional hypersingular integral equation of the type

$$Kx \equiv a(t)x(t) + \int_{-1}^{1} \frac{h(t,\tau)x(\tau)d\tau}{(\tau-t)^2} = f(t), \tag{16}$$

where $f(t) = g(t)/(t-c)$ or $f(t) = g(t)/((1-t^2)(t-c))$, $-1 < c < 1$, $g(t) \in C[-1,1]$.

Divide the interval $[-1,1]$ into two subintervals $[-1,c]$, $[c,1]$.

Let us fix a positive integer N_0. Put $h = 2/N_0$, $N_1 = \lceil(1+c)/h\rceil$, $N_2 = \lceil(1-c)/h\rceil$, $N = N_1 + N_2$.

Divide the interval $[-1,c]$ into N_1 subintervals at the points $t_k = -1 + (c+1)k/N_1$, $k = 0, 1, \ldots, N_1$.

Divide the interval $[c,1]$ into N_2 subintervals at the points $\tau_k = c + (1-c)k/N_2$, $k = 0, 1, \ldots, N_2$.

Let us introduce the nodes $\bar{t}_0 = t_0 + 1/2(N_1)^2$, $\bar{t}_k = t_k$, $k = 1, 2, \ldots, N_1 - 1$, $\bar{t}_{N_1} = t_{N_1} - 1/2(N_1)^2$; $\bar{\tau}_0 = \tau_0 + 1/2(N_2)^2$, $\bar{\tau}_k = \tau_k$, $k = 1, 2, \ldots, N_2 - 1$, $\bar{\tau}_{N_2} = 1 - 1/2(N_2)^2$.

As an approximate solution of (16), we shall seek in the form of a continuous function

$$x_N(t) = \sum_{k=0}^{N_1} \alpha_k \varphi_k(t) + \sum_{k=0}^{N_2} \beta_k \psi_k(t), \tag{17}$$

where $\varphi_k(t), k = 0, 1, \ldots, N_1, \psi_k(t), k = 0, 1, \ldots, N_2$ is a family of basis functions.

For nodes $t_k, k = 1, \ldots, N_1 - 1$, the corresponding basis elements are determined by

$$\varphi_k(t) = \begin{cases} 0, & t_{k-1} \leq t \leq t_{k-1} + \frac{1}{N_1^2}, \\ \frac{N_1^2}{(1+c)N_1-2}(t-t_{k-1}) - \frac{1}{(1+c)N_1-2}, & t_{k-1} + \frac{1}{N_1^2} \leq t \leq t_k - \frac{1}{N_1^2}, \\ 1, & t_k - \frac{1}{N_1^2} \leq t \leq t_k + \frac{1}{N_1^2}, \\ -\frac{N_1^2}{(1+c)N_1-2}(t-t_k - \frac{1}{N_1^2}) + 1, & t_k + \frac{1}{N_1^2} \leq t \leq t_{k+1} - \frac{1}{N_1^2}, \\ 0, & t_{k+1} - \frac{1}{N_1^2} \leq t \leq t_{k+1}, \\ 0, & t \in [-1,1] \setminus [t_{k-1}, t_{k+1}]. \end{cases} \tag{18}$$

For boundary nodes $t_k, k = 0$ and $k = N_1$ the corresponding basis elements are defined as

$$\varphi_0(t) = \begin{cases} 1, & -1 \leq t \leq -1 + \frac{1}{N_1^2}, \\ -\frac{N_1^2}{(1+c)N_1-2}(t+1-\frac{1}{N_1^2}) + 1, & -1 + \frac{1}{N_1^2} \leq t \leq t_1 - \frac{1}{N_1^2}, \\ 0, & t_1 - \frac{1}{N_1^2} \leq t \leq t_1, \\ 0, & [-1,1] \setminus [t_0, t_1]; \end{cases} \tag{19}$$

and

$$\varphi_{N_1}(t) = \begin{cases} 0, & -1 \leq t \leq t_{N_1-1} + \frac{1}{N_1^2}, \\ \frac{N_1^2}{(1+c)N_1-2}(t-t_{N_1-1}) - \frac{1}{(1+c)N_1-2}, & t_{N_1-1} + \frac{1}{N_1^2} \leq t \leq c - \frac{1}{N_1^2}, \\ 1, & c - \frac{1}{N_1^2} \leq t \leq c. \end{cases} \tag{20}$$

For nodes τ_k, $k = 0, 1, \ldots, N_2$, the corresponding basis elements ψ_k, $k = 0, 1, \ldots, N_2$, are determined in a the similar way: For nodes τ_k, $k = 1, \ldots, N_2 - 1$, the corresponding basis elements are determined by

$$\psi_k(t) = \begin{cases} 0, & \tau_{k-1} \leq t \leq \tau_{k-1} + \frac{1}{N_2^2}, \\ \frac{N_2^2}{(1-c)N_2-2}(t - \tau_{k-1}) - \frac{1}{(1-c)N_2-2}, & \tau_{k-1} + \frac{1}{N_2^2} \leq t \leq \tau_k - \frac{1}{N_2^2}, \\ 1, & \tau_k - \frac{1}{N_2^2} \leq t \leq \tau_k + \frac{1}{N_2^2}, \\ -\frac{N_2^2}{(1-c)N_2-2}(t - \tau_k - \frac{1}{N_2^2}) + 1, & \tau_k + \frac{1}{N_2^2} \leq t \leq \tau_{k+1} - \frac{1}{N_2^2}, \\ 0, & \tau_{k+1} - \frac{1}{N_2^2} \leq t \leq \tau_{k+1}, \\ 0, & t \in [-1,1] \setminus [\tau_{k-1}, \tau_{k+1}]. \end{cases} \quad (21)$$

For boundary nodes τ_k, $k = 0$ and $k = N_2$ the corresponding basis elements are defined as

$$\psi_0(t) = \begin{cases} 1, & c \leq t \leq c + \frac{1}{N_1^2}, \\ -\frac{N_2^2}{(1-c)N_2-2}(t - c - \frac{1}{N_2^2}) + 1, & c + \frac{1}{N_2^2} \leq t \leq \tau_1 - \frac{1}{N_2^2}, \\ 0, & \tau_1 - \frac{1}{N_2^2} \leq t \leq \tau_1, \\ 0, & [-1,1] \setminus [c, \tau_1]; \end{cases} \quad (22)$$

and

$$\psi_{N_2}(t) = \begin{cases} 0, & -1 \leq t \leq \tau_{N_2-1} + \frac{1}{N_2^2}, \\ \frac{N_2^2}{(1-c)N_2-2}(t - \tau_{N_2-1}) - \frac{1}{(1-c)N_2-2}, & \tau_{N_2-1} + \frac{1}{N_2^2} \leq t \leq 1 - \frac{1}{N_2^2}, \\ 1, & 1 - \frac{1}{N_2^2} \leq t \leq 1. \end{cases} \quad (23)$$

To simplify the description of computational scheme, we introduce the following notation:

(1) Unite the nodes $t_k, k = 0, 1, \ldots, N_1$ and $\tau_l, l = 0, 1, \ldots, N_2$, denoting them by $v_i, i = 0, 1, \ldots, N^*, N^* = N_1 + N_2$;
(2) Unite the nodes $\bar{t}_k, k = 0, 1, \ldots, N_1$ and $\bar{\tau}_l, l = 0, 1, \ldots, N_2$, denoting them by $\bar{v}_i, i = 0, 1, \ldots, N^* + 1$;
(3) Denote the family of basis functions $\{\varphi_k\}, k = 0, 1, \ldots, N_1$, $\{\psi_l\}, l = 0, 1, \ldots, N_2$ by $\{\zeta_j\}, j = 0, 1, \ldots, N^* + 1$;
(4) Denote by $\{\gamma_k\}, k = 0, 1, \ldots, N^* + 1$, unknowns $\{\alpha_i\}, i = 0, 1, \ldots, N_1$, $\{\beta_j\}, j = 0, 1, \ldots, N_2$.

Here $v_i = t_i, i = 0, 1, \ldots, N_1$, $v_{N_1+i} = \tau_i, i = 1, 2, \ldots, N_2$,

$$\gamma_i = \alpha_i, i = 0, 1, \ldots, N_1, \gamma_{N_1+1+i} = \beta_i, i = 0, 1, \ldots, N_2,$$
$$\zeta_i = \varphi_i, i = 0, 1, \ldots, N_1, \zeta_{N_1+1+i} = \psi_i, i = 0, 1, \ldots, N_2.$$

Let us recall that the points t_{N_1} and τ_0 coincide.

Applying the collocation method on the knots $\bar{v}_k, k = 0, 1, \ldots, N^* + 1$ to the Equation (16), we obtain the following system of algebraic equations for finding unknown coefficients $\{\gamma_k\}$ of the polygon (17)

$$a(\bar{v}_k)\gamma_k + \sum_{l=0}^{N^*+1} h(\bar{v}_k, v_l)\gamma_l \int_{-1}^{1} \frac{\zeta_l(\tau)}{(\tau - \bar{v}_k)^2} d\tau = f(\bar{v}_k), \quad (24)$$

$k = 0, 1, \ldots, N^* + 1$.

Using the definition of hypersingular integrals, we receive:

$$\int_{v_{k-1}}^{v_{k+1}} \frac{\zeta_k(\tau) d\tau}{(\tau - \bar{v}_k)^2} = -2 \frac{N_1^2}{(1+c)N_1 - 2} \ln((1+c)N_1 - 1), k = 1, 2, \ldots, N_1 - 1; \quad (25)$$

$$\int_{v_{k-1}}^{v_{k+1}} \frac{\zeta_{k+1}(\tau)d\tau}{(\tau - \bar{v}_k)^2} = -2\frac{N_2^2}{(1-c)N_2 - 2}\ln((1-c)N_2 - 1), k = N_1 + 2, \ldots, N^* - 1; \tag{26}$$

$$\int_{-1}^{v_1} \frac{\zeta_0(\tau)d\tau}{(\tau + 1 - \frac{1}{2N_1^2})^2} = -2N_1^2 - N_1^2 \frac{\ln(2(c+1)N_1 - 3)}{(c+1)N_1 - 2}, \tag{27}$$

$$\int_{v_{N_1-1}}^{v_{N_1}} \frac{\zeta_{N_1}(\tau)d\tau}{(\tau - v_{N_1} + \frac{1}{2N_1^2})^2} = -2N_1^2 - N_1^2 \frac{\ln(2(c+1)N_1 - 3)}{(c+1)N_1 - 2}, \tag{28}$$

$$\int_{v_{N_1}}^{v_{N_1+1}} \frac{\zeta_{N_1+1}(\tau)d\tau}{(\tau - v_{N_1} - \frac{1}{2N_2^2})^2} = -2N_2^2 - N_2^2 \frac{\ln(2(1-c)N_2 - 3)}{(1-c)N_2 - 2}, \tag{29}$$

$$\int_{v_{N^*-1}}^{1} \frac{\zeta_{N^*+1}(\tau)d\tau}{(\tau - 1 + \frac{1}{2N^2})^2} = -2N_2^2 - N_2^2 \frac{\ln(2(1-c)N_2 - 3)}{(1-c)N_2 - 2}, \tag{30}$$

$$\int_{-1}^{1} \left[\sum_{l=1}^{N^*+1} \zeta_l(\tau)\right] \frac{d\tau}{(\tau + 1 - \frac{1}{2N_1^2})^2} = -\frac{2N_1^2}{4N_1^2 - 1} + N_1^2 \frac{\ln(2(1+c)N_1 - 3)}{(1+c)N_1 - 2}, \tag{31}$$

$$\int_{-1}^{1} \left[\sum_{l=0}^{N^*} \zeta_l(\tau)\right] \frac{d\tau}{(\tau - 1 + \frac{1}{2N_2^2})^2} = -\frac{2N_2^2}{4N_2^2 - 1} + N_2^2 \frac{\ln(2(1-c)N_2 - 3)}{(1-c)N_2 - 2}, \tag{32}$$

$$\int_{-1}^{1} \left[\sum_{l=0}^{N^*+1} {}'\zeta_l(\tau)\right] \frac{d\tau}{(\tau - (c - \frac{1}{2N_1^2}))^2} = -\frac{2N_1^2}{4N_1^2 - 1} + N_1^2 \frac{\ln(2(1+c)N_1 - 3)}{(1+c)N_1 - 2}, \tag{33}$$

$$\int_{-1}^{1} \left[\sum_{l=0}^{N^*+1} {}''\zeta_l(\tau)\right] \frac{d\tau}{(\tau - (c + \frac{1}{2N_2^2}))^2} = -\frac{2N_2^2}{4N_2^2 - 1} + N_2^2 \frac{\ln(2(1+c)N_2 - 3)}{(1+c)N_2 - 2}, \tag{34}$$

$$\int_{-1}^{1} \left[\sum_{l=0}^{N^*+1} {}'''\varphi_l(\tau)\right] \frac{d\tau}{(\tau - v_k)^2} = -\frac{N_1}{(c+1)k} - \frac{N_1}{2N_1 - (c+1)k} + \frac{2N_1^2}{(1+c)N_1 - 2}\ln((c+1)N_1 - 1); \tag{35}$$

$$\int_{-1}^{1} \left[\sum_{l=0}^{N^*+1} {}''''\varphi_l(\tau)\right] \frac{d\tau}{(\tau - v_k)^2} = -\frac{N_2}{(1-c)k} - \frac{N_2}{2N_2 - (1-c)k} + \frac{2N_2^2}{(1-c)N_2 - 2}\ln((1-c)N_2 - 1). \tag{36}$$

Here $\sum_l'$, $\sum_l''$, $\sum_l'''$, $\sum_l''''$ indicates a summation over $l \neq N_1$, $l \neq N_1 + 1$, $l \neq k(1 \leq k \leq N_1 - 1)$, $l \neq k(N_1 + 2 \leq k \leq N^* - 1)$, respectively. Detailed calculations are given in [23].

We can rewrite the system (24) as

$$a(\bar{v}_k)\gamma_k - h(\bar{v}_k, \bar{v}_k)2N_1^2 \frac{\ln(N_1 - 1)}{(1+c)N_1 - 2}\gamma_k + \sum_{l=0}^{N^*+1} {}'\gamma_l h(\bar{v}_k, \bar{v}_l) \int_{-1}^{1} \zeta_l(\tau) \frac{d\tau}{(\tau - \bar{v}_k)^2}$$
$$= f(\bar{v}_k), \quad k = 1, \ldots, N_1 - 1; \tag{37}$$

$$a(\bar{v}_k)\gamma_k - h(\bar{v}_k, \bar{v}_k)2N_2^2 \frac{\ln(N_2 - 1)}{(1-c)N_2 - 2}\gamma_k + \sum_{l=0}^{N^*+1} {}'\gamma_l h(\bar{v}_k, \bar{v}_l) \int_{-1}^{1} \zeta_l(\tau) \frac{d\tau}{(\tau - \bar{v}_k)^2}$$
$$= f(\bar{v}_k), \quad k = N_1 + 2, \ldots, N^*; \tag{38}$$

$$a(\bar{v}_0)\gamma_0 - h(\bar{v}_0, \bar{v}_0)(2N_1^2 + N_1^2 \frac{\ln(2(1+c)N_1 - 3)}{(1+c)N_1 - 2})\gamma_0$$
$$+ \sum_{l=1}^{N^*+1} \gamma_l h(\bar{v}_k, \bar{v}_l) \int_{-1}^{1} \zeta_l(\tau) \frac{d\tau}{(\tau - \bar{v}_0)^2} = f(\bar{v}_0); \tag{39}$$

$$a(\bar{v}_{N_1})\gamma_{N_1} - h(\bar{v}_{N_1}, \bar{v}_{N_1})(2N_1^2 + N_1^2 \tfrac{\ln(2(1+c)N_1-3)}{(1+c)N_1-2})\gamma_{N_1}$$
$$+ \sum_{l=0}^{N^*+1} {}''\gamma_l h(\bar{v}_{N_1}, \bar{v}_l) \int_{-1}^{1} \zeta_l(\tau) \tfrac{d\tau}{(\tau-\bar{v}_{N_1})^2} = f(\bar{v}_{N_1}); \tag{40}$$

$$a(\bar{v}_{N_1+1})\gamma_{N_1+1} - h(\bar{v}_{N_1+1}, \bar{v}_{N_1+1})(2N_2^2 + N_2^2 \tfrac{\ln(2(1-c)N_2-3)}{(1-c)N_2-2})\gamma_{N_1+1}$$
$$+ \sum_{l=0}^{N^*+1} {}'''\gamma_l h(\bar{v}_{N_1+1}, \bar{v}_l) \int_{-1}^{1} \zeta_l(\tau) \tfrac{d\tau}{(\tau-\bar{v}_{N_1+1})^2} = f(\bar{v}_{N_1}); \tag{41}$$

$$a(\bar{v}_{N^*+1})\gamma_{N^*+1} - h(\bar{v}_{N^*+1}, \bar{v}_{N^*+1})(2N_2^2 + N_2^2 \tfrac{\ln(2(1-c)N_2-3)}{(1-c)N_2-2})\gamma_{N^*+1}$$
$$+ \sum_{l=0}^{N^*} \gamma_l h(\bar{v}_{N^*+1}, \bar{v}_l) \int_{-1}^{1} \zeta_l(\tau) \tfrac{d\tau}{(\tau-\bar{v}_{N^*-1})^2} = f(\bar{v}_{N^*+1}). \tag{42}$$

Here $\sum'$, $\sum''$, $\sum'''$ indicates a summation over $l \neq k$, $l \neq N_1$, $l \neq N_1+1$, respectively. The system (37)–(42) is equivalent to the system

$$(sgn\, h(v_k, v_k))\left(a(\bar{v}_k)\gamma_k - h(\bar{v}_k, \bar{v}_k)2N_1^2 \tfrac{\ln(N_1-1)}{(1+c)N_1-2}\gamma_k \right.$$
$$\left. + \sum_{l=0}^{N^*+1} {}'\gamma_l h(\bar{v}_k, \bar{v}_l) \int_{-1}^{1} \zeta_l(\tau) \tfrac{d\tau}{(\tau-\bar{v}_k)^2}\right) = (sgn\, h(t_k, t_k))f(v_k), k=1,\ldots,N_1-1; \tag{43}$$

$$(sgn\, h(v_k, v_k))\left(a(\bar{v}_k)\gamma_k - h(\bar{v}_k, \bar{v}_k)2N_2^2 \tfrac{\ln(N_2-1)}{(1-c)N_2-2}\gamma_k \right.$$
$$\left. + \sum_{l=0}^{N^*+1} {}'\gamma_l h(\bar{v}_k, \bar{v}_l) \int_{-1}^{1} \zeta_l(\tau) \tfrac{d\tau}{(\tau-\bar{v}_k)^2}\right) = (sgn\, h(t_k, t_k))f(\bar{v}_k), \quad k=N_1+2,\ldots,N^*; \tag{44}$$

$$(sgn\, h(v_0, v_0))\left(a(\bar{v}_0)\gamma_0 - h(\bar{v}_0, \bar{v}_0)(2N_1^2 + N_1^2 \tfrac{\ln(2(1+c)N_1-3)}{(1+c)N_1-2})\gamma_0 \right.$$
$$\left. + \sum_{l=1}^{N^*+1} \gamma_l h(\bar{v}_k, \bar{v}_l) \int_{-1}^{1} \zeta_l(\tau) \tfrac{d\tau}{(\tau-\bar{v}_0)^2}\right) = (sgn\, h(v_0, v_0))f(\bar{v}_0); \tag{45}$$

$$(sgn\, h(v_{N_1}, v_{N_1}))\left(a(\bar{v}_{N_1})\gamma_{N_1} - h(\bar{v}_{N_1}, \bar{v}_{N_1})(2N_1^2 + N_1^2 \tfrac{\ln(2(1+c)N_1-3)}{(1+c)N_1-2})\gamma_{N_1} \right.$$
$$\left. + \sum_{l=0}^{N^*+1} {}''\gamma_l h(\bar{v}_{N_1}, \bar{v}_l) \int_{-1}^{1} \zeta_l(\tau) \tfrac{d\tau}{(\tau-\bar{v}_{N_1})^2}\right) = (sgn\, h(v_{N_1}, v_{N_1}))f(\bar{v}_{N_1}); \tag{46}$$

$$(sgn\, h(v_{N_1+1}, v_{N_1+1}))\left(a(\bar{v}_{N_1+1})\gamma_{N_1+1} - h(\bar{v}_{N_1+1}, \bar{v}_{N_1+1})(2N_2^2 \right.$$
$$\left. +N_2^2 \tfrac{\ln(2(1-c)N_2-3)}{(1-c)N_2-2})\gamma_{N_1+1} + \sum_{l=0}^{N^*+1} {}'''\gamma_l h(\bar{v}_{N_1}, \bar{v}_l) \int_{-1}^{1} \zeta_l(\tau) \tfrac{d\tau}{(\tau-\bar{v}_{N_1+1})^2}\right) \tag{47}$$
$$= (sgn\, h(v_{N_1}, v_{N_1}))f(\bar{v}_{N_1});$$

$$(sgn\, h(v_{N^*+1}, v_{N^*+1}))\left(a(\bar{v}_{N^*+1})\gamma_{N^*+1} - h(\bar{v}_{N^*+1}, \bar{v}_{N^*+1})\right.$$
$$\left.(2N_2^2 + N_2^2 \tfrac{\ln(2(1-c)N_2-3)}{(1-c)N_2-2})\gamma_{N^*+1} + \sum_{l=0}^{N^*} \gamma_l h(\bar{v}_{N^*-1}, \bar{v}_l) \int_{-1}^{1} \zeta_l(\tau) \tfrac{d\tau}{(\tau-\bar{v}_{N^*+1})^2}\right) \tag{48}$$
$$= (sgn\, h(v_{N^*+1}, v_{N^*+1}))f(\bar{v}_{N^*+1}).$$

Here $\sum'$, $\sum''$, $\sum'''$ indicates a summation over $l \neq k$, $l \neq N_1$, $l \neq N_1+1$, respectively. Let us write the system (43)–(48) in a matrix form

$$DX = F,$$

where $D = \{d_{kl}\}$, $k, l = 0, 1, \ldots, N^*+1$, $X = (x_0, x_1, \ldots, x_{N^*+1})$, $F = (f_0, f_1, \ldots, f_{N^*+1})$. The values $\{d_{kl}\}$, $\{x_k\}$, and $\{f_k\}$ are obvious.

The diagonal elements in the left-hand side of the system of Equations (43)–(48) have the following forms

$$d_{kk} = (sgn\, h(\bar{v}_k, \bar{v}_k)) \left(a(\bar{v}_k) - h(\bar{v}_k, \bar{v}_k) 2N_1^2 \frac{\ln(N_1 - 1)}{(1+c)N_1 - 2} \right)$$

$k = 1, 2, \ldots, N_1 - 1$,

$$d_{kk} = (sgn\, h(\bar{v}_k, \bar{v}_k)) \left(a(\bar{v}_k) - h(\bar{v}_k, \bar{v}_k) 2N_1^2 \frac{\ln(N_1 - 1)}{(1+c)N_1 - 2} \right),$$

$k = N_1 + 2, \ldots, N^*$,

$$d_{00} = (sgn\, h(\bar{v}_0, \bar{v}_0)) \left(a(\bar{v}_0) - h(\bar{v}_0, \bar{v}_0)(2N_1^2 + N_1^2 \frac{\ln(2(1+c)N_1 - 3)}{(1+c)N_1 - 2}) \right),$$

$$d_{N_1,N_1} = (sgn\, h(v_{N_1}, v_{N_1})) \Big(a(\bar{v}_{N_1}) - h(\bar{v}_{N_1}, \bar{v}_{N_1})(2N_1^2$$
$$+ N_1^2 \frac{\ln(2(1+c)N_1 - 3)}{(1+c)N_1 - 2}) \Big),$$

$$d_{N_1+1,N_1+1} = (sgn\, h(v_{N_1+1}, v_{N_1+1})) \Big(a(\bar{v}_{N_1+1}) - h(\bar{v}_{N_1+1}, \bar{v}_{N_1+1})$$
$$(2N_2^2 + N_2^2 \frac{\ln(2(1-c)N_2 - 3)}{(1-c)N_2 - 2}) \Big),$$

$$d_{N^*+1,N^*+1} = (sgn\, h(v_{N^*+1}, v_{N^*+1})) \big(a(\bar{v}_{N^*+1}) \cdot h(\bar{v}_{N^*+1}, \bar{v}_{N^*+1})$$
$$((2N_2^2 + N_2^2 \frac{\ln(2(1-c)N_2 - 3)}{(1-c)N_2 - 2})) \big).$$

The cubic logarithmic norm of the matrix D is equal to

$$\Lambda_2(D) = \max \Bigg(\max_{1 \leq k \leq N_1 - 1} \left(d_{kk} + \sum_{l=0}^{N^*+1}{}' |h(\bar{v}_k, \bar{v}_l)| \int_{-1}^{1} \frac{\zeta_l(\tau) d\tau}{(\tau - \bar{v}_k)^2} \right),$$

$$\max_{N_1 + 2 \leq k \leq N^*} \left(d_{kk} + \sum_{l=0}^{N^*+1}{}' |h(\bar{v}_k, \bar{v}_l)| \int_{-1}^{1} \frac{\zeta_l(\tau) d\tau}{(\tau - \bar{v}_k)^2} \right),$$

$$\left(d_{00} + \sum_{l=1}^{N^*+1} |h(\bar{v}_0, \bar{v}_l)| \int_{-1}^{1} \frac{\zeta_l(\tau) d\tau}{(\tau - \bar{v}_0)^2} \right),$$

$$\left(d_{N_1 N_1} + \sum_{l=0}^{N^*+1}{}'' |h(\bar{v}_{N_1}, \bar{v}_l)| \int_{-1}^{1} \frac{\zeta_l(\tau) d\tau}{(\tau - \bar{v}_{N_1})^2} \right),$$

$$\left(d_{N_1+1,N_1+1} + \sum_{l=0}^{N^*+1}{}''' |h(\bar{v}_{N_1+1}, \bar{v}_l)| \int_{-1}^{1} \frac{\zeta_l(\tau) d\tau}{(\tau - \bar{v}_{N_1+1})^2} \right),$$

$$\left(d_{N^*+1,N^*+1} + \sum_{l=0}^{N^*} |h(\bar{v}_{N^*+1}, \bar{v}_l)| \int_{-1}^{1} \frac{\zeta_l(\tau) d\tau}{(\tau - \bar{v}_{N^*+1})^2} \right) \Bigg).$$

From (25)–(36) it follows that for sufficiently large N $\Lambda_2(D) < 0$ occurs. By Theorem 2 it is clear that the system (43)–(48) (and (37)–(42)) has a unique solution $x_N^*(t)$ and $\|D^{-1}\| \leq 1/|\Lambda_2(D)|$.

Let $x^*(t)$ and x_N^* be solutions of (16) and (37)–(42), respectivety.

We recall the following definitions.

Definition 3. *The class $W^r(M, [a, b])$, $r = 1, 2, \ldots$, consists of all functions $f \in C([a, b])$, which have an absolutely continuous derivative $f^{(r-1)}(x)$ and piecewise derivative $f^{(r)}(x)$ with $|f^{(r)}(x)| \leq M$.*

Definition 4. Denote by $W^r(f : f_1, f_2; M, c), r = 1, 2, \ldots$, a set of functions $f(x), x \in [a, b]$, such that $f(x) = f_1(x), x \in [a, c), f(x) = f_2(x), x \in (c, b]$, where $f_1(x) \in W^r(M, [a, c]), f_2(x) \in W^r(M, [c, b]), f_1(c) \neq f_2(c), c \in (a, b)$.

Repeating the proof presented in [24] we see that the approximation of $f(t) \in W^1((f : f_1, f_2; M, c))$ by piecewise linear functions constructed on the basis $\zeta_k(t)$, $k = 0, 1, \ldots, N^* + 1$, has the error $\frac{C}{N} \max(\omega(f_1^{(1)}, \frac{1}{N}), \omega(f_1^{(1)}, \frac{1}{N}))$ for $f(t) \in W^1((f : f_1, f_2; M, c))$, and $\frac{C}{N^2}$ for $f(t) \in W^2((f : f_1, f_2; M, c))$.

In this paper, we denote the constants that do not depend on N by C.

Let $x^*(t) \in W^2((x^* : x_1^*, x_2^*; M, c))$, and $\|x_1^{*(1)}(t)\|_C \leq M_1, t \in [a, c], \|x_2^{*(1)}(t)\|_C \leq M_2, t \in [c, b]$, $M = \max(M_1, M_2), 0 < M < \infty$, where M is a bounded constant.

Repeating the arguments given in [24], we arrive at the following statement.

Theorem 6. Let the following conditions be fulfilled:

(1) Equation (16) has the unique solution $x^*(t) \in W^2(x_1^*, x_2^*; M, c), -1 < c < 1, M = \text{const}$.
(2) For all $t \in [-1, 1]$ the function $h(t, t) \neq 0$.
(3) $\Lambda_2(D) < 0$.

Then the system of Equations (37)–(42) has a unique solution $x_N^*(t)$ and the following estimate holds: $\|x^* - x_N^*\|_1 \leq CN^{-1} \ln N$.

3.2. Nonlinear Hypersingular Integral Equations

Consider the nonlinear hypersingular integral equation:

$$a(t)x(t) + \int_{-1}^{1} \frac{h(t, \tau, x(\tau)) d\tau}{(\tau - t)^2} = f(t). \tag{49}$$

The approximate solution of the Equation (49) we shall seek as a continuous function (17) with the coefficients γ_k. The coefficients γ_k are determined by the following system of nonlinear algebraic equations

$$a(\bar{v}_k)\gamma_k + \sum_{l=0}^{N^*+1} h(\bar{v}_k, v_l, \gamma_l) \int_{-1}^{1} \frac{\zeta_l(\tau)}{(\tau - \bar{v}_k)^2} d\tau = f(\bar{v}_k), \quad k = 0, 1, \ldots, N^* + 1. \tag{50}$$

Remark 4. Note that the set $\gamma_k, k = 0, 1, \ldots, N^* + 1$, is union of sets $\alpha_k, k = 0, 1, \ldots, N_1$, and $\beta_k, k = 0, 1, \ldots, N_2$.

By computing the hypersingular integrals in (50), we can rewrite the system of Equation (50) as

$$a(\bar{v}_k)\gamma_k - h(\bar{v}_k, \bar{v}_k, \gamma_k) 2N_1^2 \frac{\ln(N_1-1)}{(1+c)N_1-2} + \sum_{l=0}^{N^*+1}{'} h(\bar{v}_k, \bar{v}_l, \gamma_l) \int_{-1}^{1} \zeta_l(\tau) \frac{d\tau}{(\tau - \bar{v}_k)^2}$$
$$= f(\bar{v}_k), \quad k = 1, \ldots, N_1 - 1; \tag{51}$$

$$a(\bar{v}_k)\gamma_k - h(\bar{v}_k, \bar{v}_k, \gamma_k) 2N_2^2 \frac{\ln(N_2-1)}{(1-c)N_2-2} + \sum_{l=0}^{N^*+1}{'} \gamma_l h(\bar{v}_k, \bar{v}_l, \gamma_l) \int_{-1}^{1} \zeta_l(\tau) \frac{d\tau}{(\tau - \bar{v}_k)^2}$$
$$= f(\bar{v}_k), \quad k = N_1 + 2, \ldots, N^*; \tag{52}$$

$$a(\bar{v}_0)\gamma_0 - h(\bar{v}_0, \bar{v}_0, \gamma_0)(2N_1^2 + N_1^2 \frac{\ln 2(1+c)N_1-3)}{(1+c)N_1-2})$$
$$+ \sum_{l=1}^{N^*+1} h(\bar{v}_k, \bar{v}_l, \gamma_l) \int_{-1}^{1} \zeta_l(\tau) \frac{d\tau}{(\tau - \bar{v}_0)^2} = f(\bar{v}_0); \tag{53}$$

$$a(\bar{v}_{N_1})\gamma_{N_1} - h(\bar{v}_{N_1}, \bar{v}_{N_1}, \gamma_{N_1})(2N_1^2 + N_1^2 \frac{\ln(2(1+c)N_1-3)}{(1+c)N_1-2})$$
$$+ \sum_{l=0}^{N^*+1}{}'' h(\bar{v}_{N_1}, \bar{v}_l) \int_{-1}^{1} \zeta_l(\tau) \frac{d\tau}{(\tau-\bar{v}_{N_1})^2} = f(\bar{v}_{N_1}); \tag{54}$$

$$a(\bar{v}_{N_1+1})\gamma_{N_1+1} - h(\bar{v}_{N_1+1}, \bar{v}_{N_1+1}, \gamma_{N_1+1})(2N_2^2 + N_2^2 \frac{\ln(2(1-c)N_2-3)}{(1-c)N_2-2})$$
$$+ \sum_{l=0}^{N^*+1}{}''' h(\bar{v}_{N_1+1}, \bar{v}_l, \gamma_l) \int_{-1}^{1} \zeta_l(\tau) \frac{d\tau}{(\tau-\bar{v}_{N_1+1})^2} = f(\bar{v}_{N_1+1}); \tag{55}$$

$$a(\bar{v}_{N^*+1})\gamma_{N^*+1} - h(\bar{v}_{N^*+1}, \bar{v}_{N^*+1}, \gamma_{N^*+1})(2N_2^2 + N_2^2 \frac{\ln(2(1-c)N_2-3)}{(1-c)N_2-2})$$
$$+ \sum_{l=0}^{N^*}{}''' h(\bar{v}_{N^*+1}, \bar{v}_l, \gamma_l) \int_{-1}^{1} \zeta_l(\tau) \frac{d\tau}{(\tau-\bar{v}_{N^*+1})^2} = f(\bar{v}_{N^*+1}). \tag{56}$$

Here $\sum'$, $\sum''$, $\sum'''$ indicate summations over $l \neq k$, $l \neq N_1$, $l \neq N_1 + 1$, respectively. The Frechet derivative on a vector $(\bar{\alpha}_0, \bar{\alpha}_1, \cdots, \bar{\alpha}_{N^*+1})$ in the space R_{N^*+1} is equal to

$$\begin{pmatrix}
a(\bar{v}_k)\gamma_k - h'_3(\bar{v}_k, \bar{v}_k, \tilde{\gamma}_k) 2N_1^2 \frac{\ln(N_1-1)}{(1+c)N_1-2} \gamma_k \\
+ \sum_{l=0}^{N^*+1}{}' h'_3(\bar{v}_k, \bar{v}_l, \tilde{\gamma}_l)\gamma_l \int_{-1}^{1} \zeta_l(\tau) \frac{d\tau}{(\tau-\bar{v}_k)^2}, \quad k = 1, \ldots, N_1-1; \\
a(\bar{v}_k)\gamma_k - h'_3(\bar{v}_k, \bar{v}_k, \tilde{\gamma}_k) 2N_2^2 \frac{\ln(N_2-1)}{(1-c)N_2-2} \gamma_k \\
+ \sum_{l=0}^{N^*+1}{}' h'_3(\bar{v}_k, \bar{v}_l, \tilde{\gamma}_l)\gamma_l \int_{-1}^{1} \zeta_l(\tau) \frac{d\tau}{(\tau-\bar{v}_k)^2}, \quad k = N_1+2, \ldots, N^*; \\
a(\bar{v}_0)\gamma_0 - h'_3(\bar{v}_0, \bar{v}_0, \tilde{\gamma}_0)\gamma_0(2N_1^2 + N_1^2 \frac{\ln(2(1+c)N_1-3)}{(1+c)N_1-2}) \\
+ \sum_{l=1}^{N^*+1} h'_3(\bar{v}_0, \bar{v}_l, \tilde{\gamma}_l)\gamma_l \int_{-1}^{1} \zeta_l(\tau) \frac{d\tau}{(\tau-\bar{v}_0)^2}; \\
a(\bar{v}_{N_1})\gamma_{N_1} - h'_3(\bar{v}_{N_1}, \bar{v}_{N_1}, \tilde{\gamma}_{N_1})\gamma_{N_1}(2N_1^2 + N_1^2 \frac{\ln(2(1+c)N_1-3)}{(1+c)N_1-2}) \\
+ \sum_{l=0}^{N^*+1} h'_3(\bar{v}_{N_1}, \bar{v}_l, \tilde{\gamma}_l)\gamma_l \int_{-1}^{1} \zeta_l(\tau) \frac{d\tau}{(\tau-\bar{v}_{N_1})^2}; \\
a(\bar{v}_{N_1+1})\gamma_{N_1+1} - h'_3(\bar{v}_{N_1+1}, \bar{v}_{N_1+1}, \tilde{\gamma}_{N_1+1})\gamma_{N_1+1}(2N_2^2 + N_2^2 \frac{\ln(2(1-c)N_2-3)}{(1-c)N_2-2}) \\
+ \sum_{l=0}^{N^*+1}{}''' h'_3(\bar{v}_{N_1+1}, \bar{v}_l, \tilde{\gamma}_l)\gamma_l \int_{-1}^{1} \zeta_l(\tau) \frac{d\tau}{(\tau-\bar{v}_{N_1+1})^2}; \\
a(\bar{v}_{N^*+1})\gamma_{N^*+1} - h'_3(\bar{v}_{N^*+1}, \bar{v}_{N^*+1}, \tilde{\gamma}_{N^*+1})\gamma_{N^*+1}(2N_2^2 + N_2^2 \frac{\ln(2(1-c)N_2-3)}{(1-c)N_2-2}) \\
+ \sum_{l=0}^{N^*} h'_3(\bar{v}_{N^*+1}, \bar{v}_l, \tilde{\gamma}_l)\gamma_l \int_{-1}^{1} \zeta_l(\tau) \frac{d\tau}{(\tau-\bar{v}_{N^*+1})^2}
\end{pmatrix}. \tag{57}$$

Here $\sum'$, $\sum''$ $\sum'''$ indicate summations over $l \neq k$, $l \neq N_1$, $l \neq N_1 + 1$, respectively. The notation $h'_3(t, \tau, u) = \frac{\delta h(t,\tau,u)}{\delta u}$ is used here.

Let the Equation (49) has the unique solution $x^*(t)$ inside the ball $B(x^*, \delta)$. We shall assume that the Frechet derivative (57) in the ball $R_{N^*+1}(x^*, \delta)$ satisfies the conditions of Theorem 5. Thus, according to statements of the Theorem 5, the solution of the system of differential equations

$$\frac{d\alpha_l(\sigma)}{d\sigma} = a(\bar{t}_l)\alpha_l(\sigma) - \sum_{k=0}^{N^*+1} h(\bar{t}_l, \bar{t}_k, \alpha_k(\sigma))N\left(\frac{1}{2l-2k+1} - \frac{1}{2l-2k-1}\right)$$
$$-f(\bar{t}_l), \quad l = 0, 1, \cdots, N^* + 1, \tag{58}$$

converges to the solution of the Equation (49).

Thus, we have proven the following statement.

Theorem 7. *Let the following conditions hold:*

(1) Equation (49) has a unique solution $x^*(t)$ inside some ball $B(x^*, \delta)$, $x^* \in W^2(x^* : x_1^*, x_2^*; M, c)$;

(2) The Frechet derivative (57) in the ball $R_{N^*+1}(x^*, \delta)$ satisfies the conditions of Theorem 5.

Then the system of Equations (51)–(56) has a unique solution inside the ball $B(x^*, \delta)$, and the solution of Equation (58) converges to this solution.

The effectiveness of the presented algorithms is illustrated by solving two hypersingular integral equations modeling aerodynamics problems.

Example 1. *Let us illustrate the effectiveness of continuous method by solving the following linear hypersingular equation*

$$\int_{-1}^{1} \frac{x(\tau)}{(\tau - t)^2} d\tau = f(\gamma_1, \gamma_2, t), \qquad (59)$$

where $f(\gamma_1, \gamma_2, t)$ is the given right-hand side of the equation:

$$\begin{aligned}f(\gamma_1, \gamma_2; t) &= \gamma_1 - \gamma_2 + (a_1 - a_2)\tfrac{1}{t} - (a_1 + \gamma_1 t)\tfrac{1}{1+t} \\ &\quad - (a_2 + \gamma_2 t)\tfrac{1}{1-t} + \gamma_1 \ln\left|\tfrac{t}{1+t}\right| + \gamma_2 \ln\left|\tfrac{1-t}{t}\right|.\end{aligned}$$

The exact solution of the equation is $x(t) = (x_1(t), x_2(t))$; $x_i(t) = a_i + \gamma_i t, i = 1, 2$.

To solve the Equation (59) numerically we use the continuous method for solving operator equations and arrive to the following evolution equation

$$\frac{d\alpha_k(\sigma)}{d\sigma} = \sum_{l=0}^{N^*+1} N\alpha_l(\sigma)\left(\frac{1}{2k+2l-1} - \frac{1}{2k+2l+1}\right) - f(\gamma_1, \gamma_2; \bar{v}_k), k = 0, 1, \ldots, N^* + 1.$$

Nodes $v_k, \bar{v}_k, k = 0, 1, \ldots, N^* + 1$, have been entered above.

In Figure 1 we show the trajectories of the exact solution of the Equation (59); its approximate solution, received with continuous method; and values of error.

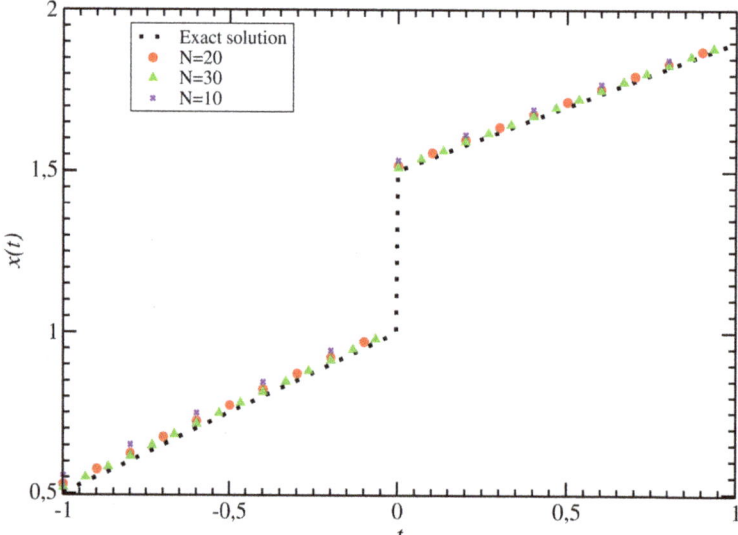

Figure 1. Numerical solutions for the linear hypersingular equation with a discontinuous right-hand side example.

Here $a_1 = 1, a_2 = 1.5, \gamma_1 = 0.5, \gamma_2 = 0.3$.

Example 2. Let us illustrate the effectiveness of the continuous method for the solutions of nonlinear hypersingular equations

$$\int_{-1}^{1} \frac{x^2(\tau)}{(\tau-t)^2} d\tau = f(\gamma_1, \gamma_2, t) \tag{60}$$

where $f(\gamma_1, \gamma_2, t)$ is the given right-hand side of the equation:

$$\begin{aligned}f(\gamma_1, \gamma_2; t) &= \gamma_1^2 + 2\gamma_1 a_1 + \gamma_1^2 t + \frac{a_1^2}{t} + (2\gamma_1 a_1 + 2\gamma_1^2 t) \ln\left|\frac{t}{1+t}\right| \\ &\quad -(a_1^2 + 2\gamma_1 a_1 t + \gamma_1^2 t^2)\frac{1}{1+t} + \gamma_2^2 - 2a_2\gamma_2 - \gamma_2^2 t - \frac{a_2^2}{t} - (a_2^2 + 2a_2\gamma_2 t \\ &\quad +\gamma_2^2 t^2)\frac{1}{1-t} + (2a_2\gamma_2 + 2\gamma_2^2 t) \ln\left|\frac{1-t}{t}\right|.\end{aligned}$$

The exact solution of the equation is $x(t) = (x_1(t), x_2(t)); x_i(t) = a_i + \gamma_i t, i = 1, 2$.

It easy to see that, if $x(t)$ is a solution of the Equation (60), then functions $-x(t), |x(t)|$ and $-|x(t)|$ are solutions of this equation too.

To solve the Equation (60) numerically we use the continuous method and receive the following evolution equation

$$\frac{d\alpha_k(\sigma)}{d\sigma} = \sum_{l=0}^{N^*+1} N\alpha_l^2(\sigma)\left(\frac{1}{2k+2l-1} - \frac{1}{2k+2l+1}\right) - f(\gamma_1, \gamma_2, \bar{v}_l),$$

$k = 0, 1, \ldots, N^* + 1$.

At first, we take $\alpha_k(0) = 0.0$ as an initial condition in order to demonstrate applicability of our method in cases of the Newton–Kantorovich method, the minimal residual method and other numerical methods; using in their construction the derivative of nonlinear operator is not applicable. Indeed, in this case the Frechet derivative (57) is not only degenerate—and, therefore, not invertable—but is an identical zero.

In Figure 2 we put $a_1 = 1, a_2 = 1.4, \gamma_1 = 0.5, \gamma_2 = -0.4$.

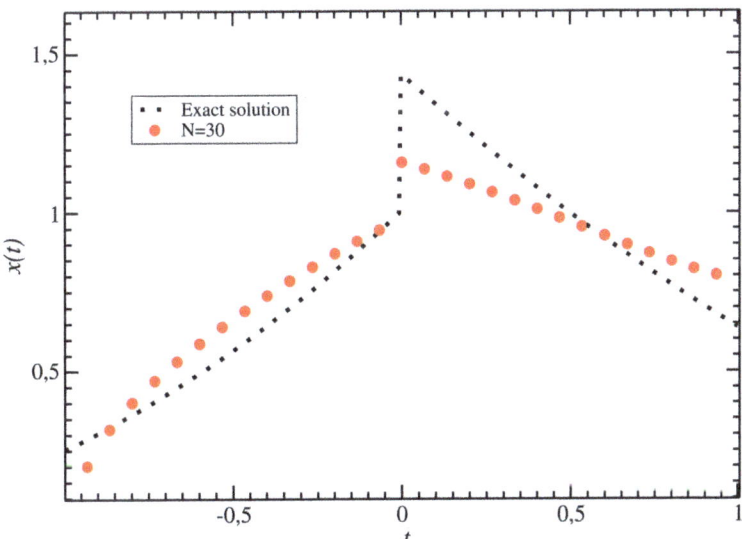

Figure 2. Numerical solution for the nonlinear hypersingular equation with a discontinuous right-hand side example.

In Figure 2 we show the trajectories of the exact solution of the Equation (60), its approximate solution, received with continuous method and values of error.

The exact solution at $t = 0$ has a jump discontinuity of $h = 0.4$. The slopes of the exact solution also change at $t = 0$. In Figure 2 we demonstrate that the numerical solution approximates the exact one at $[-1, 0)$ well. At $t = 0$ the approximate solution has a jump $\tilde{h} = 0.15$.

4. Summary and Discussion

An iterative projection method for solving linear and nonlinear hypersingular integral equations has been proposed. The method is based on the use of sufficient conditions for asymptotic stability of ODE systems. Stability conditions are expressed in terms of the logarithmic norms of the corresponding matrices. In a number of spaces often used in computational mathematics, the calculation of logarithmic norms does not cause difficulties, even for large-dimensional matrices.

What are the advantages of the presented method?

(1) The method is applicable for solving linear and nonlinear hypersingular integral equations, whose right-hand sides contain non-Riemann integrable functions.
(2) In Section 3.1 the continuous method is applied to linear hypersingular integral equations with the singularities of the second order. The conditions for the unique solvability of the constructed computing scheme are obtained and the convergence of the sequence of approximate solutions to the exact one is proven. It is shown that for linear hypersingular integral equations, the method converges for sufficiently large N and for $b(t) \neq 0, t \in [-1, 1]$.
(3) In Section 3.2 the continuous method is applied to nonlinear hypersingular integral equations with the singularities of the second order. Conditions are given for the convergence of the constructed iterative spline-collocation method to the solution of a nonlinear hypersingular integral equation. It should be noted that the method is applicable to hypersingular integral equations of the first and second kinds.

The detailed bibliography of approximation methods of hypersingular integral equations of the first and the second kinds is given in [32]. The bibliography on solving hypersingular integral equations of the first kind is presented in [45].

Mostly, papers devoted to hypersingular integral equations of the first kind focused to seek solutions in the class of functions $\sqrt{1 - t^2}\phi(t)$, where $\phi(t)$ is a smooth function. The presented method provides solutions in a general form.

The theoretical justification of the method is based on Lyapunov stability theory. It connects convergence of the method to the sign of the approximate system matrix logarithmic norm.

Said justification has advantages that allow us

1. To obtain a set of convergence conditions owing to logarithmic norm values in various spaces;
2. To determine the norm of the inverse matrix of an approximate system;
3. To determine stability boundaries for solutions with respect to variations of kernels and right-hand sides of the equations.

The major advantage of the method for nonlinear equations is as follows.

The Newton–Kantorovich method requires the Frechet derivative reversibility at each iteration step. Similar conditions are required when using other iteration methods. Our method lacks such a deficiency. It does not put any restrictions on the Frechet derivative of the nonlinear operator.

Author Contributions: Conceptualization, I.B.; Data curation, V.R.; Formal analysis, I.B. and V.R.; Funding acquisition, V.R.; Investigation, I.B. and V.R.; Methodology, V.R.; Project administration, I.B.; Resources, V.R.; Software, V.R. and A.B.; Supervision, I.B.; Visualization, A.B.; Writing—original draft, I.B.; Writing—review and editing, V.R. and A.B. All authors have read and agreed to the published version of the manuscript.

Funding: This research received no external funding.

Conflicts of Interest: The authors declare no conflict of interest.

References

1. Bogolubov N.N.; Mesheryakov V.A.; Tavkhelidze A.N. An application of Muskhelishvili's methods in the theory of elementary particles. In Proceedings of the Symposium on Solid Mechanics and Related Problems of Analysis, Metsniereba, Tbilisi, 1971; Volume 1, pp. 5–11. (In Russian)
2. Brown, D.D.; Jackson, A.D. *The Nucleon-Nucleon Interaction*; North-Holland: Amsterdam, The Netherlands, 1976.
3. Faddeev, L.; Takhtajan, L. *Hamiltonian Approach to Solution Theory*; Springer: Berlin, Germany, 1986.
4. Ioakimidis, N.I. Two methods for the numerical solution of Bueckher's singular integral equation for plane elasticity crack problems, Comput. *Methods Appl. Mech. Engrg.* **1982**, *2*, 169–177. [CrossRef]
5. Lifanov I.K.; Poltavskii, L.N.; Vainikko, G.M. *Hypersingular Integral Equations and their Applications*; Chapman Hall/CRC: Boca Raton, FL, USA; CRC Press Company: London, UK; New York, NY, USA; Washington, DC, USA, 2004.
6. Mandal, B.N.; Chakrabarti, A. *Applied Singular Integral Equations*; CRC Press: Boca Raton, FL, USA, 2011; 260p.
7. Boykov, I.V.; Boykova, A.I. Analytical methods for solution of hypersingular and polyhypersingular integral equations. *arXiv* **2019**, arXiv:1901.04880v1.
8. Gakhov, F.D. *Boundary Value Problems*; Dover Publication: Mineola, NY, USA, 1990; 561p.
9. Ivanov, V.V. *The Theory of Approximate Methods and their Application to the Numerical Solution of Singular Integral Equations*; Noordhoff International Publishing: Leiden, The Netherlands, 1976.
10. Gohberg, I.C.; Fel'dman, I.A. *Convolution Equation and Projection Methods for Their Solution*; Nauka: Moscow, Russia, 1971; English Translation: Translations of Mathematical Monographss, Volume 41; American Mathematical Society: Providence, RI, USA, 1974.
11. Golberg, M.A. Introduction to the numerical solution of Cauchy singular integral equations. In *Mathematical Concepts and Methods in Science and Engineering*; Plenum Press: New York, NY, USA, 1990; Volume 42.
12. Michlin S.G.; Prossdorf, S. *Singular Integral Operatoren*; Acad.-Verl.: Berlin, Germany, 1980.
13. Boykov, I.V. *Approximate Methods of Solution of Singular Integral Equations*; The Penza State University: Penza, Russia, 2004. (In Russian)
14. Boykov, I.V. Numerical methods for solutions of singular integral equations. *arXiv* **1973**, arXiv:1610.09611.
15. Capobiano, M.R.; Criscuolo, G.; Junghanns, P. On the numerical solution of a nonlinear integral equation of Prandtl's type. In *Operator Theory: Advances and Applications*; Birkhauser Verlag: Basel, Switzerland, 2005; Volume 160, pp. 53–79.
16. Oseledets, I.V.; Tyrtyshnicov, E.E. Approximate invention of matrices in the process of solving hypersingular integral equation. *Comput. Math. Math. Phys.* **2005**, *45*, 302–313.
17. Golberg, M.A. The convergence of several algorithms for solving integral equations with finite-part integrals, I. *Integral Equ.* **1983**, *5*, 329–340. [CrossRef]
18. Golberg, M.A. The convergence of several algorithms for solving integral equations with finite-part integrals, II. *Integral Equ.* **1985**, *9*, 267–275. [CrossRef]
19. Lifanov, I.K. *Singular Integral Equations and Discrete Vortices*; VSP: Utrecht, The Netherlands, 1996.
20. Boykov, I.V.; Zakharova Yu, F. An approximate solution to hypersingular integro-differential equations. *Univ. Proc. Volga Reg. Phys. Math. Sci. Math.* **2010**, *1*, 80–90. (In Russian)
21. Boykov, I.V.; Boykova, A.I. An approximate solution of hypersingular integral equations with odd singularities of integer order. *Univ. Proc. Volga Reg. Phys. Math. Sci. Math.* **2010**, *3*, 15–27. (In Russian)
22. Boykov, I.V.; Ventsel, E.S.; Boykova, A.I. An approximate solution of hypersingular integral equations. *Appl. Numer. Math.* **2010**, *60*, 607–628. [CrossRef]
23. Boykov, I.V.; Ventsel, E.S.; Roudnev, V.A.; Boykova, A.I. An approximate solution of nonlinear hypersingular integral equations. *Appl. Numer. Math.* **2014**, *86*, 1–21 [CrossRef]
24. Boykov, I.V.; Roudnev, V.A.; Boykova, A.I.; Baulina, O.A. New iterative method for solving linear and nonlinear hypersingular integral equations. *Appl. Numer. Math.* **2018**, *127*, 280–305. [CrossRef]
25. Kaya, A.C.; Erdogan, F. On The Solution of Integral Equations with Strongly Singular Kernels. *Quart. Appl. Math.* **1987**, *95*, 105–122. [CrossRef]
26. Eshkuvatov, Z.K.; Zulkarnain, F.S.; Nik Long, N.M.A.; Muminov, Z. Modified homotopy perturbation method for solving hypersingular integral equations of the first kind. *SpringerPlus* **2016**, *5*, 1473. [CrossRef]

27. Boykov, I.V.; Boykova, A.I.; Syomov, M.A. An approximate solution of hypersingular integral equations of the first kind. *Univ. Proc. Volga Reg. Phys. Math. Sci. Math.* **2015**, *3*, 11–27. (In Russian)
28. Boykov, I.V.; Boykova, A.I. Approximate methods for solving hypersingular integral equations of the first kind with second-order singularities on classes of functions with weights $(1-t^2)^{-1/2}$. *Univ. Proc. Volga Reg. Phys. Math. Sci. Math.* **2017**, *2*, 79–90. (In Russian)
29. Boykov, I.V.; Boykova, A.I. An approximate solution of hypersingular integral equations of the first kind with singularities of the second order on the class of functions with weight $((1-t)/(1+t))^{\pm 1/2}$. *Univ. Proc. Volga Reg. Phys. Math. Sci. Math.* **2019**, *3*, 76–92. (In Russian)
30. Lifanov, I.K.; Nenashev, A.S. Hypersingular integral equations and the theory of wire antennas. *Diff. Equ.* **2005**, *41*, 126–145. [CrossRef]
31. Lifanov, I.K.; Nenashev, A.S. Analysis of Some Computational Schemes for a Hypersingular Integral Equation on an Interval. *Diff. Equ.* **2005**, *41*, 1343–1348. [CrossRef]
32. Boykov, I.V. Analytical and numerical methods for solving hypersingular integral equations. *Dyn. Syst.* **2019**, *9*, 244–272. (In Russian)
33. Boikov, I.V. On a continuous method for solving nonlinear operator equations. *Differ. Equ.* **2012**, *48*, 1308–1314. [CrossRef]
34. Daletskii Yu, L.; Krein, M.G. *Stability of Solutions of Differential Equations in Banach Space*; Nauka: Moscow, Russia, 1970; 536p. English Translation: Translations of Mathematical Monographss, Volume 43; American Mathematical Society: Providence, RI, USA, 1974. (In Russian)
35. Lozinskii, S.M. Note on a paper by V.S. Godlevskii. *Ussr Comput. Math. Math. Phys.* **1973**, *13*, 232–234 [CrossRef]
36. Boikov, I.V. On the stability of solutions of differential and difference equations in critical cases. *Soviet Math. Dokl.* **1991**, *42*, 630–632.
37. Boikov, I.V. *Stability of Solutions of Differential Equations*; Publishing House of Penza State University: Penza, Russia, 2008; 24p. (In Russian)
38. Browder, F.E. Nonlinear mappings of the nonexpansive and accretive type in Banach spaces. *Bull. Am. Math. Soc.* **1967**, *73*, 875–882. [CrossRef]
39. Malkowski, E.; Rakočević, V. *Advanced Functional Analysis*; CRS Press, Taylor and Francis Group: Boca Raton, FL, USA, 2019.
40. Ćirić, L.; Rafiq, A.; Radenović, S.; Rajović, M. Jeong Sheok Ume On Mann implicit iterations for strongly accreative and strongly pseudo-contractive mappings. *Appl. Math. Comput.* **2008**, *198*, 128–137.
41. Todoriević, V. *Harmonic Quasiconformal Mappings and Hyperbolic Type Metrics*; Springer Nature: Cham, Switzerland, 2019
42. Todoriević, V. Subharmonic behavior and quasiconformal mappings. *Anal. Math. Phys.* **2019**, *9*, 1211–1225. [CrossRef]
43. Hadamard, J. *Lectures on Cauchy's Problem in Linear Partial Differential Equations*; Dover Publ. Inc.: New York, NY, USA, 1952.
44. Chikin, L.A. Special cases of the Riemann boundary value problems and singular integral equations. *Sci. Notes Kazan State Univ.* **1953**, *113*, 53–105. (In Russian)
45. Chan, Y.-S.; Fannjiang, A.C.; Paulino, G.H. Integral equations with hypersingular kernels-theory and applications to fracture mechanics. *Int. J. Eng. Sci.* **2003**, *41*, 683–720. [CrossRef]

© 2020 by the authors. Licensee MDPI, Basel, Switzerland. This article is an open access article distributed under the terms and conditions of the Creative Commons Attribution (CC BY) license (http://creativecommons.org/licenses/by/4.0/).

Article

On the Periodicity of General Class of Difference Equations

Osama Moaaz [1,*], **Hamida Mahjoub** [1,2] **and Ali Muhib** [1,3]

1. Department of Mathematics, Faculty of Science, Mansoura University, Mansoura 35516, Egypt; hamida.almahgoub@gmail.com (H.M.); muhib39@yahoo.com (A.M.)
2. Department of Mathematics, Faculty of Science, Benghazi, Libya
3. Department of Mathematics, Faculty of Education (Al-Nadirah), Ibb University, Ibb, Yemen
* Correspondence: o_moaaz@mans.edu.eg

Received: 27 August 2019; Accepted: 22 April 2020; Published: 1 July 2020

Abstract: In this paper, we are interested in studying the periodic behavior of solutions of nonlinear difference equations. We used a new method to find the necessary and sufficient conditions for the existence of periodic solutions. Through examples, we compare the results of this method with the usual method.

Keywords: difference equations; periodicity character; nonexistence cases of periodic solutions

1. Introduction

Difference equations are recognized as descriptions of the observed evolution of a phenomenon, where the majority of measurements of a time-evolving variable are discrete. Many mathematicians are interested of studying the qualitative behavior of difference equations motivating and fruitful as it underpins the analysis and modeling of different daily life phenomena, for example in economics, queuing theory, statistical problems, stochastic time series, probability theory, psychology, quanta in radiation, combinatorial analysis, genetics in biology, economics, electrical network, etc. Examples of difference equations that have gotten the attention of researchers see [1–40].

Grove and Ladas [9] studied the periodic character of solutions of many difference equations of higher order. Their book presented their findings along with some thought-provoking questions and many open problems and conjectures worthy of investigation. Agarwal and Elsayed [3] studied the periodicity and stability of solutions of higher order rational equation

$$w_{n+1} = a + \frac{dw_{n-l}w_{n-k}}{b - cw_{n-s}},$$

where a, b, c and d are positive real numbers. Taskara et al. [38] presented a solution and periodicity of the equation

$$w_{n+1} = \frac{p_n w_{n-k} + w_{n-(k+1)}}{q_n + w_{n-(k+1)}},$$

where p_n and q_n are periodic sequences with $(k+1)$−period and p_n is not equal to q_n. Stevic [29] studied the periodic character of equation

$$w_{n+1} = p + \frac{w_{n-(2s-1)}}{w_{n-(2l+1)s+1}},$$

where $p \geq 1$ is a real number. By a new method, Elsayed [12] and Moaaz [24] studied the existence of the solution of prime period two of equation

$$w_{n+1} = \alpha + \beta \frac{w_n}{w_{n-1}} + \gamma \frac{w_{n-1}}{w_n},$$

where α, β and γ are real numbers. Recently, Abdelrahman et al. [1] and Moaaz [25] studied the asymptotic behavior of the solutions of general equation

$$w_{n+1} = aw_{n-l} + bw_{n-k} + f(w_{n-l}, w_{n-k}),$$

where a and b are nonnegative real number.

This paper aims to shed light on the study of the existence or nonexistence of periodic solutions for difference equations. We describe and modify the new method in Elsayed [12]. Moreover, we use this new method to study the existence of periodic solutions of the general class of difference equation. Furthermore, we discuss some of the nonexistence cases of periodic solutions. Finally, through examples, we compare the results of this method with the usual method.

2. Existence and Nonexistence of a Periodic Solutions

2.1. Existence of Periodic Solutions of Period Two

Elsayed in [12] and Moaaz in [24] are established a new technique to study the existence of periodic solutions of some rational difference equation. In the following, we describe and modify this method:

Consider the difference equation

$$w_{n+1} = F(w_n, w_{n-1}, ..., w_{n-k}), \tag{1}$$

where k is positive integer. Now, we assume that Equation (1) has periodic solutions of period two

$$..., \rho, \sigma, \rho, \sigma, ...,$$

with $w_{n-(2s+1)} = \rho$ and $w_{n-2s} = \sigma$. Hence, we get that

$$\begin{cases} \rho = F(\sigma, \rho, ...); \\ \sigma = F(\rho, \sigma, ...). \end{cases} \tag{2}$$

Next, we let $\tau = \rho/\sigma$, and substitute into (2). Then, we get that

$$\begin{cases} \rho = F_1(\tau); \\ \sigma = F_2(\tau). \end{cases}$$

By using the fact $\rho - \tau\sigma = 0$, we obtain

$$F_1(\tau) - \tau F_2(\tau) = 0. \tag{3}$$

Finally, by using the relation (3), we can obtain—in most cases—the necessary and sufficient conditions that Equation (1) has periodic solutions of the prime period two.

The effectiveness of this method appears in a study the existence of periodic solutions of some difference equations with real coefficients and initial conditions (not positive only). Besides, we can study the existence of periodic solutions of some difference equations, which have never been done before due to failure while applying the usual method.

Next, we apply the new method to study the existence of periodic solutions of general equations

$$w_{n+1} = a w_{n-1} \Phi(w_n, w_{n-1}), \tag{4}$$

where a is positive real number, w_{-1}, w_0 are positive real numbers and $\Phi(u,v)$ is a homothetic function, that is there exist a strictly increasing function $G : \mathbb{R} \to \mathbb{R}$ and a homogenous function $H : \mathbb{R}^2 \to \mathbb{R}$ with degree β, such that $\Phi = G(H)$.

Remark 1. *In the following proofs, we use induction to prove the relationships. We'll only take care of the basic step of induction and the rest of the steps directly, so it was ignored.*

Theorem 1. *Assume that β is a ratios of odd positive integers and $G^{-1}(1/a)$ exists. Equation (4) has a prime period two solution $..., \rho, \sigma, \rho, \sigma, ...$ if and only if*

$$H(\tau, 1) = H(1, \tau) = \frac{A}{\sigma^\beta}, \tag{5}$$

where $\tau = \rho/\sigma$ and $A = G^{-1}(1/a)$.

Proof. We suppose that Equation (4) has a prime period two solution

$$..., \rho, \sigma, \rho, \sigma, ...$$

It follows from (4) that

$$\rho = a\rho \Phi(\sigma, \rho);$$
$$\sigma = a\sigma \Phi(\rho, \sigma).$$

Hence,

$$\Phi(\sigma, \rho) = G\left(\sigma^\beta H(1, \tau)\right) = \frac{1}{a} \tag{6}$$

and so,

$$\sigma^\beta = \frac{A}{H(1, \tau)}; \tag{7}$$

$$\rho^\beta = \frac{A \tau^\beta}{H(\tau, 1)}. \tag{8}$$

By dividing (8) by (7), we have that (5) holds.
On the other hand, let (5) holds. If we choose

$$w_{-1} = \frac{A^{1/\beta} \tau}{H^{1/\beta}(\tau, 1)} \quad \text{and} \quad w_0 = \frac{A^{1/\beta}}{H^{1/\beta}(1, \tau)},$$

for $\tau \in \mathbb{R}^+$, then we get

$$\begin{aligned} w_1 &= a w_{-1} \Phi(w_0, w_{-1}) \\ &= a \frac{A^{1/\beta} \tau}{H^{1/\beta}(\tau, 1)} G\left(H\left(\frac{A^{1/\beta}}{H^{1/\beta}(1, \tau)}, \frac{A^{1/\beta} \tau}{H^{1/\beta}(\tau, 1)}\right)\right) \\ &= a \frac{A^{1/\beta} \tau}{H^{1/\beta}(\tau, 1)} G\left(\frac{A}{H(1, \tau)} H(1, \tau)\right) \\ &= \frac{A^{1/\beta} \tau}{H^{1/\beta}(\tau, 1)} = w_{-1}. \end{aligned}$$

Similarly, we have that $w_2 = w_0$. Hence, it is followed by the induction that

$$w_{2n-1} = \frac{A^{1/\beta}\tau}{H^{1/\beta}(\tau,1)} \quad \text{and} \quad w_{2n} = \frac{A^{1/\beta}}{H^{1/\beta}(1,\tau)} \quad \text{for all } n > 0.$$

Therefore, Equation (4) has a prime period two solution, and the proof is complete. □

Consider the recursive sequence

$$w_{n+1} = f(w_{n-l}, w_{n-k}), \tag{9}$$

where the function $f(u,v) : (0,\infty)^2 \to (0,\infty)$ is continuous real function and homogenous with degree *zero*.

Theorem 2. *Assume that l odd, k even. Equation (9) has a prime period two solution* $..., \rho, \sigma, \rho, \sigma, ...$ *if and only if*

$$f(\tau,1) = \tau f(1,\tau), \tag{10}$$

where $\tau = \rho/\sigma$.

Proof. Assume that $l > k$. Since l odd and k even, we have $w_{n-l} = \rho$ and $w_{n-k} = \sigma$. From Equation (9), we get

$$\begin{aligned}
\rho &= f(\rho,\sigma) = f\left(\frac{\rho}{\sigma},1\right) \\
\sigma &= f(\sigma,\rho) = f\left(1,\frac{\rho}{\sigma}\right).
\end{aligned}$$

Since $\tau = \rho/\sigma$, we obtain

$$0 = \rho - \tau\sigma = f(\tau,1) - \tau f(1,\tau).$$

On the other hand, let (10) holds. Now, we choose

$$w_{-l+2\mu} = f(\tau,1) \quad \text{and} \quad w_{-l+2\mu+1} = f(1,\tau), \quad \mu = 0, 1, ..., (l-1)/2$$

where $\tau \in \mathbb{R}^+$. Hence, we see that

$$\begin{aligned}
w_1 &= f(w_{-l}, w_{-k}) \\
&= f(f(\tau,1), f(1,\tau)) \\
&= f(\tau f(1,\tau), f(1,\tau)) \\
&= f(\tau,1).
\end{aligned}$$

Similarly, we can prove that $w_2 = f(1,\tau)$. Hence, it is followed by the induction that

$$w_{2n-1} = f(\tau,1) \quad \text{and} \quad w_{2n} = f(1,\tau) \quad \text{for all } n > 0.$$

Therefore, Equation (9) has a prime period two solution, and the proof is complete. □

Theorem 3. *Assume that l even, k odd. Equation (9) has a prime period two solution* $..., \rho, \sigma, \rho, \sigma, ...$ *if and only if*

$$f(1,\tau) = \tau f(\tau,1), \tag{11}$$

where $\tau = \rho/\sigma$.

Proof. The proof is similar to that of proof of Theorem 2 and hence is omitted. □

Consider the difference equation

$$w_{n+1} = \gamma + \delta \frac{w_{n-1}^\beta}{g(w_n, w_{n-1})}, \qquad (12)$$

where β is a positive real number, γ, δ, w_{-1} and w_0 are arbitrary real numbers and the function $g(u,v)$ is continuous real function and homogenous with degree β

Theorem 4. *Equation (12) has a prime period two solution* $..., \rho, \sigma, \rho, \sigma, ...$ *if and only if*

$$\gamma = \delta \frac{\tau^\beta g(\tau,1) - \tau g(1,\tau)}{(\tau - 1) g(1,\tau) g(\tau,1)}, \qquad (13)$$

where $\tau = \rho/\sigma$.

Proof. Assume that there exists a prime period two solution of Equation (12) $..., \rho, \sigma, \rho, \sigma, ...$ Thus, from (12), we find $w_{n-(2r+1)} = \rho$ and $w_{n-2r} = \sigma$ for $r = 0, 1, 2, ...,$ and so

$$\rho = \gamma + \delta \frac{\rho^\beta}{g(\sigma, \rho)}$$

and

$$\sigma = \gamma + \delta \frac{\sigma^\beta}{g(\rho, \sigma)}.$$

Since $g(u,v)$ be homogenous of degree β, we get $g(u,v) = v^\beta g\left(\frac{u}{v},1\right) = u^\beta g\left(1,\frac{v}{u}\right)$ and hence,

$$\rho = \gamma + \delta \frac{\rho^\beta}{\sigma^\beta g\left(1,\frac{\rho}{\sigma}\right)}$$

$$\sigma = \gamma + \delta \frac{\sigma^\beta}{\sigma^\beta g\left(\frac{\rho}{\sigma},1\right)}.$$

Now, let $\rho = \tau \sigma$. Then, we get

$$\rho = \gamma + \delta \frac{\tau^\beta}{g(1,\tau)} \qquad (14)$$

$$\sigma = \gamma + \delta \frac{1}{g(\tau,1)}. \qquad (15)$$

By using the fact $\rho - \tau \sigma = 0$, we obtain

$$\rho - \tau \sigma = \gamma + \delta \frac{\tau^\beta}{g(1,\tau)} - \tau \left(\gamma + \delta \frac{1}{g(\tau,1)}\right)$$

$$0 = (1-\tau)\gamma + \delta \frac{\tau^\beta g(\tau,1) - \tau g(1,\tau)}{g(\tau,1) g(1,\tau)}$$

and so

$$\gamma = \delta \frac{\tau^\beta g(\tau,1) - \tau g(1,\tau)}{(\tau - 1) g(\tau,1) g(1,\tau)}.$$

Next, from (14) and (15), we see that

$$\rho = \delta \frac{\tau}{(\tau-1)} \frac{\tau^\beta g(\tau,1) - g(1,\tau)}{g(\tau,1) g(1,\tau)} \tag{16}$$

$$\sigma = \delta \frac{1}{(\tau-1)} \frac{\tau^\beta g(\tau,1) - g(1,\tau)}{g(\tau,1) g(1,\tau)}. \tag{17}$$

On the other hand, suppose that (13) holds. Let $w_{-1} = \rho$ and $w_0 = \sigma$ where ρ, σ defined as (11) and (17), respectively. Then, from (12) and (13), we find

$$\begin{aligned} w_1 &= \gamma + \delta \frac{w_{-1}^\beta}{g(w_0, w_{-1})} \\ &= \gamma + \delta \frac{\rho^\beta}{g(\sigma, \rho)} \\ &= \delta \frac{\tau^\beta g(\tau,1) - \tau g(1,\tau)}{(\tau-1) g(\tau,1) g(1,\tau)} + \delta \frac{\tau^\beta}{g(1,\tau)} = \rho. \end{aligned}$$

Similarly, we can proof that $w_2 = \sigma$. Hence, it is followed by the induction that

$$w_{2n+1} = \rho \text{ and } w_{2n} = \sigma \text{ for all } n > -1.$$

Therefore, Equation (12) has a prime period two, and the proof is complete. □

2.2. Nonexistence of Periodic Solutions of Period Two

In the following theorems, we study some general cases which there are no periodic solutions with period two of the equations

$$w_{n+1} = f(w_n, w_{n-1}) \tag{18}$$

and

$$w_{n+1} = f(w_n, w_{n-2}), \tag{19}$$

where $f \in C\left((0,\infty)^2, (0,\infty)\right)$ and w_{-1}, w_0 are positive real numbers.

Theorem 5. *Assume that $f_u > 0$ and $f_v < 0$. Then Equation (18) does not have positive period two solutions.*

Proof. On the contrary, we assume that Equation (18) has a period two distinct solution

$$\ldots, r, s, r, s, \ldots,$$

where $r \neq s$. It follows from (18) that

$$\begin{cases} r = f(s, r); \\ s = f(r, s). \end{cases} \tag{20}$$

Thus, we get

$$r f(r, s) - s f(s, r) = 0.$$

Now, we define the function

$$G_{v_0}(u) = u f(u, v_0) - v_0 f(v_0, u), \quad u > 0,$$

for $v_0 \in (0, \infty)$. Since $f > 0$, $f_u > 0$ and $f_v < 0$, we obtain

$$\frac{d}{du} G_{v_0}(u) = f(u, v_0) + u f_u(u, v_0) - v_0 f_v(v_0, u) > 0.$$

Thus, G_{v_0} is an increasing and hence G has at most one root for $u \in (0, \infty)$. But, $G(v_0) = 0$, then he only root of $G_{v_0}(w)$ is $u = v_0$. Thus, only solution of (20) is $s = r$, which is a contradiction. This completes the proof. □

Theorem 6. *Assume that $f_u > 0$ and $f_v > 0$. Then Equation (19) does not have positive period two solutions.*

Proof. The proof is similar to the proof of Theorem 5 and hence is omitted. □

Now, assume that $f_u < 0$ and $f_v > 0$. In view of [21] (Theorem 1.4.6), if Equation (18) has no solutions of prime period two, then every solution of Equation (18) converges to w^*. Therefore, we conclude the following:

Corollary 1. *Assume that $f_u < 0$ and $f_v > 0$. Then Equation (18) either every its solutions converges to w^* or has a prime period two solution.*

Corollary 2. *Assume that l and k are nonnegative integers and $w_{-\max\{l,k\}}, w_{-\max\{l,k\}+1}, ..., w_0$ are positive real numbers. The difference equation*

$$w_{n+1} = f(w_{n-l}, w_{n-k}) \tag{21}$$

does not have positive period two solutions, in the following cases:

(a) *l is even, k is odd, $f_u > 0$ and $f_v < 0$;*
(b) *l and k are even, $f_u > 0$ and $f_v > 0$.*

3. Application and Discussion

Next, we - by using Theorem 1—study the periodic character of the positive solutions of equation

$$w_{n+1} = aw_{n-1} \exp\left(\frac{-w_n w_{n-1}}{bw_n + cw_{n-1}}\right), \tag{22}$$

where $a, b, c \in (0, \infty)$. Let

$$H(u, v) = \frac{-uv}{bu + cv},$$

$G(y) = e^y$ and $\Phi(w_n, w_{n-1}) = G(H(u, v))$. From (5), if $b = c$, then (22) has a prime period two solution.

Moreover, by using Theorem 1, the discrete model with two age classes

$$w_{n+1} = w_{n-1} \exp(r - \lambda w_n - w_{n-1}), \tag{23}$$

has a prime period two solution if $\lambda = 1$.

In [10], El-Dessoky studied the periodic character of the positive solutions of equation

$$w_{n+1} = aw_{n-l} + bw_{n-k} + \frac{cw_{n-s}}{dw_{n-s} - \delta}, \tag{24}$$

where $a, b, c, d, \delta, w_{-r}, w_{-r+1}, ..., w_0$ are positive real numbers, $r = \max\{k, l, s\}$, l, k odd and s even. He is proved that the Equation (24) has no prime period two solution if $c + \delta(a + b - 1) \neq 0$. In the following, by the present method, we will find the necessary and sufficient conditions that this equation has periodic solutions of prime period two.

Corollary 3. *Equation (24) has prime period two solution if and only if $c + \delta(a + b - 1) = 0$.*

Proof. Assume that there exists a prime period two solution of Equation (24) $\dots, \rho, \sigma, \rho, \sigma, \dots$ Thus, from (24), we find
$$(1-a-b)\rho = \frac{c\sigma}{d\sigma - \delta}$$
and
$$(1-a-b)\sigma = \frac{c\rho}{d\rho - \delta}.$$

Now, let $\rho = \tau\sigma$ where $\tau \notin \{0,1\}$. Then, we get
$$d\sigma = \frac{c}{(1-a-b)\tau} + \delta$$
and
$$d\rho = \frac{c\tau}{(1-a-b)} + \delta.$$

Then, we have
$$d(\rho - \tau\sigma) = (\tau - 1)\left(\frac{c}{(1-a-b)} - \delta\right).$$

Since $\tau \neq 1$, we have
$$\frac{c}{(1-a-b)} = \delta,$$

and hence $c + \delta(a+b-1) = 0$. On the other hand, in view of [10] (Theorem 5), if $c + \delta(a+b-1) \neq 0$, then (24) has no solutions of prime period two. This completes the proof. $\square$

Example 1. *By Theorem 2, the difference equation*
$$w_{n+1} = \frac{aw_n w_{n-1}}{bw_n^2 + cw_{n-1}^2} \tag{25}$$

has periodic solutions of prime period two if and only if
$$\frac{a\tau}{b + c\tau^2} = \tau \frac{a\tau}{b\tau^2 + c}$$

and so,
$$(\tau - 1)\left(c + c\tau + c\tau^2 - b\tau\right) = 0$$

Since $p \neq q$, we have $\tau \neq 1$, and hence
$$\frac{b}{c} = \frac{1 + \tau + \tau^2}{\tau} \tag{26}$$

Now, we have $\tau > 0$, then the function $y(t) = (1 + \tau + \tau^2)/\tau$ attends its minimum value on $\mathbb{R}^+$ at $\tau_0 = 1$ and $\min_{\tau \in \mathbb{R}^+} y = y(\tau_0) = 3$, and so
$$\frac{1+\tau+\tau^2}{\tau} > \min_{\tau \in \mathbb{R}^+} y = 3 \text{ for } \tau > 0, \tau \neq 1.$$

which with (26) gives $b > 3c$. For example, $a = 3, b = 4, c = 1, w_{-1} = 0.2764$ and $w_0 = 0.7236$.

Example 2. *Consider the difference equation*
$$w_{n+1} = a + \frac{bw_{n-1}^2}{\alpha w_n^2 + \beta w_n w_{n-1} + \gamma w_{n-1}^2} \tag{27}$$

where α, β and γ are real numbers. We note that $\beta = 2$ and $f(u,v) = \alpha u^2 + \beta uv + \gamma v^2$ homogenous of degree 2. Then, Equation (27) has a prime period two solution if

$$a = b\tau \frac{\alpha + \tau\alpha + \tau\beta - \tau\gamma + \tau^2\alpha}{(\alpha\tau^2 + \beta\tau + \gamma)(\alpha + \beta\tau + \gamma\tau^2)} \tag{28}$$

Example $b = 2$, $\alpha = 0.5$, $\beta = 1.5$, $\gamma = 0.5$.
Note that, (28) implies that

$$a\left(\alpha\tau^2 + \beta\tau + \gamma\right)\left(\alpha + \beta\tau + \gamma\tau^2\right) - b\tau\left(\alpha + \tau\alpha + \tau\beta - \tau\gamma + \tau^2\alpha\right) = 0$$

and so,

$$\left(\frac{\tau^4+1}{\tau^3+\tau}\right) + \frac{a\alpha^2 - b\alpha + a\beta^2 - b\beta + a\gamma^2 + b\gamma}{a\alpha\gamma}\left(\frac{\tau}{\tau^2+1}\right) = \frac{b\alpha - a\alpha\beta - a\beta\gamma}{a\alpha\gamma}.$$

By using the facts $\frac{\tau^4+1}{\tau^3+\tau} > 1$ and $\frac{\tau}{\tau^2+1} < \frac{1}{2}$ for $\tau \in \mathbb{R}^+ \setminus \{1\}$, the condition (28) implies that

$$\begin{cases} 2(b\alpha - a\alpha\beta - a\beta\gamma) - (a\alpha^2 + 2a\alpha\gamma - b\alpha + a\beta^2 - b\beta + a\gamma^2 + b\gamma) > 0 \\ \text{and} \quad b\beta + b\alpha - a\alpha^2 - a\beta^2 - a\gamma^2 - b\gamma > 0. \end{cases}$$

Example 3. Consider the difference equation

$$w_{n+1} = a + \left(\frac{w_n}{w_{n-1}}\right)^\alpha, \tag{29}$$

where $a, \alpha \in (0, \infty)$. Now, if we define the function $f : (0, \infty)^2 \to (0, \infty)$ and

$$f(u,v) = a + \left(\frac{u}{v}\right)^\alpha,$$

then

$$\frac{\partial}{\partial u} f(u,v) = a\alpha \frac{u^{\alpha-1}}{v^\alpha} > 0;$$

$$\frac{\partial}{\partial v} f(u,v) = -a\alpha \frac{u^\alpha}{v^{\alpha+1}} < 0.$$

Thus, from Theorem 5, Equation (29) does not have positive period two solutions (Theorem 4.1 in [36]).

Example 4. Consider the May's Host Parasitoid Model

$$w_{n+1} = \frac{cw_n^2}{(1+w_n)w_{n-1}}, \tag{30}$$

where $c \in (0, \infty)$. Now, if we define the function $f : (0, \infty)^2 \to (0, \infty)$ and

$$f(u,v) = \frac{cu^2}{(1+u)v},$$

then

$$\frac{\partial}{\partial u} f(u,v) = \frac{u}{v} \frac{c}{(u+1)^2}(u+2) > 0;$$

$$\frac{\partial}{\partial v} f(u,v) = -\frac{u^2}{v^2} \frac{c}{u+1} < 0.$$

Thus, from Theorem 5, Equation (30) does not have positive period two solutions.

Author Contributions: All authors claim to have contributed equally and significantly in this paper. All authors read and approved the final manuscript.

Funding: The authors received no direct funding for this work.

Conflicts of Interest: There are no competing interests between the authors.

References

1. Abdelrahman, M.A.E.; Chatzarakis, G.E.; Li, T.; Moaaz, O. On the difference equation $w_{n+1} = aw_{n-l} + bw_{n-k} + f(w_{n-l}, w_{n-k})$. *Adv. Differ. Equ.* **2018**, *431*, 2018.
2. Abdelrahman, M.A.E, On the difference equation $zm+1 = f(zm, zm-1, ..., zm-k)$. *J. Taibah Univ. Sci.* **2019**, *13*, 1014–1021. [CrossRef]
3. Agarwal, R.P.; Elsayed, E.M. Periodicity and stability of solutions of higher order rational difference equation. *Adv. Stud. Contemp. Math.* **2008**, *17*, 181–201.
4. Ahlbrandt, C.D.; Peterson, A.C. *Discrete Hamiltonian Systems: Difference Equations, Continued Fractions, and Riccati Equations*; Kluwer Academic Publishers: Dordrecht, The Netherlands, 1996.
5. Ahmad, S. On the nonautonomous Volterra-Lotka competition equations. *Proc. Am. Math. Soc.* **1993**, *117*, 199–204. [CrossRef]
6. Allman, E.S.; Rhodes, J.A. *Mathematical Models in Biology: An Introduction*; Cambridge University Press: Cambridge, UK, 2003.
7. Andres, J.; Pennequin, D. Note on Limit-Periodic Solutions of the Difference Equation $w_{t+1} - [h(wt) + \lambda]w = rt, \lambda > 1$. *Axioms* **2019**, *8*, 19. [CrossRef]
8. Din, Q.; Elsayed, E.M. Stability analysis of a discrete ecological model. *Comput. Ecol. Softw.* **2014**, *4*, 89–103.
9. Grove E.A.; Ladas, G. *Periodicities in Nonlinear Difference Equations*; Chapman & Hall/CRC: Boca Raton, FL, USA, 2005; Volume 4.
10. El-Dessoky, M.M. On the difference equation $w_{n+1} = aw_{n-l} + bw_{n-k} + cw_{n-s}/(dw_{n-s} - e)$. *Math. Meth. Appl. Sci.* **2017**, *40*, 535–545. [CrossRef]
11. Elettreby, M.F.; El-Metwally, H. On a system of difference equations of an economic model. *Discrete Dyn. Nat. Soc.* **2013**, *6*, 405628. [CrossRef]
12. Elsayed, E.M. New method to obtain periodic solutions of period two and three of a rational difference equation. *Nonlinear Dyn.* **2015**, *79*, 241–250. [CrossRef]
13. Elsayed, E.M.; El-Dessoky, M.M. Dynamics and behavior of a higher order rational recursive sequence. *Adv. Differ. Equ.* **2012**, *2012*, 69. [CrossRef]
14. Foupouagnigni, M.; Mboutngam, S. On the Polynomial Solution of Divided-Difference Equations of the Hypergeometric Type on Nonuniform Lattices. *Axioms* **2019**, *8*, 47. [CrossRef]
15. Foupouagnigni, M.; Koepf, W.; Kenfack-Nangho, M.; Mboutngam, S. On Solutions of Holonomic Divided-Difference Equations on Nonuniform Lattices. *Axioms* **2013**, *2*, 404–434. [CrossRef]
16. Gil, M. Solution Estimates for the Discrete Lyapunov Equation in a Hilbert Space and Applications to Difference Equations. *Axioms* **2019**, *8*, 20. [CrossRef]
17. Haghighi, A.M.; Mishev, D.P. *Difference and Differential Equations with Applications in Queueing Theory*; John Wiley & Sons Inc.: Hoboken, NJ, USA, 2013.
18. Kalabusic, S.; Kulenovic, M.R.S. On the recursive sequnence $w_{n+1} = (\alpha w_{n-1} + \beta w_{n-2})/(\gamma w_{n-1} + \delta w_{n-2})$. *J. Differ. Equ. Appl.* **2003**, *9*, 701–720.
19. Kelley, W.G.; Peterson, A.C. *Difference Equations: An Introduction with Applications*, 2nd ed.; Harcour Academic: New York, NY, USA, 2001.
20. Kocic, V.L.; Ladas, G. *Global Behavior of Nonlinear Difference Equations of Higher Order with Applications*; Kluwer Academic Publishers: Dordrecht, The Netherlands, 1993.
21. Kulenovic, M.R.S.; Ladas, G. *Dynamics of Second Order Rational Difference Equations with Open Problems and Conjectures*; Chapman & Hall/CRC Press: Boca Raton, FL, USA, 2001.
22. Liu, X. A note on the existence of periodic solutions in discrete predator-prey models. *Appl. Math. Model.* **2010**, *34*, 2477–2483. [CrossRef]

23. Migda M.; Migda, J. Nonoscillatory Solutions to Second-Order Neutral Difference Equations. *Symmetry* **2018**, *10*, 207. [CrossRef]
24. Moaaz, O. Comment on new method to obtain periodic solutions of period two and three of a rational difference equation. *Nonlinear Dyn.* **2017**, *88*, 1043–1049. [CrossRef]
25. Moaaz, O. Dynamics of difference equation $w_{n+1} = f(w_{n-l}, w_{n-k})$. *Adv. Differ. Equ.* **2018**, *447*, 2018.
26. Moaaz O.; Chalishajar, D.; Bazighifan, O. Some Qualitative Behavior of Solutions of General Class of Difference Equations. *Mathematics* **2019**, *7*, 585. [CrossRef]
27. Moaaz, O.; Chatzarakis, G.E.; Chalishajar, D.; Bazighifan, O. Dynamics of General Class of Difference Equations and Population Model with Two Age Classes. *Mathematics* **2020**, *8*, 516. [CrossRef]
28. Pogrebkov, A. Hirota Difference Equation and Darboux System: Mutual Symmetry. *Symmetry* **2019**, *11*, 436. [CrossRef]
29. Stevic, S. A note on periodic character of a difference equation. *J. Differ. Equ. Appl.* **2004**, *10*, 929–932. [CrossRef]
30. Stevic, S. On the recursive sequence $w_{n+1} = \alpha + w_{n-1}^p / w_n^p$. *J. Appl. Math. Comput.* **2005**, *18*, 229–234.
31. Stevic, S. A short proof of the Cushing–Henson conjecture. *Discrete Dyn. Nat. Soc.* **2006**, *4*, 37264. [CrossRef]
32. Stevic, S. Global stability and asymptotics of some classes of rational difference equations. *J. Math. Anal. Appl.* **2006**, *316*, 60–68. [CrossRef]
33. Stevic, S. Asymptotics of some classes of higher order difference equations. *Discrete Dyn. Nat. Soc.* **2007**, *2007*, 56813. [CrossRef]
34. Stevic, S. Asymptotic periodicity of a higher order difference equation. *Discrete Dyn. Nat. Soc.* **2007**, *2007*, 13737. [CrossRef]
35. Stevic, S. Existence of nontrivial solutions of a rational difference equation. *Appl. Math. Lett.* **2007**, *20*, 28–31. [CrossRef]
36. Stevic, S. On the Recursive Sequence $y_{n+1} = A + (y_n / y_{n-1})^p$. *Discrete Dyn. Nat. Soc.* **2007**, *2007*, 34517.
37. Stevic, S.; Kent, C.; Berenaut, S. A note on positive nonoscillatory solutions of the differential equation $w_{n+1} = \alpha + w_{n-1}^p / w_n^p$. *J. Diff. Equ. Appl.* **2006**, *12*, 495-499.
38. Taskara, N.; Uslu, K.; Tollu, D.T. The periodicity and solutions of the rational difference equation with periodic coefficients. *Comput. Math. Appl.* **2011**, *62*, 1807–1813. [CrossRef]
39. Yang, C. Positive Solutions for a Three-Point Boundary Value Problem of Fractional Q-Difference Equations. *Symmetry* **2018**, *10*, 358. [CrossRef]
40. Zhou, Z.; Zou, X. Stable periodic solutions in a discrete periodic logistic equation. *Appl. Math. Lett.* **2003**, *16*, 165–171. [CrossRef]

© 2020 by the authors. Licensee MDPI, Basel, Switzerland. This article is an open access article distributed under the terms and conditions of the Creative Commons Attribution (CC BY) license (http://creativecommons.org/licenses/by/4.0/).

Article

Eigenfunction Families and Solution Bounds for Multiplicatively Advanced Differential Equations

David W. Pravica [1,2]**, Njinasoa Randriampiry** [1] **and Michael J. Spurr** [1,2,*]

[1] Department of Mathematics, East Carolina University, Greenville, NC 27858, USA; pravicad@ecu.edu (D.W.P.); randriampiryn@ecu.edu (N.R.)
[2] School of Mathematics, University of the Witwatersrand, Private Bag 3, Johannesburg P O WITS 2050, South Africa
* Correspondence: spurrm@ecu.edu; Tel.: +1-252-328-4110

Received: 4 June 2020; Accepted: 16 July 2020; Published: 21 July 2020

Abstract: A family of Schwartz functions $\mathcal{W}(t)$ are interpreted as eigensolutions of MADEs in the sense that $\mathcal{W}^{(\delta)}(t) = E\,\mathcal{W}(q^\gamma t)$ where the eigenvalue $E \in \mathbb{R}$ is independent of the advancing parameter $q > 1$. The parameters $\delta, \gamma \in \mathbb{N}$ are characteristics of the MADE. Some issues, which are related to corresponding q-advanced PDEs, are also explored. In the limit that $q \to 1^+$ we show convergence of MADE eigenfunctions to solutions of ODEs, which involve only simple exponentials and trigonometric functions. The limit eigenfunctions ($q = 1^+$) are not Schwartz, thus convergence is only uniform in $t \in \mathbb{R}$ on compact sets. An asymptotic analysis is provided for MADEs which indicates how to extend solutions in a neighborhood of the origin $t = 0$. Finally, an expanded table of Fourier transforms is provided that includes Schwartz solutions to MADEs.

Keywords: MADE; eigenfunction; convergence; Fourier transform

PACS: 34K06; 34A12; 42C40; 42A38; 33E99

1. Introduction

The introduction of a relaxing parameter $q > 1$ in differential equations was found to provide stability properties for their corresponding solutions. This is a phenomenon well-known in numerical analysis where if the Ordinary Differential Equation (ODE)

$$y'(t) = f(t, y(t))\,, \quad y(t_0) = y_0,$$

is *stiff* then one can try to use the *backward Euler method* to obtain the sequence $\{\langle t_n, y_n\rangle\}_{n=0}^{\infty}$ by first considering the algebraic equations

$$t_{n+1} = t_n + \Delta t\,, \quad y_{n+1} = y_n + f(t_{n+1}, y_{n+1}) \cdot \Delta t\,,$$

for small time-steps $\Delta t > 0$. If one can obtain y_{n+1} explicitly in terms of y_n then the iteration scheme often converges much faster, and for longer time intervals, than that provided by the *forward Euler method* [1], p. 349. That such a principle holds for ODEs as $\Delta t \to 0^+$ was established through the study of Multiplicatively Advanced Differential Equations (MADEs) as $q \to 1^+$, and will be discussed further in this article. Part of our analysis of stability will require obtaining uniform apriori bounds. This will be achieved in a somewhat general setting, and the consequences will be presented in the form of examples of advanced differential equations.

1.1. Solutions of MADEs as Eigenfunctions

In [2] solutions to equations of the form

$$y'(t) = ay(qt) + by(t), \; y(0) = 1 \text{ or } 0 \text{ (wlog)}, \tag{1}$$

were studied for $q > 1$, $a \in \mathbb{C}$, $b \in \mathbb{R}$ and $t \geq 0$. In the case that $b = 0$, with $y(0) = 0$, solutions $y(t)$ are referred to as *eigenfunctions* since $y(t) \to 0$ as $t \to \infty$. Specific asymptotic properties of solutions were obtained in Theorem 10 of [3]. Here we only consider the case that $b = 0$ and $a \in \mathbb{R}$, however the derivatives may be of higher (integer) order than in Equation (1). In addition, we extend solutions of these equations to all $t \in \mathbb{R}$ so that the eigen equation, referred to as an eigen-MADE, has a solution $y(t) \in \mathcal{S}(\mathbb{R})$ the Schwartz space of infinitely differentiable functions, with derivatives that decay faster than reciprocal polynomials (as defined in [4] section V.3). An asymptotic theory near $t = 0$ can be developed indicating that an extension to $t < 0$ is quite natural. In this way the special functions that we study are eigenfunctions in $\mathcal{L}^2(\mathbb{R})$, although not in the traditional, local ($q = 1$) sense. The significance of these functions will be demonstrated by examples, and convergence to familiar functions is obtained on compact subsets of $\mathbb{R}$, as $q \to 1^+$.

1.2. Brief Overview

The study of multiply advanced differential equations falls within the area of functional differential equations, as is studied for instance in Fox, et al. [2], Kato, et al. [3] and Dung [5]. There is also significant overlap with the area of q-difference differential equations, where the multiplicative advancement $y(t) \to y(qt)$ is referred to as a dilation and is denoted $\sigma_q(y(t)) = y(qt)$. There is a rich and active study within the area of q-difference differential equations with dilations involving $q > 1$. These are highlighted by works of: L. Di Vizio [6–8]; C. Hardouin [7]; T. Dreyfus [9,10]; A. Lastra [10–19]; S. Malek [10–22]; J. Sanz [17–19]; H. Tahara [23]; and C. Zhang [8,24]; along with further references by these researchers and others. Often these studies in q-difference differential equations overlap with the area of Gevrey asymptotics.

In the current work we continue by focusing on global solutions of a MADE on $\mathbb{R}$. In particular, we discuss several techniques for starting with a given global solution to an original MADE and then generating solutions of new related MADEs. This theme will be developed as follows: In Section 2, a known MADE solution first introduced in [25], namely $_qCos(t)$, is used to produce a simple related solution $\tilde{C}_q(t) = {_qCos}(t/\sqrt{q})$ which is an eigensolution of a MADE in the sense of the Abstract. In turn, $\tilde{C}_q(t)$ is then used to obtain a new q-advanced Airy function $Aiq(t)$ satisfying a MADE analogue of the Airy differential equation. Then $Aiq(t)$ itself is used along with convolution to generate families of functions $\phi_q(x,t)$ solving a q-advanced PDE.

In Section 3, a family of MADE solutions, under convolution and auto-correlation, are seen to produce related solutions of new MADEs. Furthermore, the least-element method in Poincare asymptotics is deployed to find natural extensions to related MADE solutions on the negative real line. A theory of asymptotic extensions to $t < 0$ is developed to clarify the notion that solutions to MADEs behave smoothly in a neigborhood of the origin. We also give conditions that ensures a natural extension to all of $\mathbb{R}$, as is needed to even consider a Fourier transform. An investigation of the inhomogeneous MADEs that these solve is begun.

In Section 4 we focus on considering solutions of MADEs as perturbations of classical solutions, and, mirroring a more direct convergence proof in Section 2, we exhibit MADE solutions which converge to a classical solution of a damped-oscillation equation—the convergence being uniform on compact subsets of $[0, \infty)$.

In Section 5, we return to the topics of convolution and auto-correlation to observe their impact when applied to MADE solutions. In this paper, we will discuss convolutions, correlations, and Fourier transforms for MADEs.

A table of Fourier transforms of global MADE solutions under study here is provided in Section 6. These will be solutions of new MADEs, for which we obtain new elements in a table of Fourier transforms. This new table mimics what is often done for Laplace transforms, in the study of linear constant coefficient ODEs.

In various theories of differential equations, convolutions provide a useful tool since general solutions can be determined from fundamental solutions, as demonstrated here in Equation (33). This is one motivation for obtaining solutions to homogenous equations, as appears in Proposition 2.

2. A Normalized Cosine Example and Extensions

From [25], consider the following Schwartz functions, for $q > 1$ and all $t \in \mathbb{R}$,

$$_q Cos(t) \equiv N_q \sum_{k=-\infty}^{\infty} \frac{(-1)^k}{q^{k^2}} \cdot \exp(-q^k |t|) \qquad (2)$$

$$_q Sin(t) \equiv sign(t) N_q \sum_{k=-\infty}^{\infty} \frac{(-1)^k}{q^{k(k-1)}} \exp(-q^k |t|) , \qquad (3)$$

where

$$\frac{1}{N_q} \equiv \sum_{k=-\infty}^{\infty} \frac{(-1)^k}{q^{k^2}} . \qquad (4)$$

Next define

$$\tilde{C}_q(t) \equiv {}_q Cos\left(\frac{t}{\sqrt{q}}\right) = N_q \sum_{k=-\infty}^{\infty} \frac{(-1)^k}{q^{k^2}} \cdot \exp\left(\frac{-q^k |t|}{\sqrt{q}}\right) . \qquad (5)$$

There are several properties that we note. In particular, the function $\tilde{C}_q(t)$ is normalized, in that the uniform bound $\|\tilde{C}_q\|_\infty = 1$ holds, after some delicate work performed in [25], for each $q > 1$. It also solves the following eigen-MADE for all $t \in \mathbb{R}$ and each $q > 1$,

$$\frac{d^2 \tilde{C}_q(t)}{dt^2} = -\tilde{C}_q(qt), \quad \tilde{C}_q(0) = 1, \quad \tilde{C}_q'(0) = 0 . \qquad (6)$$

From (6) we see that $\tilde{C}_q(t)$ satisfies an eigen-MADE in the sense of the Abstract, with $E = -1$ independently of the advancing parameter $q > 1$. Note that $_q Cos''(t) = -q \, _q Cos(qt)$ (as recorded in (10) below) does not have an eigenvalue $(-q)$ independent of q, thus we rely on $\tilde{C}_q(t)$ as the appropriate eigen-MADE solution.

Since $\tilde{C}_q(t)$ is not only $\mathcal{C}^\infty$ and bounded, but in fact Schwartz, we can obtain its Fourier transform, an operation defined for any $f \in \mathcal{L}^1(\mathbb{R})$, as

$$\hat{f}(\omega) = \mathcal{F}[f(t)](\omega) \equiv \frac{1}{\sqrt{2\pi}} \int_{-\infty}^{\infty} e^{-i\omega t} \cdot f(t) \, dt .$$

In [25] it was found that

$$\mathcal{F}[\tilde{C}_q(t)](\omega) = \frac{2(\mu_{q^2})^3 N_q}{\sqrt{2\pi}} \cdot \frac{1}{\theta(q^2; q\omega^2)} , \qquad (7)$$

where N_q was defined in Equation (4) above, and the other normalizing constant is

$$\mu_q \equiv \prod_{n=1}^{\infty} \left(1 - \frac{1}{q^n}\right) .$$

To express the Fourier transform of linear, homogeneous MADEs, we found multiple uses of the Jacobi theta function

$$\theta(q;u) \equiv \sum_{n=-\infty}^{\infty} \frac{u^n}{q^{n(n-1)/2}} = \mu_q \cdot (1+u) \cdot \prod_{n=1}^{\infty} \left(1 + \frac{u}{q^n}\right)\left(1 + \frac{1}{uq^n}\right), \qquad (8)$$

which allows the association that $N_q = \theta(q^2; -1/q)$, and which ensures that $N_q \neq 0$ for all $q > 1$, due to the product formula. It will be of significance to note that the reciprocal $1/\theta(q;u)$, for $u \geq 0$, is Schwartz when extended to be identically 0 for $u < 0$. Critical algebraic properties that we use are

$$\theta(q;q^p u) = q^{p(p+1)/2} u^p \cdot \theta(q;u), \quad \forall p \in \mathbb{Z}, \, u \in \mathbb{C}^*, \quad \text{and} \quad v \cdot \theta(q;1/v) = \theta(q;v), \quad \forall v \in \mathbb{C}^*. \qquad (9)$$

A consequence is that the only zeros of $\theta(q;u)$ are for $u = -q^p$ for all $p \in \mathbb{Z}$. This is obvious from the product definition of $\theta(q;u)$ in Equation (8).

2.1. Uniform Convergence

Using Taylor series methods as an approach paralleling that in [25] we show:

Proposition 1. *On any compact subset of $\mathbb{R}$, $\tilde{C}_q(t)$ approaches $\cos(t)$ uniformly as $q \to 1^+$.*

Proof. A given compact set is contained in an interval $[-\rho, \rho]$ for ρ sufficiently large, so it suffices to prove the theorem on $[-\rho, \rho]$.

First, recall the following results shown in [25]

$$\begin{aligned}
{}_q\text{Cos}(0) &= 1 & {}_q\text{Sin}(0) &= 0 \\
{}_q\text{Cos}'(t) &= -{}_q\text{Sin}(t) & {}_q\text{Sin}'(t) &= q \, {}_q\text{Cos}(qt) \\
{}_q\text{Cos}''(t) &= -q \, {}_q\text{Cos}(qt) & {}_q\text{Sin}''(t) &= -q^2 \, {}_q\text{Sin}(qt) \, .
\end{aligned} \qquad (10)$$

From these, by induction on the even order derivatives of ${}_q\text{Cos}(t)$, we obtain the higher order derivatives

$${}_q\text{Cos}^{(2L)}(t) = (-1)^L q^{L^2} \, {}_q\text{Cos}(q^L t), \qquad (11)$$

and

$${}_q\text{Cos}^{(2L+1)}(t) = [(-1)^L q^{L^2} \, {}_q\text{Cos}(q^L t)]' = (-1)^{L+1} q^{L^2 + L} \, {}_q\text{Sin}(q^L t) \, . \qquad (12)$$

We infer all derivatives of $\tilde{C}_q(t)$ via

$$\begin{aligned}
\tilde{C}_q^{(2L)}(t) &= [{}_q\text{Cos}(t/\sqrt{q})]^{(2L)} = (-1)^L q^{L^2} \, {}_q\text{Cos}(q^L t/\sqrt{q})(1/\sqrt{q})^{(2L)} & (13)\\
&= (-1)^L q^{L^2 - L} \, {}_q\text{Cos}(q^L t/\sqrt{q}) = (-1)^L q^{L^2 - L} \tilde{C}_q(q^L t) \, , & (14)
\end{aligned}$$

and

$$\tilde{C}_q^{(2L+1)}(t) = [(-1)^L q^{L^2 - L} \, {}_q\text{Cos}(q^L t/\sqrt{q})]' = (-1)^{L+1} q^{L^2 - 1/2} \, {}_q\text{Sin}(q^L t/\sqrt{q}) \, . \qquad (15)$$

Evaluating the derivatives of $\tilde{C}_q(t)$ at $t = 0$ yields

$$\tilde{C}_q^{(2L)}(0) = (-1)^L q^{L^2 - L} \quad \text{and} \quad \tilde{C}_q^{(2L+1)}(0) = 0 \qquad (16)$$

for all $L \geq 0$.

Next computing $P_{2N+1}[\tilde{C}_q](t)$, the $2N+1$ degree Taylor polynomial for $\tilde{C}_q(t)$ expanded about $t = 0$, gives

$$P_{2N+1}[\tilde{C}_q](t) = \sum_{p=0}^{2N+1} \frac{\tilde{C}_q^{(p)}(0)}{p!} t^p = \sum_{L=0}^{N} \frac{(-1)^L q^{L^2-L}}{(2L)!} t^{2L}, \quad (17)$$

with remainder term

$$R_{2N+1}[\tilde{C}_q](t) = \frac{\tilde{C}_q^{(2N+2)}(\xi) t^{2N+2}}{(2N+2)!} = \frac{(-1)^{N+1} q^{(N+1)^2-(N+1)} \tilde{C}_q(q^{N+1}\xi) t^{2N+2}}{(2N+2)!}, \quad (18)$$

for appropriate ξ between 0 and t. Using the sup norm $\|\tilde{C}_q\|_\infty = \|{}_qCos\|_\infty = {}_qCos(0) = 1$, along with the fact that $|t| \le \rho$, to bound from above, we obtain

$$|R_{2N+1}[\tilde{C}_q](t)| = \frac{q^{N^2+N}|\tilde{C}_q(q^{N+1}\xi)||t|^{2N+2}}{(2N+2)!} \le \frac{q^{N^2+N}\rho^{2N+2}}{(2N+2)!}.$$

Let $P_{2N+1}[\cos](t)$ and $R_{2N+1}[\cos](t)$ denote the $2N+1$ degree Taylor polynomial and remainder terms for $\cos(t)$ respectively. Then, for each $N \ge 1$ and each t with $|t| \le \rho$, one has

$$|\tilde{C}_q(t) - \cos(t)| \quad (19)$$
$$\le |\tilde{C}_q(t) - P_{2N+1}[\tilde{C}_q](t)| + |P_{2N+1}[\tilde{C}_q](t) - P_{2N+1}[\cos](t)| + |P_{2N+1}[\cos](t) - \cos(t)|$$
$$\le |R_{2N+1}[\tilde{C}_q](t)| + \left|\sum_{L=0}^{N} \frac{(-1)^L q^{L^2-L}}{(2L)!} t^{2L} - \sum_{L=0}^{N} \frac{(-1)^L}{(2L)!} t^{2L}\right| + |R_{2N+1}[\cos](t)|$$
$$\le \frac{q^{N^2+N}\rho^{2N+2}}{(2N+2)!} + \left(q^{N^2-N} - 1\right) \sum_{L=0}^{N} \frac{\rho^{2L}}{(2L)!} + \frac{\rho^{2N+2}}{(2N+2)!}$$
$$\le \frac{q^{N^2+N}\rho^{2N+2}}{(2N+2)!} + \left(q^{N^2-N} - 1\right) e^\rho + \frac{\rho^{2N+2}}{(2N+2)!}. \quad (20)$$

Now, given any $\epsilon > 0$ choose $N_0 \ge 1$ such that $\rho^{2N_0+2}/(2N_0+2)! < \epsilon/3$. Then one has $1 < \epsilon(2N_0+2)!/[3\rho^{2N_0+2}]$. Next choose $q_0 > 1$ with $1 < q_0^{N_0^2+N_0} < \epsilon(2N_0+2)!/[3\rho^{2N_0+2}]$. Then for all $1 < q < q_0$ one has

$$0 < \frac{q^{N_0^2+N_0}\rho^{2N_0+2}}{(2N_0+2)!} < \frac{q_0^{N_0^2+N_0}\rho^{2N_0+2}}{(2N_0+2)!} < \frac{\epsilon}{3} \quad \text{and} \quad 0 < \frac{\rho^{2N_0+2}}{(2N_0+2)!} < \frac{\epsilon}{3}. \quad (21)$$

Next choose $q_1 > 1$ such that $q_1^{N_0^2-N_0} - 1 < \epsilon/[3e^\rho]$. Then for all $1 < q < q_1$ one has

$$0 < \left(q^{N_0^2-N_0} - 1\right) e^\rho < \left(q_1^{N_0^2-N_0} - 1\right) e^\rho < \frac{\epsilon}{3}. \quad (22)$$

For the given ϵ, set $N = N_0$ in (19) and (20). Then for $|t| \le \rho$ and all $1 < q < \min\{q_0, q_1\}$, applying the bounds (21) and (22) to (20) gives

$$|\tilde{C}_q(t) - \cos(t)| < \frac{\epsilon}{3} + \frac{\epsilon}{3} + \frac{\epsilon}{3} = \epsilon, \quad (23)$$

verifying uniform convergence of $\tilde{C}_q(t)$ to $\cos(t)$ on $[-\rho, \rho]$ as $q \to 1^+$. □

Remark 1. Note that, alternatively, one can express Proposition 1 as

$$(\forall \mathcal{I} \subset\subset \mathbb{R} \text{ compact}) \implies \lim_{q \to 1^+} \sup\{|\tilde{C}_q(t) - \cos(t)| : t \in \mathcal{I}\} = 0 \qquad (24)$$

A similar convergence proof is given in Section 4, with details related to the novelty of the result.

2.2. *Application to PDE Example*

We are now in a position to obtain q-versions of various equations using $\tilde{C}_q(t)$ as a building block for relaxing equations. For example, define the Airy function (see page 570 in [26])

$$Ai(t) \equiv \frac{1}{\pi} \int_0^\infty \cos\left(\frac{u^3}{3} + u \cdot t\right) du, \quad t \in \mathbb{R}.$$

Some properties of this $\mathcal{C}^\infty(\mathbb{R})$ function are that $Ai(t) \to 0$ as $|t| \to \infty$, and $Ai(0) > 0$. We now show:

Proposition 2. *The q-advanced Airy function is defined here to be*

$$Aiq(t) \equiv \frac{1}{\pi} \int_0^\infty \tilde{C}_q\left(\frac{u^3}{3} + u \cdot t\right) du, \quad t \in \mathbb{R}, \qquad (25)$$

for $q > 1$. The functions $Ai(t)$ and $Aiq(t)$ satisfy the homogeneous ODE and MADE

$$Ai''(t) - t \cdot Ai(t) = 0, \quad Aiq''(t) - q^{-1/3} t \cdot Aiq\left(q^{2/3} t\right) = 0, \qquad (26)$$

respectively, for $t \geq 0$. Basic properties of $Aiq(t)$ for $q > 1$, are that $Aiq(t)$ is Schwartz with $Aiq(0) > 0$. Furthermore, for each $T > 0$, $\epsilon > 0$, and $R > T$ sufficiently large, $\exists\, q(\epsilon, T, R) > 1$ so that

$$\sup\{|Ai(t) - Aiq(t)| : |t| \leq T, \ 1 < q < q(\epsilon, T, R)\} < \epsilon. \qquad (27)$$

In other words, $Aiq(t) \to Ai(t)$ uniformly for t in compact subsets of $\mathbb{R}$, as $q \to 1^+$.

Remark 2. *Verifying convergence in Equation (27) may seem rather straight forward, due to the uniform convergence of $\tilde{C}_q(t)$ to $\cos(t)$ on compact sets. However, we need to use a careful $\epsilon/3$ argument, as demonstrated here.*

Proof. That $Aiq(t)$ is Schwartz follows from the same for $\tilde{C}_q(t)$, whereas the property $Aiq(0) > 0$ requires a manipulation of theta functions, and is shown in Appendix A. We start with the second equation in (26) since the first equation is known to hold [26]. First define the function

$$\tilde{S}_q(t) \equiv \int_0^t \tilde{C}_q(s)\, ds, \text{ so that } \tilde{S}_q(0) = 0, \ \tilde{S}_q(\pm\infty) = 0.$$

Now compute, using $v = q(u^3/3 + ut)$, and $w = q^{1/3} u$, for $t \geq 0$,

$$Aiq''(t) = \frac{1}{\pi} \int_0^\infty -u^2 \tilde{C}_q\left(q \cdot \left(\frac{u^3}{3} + u \cdot t\right)\right) du \qquad (28)$$

$$= \frac{-1}{q\pi} \int_0^\infty q(u^2 + t) \cdot \tilde{C}_q\left(q \cdot (\frac{u^3}{3} + u \cdot t)\right) du + \frac{t}{\pi} \int_0^\infty \tilde{C}_q\left(q \cdot (\frac{u^3}{3} + u \cdot t)\right) du$$

$$= \frac{-1}{q\pi} \int_0^\infty \frac{d\tilde{S}_q(v)}{dv} dv + \frac{t}{\pi} \int_0^\infty \tilde{C}_q\left(\frac{w^3}{3} + w \cdot q^{2/3} t\right) (dw/q^{1/3})$$

$$= (-\tilde{S}_q(\infty) + \tilde{S}_q(0))/(q\pi) + q^{-1/3} t \cdot Aiq\left(q^{2/3} t\right).$$

Next, to show convergence, consider any $\epsilon > 0$ and, without loss of generality, fix $T > 1$. Let t be in the interval $|t| \leq T$. Then, for any $R > T$, using integration by parts and boundedness of the sine function, we can write

$$Ai(t) - \frac{1}{\pi} \int_0^R \cos\left(\frac{u^3}{3} + u \cdot t\right) du = \frac{1}{\pi} \int_R^\infty \frac{1}{u^2 + t} \cdot \frac{d}{du} \sin\left(\frac{u^3}{3} + u \cdot t\right) du \qquad (29)$$

$$= \frac{-\sin\left(\frac{R^3}{3} + R \cdot t\right)}{\pi(R^2 + t)} - \frac{1}{\pi} \int_R^\infty \frac{-2u}{(u^2 + t)^2} \cdot \sin\left(\frac{u^3}{3} + u \cdot t\right) du$$

Thus, for all $|t| \leq T$ we can easily find $R > T$ sufficiently large so that

$$\left| Ai(t) - \frac{1}{\pi} \int_0^R \cos\left(\frac{u^3}{3} + u \cdot t\right) du \right| \leq \frac{2}{\pi \cdot (R^2 - T)}. \qquad (30)$$

The bound in Equation (30) also holds if $Ai(t)$ is replaced with $Aiq(t)$ since $|\tilde{C}_q(t)| \leq 1$ and $|\tilde{S}_q(t)| \leq 1$ for all $q > 1$. Now, fix $R > 0$ sufficiently large so that the bounds in (30), and also (30) with cos replaced by $\tilde{C}_q$, are less than $\epsilon/3$. It is essential to note that this value of R is independent of $q > 1$.

Finally, for each $t \in \mathbb{R}$, define the function

$$V_t(u) \equiv \frac{u^3}{3} + ut, \text{ so that } V_t([0, R]) = \begin{cases} [0, R^3/3 + Rt] & , t \geq 0 \\ [-2|t|^{3/2}/3, \max\{0, R^3/3 + Rt\}] & , t < 0 \end{cases}.$$

The union of these $V_t([0, R])$ over $t \in [0, R]$, is the interval $I \equiv [-2T^{3/2}/3, R^3/3 + RT]$. From the uniform convergence in Equation (24) we can choose $q(\epsilon, T, R) > 1$ so that

$$\left| \cos(V_t(u)) - \tilde{C}_q(V_t(u)) \right| < \frac{\pi \epsilon}{3 \cdot R}, \qquad (31)$$

for $|t| \leq T$, $|u| \leq R$, and $1 < q < q(\epsilon, T, R)$. This is now sufficient to verify the expression in Equation (27). $\square$

2.3. A q-Advanced PDE Example

The argument in the proof of Proposition 2 shows that knowledge of one MADE can help to generate and study other MADEs. In fact, this extends to Partial Differential Equations (PDEs). For example, consider the linear constant-coefficient Airy PDE [27]

$$\partial_t \phi(x, t) = a \partial_x^3 \phi(x, t), \quad \phi(x, 0) = f(x), \qquad (32)$$

for $x \in \mathbb{R}$, $t \in \mathbb{R}_0^+$, and constant $a > 0$. To obtain an advanced-type equation, consider the kernel function, defined for each $t > 0$,

$$\mathcal{A}q_{(t)}(x) \equiv \frac{1}{\sqrt[3]{t} A_0(q)} Aiq\left(\frac{x}{\sqrt[3]{t}}\right), \text{ for } x \in \mathbb{R}, \qquad (33)$$

for appropriate $A_0(q) \neq 0$, to be determined. For any integrable $f(x)$ and any $a \neq 0$, define,

$$\phi_q(x, t) \equiv \left[\mathcal{A}q_{(-at)} * f\right](x) = \left[f * \mathcal{A}q_{(-at)}\right](x) = \int_{-\infty}^\infty f(y) \cdot \mathcal{A}q_{(-at)}(x - y) dy, \qquad (34)$$

(compare with Equation (2.2) of [27]). Recall that the functional operation of convolution for integrable functions $g, h \in \mathcal{L}^1(\mathbb{R})$ gives a new function $g * h \in \mathcal{L}^1(\mathbb{R})$ defined by

$$g * h(x) \equiv \int_{-\infty}^{\infty} g(y) \cdot h(x-y)\, dy = \sqrt{2\pi}\, \mathcal{F}^{-1}\left[\mathcal{F}[g] \cdot \mathcal{F}[h]\right](x), \tag{35}$$

where the last equality in Equation (35) is the Convolution Theorem (see [28] Theorem IX.4). To discover the PDE that ϕ_q solves, first compute the t-partial derivative of Equation (34), to obtain

$$\partial_t \phi_q(x,t) \equiv \frac{-1}{3t} \phi_q(x,t) + \frac{-1}{3t}\int_{-\infty}^{\infty} f(y) \cdot \frac{(x-y)}{(at)^{2/3} A_0(q)} \cdot Aiq'\left(\frac{x-y}{(-at)^{1/3}}\right) dy. \tag{36}$$

Now, taking three derivatives of Equation (34) with respect to x, gives

$$\partial_x^3 \phi_q(x,t) \equiv \frac{1}{(-at)^{1/3}} \frac{\partial}{\partial x}\int_{-\infty}^{\infty} f(y) \cdot \frac{q^{-1/3}(x-y)}{-at\, A_0(q)} \cdot Aiq\left(\frac{q^{2/3}(x-y)}{(-at)^{1/3}}\right) dy \tag{37}$$

$$= \frac{-1}{atq}\phi_q\left(x, \frac{t}{q^2}\right) + \frac{-1}{atq}\int_{-\infty}^{\infty} f(y) \cdot \frac{(x-y)}{(at/q^2)^{2/3} A_0(q)} \cdot Aiq'\left(\frac{x-y}{(-at/q^2)^{1/3}}\right) dy. \tag{38}$$

By replacing $t \to q^2 t$ in Equation (38), one can verify that the q-advanced PDE, for $q > 1$ and $q^2 > 1$,

$$\partial_t \phi_q(x,t) = \frac{a q^3}{3} \cdot \partial_x^3 \phi_q\left(x, q^2 t\right), \tag{39}$$

holds. To obtain consistency with the initial data $f(x)$, first define the constant, for $q > 1$,

$$A_0(q) \equiv \int_{-\infty}^{\infty} Aiq(t)\, dt, \tag{40}$$

which is finite since $Aiq(t)$ is Schwartz (thus integrable) for $q > 1$. Then we require,

The q-Airy Hypothesis: Given $q > 1$, the expression in Equation (40) does not vanish, ie. $A_0(q) \neq 0$. In Appendix B we show that the q-Airy Hypothesis holds for all $q > 1$. Then

$$\begin{cases} f(x) \text{ is continuous, integrable, and bounded} \\ \quad \text{and} \\ \text{the } q\text{-Airy Hypothesis holds} \end{cases} \implies (\forall x \in \mathbb{R})\left(\lim_{t \to 0^+} \phi_q(x,t) = f(x)\right), \tag{41}$$

where convergence in Equation (41) is pointwise, and is shown in Appendix C using a mollifier-type argument. If, in addition, we have $f \in C^1 \cap \mathcal{L}^1$ and $f' \in \mathcal{L}^\infty$, then convergence in Equation (41) becomes uniform.

3. Solutions of MADEs and Natural Extensions

Define the family of Dirichlet-type functions for $t \in \mathbb{R}_0^+$, and $q > 1$, as introduced in [29],

$$f_{\mu,\lambda}(t) \equiv \sum_{m=-\infty}^{\infty} (-1)^m \frac{e^{-q^m t}}{q^{m(m-\mu)/\lambda}}. \tag{42}$$

For each $\mu \in \mathbb{Q}$ and $\lambda \in \mathbb{Q}^+$, the corresponding function solves the eigen-MADE

$$\partial_t^\delta f_{\mu,\lambda}(t) = (-1)^{\gamma+\delta}\, q^{\gamma(\gamma+\mu)/\lambda}\, f_{\mu,\lambda}\left(q^\gamma t\right). \tag{43}$$

Here $\lambda/2 = \gamma/\delta \in \mathbb{Q}^+$ is in reduced form with $\gamma, \delta \in \mathbb{N}$. The function $f_{\mu,\lambda}(t)$ has eigenvalue $E = (-1)^{\gamma+\delta} q^{\gamma(\gamma+\mu)/\lambda}$ and can be normalized so that the function $g_{\mu,\lambda}(t) \equiv f_{\mu,\lambda}\left(t/q^{\gamma(\gamma+\mu)/(\delta\lambda)}\right)$ solves the q-advanced eigen equation

$$\partial_t^\delta g_{\mu,\lambda}(t) = (-1)^{\gamma+\delta} g_{\mu,\lambda}(q^\gamma t) . \tag{44}$$

for $t > 0$. In this manner the q-dependence of the eigenvalue can be removed. Note that the sign of the eigenvalue $(-1)^{\gamma+\delta}$ can dramatically affect the behavior of the solution.

3.1. Flat Solutions of MADEs

In [29] we found special conditions under which $f_{\mu,\lambda}(t)$ extends to all $t \in \mathbb{R}$, so that

$$F_{\mu,\lambda}(t) \equiv \begin{cases} f_{\mu,\lambda}(t) & , t \geq 0 \\ 0 & , t < 0 \end{cases} \tag{45}$$

gives a Schwartz solution to an associated MADE to all $t \in \mathbb{R}$. The essential condition is that $f_{\mu,\lambda}^{(n)}(0^+) = 0$ for all $n \in \mathbb{N}_0$, which is a property called *flatness*, at $t = 0$. It was shown in [29] that

$$f_{\mu,\lambda}(t) \text{ is flat at } t = 0 \iff \mu \text{ is an odd integer and } \lambda \text{ is an even integer} .$$

This condition for flatness can be expressed as

$$\mu = 2M+1 \text{ (odd)}, M \in \mathbb{Z} \quad \text{and} \quad \lambda = 2N \text{ (even)}, N \in \mathbb{N}. \tag{46}$$

Then, for $\langle \mu, \lambda \rangle$ as in Equation (46), $F_{\mu,\lambda}(t)$ all solve first-order MADEs:

$$\partial_t F_{\mu,\lambda}(t) = (-1)^{N+1} q^{(N+2M+1)/2} F_{\mu,\lambda}(q^N t) ,$$

for $t \in \mathbb{R}$ and $q > 1$. See examples in Figure 1. Furthermore, the Fourier transform has a special form:

$$\mathcal{F}[F_{2M+1,2N}](\omega) = \frac{(-1)^M \mu_{q^{1/N}}^3}{\sqrt{\pi}} \cdot \frac{q^{M(M+1)/(2N)}}{i\omega} \times \left[\frac{1}{N}\sum_{j=0}^{N-1} \frac{1}{\theta(q^{1/N}, z_j(\omega)/q^{(M+1)/N})}\right],$$

where for each $j \in \{0, 1, 2, \ldots, N-1\}$, the points of valuation of the theta function require,

$$z_j(\omega) = -|\omega|^{1/N} \cdot e^{3\pi i/(2N)} \cdot e^{i[\arg(\omega)]/N} \cdot \rho^j ,$$

for $\rho \equiv e^{i2\pi/N}$, and $\{z_j\}$ are the N distinct solutions of $(-z_j)^N = -i\omega$.

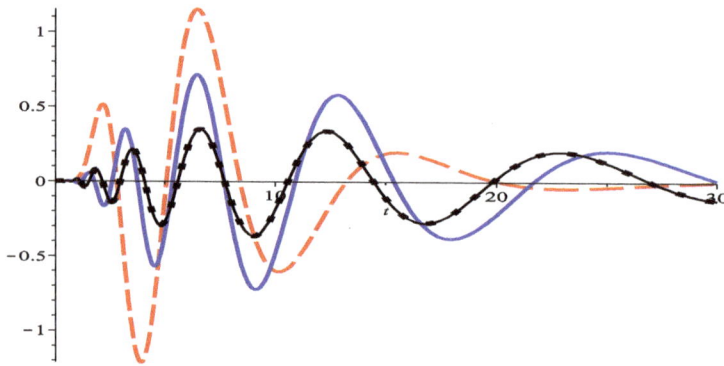

Figure 1. Three Flat Functions: Normalized plots of first-order MADE solutions that are flat at $t = 0$, (1) $f_{1,2}(t)$ (dashed red), (2) $f_{1,4}(t)$ (solid blue), (3) $f_{1,6}(t)$ (dotted black line) all for $q = 1.3$.

3.2. A Non-Trivial Extension of a MADE Solution

Now consider the situation where $\exists n_* \in \mathbb{N}_0$ where $f_{\mu,\lambda}^{(n_*)}(0^+) \neq 0$. Then an extension of $f_{\mu,\lambda}(t)$ to the region $t < 0$ is not so clear. However, by truncating the series in Equation (42) an asymptotic exponential-series is obtainable that provides, what appears to be, a smooth extension to the $t < 0$ region. However, extending in this manner does not lead to a homogeneous, eigen-MADE in the region $t < 0$. This is demonstrated with a specific example.

We begin by recalling the Airy equation as given in Proposition 2

$$y''(t) - t\, y(t) = 0. \tag{47}$$

However, taking the derivative of this equation gives a generalization

$$y'''(t) - y(t) = t\, y'(t), \tag{48}$$

where the right hand side is expected to be small for $t \simeq 0$. Hence a solution to the constant coefficient equation

$$y'''(t) - y(t) = 0, \tag{49}$$

see Section 4, may be considered to be an approximate solution to the Airy equation near the origin. For example, the function

$$y(t) = (2/\sqrt{3})\, e^{-t/2} \sin(\sqrt{3}t/2), \tag{50}$$

solves (49) with initial conditions

$$y(0) = 0, \quad y'(0) = 1, \quad y''(0) = -1. \tag{51}$$

Now we consider a q-relaxed version of (50) in the form of a solution to the MADE

$$\eta'''(t) - q^3\, \eta(qt) = 0, \tag{52}$$

with parameter $q > 1$. Note that (52) is a multiplicatively advanced relaxed version of the approximate Airy ODE (49) for $q \simeq 1^+$. From Equations (42) and (43), a particular solution of Equation (52) is $\eta(t) = f_{1,2/3}(t)$ for $t \geq 0$. To extend $\eta(t)$ to all of $t \in \mathbb{R}$ in a $\mathcal{C}^\infty$ fashion, we find that

$$\mathcal{W}_{1,2/3}(t) \equiv \begin{cases} f_{1,2/3}(t) & ,\text{ for } t \geq 0 \\ (-1)f_{1,2/3}\left(e^{2\pi i/3}t\right) + (-1)f_{1,2/3}\left(e^{4\pi i/3}t\right) & ,\text{ for } t < 0 \end{cases} \tag{53}$$

is a Schwartz function, where $f_{1,2/3}(z)$ is analytic for $\Re(z) > 0$ and bounded for $\Re(z) = 0$. Although there is no unique solution to MADEs in general, the function $\mathcal{W}_{1,2/3}(t)$ constructed in Equation (53) will be called *canonical*, and it solves the MADE in Equation (52) for all $t \in \mathbb{R}$.

3.3. Asymptotic Analysis of an Extension

There is an alternate continuous way to extend $\eta(t)$ to the region $t_* < t < 0$, for $t_* < 0$ defined below, in terms of $q > 1$. Define the constant $\mathcal{C}_q^+$ so that

$$\frac{1}{\mathcal{C}_q^+} \equiv -\sum_{k=-\infty}^{\infty} \frac{(-1)^k q^k}{q^{3k(k-1)/2}} = -\theta(q^3; -q), \tag{54}$$

where the last equality follows from (8). Note that $\theta(q^3; -q)$ is non-zero for real $q > 1$ by (9), whence $\mathcal{C}_q^+$ is well-defined and finite. For $t \geq 0$ the function $\eta(t)$ is defined as

$$\eta(t) \equiv \mathcal{C}_q^+ \sum_{k=-\infty}^{\infty} \frac{(-1)^k e^{-q^k t}}{q^{3k(k-1)/2}} = \frac{f_{1,2/3}(t)}{-\theta(q^3; -q)} = \frac{f_{1,2/3}(t)}{f'_{1,2/3}(0)}. \tag{55}$$

Now, for $t \geq 0$, $\eta(t)$ solves (52) with initial conditions

$$\eta(0) = 0, \quad \eta'(0) = 1, \quad \eta''(0) = -q. \tag{56}$$

However, for each $t < 0$ the function $\eta(t)$ diverges, due to the rapid growth of $e^{-q^k t} = e^{q^k|t|}$, in k, as compared to that of $q^{3k(k-1)/2}$ in the summands of (55), as k approaches infinity. Thus, for each $t < 0$ the function $\eta(t)$ is not defined.

To remedy this, while keeping the same summands as in (55), we truncate the upper limit of summation in (55). Thus, for all $t \in \mathbb{R}$ we define the asymptotic extension $\tilde{\eta}(t)$ of $\eta(t)$ by

$$\tilde{\eta}(t) \equiv \tilde{c}(q, t) \sum_{k=-\infty}^{N(q;t)} \frac{(-1)^k e^{-q^k t}}{q^{3k(k-1)/2}}, \tag{57}$$

where the integer upper limit of the sum, and the normalizing coefficient, are defined to be

$$N(q, t) = \begin{cases} \infty & , t \geq 0 \\ N_*(q, t) & , t < 0 \end{cases}, \quad \tilde{c}(q, t) = \begin{cases} \mathcal{C}_q^+ & , t \geq 0 \\ \mathcal{C}_q^- \equiv \left(-\sum_{k=-\infty}^{\lfloor N_*(q,t) \rfloor} \frac{(-1)^k q^k}{q^{3k(k-1)/2}}\right)^{-1} & , t < 0 \end{cases}. \tag{58}$$

Since it will follow from the definition below that $N_*(q, t) \to \infty$ as $t \to 0^-$, continuity for $\tilde{\eta}(t)$ is achieved at $t = 0$. However, as a solution to a MADE, we have that $\tilde{\eta}(t) \in \mathcal{D}'$, where $\mathcal{D}'$ is the space of distributions, dual to $\mathcal{D} \equiv \mathcal{C}_0^\infty(\mathbb{R})$, the set of compactly supported, infinitely differentiable functions. In fact, since

$$\tilde{\eta}'''(t) - q^3 \tilde{\eta}(qt) = \tilde{f}(t), \tag{59}$$

where $\tilde{f} \in \mathcal{D}'$, with $\text{supp}(\tilde{f}) \subset (-\infty, 0]$, we have that $\tilde{\eta}(t)$ is a weak solution (as defined in [4] p. 149) to the inhomogeneous extension of (52).

For $t < 0$, a best choice for $N_*(q, t)$ is chosen to be the k value at which a local minimum for the function

$$\mathcal{T}(k, |t|) \equiv \frac{e^{q^k|t|}}{q^{3k(k-1)/2}} = e^{h(k,|t|)}, \tag{60}$$

exists, where the exponent function is defined to be

$$h(k, |t|) \equiv q^k |t| - \ln(q)(3k(k-1)/2). \tag{61}$$

The choice of truncation $N_*(q)$ presented here is made based on the least-term approximation from Poincaré asymptotics, as presented on p. 94 of Bender and Osrzag [26]:

> "We look over the individual terms in the asymptotic series; ...For every given value of ...[t]... we locate the smallest term. We then add all the preceding terms in the asymptotic series up to but *not* including the smallest term."

Traditionally this rule gives a good estimate of the actual function, which is often the solution of a differential equation. In our case the rule above can only be applied for $t < 0$ sufficiently close to the origin, which for this function turns out to be

$$|t| < 3/(e\sqrt{q}\ln(q)).$$

This is a consequence of the following more general result.

Proposition 3. *For $\mu, \lambda \in \mathbb{R}$ with $\lambda > 0$, define the following function on $t \in \mathbb{R}$*

$$\tilde{f}(t) = \sum_{k=-\infty}^{N(q,t)} \frac{a_k e^{-q^k t}}{q^{k(k-\mu)/\lambda}}, \quad \text{where:} \quad N(q,t) = \begin{cases} \infty, & t \geq 0 \\ N_*(q,t), & t_* < t < 0 \end{cases}, \quad (62)$$

for any bounded sequence $\{a_k\} \in \ell^\infty$. Define the exponential growth portion of the summands as

$$\mathcal{T}_{\mu,\lambda}(k,|t|) \equiv \frac{e^{q^k|t|}}{q^{k(k-\mu)/\lambda}} = e^{h_{\mu,\lambda}(k,|t|)}, \quad \text{where:} \quad h_{\mu,\lambda}(k,|t|) \equiv q^k|t| - \ln(q) \cdot \frac{k(k-\mu)}{\lambda}. \quad (63)$$

Then, define two constants, for fixed $q > 1$,

$$t_* \equiv \frac{-2}{\lambda e q^{\mu/2} \ln(q)} < 0 \quad \text{and} \quad N_*(q,t_*) \equiv \frac{1}{\ln(q)} + \frac{\mu}{2}. \quad (64)$$

For $t \in (t_*, 0)$, the function $N_*(q,t)$ exists uniquely as the local minimum of $\mathcal{T}_{\mu,\lambda}(k,|t|)$.

Remark 3. *The coefficients a_k in Equation (62) play no part in the following analysis. However, if they decay as $|k| \to \infty$, or if they change sign, then the asymptotic behavior may be different than what is derived here.*

Proof. Differentiating the exponent $h_{\mu,\lambda}(k,|t|) = \ln[\mathcal{T}_{\mu,\lambda}(k,|t|)]$ in (63) with respect to k gives the critical point condition

$$\ln(q) q^k |t| - \ln(q)(2k - \mu)/\lambda = 0 \iff |t|q^k - (2k-\mu)/\lambda = 0$$
$$\iff q^k = (2k-\mu)/(\lambda |t|). \quad (65)$$

Taking a second derivative of $h_{\mu,\lambda}(k,|t|)$ with respect to k gives the inflection point condition

$$\ln^2(q) q^k |t| - \ln(q)(2/\lambda) = 0 \iff \ln(q) q^k |t| - (2/\lambda) = 0$$
$$\iff k = \frac{\ln[2/(\lambda|t|\ln(q))]}{\ln(q)}. \quad (66)$$

Interpreting the middle critical point condition in (65) as the intersection of the concave up function $|t|q^k$ with the fixed line $(2k-\mu)/\lambda$ reveals three possibilities:

Case 1: There are two critical points $k_1 < k_2$ with an intervening inflection point $k_3 \in (k_1, k_2)$ for $|t|$ and q sufficiently small. By the first derivative test, a local maximum occurs at k_1 while the desired local minimum then occurs at k_2.

Case 2: An edge case occurs, in which the two critical points coalesced to one point equaling the inflection point, $k_1 = k_2 = k_3$. There is no local minimum for $h_{\mu,\lambda}(k, |t|)$ in this setting.

Case 3: There are no critical points when either $|t|$ or q is too large, resulting in no local minimum for $h_{\mu,\lambda}(k, |t|)$ in this setting.

Thus, the edge case, Case 2, marks the transition at which a local minimum of the summand $\mathcal{T}_{\mu,\lambda}(k, |t|)$ occurs, and hence Case 2 marks the transition at which an asymptotic phenomena for the index k occurs. To quantify this point of transition, we note that the edge case, Case 2, where the inflection point equals the critical point, implies that the solution of (65) also simultaneously solves (66) in this setting. Substituting the expression for q^k in (65) into (66) gives

$$\ln(q) \cdot \frac{2k - \mu}{\lambda} - \frac{2}{\lambda} = 0 \iff k = \frac{1}{\ln(q)} + \frac{\mu}{2}. \tag{67}$$

Then substituting the value of $k = 1/\ln(q) + \mu/2$ as obtained in (67) into the value of k in Equation (65) gives the value of $|t| = |t_*|$ that corresponds to this transition as

$$|t_*| = \frac{2}{\lambda \, e \, q^{\mu/2} \ln(q)}. \tag{68}$$

Thus, we saw that Case 2 holding implies that

$$|t| = 2/(\lambda \, e \, q^{\mu/2} \ln(q)) = |t_*| \quad \text{and} \quad k_1 = k_2 = k_3 = (\mu/2) + (1/\ln(q)).$$

Conversely, if $|t| = 2/(\lambda \, e \, q^{\mu/2} \ln(q)) = |t_*|$, then (66) holds if and only if

$$\ln(q) \, q^k \cdot 2/(\lambda \, e \, q^{\mu/2} \ln(q)) - (2/\lambda) = 0 \iff q^k = e \, q^{\mu/2} = q^{1/\ln(q)} q^{\mu/2}$$
$$\iff k = (\mu/2) + (1/\ln(q)). \tag{69}$$

Furthermore, observe that since $y = \exp(x - 1)$ is concave up with tangent line $y = x$ at $x = 1$ then the inequality $\exp(x - 1) \geq x$ holds for all x and equality holds if and only if $x = 1$. Replacing x by $(k - \mu/2) \ln(q)$ in our inequality gives

$$\frac{q^{k-\mu/2}}{e} \geq \left(k - \frac{\mu}{2}\right) \ln(q) \quad \text{with equality holding iff} \quad \left(k - \frac{\mu}{2}\right) \ln(q) = 1. \tag{70}$$

Multiplying the inequality on the left through by $2/(\lambda \ln(q))$ gives

$$\frac{2q^k}{\lambda \, e \, q^{\mu/2} \ln(q)} = q^k |t_*| \geq \frac{2k - \mu}{\lambda} \quad \text{with equality holding iff} \quad k = \frac{\mu}{2} + \frac{1}{\ln(q)}, \tag{71}$$

whence (65) also holds at the same value of $k = \mu/2 + 1/\ln(q)$. Thus, the critical points and the inflection point coalesced to the common value $k = \mu/2 + 1/\ln(q)$ and Case 2 holds. We see that Case 2 holding is equivalent to $-t = t_* = 2/(\lambda \, e \, q^{\mu/2} \ln(q))$ holding. Furthermore, one sees that Case 1 holds when $|t| < |t_*|$, and a local minimum is obtained. Thus, the asymptotic phenomena occurs for $|t| < |t_*|$ where for the upper index limit $N_*(q, t)$ we take the larger of the two solutions to the transcendental equation for k_* in Equation (65):

$$q^{k_*} = \frac{2k_* - \mu}{\lambda |t|}. \tag{72}$$

Then, for $|t| < |t_*| = 2/(\lambda \, e \, q^{\mu/2} \ln(q))$ sufficiently small, $\mathcal{T}_{\mu,\lambda}(k, |t|)$ has a local minimum at $N_*(q, t) = k_*$, which can be found by taking a seed point greater than the value $\ln(2/(|t|\lambda \ln(q)))/\ln(q)$ of the inflection point and utilizing Newton's method. □

3.4. Special Case of the Derivative of an Airy Approxiamtion

We return to considering the special case that $\mu = 1$, $\lambda = 2/3$ and $a_k = (-1)^k$. However, rather than illustrating a graph of the above phenomena for $f_{1,2/3}(t)/f'_{1,2/3}(0)$, we instead illustrate the behavior for its derivative

$$\phi(t) \equiv f'_{1,2/3}(t)/f'_{1,2/3}(0) = f_{1,5/3}(t)/f_{1,5/3}(0),$$

in Figure 2 below. In this setting, $\mu = 5/3$, $\lambda = 2/3$, and the asymptotic extension of $\phi(t)$ is

$$\tilde{\phi}(t) \equiv C_q^{-1} \sum_{k=-\infty}^{N(q,t)} \frac{(-1)^k e^{-q^k t}}{q^{3k(k-5/3)/2}}, \quad \text{where: } N(q,t) = \begin{cases} \infty, & t \geq 0 \\ N_*(q,t), & t_* < t < 0 \end{cases}, \tag{73}$$

where for $t < 0$, we compute, using $q = 1.2$, $\mu = 1$, and $\lambda = 2/3$,

$$t_* \equiv \frac{-2}{\lambda e q^{\mu/2} \ln(q)} \simeq -5.081 \quad \text{and} \quad N_*(q,t_*) \equiv \frac{1}{\ln(q)} + \frac{\mu}{2} \simeq 5.985. \tag{74}$$

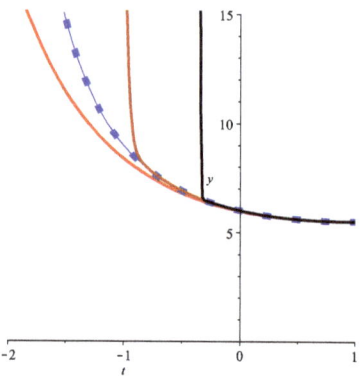

Figure 2. (**Left**) Asymptotic extension $\tilde{\phi}(t)$ from Equation (73) for $\phi(t)$ (solid red) together with a similarly constructed asymptotic extension for $-\chi_{(-\infty,0]}(t)\mathcal{W}_{1,5/3}(t)/f'_{1,2/3}(0)$ (dashed blue) both for $q = 1.2$. (**Right**) Plots of $e^t \check{K}(t)$ where the functions $\check{K}(t)$ are defined in Equation (76) for $q = 1.2$. Failure of the asymptotic extension is found to be around $t = -1$, as compared to the computed value of $t_* = -1.8$. The upper-sum limits, from left to right, are $N_* = 6, 10 (dotted), 20, 30$.

For $t \in (-t_*, 0)$ the function $N_*(q, t) = N_*(1.2, t)$ is the k value giving the larger of the two solutions to the transendental equation:

$$q^k = 3(2k - 5/3)/(2|t|), \tag{75}$$

which is the analogue of (65) and (72). The asymptotic extension $\tilde{\phi}(t)$ is given by the solid red graph in Figure 2 (Left). Defining the function

$$\mathcal{W}'_{1,2/3}(t) \equiv \mathcal{W}_{1,5/3}(t),$$

the dotted blue graph in Figure 2 (Left) is the asymptotic extension of $-\chi_{(-\infty,0]}(t)\mathcal{W}_{1,5/3}(t)$ to $\mathbb{R}$. The asymptotic extension of the derivative $f'_{1,2/3}(t)/f'_{1,2/3}(0)$ (rather than the original function $f_{1,2/3}(t)/f_{1,2/3}(0)$) is used due to non-vanishing at $t = 0$ as well as due to its comparatively flatter derivative.

From Figure 2 the asymptotic expansion is valid to around $t \sim -2$, using $N_* \sim 10$, rather than $t \sim -5$, using $N_* \sim 6$, contrary to what was expected from Equation (74). This is due to the alternation $a_k = (-1)^k$ since cancelations require a more careful analysis. This is not done here, but the next example considers a comparatively simple case, which gives a better comparison.

3.5. An even Simpler Example of MADE Asymptotics

In this section, we motivate a simpler type of asymptotic extension, distinct from Section 3.4, using two examples.

To begin, we recall a MADE that was studied in [30], for $q > 1$ and $t \geq 0$,

$$\partial_t \tilde{K}(t) = -q\tilde{K}(qt), \quad \tilde{K}(t) \equiv \sum_{j=-\infty}^{N(q;t)} \frac{e^{-q^j t}}{q^{j(j-1)/2}}, \tag{76}$$

where for $t \geq 0$ set $N(q;t) = \infty$. Here we consider the extension to negative values of the parameter. Then, for $t_* < t < 0$, we will choose the constant $N(q;t) = N_*(q, t_*)$. To use the asymptotic analysis, note that $a_k = 1$, $\mu = 1$ and $\lambda = 2$. Thus, we obtain an approximate MADE solution extension to the region $t < 0$. Start by defining

$$\mathcal{T} \equiv \frac{e^{q^j |t|}}{q^{j(j-1)/2}} = e^h, \quad \text{where:} \quad h \equiv q^j |t| - \ln(q)(j(j-1)/2).$$

Differentiating h with respect to j gives the critical condition

$$\ln(q) q^j |t| - \ln(q)(2j-1)/2 = 0 \quad \Longleftrightarrow \quad q^j = (2j-1)/(2|t|).$$

The second derivative gives the inflection condition

$$\ln^2(q) q^j |t| - \ln(q) = 0 \quad \Longleftrightarrow \quad j = -\ln[|t|\ln(q)]/(\ln(q)).$$

Combining these expressions to eliminate $q^j |t|$ gives

$$\ln(q)\left(\ln(q)(2j-1)/2\right) - \ln(q) = 0 \quad \Longleftrightarrow \quad j_* = \frac{1}{\ln(q)} + \frac{1}{2} \equiv N_*(q, t_*),$$

from Equation (67) which then results in

$$t_* = -(2j_* - 1)/(2q^{j_*}),$$

from Equation (68). For $t_* < t < 0$, we have $N(q,t) > N(q, t_*) = j_*$. By inspection, Figure 2 (Right) indicates that we maintain a good asymptotic expansion by letting all $N(q,t) = N(q,t_*) = j_*$. In particular, for $q = 1.2$ our rule suggests $j \leq \lfloor j_* \rfloor = N_* \sim 6$, which is expected to be valid for $t \in (-1.8, 0)$. The Right of Figure 2 indicates a good match for $t \in (-1, 0)$, using $N_* \sim 10$.

Finally, we return to Equation (57), and consider the slightly different series, for all $t > t_*$ (where $t_* < 0$)

$$\tilde{\eta}_*(t) \equiv \tilde{c}_*(q,t) \sum_{k=-\infty}^{\lfloor N(q;t) \rfloor} \frac{(-1)^k e^{-q^k t}}{q^{3k(k-1)/2}} + \tilde{c}_*(q,t) \sum_{k=-\infty}^{\lfloor N_*(q;t_*) \rfloor} \frac{(-1)^k}{q^{3k(k-1)/2}} \cdot \chi_{(t_*,0)}(t), \tag{77}$$

where now the integer upper-sum limit, and the normalizing coefficient, are defined to be, respectively

$$N(q,t) = \begin{cases} \infty & , t \geq 0 \\ N_*(q, t_*) & , t_* < t < 0 \end{cases}, \quad \tilde{c}(q,t) = \begin{cases} C_q^+ & , t \geq 0 \\ C_q^- \equiv \left(-\sum_{k=-\infty}^{\lfloor N_*(q,t) \rfloor} \frac{(-1)^k q^k}{q^{3k(k-1)/2}} \right)^{-1} & , t_* < t < 0 \end{cases} \tag{78}$$

The function $\tilde{\eta}(t)$ is differentiable for $t \in (t_*, \infty)$, and solves an inhomogeneous MADE

$$\tilde{\eta}_*'''(t) - q^3 \tilde{\eta}_*(qt) = \tilde{f}_*(t), \tag{79}$$

where $\tilde{f}_* \in \mathcal{D}'$ is derived in Appendix D. Note that $\tilde{f}_*(t)$ is distinct from $\tilde{f}(t)$ for $t > t_*$ in Equation (59), and the corresponding weak solution $\tilde{\eta}_*(t)$ is much easier to compute than $\tilde{\eta}(t)$, with little consequence to the asymptotics.

4. Convergence of MADEs to Classical Solutions

In this section, we present another example where we can study convergence of a MADE solution to its classical analogue. This requires an apriori uniform bound in a fixed neighborhood of $t = 0$ for all $q > 1$ sufficiently small. Obtaining a uniform-in-q bound for general $f_{\mu,\lambda}(t)$ is rather deep, and complicated by the presence of the alternation $(-1)^m$ in Equation (42). Here we study a series without this alternating factor, which defines a function that behaves like a damped oscillation. The details are more challenging than what appears in the proof of Proposition 1, so a full analysis is provided.

Consider the following linear third-order MADE

$$f^{(3)}(t) = q^3 f(qt) \tag{80}$$

for $q > 1$, on the interval $t \in [0, \infty)$, satisfying the initial conditions

$$f(0) = 0, \quad f'(0) = 1, \quad f''(0) = -q. \tag{81}$$

For small $q > 1$, as $q \to 1^+$, Equations (80) and (81) can be considered to be a perturbation of the classical analogue, which is the ODE

$$g^{(3)}(t) = g(t) \tag{82}$$

with initial conditions

$$g(0) = 0, \quad g'(0) = 1 \quad g''(0) = -1 \tag{83}$$

obtained by setting $q = 1$ in (80) and (81). One can check directly that (82) and (83) is solved uniquely by

$$g(t) = 2 \cdot \exp(-t/2) \cdot \sin(\sqrt{3}t/2)/\sqrt{3}. \tag{84}$$

Now, using techniques mirroring those of Theorem 3.2 of [29], a particular solution to (80) is

$$\tilde{f}(t) = \sum_{k=-\infty}^{\infty} \frac{e^{-q^k t/2} \sin(\sqrt{3} q^k t/2)}{q^{k(k-1)/(2/3)}}, \tag{85}$$

for $t \geq 0$. Note that the expression in Equation (85) does not have the alternation $(-1)^k$, unlike the expression in Equation (55) for $\eta(t)$, and this will allow a sharp bound on $\tilde{f}(t)$ for all $t \geq 0$, independent of $q > 1$.

The first derivative of $\tilde{f}(t)$ is seen to be

$$\begin{aligned}\tilde{f}'(t) &= \sum_{k=-\infty}^{\infty} \frac{q^k e^{-q^k t/2}[(-1/2)\sin(\sqrt{3}q^k t/2) + (\sqrt{3}/2)\cos(\sqrt{3}q^k t/2)]}{q^{k(k-1)/(2/3)}} \\ &= \sum_{k=-\infty}^{\infty} \frac{e^{-q^k t/2} \sin(\sqrt{3}q^k t/2 + 2\pi/3)}{q^{k(k-1-2/3)/(2/3)}}\end{aligned} \tag{86}$$

where the fact that:

$$\frac{-\sin(x) + \sqrt{3}\cos(x)}{2} = \cos(2\pi/3)\sin(x) + \sin(2\pi/3)\cos(x) = \sin(x + 2\pi/3),$$

was used explicitly to obtain the last equality in (86). Using this identity implicitly, we obtain:

$$\tilde{f}^{(2)}(t) = \sum_{k=-\infty}^{\infty} \frac{q^k e^{-q^k t/2} \sin(\sqrt{3} q^k t/2 + 4\pi/3)}{q^{k(k-1-2/3)/(2/3)}} \tag{87}$$

$$= \sum_{k=-\infty}^{\infty} \frac{e^{-q^k t/2} \sin(\sqrt{3} q^k t/2 + 4\pi/3)}{q^{k(k-1-4/3)/(2/3)}}$$

and finally we verify:

$$\tilde{f}^{(3)}(t) = \sum_{k=-\infty}^{\infty} \frac{e^{-q^k t/2} \sin(\sqrt{3} q^k t/2 + 6\pi/3)}{q^{k(k-1-6/3)/(2/3)}} \tag{88}$$

$$= \sum_{k=-\infty}^{\infty} \frac{e^{-q^{k-1}(qt)/2} \sin(\sqrt{3} q^{k-1} qt/2)}{q^{[\{(k-1)+1\}][\{(k-1)-1\}-1]/(2/3)}}$$

$$= \sum_{m=-\infty}^{\infty} \frac{e^{-q^m(qt)/2} \sin(\sqrt{3} q^m(qt)/2)}{q^{[m+1][\{m-1\}-1]/(2/3)}}$$

$$= q^3 \sum_{m=-\infty}^{\infty} \frac{e^{-q^m(qt)/2} \sin(\sqrt{3} q^m(qt)/2)}{q^{m(m-1)/(2/3)}} = q^3 \tilde{f}(qt) \ . \tag{89}$$

A re-indexing $m = k - 1$ was used to move from (88) to (89). Note that (89) gives that (80) holds. From (85)–(87), one sees that

$$\tilde{f}(0) = \sum_{k=-\infty}^{\infty} \frac{\sin(0)}{q^{k(k-1)/(2/3)}} = 0 \ , \tag{90}$$

$$\tilde{f}'(0) = \sum_{k=-\infty}^{\infty} \frac{\sin(2\pi/3)}{q^{k(k-5/3)/(2/3)}} = \frac{\sqrt{3}}{2} \sum_{k=-\infty}^{\infty} \frac{q^k}{(q^3)^{k(k-1)/2}} = \frac{\sqrt{3}}{2} \theta(q^3; q) \tag{91}$$

$$\tilde{f}^{(2)}(0) = \sum_{k=-\infty}^{\infty} \frac{\sin(4\pi/3)}{q^{k(k-7/3)/(2/3)}} = \frac{-\sqrt{3}}{2} \sum_{k=-\infty}^{\infty} \frac{(q^2)^k}{(q^3)^{k(k-1)/2}} = \frac{-\sqrt{3}}{2} \theta(q^3; q^2) \ , \tag{92}$$

where the last equalities of (91) and (92) are obtained from (8).

Normalizing $\tilde{f}(t)$ by $\tilde{f}'(0) = (\sqrt{3}/2)\theta(q^3; q)$ to obtain

$$f(t) = \tilde{f}(t)/\tilde{f}'(0) \ , \tag{93}$$

one sees that $f(t)$ now satisfies the MADE (80) along with the initial conditions (81). The last initial condition follows from the fact that

$$f^{(2)}(0) = \frac{\tilde{f}^{(2)}(0)}{\tilde{f}^{(1)}(0)} = \frac{-(\sqrt{3}/2)\theta(q^3; q^2)}{(\sqrt{3}/2)\theta(q^3; q)} = \frac{-\theta(q^3; q^2)}{\theta(q^3; q)} = -q \ , \tag{94}$$

where the last equality in (94) follows from the next lemma.

Lemma 1. *For $q > 1$ the Jacobi theta function (8) satisfies*

$$\frac{\theta(q^3; q^2)}{\theta(q^3; q)} = q = \frac{\theta(q^3; q^2)}{\theta(q^3; -q)} \ . \tag{95}$$

Proof. For the first equality in (95) one can write

$$\theta(q^3; q^2) = \theta(q^3; q^3(1/q)) = q^3(1/q)\theta(q^3; 1/q) = q\left[q\theta(q^3; 1/q)\right] = q\left[\theta(q^3; q)\right] \ , \tag{96}$$

where the second equality is obtained from Equation (9) with $u = (1/q)$, and the last equality is the reciprocal identity in Equation (9) with $v = q$. Dividing (96) by $\theta(q^3; q)$ gives (95). For the second equality in (95), let $u = (-1/q)$ and $v = -q$ in Equation (9). Then as above $\theta(q^3; -q^2) = q\,\theta(q^3; -q)$. The lemma is shown. □

In addition to the last equality of (94) being proven by the first equality in (95) in Lemma 1, the second equality of (95) proves that the second derivative of $\mathcal{W}_{1,2/3}(t)/\mathcal{W}'_{1,2/3}(0)$ at $t = 0$ equals $-q$.

The following theta function bound will also be helpful.

Lemma 2. *For $q > 1$ the Jacobi theta function (8) satisfies*

$$\frac{\theta(q^3; 1)}{\theta(q^3; q)} \leq 1 + \frac{1}{q^3} < 2 . \qquad (97)$$

Proof. Observe that

$$\theta(q^3; 1) = \mu_{q^3} \prod_{n=0}^{\infty} \left[\left(1 + \frac{1}{q^{3n}}\right) \left(1 + \frac{1}{q^{3(n+1)}}\right) \right]$$

$$= \mu_{q^3} \left[\prod_{n=0}^{\infty} \left(1 + \frac{1}{q^{3n}}\right) \right] \left(1 + \frac{1}{q^3}\right) \left[\prod_{n=0}^{\infty} \left(1 + \frac{1}{q^{6+3n}}\right) \right] , \qquad (98)$$

while

$$\theta(q^3; q) = \mu_{q^3} \prod_{n=0}^{\infty} \left[\left(1 + \frac{q}{q^{3n}}\right) \left(1 + \frac{1}{qq^{3(n+1)}}\right) \right]$$

$$= \mu_{q^3} \left[\prod_{n=0}^{\infty} \left(1 + \frac{q}{q^{3n}}\right) \right] \left[\prod_{n=0}^{\infty} \left(1 + \frac{1}{q^{4+3n}}\right) \right] . \qquad (99)$$

Comparing each factor in the square brackets in (98) with the corresponding factor in the square brackets of (99) one sees that for all $n \geq 0$

$$\left(1 + \frac{1}{q^{3n}}\right) \leq \left(1 + \frac{q}{q^{3n}}\right) \quad \text{and} \quad \left(1 + \frac{1}{q^{6+3n}}\right) \leq \left(1 + \frac{1}{q^{4+3n}}\right) , \qquad (100)$$

from which one concludes that

$$\frac{\theta(q^3; 1)}{1 + 1/q^3} \leq \theta(q^3; q) , \qquad (101)$$

giving the left inequality in (97). The right inequality in (97) holds via the assumption that $q > 1$. □

Next we compute all derivatives of $g(t) = 2\exp(-t/2)\sin(\sqrt{3}t/2)/\sqrt{3}$ at $t = 0$ and of $f(t) = \tilde{f}(t)/\tilde{f}'(0)$ at $t = 0$, in preparation for the computation of the Taylor series expansion at $t = 0$ for both $g(t)$ and $f(t)$. From (82) we immediately have that for $k \geq 0$ and $j = 0, 1, 2$

$$g^{(3k+j)}(t) = g^{(j)}(t) . \qquad (102)$$

From (102) and (83) one concludes that for $k \geq 0$

$$g^{(3k)}(0) = g(0) = 0 , \quad g^{(3k+1)}(0) = g'(0) = 1 , \quad g^{(3k+2)}(0) = g''(0) = -1 . \qquad (103)$$

The analogous results for $f(t) = \tilde{f}(t)/\tilde{f}'(0)$ are obtained in the following lemma.

Lemma 3. *For $t \geq 0$ and $q > 1$, let $f(t) = \tilde{f}(t)/\tilde{f}'(0)$ with $\tilde{f}(t)$ given by* (85). *Then for $k \geq 0$ and $j = 0, 1, 2$ one has*

$$f^{(3k+j)}(t) = \left(q^3\right)^{k(k+1)/2} q^{jk} f^{(j)}(q^k t) . \tag{104}$$

Furthermore, at $t = 0$ one has

$$f^{(3k)}(0) = 0, \quad f^{(3k+1)}(0) = \left(q^3\right)^{k(k+1)/2} q^k, \quad f^{(3k+2)}(0) = -\left(q^3\right)^{k(k+1)/2} q^{2k} q . \tag{105}$$

Proof. We first establish (104) for the case that $j = 0$ by induction on k. So for $j = 0$ note that (104) holds as a tautology for $k = 0$, and for $k = 1$ it holds by (89). Assume that $f^{(3k)}(t) = \left(q^3\right)^{k(k+1)/2} f(q^k t)$ for fixed k. Then

$$\begin{aligned}
f^{3(k+1)}(t) &= f^{(3k+3)}(t) = \left[f^{(3k)}(t)\right]^{(3)} = \left[\left(q^3\right)^{k(k+1)/2} f(q^k t)\right]^{(3)} \tag{106} \\
&= \left(q^3\right)^{k(k+1)/2} q^3 f(qq^k t) q^{3k} = \left(q^3\right)^{(k+1)(k+2)/2} f(q^{k+1} t) , \tag{107}
\end{aligned}$$

where: the inductive hypothesis gives the rightmost equality in (106), and that (89) along with the chain rule gives the first equality in (107). Thus, the $j = 0$ case holds for all k. Now differentiate the expression $f^{(3k)}(t) = \left(q^3\right)^{k(k+1)/2} f(q^k t)$ either $j = 1$ or $j = 2$ times to obtain (104) in all remaining cases. Evaluating (104) at $t = 0$ and relying on (90)–(94) gives (105). □

Next, the $3N + 2$-degree Taylor polynomials $P_{3N}[g](t), P_{3N}[f](t)$ of g and f, respectively, expanded about $t = 0$ are given by

$$\begin{aligned}
P_{3N+2}[g](t) &= \sum_{n=0}^{3N+2} \frac{g^{(n)}(0)}{n!} t^n = \sum_{k=0}^{N} \frac{g^{(3k)}(0)}{(3k)!} t^{3k} + \sum_{k=0}^{N} \frac{g^{(3k+1)}(0)}{(3k+1)!} t^{3k+1} \\
&\quad + \sum_{k=0}^{N} \frac{g^{(3k+2)}(0)}{(3k+2)!} t^{3k+2} \\
&= \sum_{k=0}^{N} \frac{1}{(3k+1)!} t^{3k+1} + \sum_{k=0}^{N} \frac{-1}{(3k+2)!} t^{3k+2} \tag{108} \\
P_{3N+2}[f](t) &= \sum_{n=0}^{3N+2} \frac{f^{(n)}(0)}{n!} t^n = \sum_{k=0}^{N} \frac{f^{(3k)}(0)}{(3k)!} t^{3k} + \sum_{k=0}^{N} \frac{f^{(3k+1)}(0)}{(3k+1)!} t^{3k+1} \\
&\quad + \sum_{k=0}^{N} \frac{f^{(3k+2)}(0)}{(3k+2)!} t^{3k+2} \\
&= \sum_{k=0}^{N} \frac{q^{3k(k+1)/2} q^k}{(3k+1)!} t^{3k+1} + \sum_{k=0}^{N} \frac{-q^{3k(k+1)/2} q^{2k} q}{(3k+2)!} t^{3k+2} \tag{109}
\end{aligned}$$

where (108) follows from (103), and (109) follows from (105). For $t \geq 0$, these have respective remainder terms

$$\begin{aligned}
R_{3N+2}[g](t) &= \frac{g^{(3N+3)}(\xi)}{(3N+3)!} t^{3N+3} = \frac{g(\xi)}{(3N+3)!} t^{3N+3} , \tag{110} \\
R_{3N+2}[f](t) &= \frac{f^{(3N+3)}(\zeta)}{(3N+3)!} t^{3N+3} = \frac{q^{3(N+1)(N+2)/2} f(q^{N+1} \zeta)}{(3N+3)!} t^{3N+3} \tag{111}
\end{aligned}$$

for some $\xi \in [0, t]$ and $\zeta \in [0, t]$. The goal of uniform convergence on compact subsets is now obtained in the following proposition.

Proposition 4. *Let S be any compact set contained in $[0, \infty)$. Then $f(t)$ converges uniformly to $g(t)$ on S as $q \to 1^+$, where $f(t)$ is given by both (93) and (85), while $g(t)$ is given by (84).*

Proof. Without loss of generality, there is a $\rho > 0$ such that $S \subseteq [0, \rho]$, and it is sufficient to prove uniform convergence on $[0, \rho]$. For $t \in [0, \rho]$, from the triangle inequality one has

$$\begin{aligned}|f(t) - g(t)| &\leq |f(t) - P_{3N+2}[f](t)| + |P_{3N+2}[f](t) - P_{3N+2}[g](t)| \\ &\quad + |P_{3N+2}[g](t) - g(t)| \end{aligned} \quad (112)$$

$$\begin{aligned}&= |R_{3N+2}[f](t)| + |P_{3N+2}[f](t) - P_{3N+2}[g](t)| \\ &\quad + |R_{3N+2}[g](t)|\end{aligned} \quad (113)$$

Now for $0 \leq t \leq \rho$ and relying on (111), one starts with (114) to see

$$|R_{3N+2}[f](t)| = \left| \frac{q^{3(N+1)(N+2)/2} f(q^{N+1}\zeta)}{(3N+3)!} t^{3N+3} \right| \quad (114)$$

$$\leq \frac{q^{3(N+1)(N+2)/2} \rho^{3N+3}}{(3N+3)!} \left| f(q^{N+1}\zeta) \right|$$

$$= \frac{q^{3(N+1)(N+2)/2} \rho^{3N+3}}{(3N+3)!} \left| \frac{1}{\tilde{f}(0)} \tilde{f}(q^{N+1}\zeta) \right| \quad (115)$$

$$= \frac{q^{3(N+1)(N+2)/2} \rho^{3N+3}}{(3N+3)!} \times \quad (116)$$

$$\left| \frac{1}{(\sqrt{3}/2)\theta(q^3;q)} \sum_{k=-\infty}^{\infty} \frac{e^{-q^k q^{N+1}\zeta/2} \sin(\sqrt{3}q^k q^{N+1}\zeta/2)}{q^{k(k-1)/(2/3)}} \right|$$

$$\leq \frac{q^{3(N+1)(N+2)/2} \rho^{3N+3}}{(3N+3)!} \frac{2}{\sqrt{3}\theta(q^3;q)} \sum_{k=-\infty}^{\infty} \frac{1}{(q^3)^{k(k-1)/(2)}} \quad (117)$$

$$= \frac{q^{3(N+1)(N+2)/2} \rho^{3N+3}}{(3N+3)!} \frac{2}{\sqrt{3}\theta(q^3;q)} \theta(q^3;1) \quad (118)$$

$$< \frac{q^{3(N+1)(N+2)/2} \rho^{3N+3}}{(3N+3)!} \frac{4}{\sqrt{3}}, \quad (119)$$

where: moving to (115) is obtained via (93); (116) follows from (85) and (91); the equality in (118) is obtained by (8); and the inequality in (119) is given by (97) in Lemma 2. Similarly, from (110) and (84), one has

$$|R_{3N+2}[g](t)| = \left| \frac{g(\zeta)}{(3N+3)!} t^{3N+3} \right| \quad (120)$$

$$\leq \frac{\rho^{3N+3}}{(3N+3)!} \left| 2\exp(-\zeta/2)\sin(\sqrt{3}\zeta/2)/\sqrt{3} \right| \leq \frac{2\rho^{3N+3}}{\sqrt{3}(3N+3)!}.$$

Also, from (108) and (109) if we let: $\Delta P[f,g](t) \equiv P_{3N+2}[f](t) - P_{3N+2}[g](t)$, then

$$\begin{aligned} |\Delta P[f,g](t)| &= \left| \sum_{k=0}^{N} \frac{q^{3k(k+1)/2} q^k - 1}{(3k+1)!} t^{3k+1} + \sum_{k=0}^{N} \frac{-q^{3k(k+1)/2} q^{2k} q + 1}{(3k+2)!} t^{3k+2} \right| \\ &\leq \sum_{k=0}^{N} \frac{q^{3k(k+1)/2} q^k - 1}{(3k+1)!} \rho^{3k+1} + \sum_{k=0}^{N} \frac{q^{3k(k+1)/2} q^{2k} q - 1}{(3k+2)!} \rho^{3k+2} \\ &\leq \left[q^{3N(N+1)/2} q^{2N} q - 1 \right] \left[\sum_{k=0}^{N} \frac{\rho^{3k+1}}{(3k+1)!} + \sum_{k=0}^{N} \frac{\rho^{3k+2}}{(3k+2)!} \right] \\ &\leq \left[q^{3N(N+1)/2} q^{2N} q - 1 \right] e^{\rho} \, . \end{aligned} \qquad (121)$$

Applying (118), (121), and (120) to (113) one has that for $N \geq 0$

$$|f(t) - g(t)| \leq \frac{q^{3(N+1)(N+2)/2} \rho^{3N+3}}{(3N+3)!} \frac{4}{\sqrt{3}} \qquad (122)$$
$$+ \left[q^{3N(N+1)/2} q^{2N} q - 1 \right] e^{\rho} + \frac{2\rho^{3N+3}}{\sqrt{3}(3N+3)!} \, .$$

Now, given $\epsilon > 0$, choose N_0 sufficiently large such that one has $4\rho^{3N_0+3}/\left[\sqrt{3}(3N_0+3)!\right] < \epsilon/3$. Then

$$1 < (\epsilon/3) \left[\sqrt{3}(3N_0+3)!\right]/\left[4\rho^{3N_0+3}\right] \quad \text{and} \quad 1 < 1 + \epsilon/\left[3e^{\rho}\right] \, .$$

Pick $q_0 > 1$ so that

$$q_0^{3(N_0+1)(N_0+2)/2} < (\epsilon/3) \left[\sqrt{3}(3N_0+3)!\right]/\left[4\rho^{3N_0+3}\right] \qquad (123)$$

and

$$q_0^{3N_0(N_0+1)/2} q_0^{2N_0} q_0 < 1 + \epsilon/\left[3e^{\rho}\right] \, .$$

Then for $1 < q < q_0$ one has

$$q^{3(N_0+1)(N_0+2)/2} < (\epsilon/3) \left[\sqrt{3}(3N_0+3)!\right]/\left[4\rho^{3N_0+3}\right] \qquad (124)$$

and

$$q^{3N_0(N_0+1)/2} q^{2N_0} q < 1 + \epsilon/\left[3e^{\rho}\right] \, ,$$

whence for $1 < q < q_0$

$$\frac{2\rho^{3N_0+3}}{\sqrt{3}(3N_0+3)!} < q^{3(N_0+1)(N_0+2)/2} \frac{4\rho^{3N_0+3}}{\sqrt{3}(3N_0+3)!} < (\epsilon/3) \qquad (125)$$

and

$$\left[q^{3N_0(N_0+1)/2} q^{2N_0} q - 1 \right] e^{\rho} < \epsilon/3 \, .$$

Applying (125) to (122) with N taken to be N_0 one has that for $1 < q < q_0$

$$|f(t) - g(t)| \leq \epsilon/3 + \epsilon/3 + \epsilon/3 = \epsilon \, . \qquad (126)$$

So $f(t)$ approaches $g(t)$ uniformly on $[0, \rho]$ as $q \to 1^+$, and the proposition is proven. □

5. Convolutions, Correlations and Bounds

Here we briefly demonstrate that solutions of MADEs beget new solutions of different MADEs.

5.1. Distinction between Convolutions and Correlations

Let $f, g \in \mathcal{L}^1(\mathbb{R})$ and recall the standard definitions:

$$\text{Convolution between } f \text{ and } g \equiv [f * g](t) = \int_{-\infty}^{\infty} f(s) \cdot g(t-s)\, ds \tag{127}$$

$$\text{Correlation between } f \text{ and } g \equiv [f \star g](t) = \int_{-\infty}^{\infty} f(s) \cdot g(t+s)\, ds \tag{128}$$

Proposition 5. *Consider $f, g \in \mathcal{S}(\mathbb{R})$, which solve the following MADEs*

$$f^{(a)} = c_f \cdot f(qt), \quad g^{(b)} = c_g \cdot g(qt), \tag{129}$$

respectively, for $q > 1$, $a, b \in \mathbb{N}$, and $c_f \neq 0$, $c_g \neq 0$. Then the correlation and convolution solve the following higher-order MADEs

$$[f * g]^{(a+b)}(t) = \frac{c_f \cdot c_g}{q} [f * g](qt), \quad [f \star g]^{(a+b)}(t) = (-1)^a \frac{c_f \cdot c_g}{q} [f \star g](qt),$$

and $[f * g], [f \star g] \in \mathcal{S}(\mathbb{R})$.

Proof. The fact that convolution and correlation preserve the Schwartz property follows from Theorem 3.3 of [31]. The MADE equations easily follow from repeated applications of integration by parts, use of Equation (129), and a change of variables. □

5.2. Auto-Correlation

It was shown in Theorem 7 of [25] that the auto-correlation of $\mathcal{W}_{-1,2}(t) = F_{-1,2}(t)$, as defined in (45) for $\mu = -1$ and $\lambda = 2$, gives $\mathcal{W}_{0,1}(t) = f_{0,1}(|t|)$, as defined in (42) for $\mu = 0$ and $\lambda = 1$, in the sense that

$$[\mathcal{W}_{-1,2} \star \mathcal{W}_{-1,2}](t) \equiv \int_{-\infty}^{\infty} \mathcal{W}_{-1,2}(u) \cdot \mathcal{W}_{-1,2}(u+t)\, du$$

$$= \frac{-\mu_q^4}{2\mu_{q^2}^2} \cdot \mathcal{W}_{-2,1}(-t) = \frac{-\mu_q^4}{2\mu_{q^2}^2} \cdot \mathcal{W}_{-2,1}(t) = \frac{+\mu_q^4}{2\mu_{q^2}^2} \cdot \mathcal{W}_{0,1}\left(\frac{t}{q}\right),$$

where $\mathcal{W}_{-2,1}(t) = f_{-2,1}(|t|)$, as defined in (42) for $\mu = -2$ and $\lambda = 1$. Using this result, along with the Cauchy-Schwartz inequality it was shown in Proposition 4 of [25] that

$$0 < \|\mathcal{W}_{0,1}\|_\infty = \mathcal{W}_{0,1}(0) = \theta(q^2; -1/q) < 1, \quad \forall q > 1.$$

This important bound allows one to obtain uniform convergence of the normalized function $\mathcal{W}_{0,1}(t)/\mathcal{W}_{0,1}(0) \to \cos(t)$, as $q \to 1^+$.

5.3. Cross-Correlation

Let us consider an example that involves different MADE solutions, to obtain a new MADE. Knowing the Fourier transform of these functions allows us to easily derive properties of the resulting function. Compute, using Plancherel's Lemma,

$$
\begin{aligned}
[\mathcal{W}_{-1,2} \star \mathcal{W}_{0,1}](t) &\equiv \int_{-\infty}^{\infty} \mathcal{W}_{-1,2}(u) \cdot \mathcal{W}_{0,1}(u+t)\, du \\
&= \int_{-\infty}^{\infty} e^{-i\omega t} \mathcal{F}[\mathcal{W}_{-1,2}](\omega) \cdot \mathcal{F}[\mathcal{W}_{0,1}](\omega)\, d\omega .
\end{aligned}
\tag{130}
$$

Now, to simplify the integrand in (130), we use the Fourier transforms from [25,32] respectively, to write:

$$
\begin{aligned}
\mathcal{F}[\mathcal{W}_{-1,2}](\omega) \cdot \mathcal{F}[\mathcal{W}_{0,1}](\omega) &= \frac{i\mu_q^3}{\sqrt{2\pi}\,\omega\, \theta(q;i\omega)} \times \frac{2(\mu_{q^2})^3}{\sqrt{2\pi}\,\theta(q^2;\omega^2)} \\
&= \frac{i(\mu_q \cdot \mu_{q^2})^3}{\pi} \times \frac{1}{\omega\, \theta(q;i\omega)\, \theta(q^2;\omega^2)} \\
&= \frac{(\mu_q^2 \cdot \mu_{q^2})^2}{\pi} \times \frac{1}{(-i\omega)\, \theta(q;-i\omega)\, \theta^2(q;i\omega)} .
\end{aligned}
\tag{131}
$$

The equality in (131) follows from the fact that $\theta(q^2;\omega^2) = \theta(q;i\omega)\,\theta(q;-i\omega)$ and uses the definition of the Jacobi theta function in Equation (8). The consequence is that there are simple poles when $\omega = -iq^k$ for $k \in \mathbb{Z}$, but double poles at $\omega = iq^k$. Computing the integral in (130) using residue theory, requires a careful consideration of the position of these poles off the real axis.

For $t \geq 0$ the contour for ω must traverse the lower-half plane, encompassing the simple poles $\omega = -iq^k$. Consequently, residue theory and Equation (9) gives

$$
[\mathcal{W}_{-1,2} \star \mathcal{W}_{0,1}](t) = C_q \sum_{k=-\infty}^{\infty} (-1)^k \frac{e^{-q^k t}}{q^{3k(k+1)/2}} = C_q \cdot f_{-1,2/3}(t) ,
$$

which solves the eigen-MADE

$$
f^{(3)}_{-1,2/3}(t) = f_{-1,2/3}(qt) .
$$

6. Expanded Table of Fourier Transforms

In this final section we establish a short table of Fourier transforms for solutions of MADEs and their relations to Jacobi theta functions. Included are well-established results, along with new functions. The positive constants K_1 and K_2 are generic, but estimates are not presented here.

The introduction of new functions are as follows: For $K(t)$ see [32] for decay constants K_1 and K_2 in Table 1; The functions $_q Cos(t)$ and $_q Sin(t)$ are closely related to $\tilde{C}_q(t)$ and $\tilde{S}_q(t)$, respectively, introduced in [25], where constants K_3 and K_4 are obtained; The q-Bessel functions, related to $\mathcal{J}(t)$, were introduced in [33], along with decay constants K_5 and K_6; Flat wavelets $F(t)$ have Fourier transforms that are averages of theta functions, first derived in [29], along with constants K_7 and K_8. The functions $K * \tilde{C}_q(t)$ and $\mathcal{W}_{\mu,\lambda}(t)$, have Fourier transforms that involve theta functions, which can be used to obtain decay parameters K_9 and K_{10}.

Note that similar tables for Laplace transforms are quite extensive, since applications only require control of function growth on $\mathbb{R}^+$. Here we are concerned with globally defined functions on $\mathbb{R}$ for which a Fourier transform can be defined.

Table 1. Table of Fourier transforms with solutions of ODEs and MADEs.

Global Function	Property	Differential Equation	$f(0)$	$f(\pm\infty)$ decay Rate	Fourier Transf. (Modulo Coef.)
$f(t) = e^{-t^2/2}$	Entire Schwartz	$-f''(t) + t^2 f(t) = f(t)$	1	Gaussian	$e^{-x^2/2}$
$f(t) = e^{-|t|}$	$C^0 \cap \mathcal{L}^p$ $1 \le p \le \infty$	$f'(t) + f(t) = -2\delta(t)$	1	exponential	$(1+x^2)^{-1}$
$e^{i(x-x_0)t} =$ $\exp[i(x-x_0)t]$	$C^0 \cap \mathcal{L}^\infty$	$\partial_t \exp[i(x-x_0)t] =$ $i(x-x_0)\exp[i(x-x_0)t]$	1	undefined	$\delta_0(x-x_0) = \delta_{x_0}(x)$
$j_0(t) = \frac{\sin(t)}{t}$	$C^\infty \cap \mathcal{L}^p$ $1 < p \le \infty$	$j_0''(t) + \frac{2}{t}j_0'(t) = -j_0(t)$	1	$1/|t|$	$\chi_{[-1,1]}(x)$
$Ai(t) =$ $\int_0^\infty \cos(\frac{u^3}{3}+ut)\,du$	$C^\infty \cap \mathcal{L}^p$ $4 < p \le \infty$	$Ai''(t) = t \cdot Ai(t)$	$Ai(0)$ smooth	$1/|t|^{1/4}$	$e^{ikx^3/3}$
$\cos(t)$ $\sin(t)$	$C^0 \cap \mathcal{L}^\infty$	$\cos''(t) + \cos(t) = 0$ $\sin''(t) + \sin(t) = 0$	1	undefined	$\delta_1(x) \pm \delta_{-1}(x)$
$K(t) \equiv F_{-1,2}(t)$	Schwartz wavelet	$K'(t) = K(qt)$	0 flat	$|t|^{-K_1 \ln|t|+K_2}$	$\frac{1}{ix\theta(q;ix)}$
$\tilde{C}_q(t) =$ $f_{0,1}(\frac{|t|}{\sqrt{q}})/f_{0,1}(0)$	Schwartz wavelet	$\tilde{C}_q''(t) + \tilde{C}_q(qt) = 0$	1 smooth	$|t|^{-K_3 \ln|t|+K_4}$	$\frac{1}{\theta(q^2;qx^2)}$
$\tilde{S}_q(t) = \int_0^t \tilde{C}_q(u)du$	Schwartz wavelet	$\tilde{S}_q''(t) + q^{-1}\tilde{S}_q(qt) = 0$	0 smooth	$|t|^{-K_3 \ln|t|+K_4}$	$\frac{-iq^3 x}{\theta(q^2;q^3x^2)}$
$CiS_q(t) =$ $\tilde{C}_q(t) + i\tilde{S}_q(t)$	Schwartz wavelet	$\partial_t^2 CiS_q(t) = -CiS_q(qt)$	1 smooth	$|t|^{-K_3 \ln|t|+K_4}$	$\frac{1}{\theta(q^2;qx^2)} + \frac{q^3 x}{\theta(q^2;q^3x^2)}$
$\mathcal{J}(t) = \frac{\tilde{S}_q(t)}{t}$	Schwartz	$\mathcal{J}''(t) + \frac{2}{t}\mathcal{J}'(t) = -\mathcal{J}(qt)$	$1/q$	$|t|^{-K_5 \ln|t|+K_6}$	$\int_{-\infty}^{x} \frac{\omega\,d\omega}{\theta(q^2;q^3\omega^2)}$
$F(t) =$ $F_{2M+1,2N}(t)$	Schwartz wavelet	$a = N+1$ $b = (N+2M+1)/2$ $F'(t) = (-1)^a q^b F(q^N t)$	0 flat	$|t|^{-K_7 \ln|t|+K_8}$	$(-z_j)^N = -ix$ $\frac{-i}{xN}\sum_{j=1}^{N} \frac{z_j^{M+1}}{\theta(q^{1/N};z_j)}$
$M_1(t) =$ $[K * \tilde{C}_q](t)$	Schwartz wavelet	$M_1'''(t) = -q^{-1}M_1(qt)$	0 smooth	$|t|^{-K_9 \ln|t|+K_{10}}$	$\frac{1}{ix\,\theta(q;ix)\,\theta(q^2;qx^2)}$
$M_2(t) =$ $\mathcal{W}_{1,2/3}(t)$	Schwartz wavelet	$M_2'''(t) = q^3 M_2(qt)$	0 smooth	$|t|^{-K_9 \ln|t|+K_{10}}$	$\frac{x^2}{\theta(q^3;-ix^3)}$
$Aiq(t) =$ $\int_0^\infty \tilde{C}_q(\frac{u^3}{3}+ut)\,du$	Schwartz	$Aiq''(t) =$ $q^{-1/3}t \cdot Aiq(q^{2/3}t)$	$Aiq(0)$ smooth	$|t|^{-K_3 \ln|t|+K_4}$	$\int_0^\infty \frac{e^{ix^3/(3k^2)}}{\theta(q^2;qk^2)}\,dk$

Author Contributions: Conceptualization, D.W.P., N.R. and M.J.S.; investigation, D.W.P., N.R. and M.J.S.; writing–original draft preparation, D.W.P., N.R. and M.J.S.; writing–review and editing, D.W.P., N.R. and M.J.S. All authors have read and agreed to the published version of the manuscript.

Funding: This research received no external funding. This research was supported in part by the ECU Mathematics Department and by the ECU Thomas Harriot College of Arts and Sciences.

Acknowledgments: The authors would like to thank the reviewers for very helpful comments and suggestions that aided in the completion of this project.

Conflicts of Interest: The authors declare no conflict of interest.

Abbreviations

The following abbreviations are used in this manuscript:

ODE   Ordinary Differential Equation
PDE   Partial Differential Equation
MADE Multiplicatvely Advanced Differential Equation

Appendix A. Normalization in Terms of Theta Functions

The normalization for $\tilde{C}_q(t)$ in Equation (4) involves a theta function, so that

$$\frac{1}{N_q} \equiv \sum_{k=-\infty}^{\infty} \frac{(-1)^k}{q^{k^2}} = \sum_{k=-\infty}^{\infty} \frac{(-1/q)^k}{(q^2)^{k(k-1)/2}} = \theta\left(q^2; \frac{-1}{q}\right). \tag{A1}$$

The last expression in Equation (A1) does not vanish for $q > 1$ due to the product formula in Equation (8). Similarly, we can show that $Aiq(0) \neq 0$. Indeed, from the definition, note that using the change of variables $w = q^{(k-1/2)/3}u$,

$$\begin{aligned}
Aiq(0) &= \frac{1}{\pi} \int_0^\infty \tilde{C}_q(u^3)\, du \\
&= \frac{N_q}{\pi} \int_0^\infty \sum_{k=-\infty}^{\infty} \frac{(-1)^k}{q^{k^2}} e^{-q^k u^3/\sqrt{q}}\, du \\
&= \frac{N_q}{\pi} \sum_{k=-\infty}^{\infty} \frac{(-1)^k q^{1/6}}{q^{k^2} q^{k/3}} \int_0^\infty e^{-w^3}\, dw \\
&= \frac{q^{1/6} N_q}{\pi} \cdot \int_0^\infty e^{-w^3}\, dw \cdot \sum_{k=-\infty}^{\infty} \frac{(-1/q^{1/3})^k}{(q^2)^{k^2/2}} \\
&= \frac{q^{1/6}}{\pi} \cdot \int_0^\infty e^{-w^3}\, dw \cdot \left[\frac{\theta(q^2; -1/q^{4/3})}{\theta(q^2; -1/q)}\right],
\end{aligned}$$

and the final expression clearly does not vanish for any $q > 1$.

Appendix B. Establishing the q-Airy Hypothesis for $q > 1$

To compute $A_0(q)$ explicitly, we will find the Fourier transform of $Aiq(t)$ and then find its value at the origin. This requires a careful change of variables. To begin, we combine definition Equation (25) and the inverse Fourier transform of formula in Equation (7) giving

$$Aiq(x) = \frac{2(\mu_{q^2})^3 N_q}{2\pi\,\pi} \cdot \int_0^\infty \int_{-\infty}^\infty \frac{\exp\left(ik(t^3/3 + xt)\right)}{\theta(q^2; q k^2)}\, dk\, dt\,.$$

To handle the double integral note that the odd power of both the k and the t variables allows the following rearrangement

$$\int_0^\infty \int_0^\infty \frac{\exp\left(ik(t^3/3 + xt)\right) + \exp\left(i(-k)(t^3/3 + xt)\right)}{\theta(q^2; q k^2)}\, dk\, dt = \int_0^\infty \int_{-\infty}^\infty \frac{\exp\left(ik(t^3/3 + xt)\right)}{\theta(q^2; q k^2)}\, dt\, dk\,.$$

We can now obtain the Fourier transform

$$\begin{aligned}
\mathcal{F}[Aiq(x)](\omega) &= \frac{2(\mu_{q^2})^3 N_q}{(2\pi)^{3/2}\,\pi} \cdot \int_{-\infty}^\infty \int_0^\infty \int_{-\infty}^\infty e^{-ix(\omega-kt)} \cdot \frac{\exp\left(ik(t^3/3)\right)}{\theta(q^2; q k^2)}\, dt\, dk\, dx \\
&= \frac{2(\mu_{q^2})^3 N_q}{\sqrt{2\pi}\,\pi} \cdot \int_0^\infty \frac{\exp\left(i\omega^3/(3k^2)\right)}{\theta(q^2; q k^2)}\, dk\,.
\end{aligned}$$

Finally, computing at $\omega = 0$ gives the final result

$$A_0(q) = \mathcal{F}[Aiq(t)](0) = \int_{-\infty}^\infty Aiq(t)\, dt = \frac{2(\mu_{q^2})^3 N_q}{\sqrt{2\pi}\,\pi\sqrt{q}} \cdot \int_0^\infty \frac{dk}{\theta(q^2; k^2)} > 0,$$

which is clearly a finite, positive, non-zero quantity, for each $q > 1$.

Appendix C. Mollifier Argument for Airy PDE Initial Profile

Let us first make clear the importance of normalization. Indeed, observe that if $A_0(q) \neq 0$, then the change of variables $u = y/\sqrt[3]{t}$ for $t > 0$, gives

$$\lim_{t \to 0^+} \frac{1}{\sqrt[3]{t} A_0(q)} \int_{-\infty}^{\infty} Aiq\left(\frac{y}{\sqrt[3]{t}}\right) dy = \lim_{t \to 0^+} \frac{1}{A_0(q)} \int_{-\infty}^{\infty} Aiq(u)\, du = \frac{A_0(q)}{A_0(q)} = 1.$$

Thus, explicitly, for each fixed $q > 1$ and $x \in \mathbb{R}$, and $\phi_q(x,t)$ as in Equation (34)

$$|\phi_q(x,0) - f(x)| = \left| \lim_{t \to 0^+} \frac{1}{\sqrt[3]{t} A_0(q)} \int_{-\infty}^{\infty} Aiq\left(\frac{y}{\sqrt[3]{t}}\right) \cdot (f(x-y) - f(x))\, dy \right|$$

$$\leq \lim_{t \to 0^+} \frac{1}{|A_0(q)|} \int_{-\infty}^{\infty} |Aiq(u)| \cdot \left| f\left(x - u\sqrt[3]{t}\right) - f(x) \right| du. \tag{A2}$$

At this point we use the Schwartz property of $Aiq(u)$, along with the integrability and continuity of $f(x)$, to argue that the expression above is arbitrarily close to 0. This will be done in two parts.

Given $\epsilon > 0$, choose $R_\epsilon > 0$ so that

$$\int_{|u| \geq R_\epsilon} |Aiq(u)|\, du \leq \frac{|A_0(q)|}{4\|f\|_\infty} \cdot \epsilon.$$

Note that this estimate is independent of $t > 0$, so with $q > 1$ and $x \in \mathbb{R}$ fixed, the choice of R_ϵ will determine a bound that is needed on t, near 0.

Now, consider the region $|u| \leq R_\epsilon$. Since $f \in \mathcal{C}^1(\mathbb{R})$, $f(x)$ is continuous, so given $\epsilon > 0$ and $x \in \mathbb{R}$, $\exists \delta_{\epsilon,x} > 0$ so that

$$|x - y| < \delta_{\epsilon,x} \quad \Longrightarrow \quad |f(x) - f(y)| < \frac{\epsilon}{2} \cdot \left(\frac{|A_0(q)|}{\|Aiq\|_1} \right).$$

Thus, we require $|u\sqrt[3]{t}| \leq R_\epsilon \sqrt[3]{t} < \delta_{\epsilon,x}$ so that

$$t \in (0, (\delta_{\epsilon,x}/R_\epsilon)^3) \quad \Longrightarrow \quad |\phi_q(x,0) - f(x)| < \epsilon, \tag{A3}$$

which establishes pointwise convergence. However, if $f \in \mathcal{C}^1 \cap \mathcal{L}^1$ and $f' \in \mathcal{L}^\infty$, returning to Equation (A2) for $t > 0$, we obtain uniform convergence as follows:

$$\frac{1}{|A_0(q)|} \int_{-R_\epsilon}^{R_\epsilon} |Aiq(u)| \cdot \frac{\left| f\left(x - u\sqrt[3]{t}\right) - f(x) \right|}{|u\sqrt[3]{t}|} \cdot |u\sqrt[3]{t}|\, du \leq \frac{R_\epsilon \sqrt[3]{t}}{|A_0(q)|} \cdot \|f'\|_\infty \cdot \int_{-R_\epsilon}^{R_\epsilon} |Aiq(u)|\, du$$

Now, clearly, the condition in Equation (A3) can be achieved.

Thus, we verified that the solution to the q-advanced PDE in Equation (34) has the property that a continuous, bounded and integrable initial profile $f(x)$ is recovered at $t = 0$, as indicated in Equation (41).

Appendix D. Derivation of Inhomogeneous MADE

Using the characteristic function $\chi_S(t)$, and delta function centered at the origin $\delta_0(t)$, express the function in Equation (57) as

$$\tilde{\eta}(t) = C_q^- \cdot \left(\sum_{k=-\infty}^{\lfloor N_* \rfloor} (-1)^k \frac{e^{-q^k t} - 1}{q^{3k(k-1)/2}} \right) \cdot \chi_{(t_*,0)}(t) + C_q^+ \cdot \left(\sum_{k=-\infty}^{\infty} \frac{(-1)^k e^{-q^k t}}{q^{3k(k-1)/2}} \right) \cdot \chi_{[0,\infty)}(t),$$

for $N_* = N_*(q, t_*)$ fixed. Note that $\mathcal{C}_q^+$ is defined in Equation (54), and $\mathcal{C}_q^-$ is defined in Equation (78), so that $\tilde{\eta}(0^+) = \tilde{\eta}(0^-) = 0$, and $\tilde{\eta}'(0^+) = \tilde{\eta}'(0^-) = 1$. Thus, the first derivative is continuous, and the second derivative is bounded. However, the third derivative results in the appearance of a distribution,

$$\begin{aligned}\tilde{\eta}'''(t) &= q^3 \tilde{\eta}(qt) + \left[q^3 \mathcal{C}_q^- \cdot \left(\sum_{k=-\infty}^{\lfloor N_* \rfloor} (-1)^k \cdot \frac{q^{2k}}{q^{3k(k-1)/2}} \right) \cdot \chi_{(t_*, 0)}(t) \right.\\ &\quad \left. - \mathcal{C}_q^- \cdot \left(\sum_{k=-\infty}^{\lfloor N_* \rfloor} \frac{(-1)^k q^{2k}}{q^{3k(k-1)/2}} \right) \cdot \delta_0(t) + \mathcal{C}_q^+ \cdot \left(\sum_{k=-\infty}^{\infty} \frac{(-1)^k q^{2k}}{q^{3k(k-1)/2}} \right) \cdot \delta_0(t) \right]\\ &= q^3 \tilde{\eta}(qt) + [\tilde{f}_*(t)] ,\end{aligned}$$

which is an inhomogeneous MADE for all $t > t_*$, and which defines $\tilde{f}_*(t)$, by inspection of the quantity in the square brackets. The last three terms on the right hand side vanish as $q \to 1^+$, where $N_* \to \infty$ in a manner described after the proof of Proposition 3.

References

1. Burden, R.L.; Fairs, D.J.; Burden, A.M. *Numerical Analysis*; 10E. Cengage Learning: Boston, MA, USA, 2016.
2. Fox, L.; Mayers, D.F.; Ockendon, J.R.; Tayler, A.B. On a Functional Differential Equation. *IMA J. Appl. Math.* **1971**, *8*, 271–307. [CrossRef]
3. Kato, T.; McLeod, J.B. The Functional-Differential Equation $y'(x) = ay(\lambda x) + by(x)$. *Bull. Am. Math. Soc.* **1971**, *77*, 891–937.
4. Reed, M.; Simon, B. *Functional Analysis I*; Academic Press: New York, NY, USA, 1975.
5. Dung, N.T. Asymptotic behavior of linear advanced differential equations. *Acta Math. Sci.* **2015**, *35*, 610–618. [CrossRef]
6. Di Vizio, L. An ultrametric version of the Maillet-Malgrange theorem for nonlinear q-difference equations. *Proc. Am. Math. Soc.* **2008**, *136*, 2803–2814. [CrossRef]
7. Di Vizio, L.; Hardouin, C. Descent for differential Galois theory of difference equations: Confluence and q-dependence. *Pac. J. Math.* **2012**, *256*, 79–104. [CrossRef]
8. Di Vizio, L.; Zhang, C. On q-summation and confluence. *Ann. Inst. Fourier (Grenoble)* **2009**, *59*, 347–392. [CrossRef]
9. Dreyfus, T. Building meromorphic solutions of q-difference equations using a Borel-Laplace summation. *Int. Math. Res. Not. IMRN* **2015**, *15*, 6562–6587. [CrossRef]
10. Dreyfus, T.; Lastra, A.; Malek, S. On the multiple-scale analysis for some linear partial q-difference and differential equations with holomorphic coefficients. *Adv. Differ. Equ.* **2019**, *2019*, 326. [CrossRef]
11. Lastra, A.; Malek, S. On q-Gevrey asymptotics for singularly perturbed q-difference-differential problems with an irreguular singularity. *Abstr. Appl. Anal.* **2012**, *2012*, 860716. [CrossRef]
12. Lastra, A.; Malek, S. On parametric Gevrey asymptotics for singularly perturbed partial differential equations with delays. *Abstr. Appl. Anal.* **2013**, *2013*, 723040. [CrossRef]
13. Lastra, A.; Malek, S. Parametric Gevrey asymptotics for some Cauchy problems in quasiperiodic function spaces. *Abstr. Appl. Anal.* **2014**, *2014*, 153169. [CrossRef]
14. Lastra, A.; Malek, S. On parametric multilevel q-Gevrey asymptotics for some linear q-difference-differential equations. *Adv. Differ. Equ.* **2015**, *2015*, 344. [CrossRef]
15. Lastra, A.; Malek, S. On multiscale Gevrey and q-Gevrey asymptotics for some linear q-difference differential initial value Cauchy problems. *J. Differ. Equ. Appl.* **2017**, *23*, 1397–1457. [CrossRef]
16. Lastra, A.; Malek, S. On a q-Analog of Singularly Perturbed Problem of Irregular Type with Two Complex Time Variables. *Mathematics* **2019**, *2019*, 924. [CrossRef]
17. Lastra, A.; Malek, S.; Sanz, J. On q-asymptotics for linear q-difference-differential equations with Fuchsian and irreguular singularities. *J. Differ. Equ.* **2012**, *252*, 5185–5216. [CrossRef]
18. Lastra, A.; Malek, S.; Sanz, J. On q-asymptotics for q-difference-differential equations with Fuchsian and irregular singularities, Formal and analytic solutions of differential and difference equations. *Pol. Acad. Sci. Inst. Math.* **2012**, *97*, 73–90.

19. Lastra, A.; Malek, S.; Sanz, J. On Gevrey solutions of threefold singular nonlinear partial differential equations. *J. Differ. Equ.* **2013**, *255*, 3205–3232. [CrossRef]
20. Malek, S. On complex singularity analysis for linear partial q-difference-differential equations using nonlinear differential equations. *J. Dyn. Control Syst.* **2013**, *19*, 69–93. [CrossRef]
21. Stéphane, M. On parametric Gevrey asymptotics for a q-analog of some linear initial value problem. *Funkcial. Ekvac.* **2017**, *60*, 21–63.
22. Malek, S. On a Partial q-Analog of a Singularly Perturbed Problem with Fuchsian and Irregular Time Singularity. *Abstr. Appl. Anal.* **2020**, *2020*, 7985298. [CrossRef]
23. Tahara, H. q-analogues of Laplace and Borel transforms by means of q-exponentials. *Ann. Inst. Fourier Grenoble* **2017**, *67* 1865–1903. [CrossRef]
24. Zhang, C. Analytic continuation of solutions of the pantograph equation by means of θ-modular forms. *arXiv* **2012**, arXiv:1202.0423.
25. Pravica, D.W.; Randriampiry, N.; Spurr, M.J. Reproducing kernel bounds for an advanced wavelet frame via the theta function. *Appl. Comput. Harmon. Anal.* **2012**, *33*, 79–108. [CrossRef]
26. Bender, C.M.; Orszag S.A. *Asymptotic Analysis*; Advanced Mathematical Methods for Scientists and Engineers; Springer: New York, NY, USA, 1999.
27. Craig, W.; Goodman, J. Linear Dispersive Equations of Airy Type. *J. Differ. Equ.* **1990**, *87*, 38–61. [CrossRef]
28. Reed, M.; Simon, B. *Fourier Analysis, Self-Adjointness II*; Academic Press: New York, NY, USA, 1975.
29. Pravica, D.W.; Randriampiry, N.; Spurr, M.J. Solutions of a class of multiplicatively advanced differential equations. *C.R. Acad. Sci. Paris Ser. I* **2018**, *356*, 776–817. [CrossRef]
30. Pravica, D.; Spurr, M. Analytic Continuation into the Future. *Discret. Contin. Dyn. Syst.* **2003**, *2003*, 709–716.
31. Stein, E.; Weiss, G. *Introduction to Fouier Analysis on Euclidian Spaces*; Princeton University Press: Princeton, NJ, USA, 1971.
32. Pravica, D.; Randriampiry, N.; Spurr, M. Applications of an advanced differential equation in the study of wavelets. *Appl. Comput. Harmon. Anal.* **2009**, *27*, 2–11. [CrossRef]
33. Pravica, D.; Randriampiry, N.; Spurr, M. On q-advanced spherical Bessel functions of the first kind and perturbations of the Haar wavelet. *Appl. Comput. Harmon. Anal.* **2018**, *44*, 350–413. [CrossRef]

 © 2020 by the authors. Licensee MDPI, Basel, Switzerland. This article is an open access article distributed under the terms and conditions of the Creative Commons Attribution (CC BY) license (http://creativecommons.org/licenses/by/4.0/).

Article

On the Regularized Asymptotics of a Solution to the Cauchy Problem in the Presence of a Weak Turning Point of the Limit Operator

Alexander Yeliseev

National Research University «MPEI», Krasnokazarmennaya 14, 111250 Moscow, Russia ;Yeliseevag@mpei.ru; Tel.: +8-495-362-71-31

Received: 11 June 2020; Accepted: 15 July 2020; Published: 23 July 2020

Abstract: An asymptotic solution of the linear Cauchy problem in the presence of a "weak" turning point for the limit operator is constructed by the method of S. A. Lomov regularization. The main singularities of this problem are written out explicitly. Estimates are given for ε that characterize the behavior of singularities for $\varepsilon \to 0$. The asymptotic convergence of a regularized series is proven. The results are illustrated by an example. Bibliography: six titles.

Keywords: singular Cauchy problem; asymptotic series; regularization method; turning point

1. Introduction

In this paper, the regularization method of S. A. Lomov [1] is used to construct a regularized asymptotic solution of a singularly-perturbed inhomogeneous Cauchy problem on the entire interval [0, T] in the presence of a spectral singularity in the form of a "weak" turning point for the limit operator.

We note the paper [2] devoted to the construction of the asymptotic behavior of the solution of singularly-perturbed Cauchy problems for integro-differential equations in the presence of spectral features of the limit operator. The point $\varepsilon = 0$ for a singularly-perturbed Cauchy problem is singular in the sense that in the classical theorems, the existence of a solution to the Cauchy problem does not take place at this point. Therefore, in the solution of singularly-perturbed problems, essentially, singularities arise that describe the irregular dependence of the solution on ε. The description of these singularities is the main problem of the regularization method. Under the conditions of spectrum stability, essentially, singularities are described using exponentials of the form $e^{\frac{1}{\varepsilon}\varphi(t)}$, where $\varphi(t)$, is smooth, in the general case, complex function of a real variable t. For solutions of linear homogeneous equations, such singularities were singled out by Liouville [3].

If the stability conditions are violated—for example, if the points of the spectrum intersect at one or more points t—the description is more complicated. In [4], singularities are presented in the case of a simple turning point, when the only eigenvalue of the operator $A(t)$ has the form

$$\lambda(t) = t^{k_0}(t-t_1)^{k_1}\ldots(t-t_m)^{k_m}a(t), a(t) \neq 0, k_0 + k_1 + \ldots + k_m = n.$$

Moreover, it is assumed that the operator $A(t)$ has a diagonal form for any $t \subset [0, T]$. Additionally, in [5], a rational simple turning point was considered, an irrational simple turning point was considered in [6]. Significantly special singularities from a mathematical point of view are special functions that describe the irregular dependence of the solution on ε for $\varepsilon \to 0$, and from the point of view of the hydrodynamics of the boundary layer function generated by the spectral singularity of the point $\lambda(t)$. The question of essentially special singularities is related to how the solution of a singularly-perturbed Cauchy problem inherits the smoothness properties of the coefficients of the equation. In particular, the coefficients of the equation depend analytically on the parameter ε. In the presence of a singular point $\varepsilon = 0$, analyticity at this point is inherited by solving the problem of a singularly-perturbed Cauchy problem not as is known from classical existence theorems: a singular point and a certain character of the spectrum of the operator $A(t)$ generate significantly in the solution singularities, highlighting that we have the right to calculate that the rest of the solution will already be analytic in some neighborhood of the value $\varepsilon = 0$ if on $h(t)$ and $A(t)$ to impose certain restrictions (infinite differentiability with respect to t is not enough!). Let us explain the words "the rest of the solution" with the simplest example of a scalar problem

$$\varepsilon \dot{u}(t,\varepsilon) = a(t)u(t,\varepsilon) + h(t), u(0,\varepsilon) = u^0. \tag{1}$$

If $a(t) < 0$, then the solution to this problem has the following structure provided $a(t) = tk(t), k(t) < 0$

$$u(t,\varepsilon) = f(t,\varepsilon)e^{\frac{1}{\varepsilon}\int_0^t a(s)ds} + g(t,\varepsilon)e^{\frac{1}{\varepsilon}\int_0^t a(s)ds}\int_0^t e^{-\frac{1}{\varepsilon}\int_0^s a(s_1)ds_1} ds + y(t,\varepsilon)$$

where the functions $f(t,\varepsilon), g(t,\varepsilon), y(t,\varepsilon)$ are analytic in ε, if on $k(t), h(t)$ problems (1) impose certain requirements. This paper continues the study of the turning points [3,4], namely, "weak", the turning point of the regularization method.

The simplest case of a weak turning point is the point of the first order, i.e., $\lambda_2(t) - \lambda_1(t) = ta(t), a(t) \neq 0$. The solution to a singularly-perturbed Cauchy problem in this case is described in [7]. It is assumed that the eigenspaces corresponding to the eigenvalues $\lambda_1(t), \lambda_2(t)$ are one-dimensional. In this paper, we consider the general case of a weak turning point. The definition of a "weak" point for the limit operator will be given below in the statement of the problem.

2. Statement of the Problem, Description of the Main Singularities of the Problem

1^0. **Statement of the problem.** Let a singularly-perturbed Cauchy problem be given

$$\varepsilon \dot{u}(t,\varepsilon) = A(t)u(t,\varepsilon) + h(t), u(t,\varepsilon) = u^0. \tag{2}$$

and conditions are met

(1) $h(t) \in C^\infty([0,T], R^n)$;
(2) $A(t) \in C^\infty([0,T], \mathscr{L}(R^n, R^n))$ having a smooth spectrum
   $\lambda_i(t) \in C^\infty([0,T])$, $i = 1,2$;
(3) $A(t) = \lambda_1(t)P_1(t) + \lambda_2(t)P_2(t)$, $P_1(t) + P_2(t) = I$,
(4) The condition of a weak point

$\lambda_2(t) - \lambda_1(t) = t^{k_0}(t-t_1)^{k_1}\ldots(t-t_m)^{k_m}a(t), a(t) \neq 0, k_0 + k_1 + \ldots + k_m = n,$

$\lambda_2(t) \neq \lambda_1(t) \; \forall t \in (0,t_1) \cup (t_1,t_2) \cup \ldots \cup (t_{m-1},t_m) \cup (t_m, T]$;

moreover, the geometric multiplicity of the eigenvalues is equal to the algebraic

for any $t \in [0,T]$;
(5) $\lambda_i(t) \neq 0$, $Re\lambda_i(t) \leq 0$, $\forall t \in [0,T]$.

2^0. **Description of the space without resonant solutions. The formalism of the regularization method.** In presenting the regularization method for solving problem (2), we will use the

Lagrange–Sylvester interpolation polynomials, which describe differentiable functions f (t) defined at the point $t_0, t_1, ..., t_m$ together with their derivatives. They have the form:

$$K(t)f(t) = \sum_{j=0}^{m} \sum_{i=0}^{k_j-1} K_{j,i}(t) f^{(i)}(t_j) \tag{3}$$

where $K_{j,i}(t)$ are polynomials with the property $\frac{d^s}{dt^s} K_{j,i}(t)|_{t=t_k} = \delta_j^k \delta_i^s$, $j, k = \overline{0, m}$, $i, s = \overline{0, k_j - 1}$. The singularities $J_1(t,\varepsilon), J_2(t,\varepsilon)$ of this problem (2) are found from the solution of the Cauchy problem.

$$\begin{cases} \varepsilon \dot{J}_1(t,\varepsilon) = \lambda_1(t) J_1(t,\varepsilon) + \varepsilon K(t) J_2(t,\varepsilon) \\ \varepsilon \dot{J}_2(t,\varepsilon) = \lambda_2(t) J_2(t,\varepsilon) + \varepsilon K(t) J_1(t,\varepsilon) \\ J_1(0,\varepsilon) = 1, J_2(0,\varepsilon) = 1. \end{cases} \tag{4}$$

Here $K(t) = \sum_{j=0}^{m} \sum_{i=0}^{k_j-1} K_{j,i}(t)$. Proof of the existence of a solution to the system (4) and decision evaluations are made in the annex. The solutions of system (4) generate a series of functions that describe the singularities of problem (2).

$$\varphi_i(t) = \frac{1}{\varepsilon} \int_0^t \lambda_i(s) ds, \quad \sigma_{i,0}(t,\varepsilon) = e^{\varphi_i(t)}, \quad i = 1,2$$

$$\sigma_{1,1}^{(j_1,i_1)}(t,\varepsilon) = e^{\varphi_1(t)} \int_0^t e^{\Delta \varphi(s_1)} K_{j_1,i_1}(s_1) ds_1,$$

$$\sigma_{2,1}^{(j_1,i_1)}(t,\varepsilon) = e^{\varphi_2(t)} \int_0^t e^{-\Delta \varphi(s_1)} K_{j_1,i_1}(s_1) ds_1,$$

$$\sigma_{1,p}^{(j_1,i_1,...,j_p,i_p)}(t,\varepsilon) = e^{\varphi_1(t)} \int_0^t e^{\Delta \varphi(s_1)} K_{j_p,i_p}(s_1) \int_0^{s_1} e^{-\Delta \varphi(s_2)} K_{j_{p-1},i_{p-1}}(s_2) ... \tag{5}$$

$$\int_0^{s_{p-1}} e^{(-1)^{p-1} \Delta \varphi(s_p)} K_{j_1,i_1}(s_p) ds_p ... ds_1,$$

$$\sigma_{2,p}^{(j_1,i_1,...,j_p,i_p)}(t,\varepsilon) = e^{\varphi_2(t)} \int_0^t e^{-\Delta \varphi(s_1)} K_{j_p,i_p}(s_1) \int_0^{s_1} e^{\Delta \varphi(s_2)} K_{j_{p-1},i_{p-1}}(s_2) ...$$

$$\int_0^{s_{p-1}} e^{(-1)^p \Delta \varphi(s_p)} K_{j_1,i_1}(s_p) ds_p ... ds_1$$

where p is the number of integrals, $j_s = \overline{0,m}$, $i_s = \overline{0, k_s - 1}$, $\Delta \varphi(t) = \int_0^t (\lambda_2(s) - \lambda_1(s)) ds$.

Note that $\sigma_{s,p}^{(j_1,i_1,...,j_p,i_p)}(t,\varepsilon)$ satisfy the system.

$$\begin{cases} \varepsilon \dot{\sigma}_{1,p}^{(j_1,i_1,...,j_p,i_p)}(t,\varepsilon) = \\ \quad \lambda_1(t) \sigma_{1,p}^{(j_1,i_1,...,j_p,i_p)}(t,\varepsilon) + \varepsilon K_{j_p,i_p} \sigma_{2,p-1}^{(j_1,i_1,...,j_{p-1},i_{p-1})}(t,\varepsilon), \\ \varepsilon \dot{\sigma}_{2,p}^{(j_1,i_1,...,j_p,i_p)}(t,\varepsilon) = \\ \quad \lambda_2(t) \sigma_{2,p}^{(j_1,i_1,...,j_p,i_p)}(t,\varepsilon) + \varepsilon K_{j_p,i_p} \sigma_{1,p-1}^{(j_1,i_1,...,j_{p-1},i_{p-1})}(t,\varepsilon). \end{cases} \tag{6}$$

Instead of the desired solution $u(t,\varepsilon)$ of problem (2), we study the vector function $z(t,\sigma,\varepsilon)$ such that its restriction coincides with the desired solution.

$$z(t,\sigma,\varepsilon)\Big|_{\sigma = \sigma_{s,p}^{(j_1,i_1,...,j_p,i_p)}(t,\varepsilon)} = u(t,\varepsilon), \quad s = 1,2, \quad p = \overline{0,\infty} \tag{7}$$

In view of (2), (5), (6), we can write the problem for $z(t, \sigma, \varepsilon)$. Using the complex differentiation formula

$$\frac{dz}{dt} = \dot{z} + \sum_{s=1}^{2} \sum_{p=0}^{\infty} \sum_{j_1,\ldots,j_p=0}^{m} \sum_{i_1,\ldots,i_p=0}^{k_1-1,\ldots,k_p-1} \left(\frac{\lambda_s}{\varepsilon} \sigma_{s,p}^{(j_1,i_1,\ldots,j_p,i_p)} + K_{j_p,i_p} \sigma_{3-s,p-1}^{(j_1,i_1,\ldots,j_{p-1},i_{p-1})} \right) \frac{\partial z}{\partial \sigma_{s,p}^{(j_1,i_1,\ldots,j_p,i_p)}}, \tag{8}$$

we get task for the extended function $z(t, \sigma, \varepsilon)$.

$$\begin{cases} A(t)z - \sum_{s=1}^{2} \sum_{p=0}^{\infty} \sum_{j_1,\ldots,j_p=0}^{m} \sum_{i_1,\ldots,i_p=0}^{k_1-1,\ldots,k_p-1} (\lambda_s \sigma_{s,p}^{(j_1,i_1,\ldots,j_p,i_p)} - \\ -K_{j_p,i_p} \sigma_{3-s,p-1}^{(j_1,i_1,\ldots,j_{p-1},i_{p-1})}) \frac{\partial z}{\partial \sigma_{s,p}^{(j_1,i_1,\ldots,j_p,i_p)}} = \varepsilon \dot{z} - h(t), \\ z(0, 0, \varepsilon) = u^0. \end{cases} \tag{9}$$

By convention, we assume that if the term containing $p - 1 < 0$ in the index, then this term is equal to zero. To solve this problem, we introduce the space of non-resonant solutions $\hat{E}$.

$$\hat{E} = \bigoplus_{s=1}^{2} \bigoplus_{p=0}^{\infty} \bigoplus_{j_1,\ldots,j_p=0}^{m} \bigoplus_{i_1,\ldots,i_p=0}^{k_1-1,\ldots,k_p-1} E \otimes \{\sigma_{s,p}^{(j_1,i_1,\ldots,j_p,i_p)}\} \oplus E$$

The element $\hat{z} \in \hat{E}$ has the form

$$\hat{z} = \sum_{s=1}^{2} \sum_{p=0}^{\infty} \sum_{j_1,\ldots,j_p=0}^{m} \sum_{i_1,\ldots,i_p=0}^{k_1-1,\ldots,k_p-1} z_{s,p}^{(j_1,i_1,\ldots,j_p,i_p)} \otimes \{\sigma_{s,p}^{(j_1,i_1,\ldots,j_p,i_p)}\} + w$$

where $z_{s,p}^{(j_1,i_1,\ldots,j_p,i_p)}, w \in E$.

Here $\oplus$ is the symbol of the direct sum of linear spaces; $\otimes$ is the symbol of the tensor product. We introduce the operators generated by problem (9).

$$\begin{cases} \mathscr{L}_0 = \bigoplus_{s=1}^{2} \bigoplus_{p=0}^{\infty} \bigoplus_{j_1,\ldots,j_p=0}^{m} \bigoplus_{i_1,\ldots,i_p=0}^{k_1-1,\ldots,k_p-1} (A(t) - \lambda_s(t)) \otimes \{\sigma_{s,p}^{(j_1,i_1,\ldots,j_p,i_p)} \\ \frac{\partial}{\partial \sigma_{s,p}^{(j_1,i_1,\ldots,j_p,i_p)}}\} \oplus A(t) \\ \mathscr{L}_1 = \bigoplus_{s=1}^{2} \bigoplus_{p=0}^{\infty} \bigoplus_{j_1,\ldots,j_p=0}^{m} \bigoplus_{i_1,\ldots,i_p=0}^{k_1-1,\ldots,k_p-1} \bigoplus_{j_{p+1}=0}^{m} \bigoplus_{i_{p+1}=0}^{k_{p+1}-1} K_{j_{p+1},i_{p+1}}(t) \\ \otimes \{\sigma_{3-s,p}^{(j_1,i_1,\ldots,j_p,i_p)} \frac{\partial}{\partial \sigma_{s,p}^{(j_1,i_1,\ldots,j_{p+1},i_{p+1})}}\} \\ Gz = z(0,0,\varepsilon). \end{cases} \tag{10}$$

Operator actions are recorded as

$$\begin{cases} \mathscr{L}_0 \hat{z}(t) = \sum_{s=1}^{2} \sum_{p=0}^{\infty} \sum_{j_1,\ldots,j_p=0}^{m} \sum_{i_1,\ldots,i_p=0}^{k_1-1,\ldots,k_p-1} (A(t) - \lambda_s(t)) z_{s,p}^{(j_1,i_1,\ldots,j_p,i_p)}(t) \\ \otimes \{\sigma_{s,p}^{(j_1,i_1,\ldots,j_p,i_p)}\} + A(t)w \\ \mathscr{L}_1 \hat{z}(t) = \sum_{s=1}^{2} \sum_{p=0}^{\infty} \sum_{j_1,\ldots,j_p=0}^{m} \sum_{i_1,\ldots,i_p=0}^{k_1-1,\ldots,k_p-1} (\sum_{j_{p+1}=0}^{m} \sum_{i_{p+1}=0}^{k_{p+1}-1} K_{j_{p+1},i_{p+1}}(t) \\ z_{3-s,p+1}^{(j_1,i_1,\ldots,j_{p+1},i_{p+1})}(t)) \otimes \{\sigma_{s,p}^{(j_1,i_1,\ldots,j_p,i_p)}\} \\ G\hat{z} = z(0,0,\varepsilon). \end{cases} \quad (11)$$

In addition, we introduce spectral projectors

$$\begin{cases} \hat{P}_{k,s,p}^{(j_1,i_1,\ldots,j_p,i_p)}(t) = P_k(t) \otimes \{\sigma_{s,p}^{(j_1,i_1,\ldots,j_p,i_p)} \frac{\partial}{\partial \sigma_{s,p}^{(j_1,i_1,\ldots,j_p,i_p)}}\} \\ \hat{\pi}_{k,s,p}^{(j_0,i_0,j_1,i_1,\ldots,j_p,i_p)}(t) = (-1)^{i_0} P_k(t) < \delta^{(i_0)}(t-t_{j_0}), P_k(t) \cdot > \otimes \\ \{\sigma_{s,p}^{(j_1,i_1,\ldots,j_p,i_p)} \frac{\partial}{\partial \sigma_{s,p}^{(j_1,i_1,\ldots,j_p,i_p)}}\} \\ \hat{P}_0(t) = \sum_{s=1}^{2} \sum_{p=0}^{\infty} \sum_{j_1,\ldots,j_p=0}^{m} \sum_{i_1,\ldots,i_p=0}^{k_1-1,\ldots,k_p-1} \hat{P}_{s,s,p}^{(j_1,i_1,\ldots,j_p,i_p)}(t) - \\ \text{kernel projection operator.} \mathscr{L}_0 \\ \hat{\pi}_0^{(j_0,i_0)}(t) = \sum_{s=1}^{2} \sum_{p=0}^{\infty} \sum_{j_1,\ldots,j_p=0}^{m} \sum_{i_1,\ldots,i_p=0}^{k_1-1,\ldots,k_p-1} \hat{\pi}_{3-s,s,p}^{(j_0,i_0,j_1,i_1,\ldots,j_p,i_p)}(t) \\ j_0 = \overline{0,m}, \quad i_0 = \overline{0, k_{j_0} - 1}. \end{cases} \quad (12)$$

The action of projectors on the element $\hat{z} \in \hat{E}$ is written as

(a) $\hat{P}_{k,s,p}^{(j_1,i_1,\ldots,j_p,i_p)}(t)\hat{z}(t) = (\lambda_k(t) - \lambda_s(t)) P_k(t) z_{s,p}^{(j_1,i_1,\ldots,j_p,i_p)}(t) \otimes \sigma_{s,p}^{(j_1,i_1,\ldots,j_p,i_p)}$

(b) $\hat{\pi}_{k,s,p}^{(j_0,i_0,j_1,i_1,\ldots,j_p,i_p)}(t)\hat{z}(t) = P_k(t)(\frac{d}{dt})^{i_0} P_k(t) z_{s,p}^{(j_1,i_1,\ldots,j_p,i_p)}(t)|_{t=t_{j_0}} \otimes \sigma_{s,p}^{(j_1,i_1,\ldots,j_p,i_p)}$

Using the operators (10), problem (9) can be rewritten in the space $\hat{E}$ as follows:

$$\begin{cases} \mathscr{L}_0 \hat{z} = \varepsilon \mathscr{L}_1 \hat{z} + \varepsilon \dot{\hat{z}} - h(t), \\ G\hat{z} = u^0. \end{cases} \quad (13)$$

Problem (13) is regular with respect to ε. Therefore, the solution (13) will be defined as a regular series in powers of ε, i.e.,

$$\hat{z} = \sum_{k=0}^{\infty} \varepsilon^k \hat{z}_k \quad (14)$$

Substituting series (14) into problem (13), we obtain the following series of iterative problems:

$$\begin{cases} \mathscr{L}_0 \hat{z}_0 = -h(t), \quad G\hat{z}_0 = u^0. \\ \mathscr{L}_0 \hat{z}_k = \mathscr{L}_1 \hat{z}_{k-1} + \dot{\hat{z}}_{k-1}, \quad k = \overline{1,\infty} \\ G\hat{z}_k = 0. \end{cases} \quad (15)$$

3. Solvability of Iterative Problems

In order to solve the iterative problems (15) we formulate solvability theorem for equations of the form $\mathscr{L}_0(t)\hat{z} = \hat{h}(t)$ in the space $\hat{E}$. The following normal solvability theorem holds.

Theorem 1. *Let $\hat{E}$ contain the equation*

$$\mathscr{L}_0 \hat{z} = \hat{h}(t). \qquad (16)$$

and conditions (1)–(5) of problem (2) are satisfied.

Then Equation (16) is solvable in $\hat{E}$ if and only if

(1) $\hat{P}_0(t)\hat{h}(t) = 0, \forall t \in [0, T]$;
(2) $\hat{\pi}_0^{(j_0,i_0)}(t)\hat{h}(t) = 0$, $j_0 = \overline{0, m}$, $i_0 = \overline{0, k_{j_0} - 1}$.

Proof. Let the equation be solvable. We approach the equation by the operator $\hat{P}_0(t)$. Since $\hat{P}_0(t)\mathscr{L}_0(t) = 0$, so $\hat{P}_0(t)\hat{h}(t) = 0$. It follows that

$$\hat{h}(t) = \sum_{s=1}^{2} \sum_{p=0}^{\infty} \sum_{j_1,\ldots,j_p=0}^{m} \sum_{i_1,\ldots,i_p=0}^{k_1-1,\ldots,k_p-1} P_{3-s}(t) h_{s,p}^{(j_1,i_1,\ldots,j_p,i_p)}(t) \otimes \{\sigma_{s,p}^{(j_1,i_1,\ldots,j_p,i_p)}\} + h_0(t). \qquad (17)$$

We act with the operator $\hat{\pi}_0$. Then as

$$\hat{\pi}_0^{(j_0,i_0)}(t)\mathscr{L}_0 \hat{z} =$$
$$\sum_{s=1}^{2} \sum_{p=0}^{\infty} \sum_{j_1,\ldots,j_p=0}^{m} \sum_{i_1,\ldots,i_p=0}^{k_1-1,\ldots,k_p-1} P_{3-s}(t_{j_0}) (\tfrac{d}{dt})^{i_0} ((\lambda_{3-s}(t) - \lambda_s(t)))$$
$$P_{3-s}(t) z_{s,p}^{(j_1,i_1,\ldots,j_p,i_p)}(t))|_{t=t_{j_0}} \otimes \{\sigma_{s,p}^{(j_1,i_1,\ldots,j_p,i_p)}\} = 0,$$

then

$$\hat{\pi}_0^{(j_0,i_0)}(t)\hat{h}(t) = \sum_{s=1}^{2} \sum_{p=0}^{\infty} \sum_{j_1,\ldots,j_p=0}^{m} \sum_{i_1,\ldots,i_p=0}^{k_1-1,\ldots,k_p-1} P_{3-s}(t) (\tfrac{d}{dt})^{i_0} (P_{3-s}(t)$$
$$h_{s,p}^{(j_1,i_1,\ldots,j_p,i_p)}(t))|_{t=t_{j_0}} \otimes \{\sigma_{s,p}^{(j_1,i_1,\ldots,j_p,i_p)}\} = 0 \qquad (18)$$

Sufficiency is obvious. □

As a result, we get a solution

$$\hat{z}(t) = \hat{P}_0(t)\hat{z}(t) + \sum_{s=1}^{2} \sum_{p=0}^{\infty} \sum_{j_1,\ldots,j_p=0}^{m} \sum_{i_1,\ldots,i_p=0}^{k_1-1,\ldots,k_p-1}$$
$$(A(t) - \lambda_s(t))^{-1} P_{3-s}(t) h_{s,p}^{(j_1,i_1,\ldots,j_p,i_p)}(t) \otimes \{\sigma_{s,p}^{(j_1,i_1,\ldots,j_p,i_p)}\} + A^{-1}(t)h_0(t). \qquad (19)$$

Here $\hat{P}_0(t)\hat{z}(t)$ is an arbitrary vector from the kernel of the operator $\mathscr{L}_0(t)$.

Theorem 2. *Let the task be given in $\hat{E}$*

$$\mathscr{L}_0 \hat{z} = 0, G\hat{z} = 0. \qquad (20)$$

and the conditions of Theorem 1. are satisfied. Then, when

$$\begin{cases} \hat{P}_0(t)(\mathscr{L}_1 \hat{z} + \dot{\hat{z}}) = 0 \\ \hat{\pi}_0^{(j_0,i_0)}(t)(\mathscr{L}_1 \hat{z} + \dot{\hat{z}}) = 0, \\ j_0 = \overline{0,m}, \; i_0 = \overline{0, k_{j_0} - 1}. \end{cases} \qquad (21)$$

the solution of problem (20) is unique and identically equal to zero.

Proof. The solution of the equation from system (20) can be written as

$$\hat{z}_0 = \sum_{s=1}^{2} \sum_{p=0}^{\infty} \sum_{i_1,\ldots,i_p=0}^{m} \sum_{i_1,\ldots,i_p=0}^{k_1-1,\ldots,k_p-1} P_s(t) z_{s,p}^{(j_1,i_1,\ldots,j_p,i_p)}(t) \otimes \{\sigma_{s,p}^{(j_1,i_1,\ldots,j_p,i_p)}\} \qquad (22)$$

We calculate

$$\mathscr{L}_1 \dot{z}_0 + \dot{z}_0 = \sum_{s=1}^{2} \sum_{p=0}^{\infty} \sum_{j_1,\ldots,j_p=0}^{m} \sum_{i_1,\ldots,i_p=0}^{k_0-1,\ldots,k_p-1} [\frac{d}{dt}(P_s(t) z_{s,p}^{(j_1,i_1,\ldots,j_p,i_p)}(t)) + \\ + \sum_{j_{p+1}=0}^{m} \sum_{i_{p+1}=0}^{k_{p+1}-1} K_{j_{p+1},i_{p+1}}(t) P_{3-s}(t) z_{3-s,p+1}^{(j_1,i_1,\ldots,j_{p+1},i_{p+1})}(t))] \otimes \sigma_{s,p}^{(j_1,i_1,\ldots,j_p,i_p)}. \tag{23}$$

where $P_s(t) z_{s,p}^{(j_1,i_1,\ldots,j_p,i_p)}(t)$ is an arbitrary eigenvector of the operator $A(t)$.

We submit (22) to the initial condition. Moreover, we take into account that $\sigma_{s,p}^{(j_1,i_1,\ldots,j_p,i_p)}(0,\varepsilon) = 0$, $p \geq 1$. Then we have $P_s(0) z_{s,0}(0) = 0$, $s = 1,2$.

As $\hat{P}_0(t)(\mathscr{L}_1 \hat{z}(t) + \dot{\hat{z}}(t)) = 0$, from here we get a series of Cauchy problems

$$\begin{cases} p = 0 \\ \frac{d}{dt}(P_s(t) z_{s,0}(t)) = \dot{P}_s(t)(P_s(t) z_{s,0}(t)) \\ P_s(0) z_{s,0}(0) = 0, s = 1,2 \\ \text{при } p \geq 1 \\ \frac{d}{dt}(P_s(t) z_{s,p}^{(j_1,i_1,\ldots,j_p,i_p)}(t)) = \dot{P}_s(t)(P_s(t) z_{s,p}^{(j_1,i_1,\ldots,j_p,i_p)}(t)) \\ P_s(t_{j_p}) z_{s,p}^{(j_1,i_1,\ldots,j_p,i_p)}(t_{j_p}) = ?\text{not defined at the moment} \end{cases} \tag{24}$$

To solve the arising Cauchy problems, we introduce resolving operators

$$\begin{cases} \frac{d}{dt} U_s(t,\tau) = \dot{P}_s(t) U_s(t,\tau) \\ U_s(t,t) = I, s = 1,2 \end{cases} \tag{25}$$

The solution for $p = 0$ will be $P_s(t) z_{s,0}(t) = U_s(t,0) P_s(0) z_{s,0}(0) \equiv 0$. To determine the initial conditions for Cauchy problems (24) $p \geq 1$, we calculate

$$\hat{\pi}_0^{(j_0,i_0)}(t)(\mathscr{L}_1 \hat{z} + \dot{\hat{z}}) = 0, \quad j_0 = \overline{0,m}, \quad i_0 = \overline{0,k_{j_0}-1}.$$

$$\begin{cases} \sum_{j_{p+1}=0}^{m} \sum_{i_{p+1}=0}^{k_{p+1}-1} P_s(t_{j_0})(\frac{d}{dt})^{i_0}(K_{j_{p+1},i_{p+1}}(t) P_s(t) z_{s,p+1}^{(j_1,i_1,\ldots,j_{p+1},i_{p+1})}(t))|_{t=t_{j_0}} \\ = P_s(t_{j_0})(\frac{d}{dt})^{i_0}(\dot{P}_s(t) P_{3-s}(t) z_{3-s,p}^{(j_1,i_1,\ldots,j_p,i_p)}(t))|_{t=t_{j_0}}, p \geq 0 \\ j_0 = \overline{0,m}, \quad i_0 = \overline{0,k_0-1}. \end{cases} \tag{26}$$

From system (26) we obtain the initial conditions for the remaining Cauchy problems. To do this, sequentially sorting i_p, we obtain

$$\begin{cases} p = 0 \\ j_0 = \overline{0,m}, i_0 = 0, s = 1,2 \\ P_s(t_{j_0}) z_{s,1}^{(j_0,0)}(t_{j_0}) = \\ \dot{P}_s(t_{j_0}) P_{3-s}(t_{j_0}) z_{3-s,0}(t_{j_0})) = 0 \end{cases} \tag{27}$$

From here when

$$p = 1 \; j_1 = \overline{0,m}, \; i_1 = 0, s = 1,2, \\ P_s(t) z_{s,1}^{(j_1,0)}(t) = U_s(t,t_{j_1}) P_s(t_{j_1}) z_{s,1}^{(j_1,0)}(t_{j_1}) \equiv 0.$$

$$\begin{cases} p=0 \\ j_0=\overline{0,m},\ i_0=1,\ s=1,2 \\ P_s(t_{j_0})z_{s,1}^{(j_0,1)}(t_{j_0})= \\ -P_s(t_{j_0})\frac{d}{dt}(P_s(t)z_{s,0}(t))|_{t=t_{j_0}}+ \\ +P_s(t_{j_0})\frac{d}{dt}(\dot{P}_s(t)P_{3-s}(t)z_{3-s,0}(t))|_{t=t_{j_0}}=0 \end{cases} \quad (28)$$

$$p=1\ j_1=\overline{0,m},\ i_1=1, s=1,2,$$
$$P_s(t)z_{s,1}^{(j_1,1)}(t)=U_s(t,t_{j_1})P_s(t_{j_1})z_{s,1}^{(j_1,1)}(t_{j_1})\equiv 0.$$

$$\begin{cases} p=0 \\ j_0=\overline{0,m},\ i_0=n,\ n=\overline{0,k_0-1}, s=1,2 \\ P_s(t_{j_0})z_{s,1}^{(j_0,n)}(t_{j_0})= \\ -\sum_{j_1=0}^{m}\sum_{i=0}^{n-1}C_n^i P_s(t_{j_0})(\frac{d}{dt})^{n-i}(P_s(t)z_{s,0}(t))|_{t=t_{j_0}}+ \\ +P_s(t_{j_0})(\frac{d}{dt})^n(\dot{P}_s(t)P_{3-s}(t)z_{3-s,0}(t))|_{t=t_{j_0}}=0 \end{cases} \quad (29)$$

From here when

$$p=1\ j_1=\overline{0,m},\ i_1=n,$$
$$P_s(t)z_{s,1}^{(j_1,n)}(t)=U_s(t,t_{j_1})P_s(t_{j_1})z_{s,1}^{(j_1,n)}(t_{j_1})\equiv 0.$$

Having considered the case $p=1$ (recall that $p=1$ means the order of multiple singular integrals), we pass to the case $p=2$. Since the initial conditions for p are expressed in terms of the initial conditions for $p-1$, by induction we prove that the initial conditions are equal to zero for any p. From there,

$$P_s(t)z_{s,p}^{(j_1,i_1,\ldots,j_p,i_p)}(t))=U_s(t,t_{j_p})P_s(t_{j_p})z_{s,p}^{(j_1,i_1,\ldots,j_p,i_p)}(t_{j_p}))\equiv 0.$$

Therefore, the solution of problem (20) is identically equal to zero. □

4. Construction of a Formal Asymptotic Solution

We apply Theorems I and II to solve iterative problems (15). We write the problem at the iterative step
ε^0

$$\mathcal{L}_0\hat{z}_0=-h(t), G\hat{z}_0=u^0. \quad (30)$$

Or component-wise

$$\begin{cases} (A(t)-\lambda_s(t))z_{s,p,0}^{(j_1,i_1,\ldots,j_p,i_p)}(t)=0, \\ A(t)w_0(t)=-h(t), \\ z_{1,0,0}(0)+z_{2,0,0}(0)+w_0(0)=u^0 \\ z_{s,p,0}^{(j_1,i_1,\ldots,j_p,i_p)}(t_{j_p}), p\geq 1, s=1,2\ \text{(determined in the decision process} \\ \text{iterative tasks)} \end{cases}$$

Solution (30) can be written as

$$\hat{z}_0=\sum_{s=1}^{2}\sum_{p=0}^{\infty}\sum_{j_1,\ldots,j_p=0}^{m}\sum_{i_1,\ldots,i_p=0}^{k_1-1,\ldots,k_p-1}P_s(t)z_{s,p,0}^{(j_1,i_1,\ldots,j_p,i_p)}(t)\otimes\{\sigma_{s,p}^{(j_1,i_1,\ldots,j_p,i_p)}(t,\varepsilon)\} \quad (31)$$
$$-A^{-1}(t)h(t).$$

where $P_s(t)z_{s,p,0}^{(j_1,i_1,\ldots,j_p,i_p)}(t)$ is an arbitrary eigenvector of the operator $A(t)$.

We obey (31) the initial condition. Moreover, we take into account that $\sigma_{s,p}^{(j_1,i_1,\ldots,j_p,i_p)}(0,\varepsilon) = 0$, $p \geq 1$. Then we have $P_1(0)z_{1,0,0}(0) + P_2(0)z_{2,0,0}(0) - A^{-1}(0)h(0) = u^0$.
From here $P_s(0)z_{s,0,0}(0) = P_s(0)u^0 + \frac{P_s(0)h(0)}{\lambda_s(0)}$, $s = 1,2$.

Initial conditions for $P_s(t_{j_p})z_{s,p,0}^{(j_1,i_1,\ldots,j_p,i_p)}(t_{j_p})$ are determined from the solvability conditions of the iterative system at the first iterative step. Thus, at the zero iteration step, we obtained

$$\begin{cases} \hat{z}_0 = \sum_{s=1}^{2} \sum_{p=0}^{\infty} \sum_{j_1,\ldots,j_p=0}^{m} \sum_{i_1,\ldots,i_p=0}^{k_1-1,\ldots,k_p-1} P_s(t)z_{s,p,0}^{(j_1,i_1,\ldots,j_p,i_p)}(t) \\ \otimes \{\sigma_{s,p}^{(j_1,i_1,\ldots,j_p,i_p)}\} - A^{-1}(t)h(t). \\ P_s(0)z_{s,0,0}(0) = P_s(0)u^0 + \frac{P_s(0)h(0)}{\lambda_s(0)}, \quad s = 1,2 \end{cases} \quad (32)$$

The task in the first iterative step ε has the form

$$\begin{cases} \mathscr{L}_0 \hat{z}_0 = \dot{\hat{z}}_0 + \mathscr{L}_1 \hat{z}_0, \\ G\hat{z}_1 = 0. \end{cases} \quad (33)$$

solvable in $\hat{E}$ if the right-hand side satisfies the conditions of Theorem I. First, we calculate

$$\mathscr{L}_1 \hat{z}_0 + \dot{\hat{z}}_0 = \sum_{s=1}^{2} \sum_{p=0}^{\infty} \sum_{j_1,\ldots,j_p=0}^{m} \sum_{i_1,\ldots,i_p=0}^{k_0-1,\ldots,k_p-1} [\frac{d}{dt}(P_s(t)z_{s,p,0}^{(j_1,i_1,\ldots,j_p,i_p)}(t)) + \\ + \sum_{j_{p+1}=0}^{m} \sum_{i_{p+1}=0}^{k_{p+1}-1} K_{j_{p+1},i_{p+1}}(t)z_{3-s,p+1,0}^{(j_1,i_1,\ldots,j_{p+1},i_{p+1})}(t)] \otimes \sigma_{s,p}^{(j_1,i_1,\ldots,j_p,i_p)}(t,\varepsilon) \\ - \frac{d}{dt}A^{-1}(t)h(t). \quad (34)$$

By writing (33) at the first iteration step by components and taking into account (34), we obtain a series of problems:

$$\begin{cases} (A(t) - \lambda_s(t))z_{s,p,1}^{(j_1,i_1,\ldots,j_p,i_p)}(t) = \frac{d}{dt}(P_s(t)z_{s,p,0}^{(j_1,i_1,\ldots,j_p,i_p)}(t)) + \\ + \sum_{j_{p+1}=0}^{m} \sum_{i_{p+1}=0}^{k_{p+1}-1} K_{j_{p+1},i_{p+1}}(t)P_{3-s}(t)z_{3-s,p+1,0}^{(j_1,i_1,\ldots,j_{p+1},i_{p+1})}(t), \\ z_{1,0,1}(0) + z_{2,0,1}(0) = ((A^{-1}(t)\frac{d}{dt})^2 \int_0^t h(s)ds)(0), \\ z_{s,p,1}(t_{j_p}), p \geq 1, s = 1,2 \text{ (determined in the decision process iterative tasks).} \end{cases} \quad (35)$$

From the solvability conditions (35) and taking into account (32), we obtain a series of Cauchy problems

$$\begin{cases} p = 0 \\ \frac{d}{dt}(P_s(t)z_{s,0,0}(t)) = \dot{P}_s(t)(P_s(t)z_{s,0,0}(t)) \\ P_s(0)z_{s,0,0}(0) = P_s(0)u^0 + \frac{P_s(0)h(0)}{\lambda_s(0)}, s = 1,2 \\ \text{при } p \geq 1 \\ \frac{d}{dt}(P_s(t)z_{3,p,0}^{(j_1,i_1,\ldots,j_p,i_p)}(t)) = \dot{P}_s(t)(P_s(t)z_{s,p,0}^{(j_1,i_1,\ldots,j_p,i_p)}(t)) \\ P_s(t_{j_p})z_{s,p,0}^{(j_1,i_1,\ldots,j_p,i_p)}(t_{j_p}) = ?\text{not defined at the moment} \end{cases} \quad (36)$$

To determine the initial conditions for the Cauchy problems (36) $p \geq 1$, we calculate

$$\hat{\pi}_0^{(j_0,i_0)}(t)(\mathscr{L}_1 \hat{z}_0 + \dot{\hat{z}}_0) = 0, \quad j_0 = \overline{0,m}, \quad i_0 = \overline{0,k_{j_0}-1}.$$

Then we get

$$\sum_{j_{p+1}=0}^{m} \sum_{i_{p+1}=0}^{k_{p+1}-1} P_s(t_{j_0}) (\tfrac{d}{dt})^{i_0} (K_{j_{p+1},i_{p+1}}(t) P_s(t) z_{s,p+1,0}^{(j_1,i_1,\ldots,j_{p+1},i_{p+1})}(t)))|_{t=t_{j_0}}$$
$$= P_s(t_{j_0}) (\tfrac{d}{dt})^{i_0} (\dot{P}_s(t) P_{3-s}(t) z_{3-s,p,0}^{(j_1,i_1,\ldots,j_p,i_p)}(t)))|_{t=t_{j_0}} \qquad (37)$$

Going over i_0 sequentially for a fixed p, we get

$$\begin{cases}
j_0 = \overline{0,m}, \ i_0 = 0, \ s = 1,2 \\[4pt]
P_s(t_{j_0}) z_{s,p+1,0}^{(j_1,i_1,\ldots,j_p,i_p,j_0,0)}(t_{j_0}) = \\
\dot{P}_s(t_{j_0}) P_{3-s}(t_{j_0}) z_{3-s,p,0}^{(j_1,i_1,\ldots,j_p,i_p)}(t_{j_0})) \\[6pt]
j_0 = \overline{0,m}, \ i_0 = 1, \ s = 1,2 \\[4pt]
P_s(t_{j_0}) z_{s,p+1,0}^{(j_1,i_1,\ldots,j_p,i_p,j_0,1)}(t_{j_0}) = \\
-P_s(t_{j_0}) \tfrac{d}{dt} (P_s(t) z_{s,p,0}^{(j_1,i_1,\ldots,j_p,i_p)}(t))|_{t=t_{j_0}} + \\
+P_s(t_{j_0}) \tfrac{d}{dt} (\dot{P}_s(t) P_{3-s}(t) z_{3-s,p,0}^{(j_1,i_1,\ldots,j_p,i_p)}(t))|_{t=t_{j_0}} \\[6pt]
j_0 = \overline{0,m}, \ i_0 = n, \ n = \overline{0,k_0-1}, s = 1,2 \\[4pt]
P_s(t_{j_0}) z_{s,p+1,0}^{(j_1,i_1,\ldots,j_p,i_p,j_0,n)}(t_{j_0}) = \\
- \sum_{j_{p+1}=0}^{m} \sum_{i=0}^{n-1} C_n^i P_s(t_{j_0}) (\tfrac{d}{dt})^{n-i} (P_s(t) z_{s,p+1,0}^{(j_1,i_1,\ldots,j_p,i_p,j_0,i)}(t))|_{t=t_{j_0}} + \\
+ P_s(t_{j_0}) (\tfrac{d}{dt})^n (\dot{P}_s(t) P_{3-s}(t) z_{3-s,p,0}^{(j_1,i_1,\ldots,j_p,i_p)}(t))|_{t=t_{j_0}}
\end{cases} \qquad (38)$$

Since the initial conditions for $p+1$ are expressed in terms of the initial conditions for p, we thereby prove by induction that the initial conditions are defined for any p.

After determining the initial conditions from system (38), we obtain solutions to system (36).

$$\begin{cases}
P_s(t) z_{s,0,0}(t) = U_s(t,0)(P_s(0)u^0 + \tfrac{P_s(0)h(0)}{\lambda_s(0)}), \\
P_s(t) z_{s,p,0}^{(j_1,i_1,\ldots,j_p,i_p)}(t) = U_s(t,t_{j_p}) P_s(t_{j_p}) z_{s,p,0}^{(j_1,i_1,\ldots,j_p,i_p)}(t_{j_p}) \\
s = 1,2, \ p = \overline{0,\infty}, \ j_p = \overline{0,m}, \ i_p = \overline{0, k_{j_p}-1}
\end{cases} \qquad (39)$$

Custom vectors $P_s(t) z_{s,p,1}^{(j_1,i_1,\ldots,j_p,i_p)}(t)$ and the remaining initial conditions are on the second iteration step. By this scheme, all terms of the solution to problem (30) are found.

5. Evaluation of the Remainder Term

Let the terms of the double series (14) as a result of solving iterative problems be defined for $0 \leqslant q \leqslant n, 0 \leqslant p \leqslant r$, здесь q—iterative step in ε, а p—orders of singular integrals. We write the relation for the remainder $R_{n,r}(t,\varepsilon)$:

We rewrite the series (14) in the form

$$\hat{z}(t,\varepsilon) = \sum_{q=0}^{n} \varepsilon^q \sum_{s=1}^{2} \sum_{p=0}^{r} \sum_{j_1,\ldots,j_p=0}^{m} \sum_{i_1,\ldots,i_p=0}^{k_1-1,\ldots,k_p-1} z_{s,p,q}^{(j_1,i_1,\ldots,j_p,i_p)}(t) \sigma_{s,p}^{(j_1,i_1,\ldots,j_p,i_p)}(t,\varepsilon) + $$
$$+ \sum_{q=0}^{n} \varepsilon^q w_q(t) + \varepsilon^{n+1} \cdot R_{n,r}(t,\varepsilon) \qquad (40)$$

Substituting (40) into (2) and taking into account the iterative problems, we obtain the problem for the remainder term $R_{n,r}(t,\varepsilon)$,

$$\begin{cases} \varepsilon \dot{R}_{n,r}(t,\varepsilon) - A(t)R_{n,r}(t,\varepsilon) = -H(t,\varepsilon), \\ R_{n,r}(0,\varepsilon) = 0, \end{cases} \quad (41)$$

where

$$H(t,\varepsilon) = \sum_{s=1}^{2}\Big[\sum_{p=0}^{r-1}\sum_{j_1,\ldots,j_p=0}^{m}\sum_{i_1,\ldots,i_p=0}^{k_1-1,\ldots,k_p-1}(\dot{z}_{s,p,n}^{(j_1,i_1,\ldots,j_p,i_p)}(t)+ \\ \sum_{j_{p+1}=0}^{m}\sum_{i_{p+1}=0}^{k_{p+1}-1}K_{j_{p+1},i_{p+1}}(t)z_{3-s,p+1,n}^{(j_1,i_1,\ldots,j_{p+1},i_{p+1})}(t))\sigma_{s,p}^{(j_1,i_1,\ldots,j_p,i_p)}(t,\varepsilon)+ \\ \sum_{j_1,\ldots,j_r=0}^{m}\sum_{i_1,\ldots,i_r=0}^{k_1-1,\ldots,k_r-1}\dot{z}_{s,r,n}^{(j_1,i_1,\ldots,j_r,i_r)}(t)\sigma_{s,r}^{(j_1,i_1,\ldots,j_r,i_r)}(t,\varepsilon)\Big] + \dot{w}_n(t) \quad (42)$$

As follows from conditions (5) on the spectrum in problem (2) and estimates of the integrals $\sigma_{s,p}^{(j_1,i_1,\ldots,j_p,i_p)}(t,\varepsilon)$ in the lemma 1, the right-hand side of (41) has the estimate

$$\|H(t,\varepsilon)\|_{C[0,T]} \leqslant C, \quad \forall (t,\varepsilon) \in [0,T] \times (0,\varepsilon_0].$$

We write the solution (41) in the form:

$$R_{k,m} = \frac{1}{\varepsilon}\int_0^t U_\varepsilon(t,s)H(s,\varepsilon)\,ds,$$

where $U_\varepsilon(t,s)$ is the resolving operator, which is a solution to the Cauchy problem:

$$\varepsilon \dot{U}_\varepsilon(t,s) = A(t)U_\varepsilon(t,s), \quad U_\varepsilon(t,s)\Big|_{s=t} = I.$$

It follows from the conditions 5) on the spectrum in problem (2) that $U_\varepsilon(t,s)$ limited to $[0,T] \times [0,t], \varepsilon \in (0,\varepsilon_0]$:

$$\|U_\varepsilon(t,s)\|_{C[0,T]} \leq C.$$

Therefore, from the relation

$$R_{k,m} = -U_\varepsilon(t,s)A^{-1}(s)H(s,\varepsilon)\Big|_0^t + \int_0^t U_\varepsilon(t,s)\frac{d}{ds}A^{-1}(s)H(s,\varepsilon)\,ds = \\ = -A^{-1}(t)H(t,\varepsilon) + U_\varepsilon(t,0)A^{-1}(0)H(0,\varepsilon) + \int_0^t U_\varepsilon(t,s)\frac{d}{ds}A^{-1}(s)H(s,\varepsilon)\,ds,$$

we get:

$$\|R_{n,r}\|_{C[0,T]} \leqslant C.$$

From these estimates, we move on to the following.

Theorem 3. *On estimating the remainder (asymptotic convergence).*

Let Cauchy problem (2) be given and conditions (1) ÷ (5) be satisfied. Then the estimate is correct

$$\Big\|u(t,\varepsilon) - \sum_{q=0}^{n}\varepsilon^q\sum_{s=1}^{2}\sum_{p=0}^{r}\sum_{j_1,\ldots,j_p=0}^{m}\sum_{i_1,\ldots,i_p=0}^{k_1-1,\ldots,k_p-1}z_{s,p,q}^{(j_1,i_1,\ldots,j_p,i_p)}(t) \\ \sigma_{s,p}^{(j_1,i_1,\ldots,j_p,i_p)}(t,\varepsilon) + \sum_{q=0}^{n}\varepsilon^q w_q(t)\Big\|_{C[0,T]} \leqslant C \cdot \varepsilon^{n+1}, \quad (43)$$

where $\mathbb{C} \geqslant 0$ is a constant independent of ε, a $z_{s,p,q}^{(j_1,i_1,\ldots,j_p,i_p)}(t), w_q(t)$ obtained from solving iterative problems for $0 \leqslant q \leqslant n, 0 \leqslant p \leqslant r$.

Theorem 4. *About the passage to the limit.*

Let problem (2) be given and the conditions (1) ÷ (5). Then:

(a)   *If* $\operatorname{Re} \lambda_i \leq -\delta < 0$, *then*

$$\lim_{\varepsilon \to 0} u(t,\varepsilon) = -A^{-1}(t)h(t), \qquad t \in [\delta_0, T], \quad \delta_0 > 0\text{— arbitrarily small;}$$

(b)   *If* $\operatorname{Re} \lambda_i \leqslant 0$, *then* $\forall \, \varphi(t) \in C^\infty[0,T]$

$$\lim_{\varepsilon \to 0} \int_0^T \left(u(t,\varepsilon) + A^{-1}(t)h(t) \right) \varphi(t) dt = 0.$$

Proof.

(a)   The statement of this section directly follows from estimates of the integrals $\sigma_{s,p}^{(j_1,i_1,\ldots,j_p,i_p)}(t,\varepsilon)$ in the Lemma 1.

(b)   In this case $\sigma_{s,p}^{(j_1,i_1,\ldots,j_p,i_p)}(t,\varepsilon)$ are rapidly oscillating functions and the proof of the passage to the limit in the weak sense follows from the Riemann–Lebesgue lemma. □

6. Application

$$\begin{cases} \varepsilon \dot{J}(t) = \begin{pmatrix} \lambda_1(t) & 0 \\ 0 & \lambda_2(t) \end{pmatrix} J(t) + \varepsilon K(t) \begin{pmatrix} 0 & 1 \\ 1 & 0 \end{pmatrix} J(t), \\ J(0) = \begin{pmatrix} 1 \\ 1 \end{pmatrix}. \end{cases} \qquad (44)$$

where $J(t) = \begin{pmatrix} J_1(t) \\ J_2(t) \end{pmatrix}$ is a vector-function. The system (44) in the general case is not explicitly solved. We find the solution (44) by the method of successive approximations.

Lemma 1. *The solution (44) is represented as a uniformly converging series on* $[0,T] \times (0, \varepsilon_0]$, *which admits an estimate*

(a)   *If* $\operatorname{Re} \lambda_i \leq -\delta < 0$, *then*

$$\|J\|_{C[0,T]} \leqslant e^{-\delta t/\varepsilon} C;$$

(b)   *If* $\operatorname{Re} \lambda_i \leqslant 0$, *then*

$$\|J\|_{C[0,T]} \leqslant C,$$

where $C > 0$ *is a constant independent of* ε.

Proof. Solving (44) by the method of successive approximations, we obtain:

$$J(t) = \exp\left(\tfrac{1}{\varepsilon}\Lambda_0^t\right) + \exp\left(\tfrac{1}{\varepsilon}\Lambda_0^t\right) \int_0^t K(s) \exp\left(-\tfrac{1}{\varepsilon}\Lambda_0^s\right) \cdot T \cdot \exp\left(\tfrac{1}{\varepsilon}\Lambda_0^s\right) ds +$$

$$+ \exp\left(\tfrac{1}{\varepsilon}\Lambda_0^t\right) \int_0^t K(s) \exp\left(-\tfrac{1}{\varepsilon}\Lambda_0^s\right) \cdot T \cdot \exp\left(\tfrac{1}{\varepsilon}\Lambda_0^s\right) \int_0^s K(s_1) \exp\left(-\tfrac{1}{\varepsilon}\Lambda_0^{s_1}\right) \cdot T \cdot$$

$$\cdot \exp\left(\tfrac{1}{\varepsilon}\Lambda_0^{s_1}\right) ds_1\, ds + \ldots,$$

$$\text{here} \quad T = \begin{pmatrix} 0 & 1 \\ 1 & 0 \end{pmatrix},\ \Lambda_0^t = \begin{pmatrix} \varphi_1(t) & 0 \\ 0 & \varphi_2(t) \end{pmatrix}.$$

Using property

$$T \cdot \exp\left(\frac{1}{\varepsilon}\begin{pmatrix} \varphi_1(t) & 0 \\ 0 & \varphi_2(t) \end{pmatrix}\right) = \exp\left(\frac{1}{\varepsilon}\begin{pmatrix} \varphi_2(t) & 0 \\ 0 & \varphi_1(t) \end{pmatrix}\right) \cdot T,$$

we get

$$J(t) = \exp\left(\tfrac{1}{\varepsilon}\Lambda_0^t\right) + \exp\left(\tfrac{1}{\varepsilon}\Lambda_0^t\right) \int_0^t K(s) \exp\left(\tfrac{1}{\varepsilon}\triangle_0^s\right) T\, ds + \exp\left(\tfrac{1}{\varepsilon}\Lambda_0^t\right)$$

$$\int_0^t K(s) \exp\left(\tfrac{1}{\varepsilon}\triangle_0^s\right) \int_0^s K(s_1) \exp\left(-\tfrac{1}{\varepsilon}\triangle_0^{s_1}\right) ds_1\, ds + \exp\left(\tfrac{1}{\varepsilon}\Lambda_0^t\right) \int_0^t K(s) \qquad (45)$$

$$\exp\left(\tfrac{1}{\varepsilon}\triangle_0^s\right) \int_0^s K(s_1) \exp\left(-\tfrac{1}{\varepsilon}\triangle_0^{s_1}\right) \cdot \int_0^{s_1} K(s_2) \exp\left(\tfrac{1}{\varepsilon}\triangle_0^{s_2}\right) T\, ds_2\, ds_1\, ds + \ldots,$$

here

$$\triangle_0^t = \begin{pmatrix} \varphi_2(t) - \varphi_1(t) & 0 \\ 0 & \varphi_1(t) - \varphi_2(t) \end{pmatrix}.$$

Component by component (45) looks like

$$J_1(t) = \exp\left(\tfrac{1}{\varepsilon}\varphi_1(t)\right) + \exp\left(\tfrac{1}{\varepsilon}\varphi_1(t)\right) \int_0^t K(s) \exp\left(\tfrac{1}{\varepsilon}\Delta\varphi(s)\right) ds +$$

$$+ \exp\left(\tfrac{1}{\varepsilon}\varphi_1(t)\right) \int_0^t K(s) \exp\left(\tfrac{1}{\varepsilon}\Delta\varphi(s)\right) \int_0^s K(s_1) \exp\left(-\tfrac{1}{\varepsilon}\Delta\varphi(s_1)\right) ds_1\, ds + \ldots,$$

$$J_2(t) = \exp\left(\tfrac{1}{\varepsilon}\varphi_2(t)\right) + \exp\left(\tfrac{1}{\varepsilon}\varphi_2(t)\right) \int_0^t K(s) \exp\left(-\tfrac{1}{\varepsilon}\Delta\varphi(s)\right) ds + \qquad (46)$$

$$+ \exp\left(\tfrac{1}{\varepsilon}\varphi_2(t)\right) \int_0^t K(s) \exp\left(-\tfrac{1}{\varepsilon}\Delta\varphi(s)\right) \int_0^s K(s_1) \exp\left(\tfrac{1}{\varepsilon}\Delta\varphi(s_1)\right) ds_1\, ds + \ldots$$

The uniform convergence of the series (46) follows from the estimates: $k_0 + k_1 + \ldots + k_m = n$

(a) $\operatorname{Re}\lambda_i \leqslant -\delta < 0, M = \max|M_{j,i}|,\ M_{j,i} = \max|K_{j,i}(t)|\quad t \in [0, T]$.

$$\left|e^{\varphi_1(t)/\varepsilon}\right| \leq e^{-\delta t/\varepsilon},$$

$$\left|e^{\varphi_1(t)/\varepsilon} \int_0^t K(s) e^{(\varphi_2(s)-\varphi_1(s))/\varepsilon}\, ds\right| \leq Mn \int_0^t e^{\int_s^t \operatorname{Re}\lambda_1(s_1)ds_1/\varepsilon + \int_0^s \operatorname{Re}\lambda_2(s_2)ds_2/\varepsilon}\, ds \leq$$

$$\leq Mn \int_0^t e^{-\delta(t-s+s)/\varepsilon}\, ds = e^{-\delta t/\varepsilon} \cdot Mnt,$$

....

$$\left| e^{\varphi_1(t)/\varepsilon} \int_0^t K(s_1) e^{\Delta\varphi(s_1)/\varepsilon} \int_0^{s_1} K(s_2) e^{-\Delta\varphi(s_2)/\varepsilon} \ldots \int_0^{s_{p-1}} K(s_p) e^{(-1)^p \Delta\varphi(s_n)/\varepsilon} \, ds_p \ldots ds_1 \right| \leq$$

$$\leq (Mn)^p \int_0^t \int_0^{s_1} \ldots \int_0^{s_{p-1}} e^{(\varphi_1(t)-\varphi_1(s_1)+\ldots+(-1)^{p+1}\varphi_1(s_p))/\varepsilon} \cdot$$

$$\cdot e^{(\varphi_2(s_1)-\varphi_2(s_2)+\ldots+(-1)^p \varphi_2(s_p))/\varepsilon} \, ds_1 \ldots ds_p \leq e^{-\delta t/\varepsilon} \cdot \frac{(Mnt)^p}{p!}$$

In this way

$$|J_1(t,\varepsilon)| \leq e^{-\delta t/\varepsilon} \cdot e^{Mnt} \leq e^{MnT} \cdot e^{-\delta t/\varepsilon};$$

similarly

$$|J_2(t,\varepsilon)| \leq e^{MnT} \cdot e^{-\delta t/\varepsilon}.$$

(b) $\operatorname{Re}\lambda_i \leq 0, \quad t \in [0,T]$

$$|J_i(t)| \leq e^{MnT}, \quad i = 1, 2.$$

Therefore, the series (46) converge uniformly in ε и t on $[0,T] \times (0,\varepsilon_0]$. In addition, it is easy to verify that the rows withstand operator action $\varepsilon \frac{d}{dt}$ to any degree. □

7. Example

The simplest case of a weak turning point is the point of the first order, i.e., $\lambda_2(t) - \lambda_1(t) = ta(t), a(t) \neq 0$. The solution to a singularly perturbed Cauchy problem in this case is described in [6]. Here we give a solution to the Cauchy problem

$$\varepsilon \dot{u}(t,\varepsilon) = A(t)u(t,\varepsilon) + h(t), u(t,\varepsilon) = u^0. \tag{47}$$

and conditions are met:

(1)   Conditions (1) ÷ (3) from (2);
(2)   Weak point condition

$$\lambda_2(t) - \lambda_1(t) = t(t-1)a(t), \ a(t) \neq 0, \lambda_2(t) \neq \lambda_1(t) \ \forall t \in (0,1) \cup (1,T];$$

moreover, the geometric multiplicity of the eigenvalues is algebraic for any $t \in [0,T]$;

(3) $\lambda_i(t) \neq 0, \ \operatorname{Re}\lambda_i(t) = 0 \ \forall t \in [0,T]$.

The Lagrange–Sylvester interpolation polynomial for function f(t) given at the nodes $t_0 = 0, t_1 = 1$ has the form $K(t)f(t) = (1-t(f(0) + tf(1), K_0(t) = 1 - t, K_1(t) = t$.

Singularities are described as

$$\varphi_i(t) = \tfrac{1}{\varepsilon}\int_0^t \lambda_i(s)ds,$$
$$\sigma_{i,0}(t,\varepsilon) = e^{\varphi_i(t)}, \quad i = 1,2$$
$$\sigma_{1,1}^{(0)}(t,\varepsilon) = e^{\varphi_1(t)} \int_0^t e^{\Delta\varphi(s_1)} K_0(s_1) ds_1,$$
$$\sigma_{1,1}^{(1)}(t,\varepsilon) = e^{\varphi_1(t)} \int_0^t e^{\Delta\varphi(s_1)} K_1(s_1) ds_1,$$
$$\sigma_{2,1}^{(0)}(t,\varepsilon) = e^{\varphi_2(t)} \int_0^t e^{-\Delta\varphi(s_1)} K_0(s_1) ds_1,$$
$$\sigma_{2,1}^{(1)}(t,\varepsilon) = e^{\varphi_2(t)} \int_0^t e^{-\Delta\varphi(s_1)} K_1(s_1) ds_1, \tag{48}$$
$$\sigma_{1,p}^{(j_1,\ldots,j_p)}(t,\varepsilon) = e^{\varphi_1(t)} \int_0^t e^{\Delta\varphi(s_1)} K_{j_p}(s_1) \int_0^{s_1} e^{-\Delta\varphi(s_2)} K_{j_{p-1}}(s_2)\ldots$$
$$\int_0^{s_{p-1}} e^{(-1)^{p-1}\Delta\varphi(s_p)} K_{j_1}(s_p) ds_p \ldots ds_1,$$
$$\sigma_{2,p}^{(j_1,\ldots,j_p)}(t,\varepsilon) = e^{\varphi_2(t)} \int_0^t e^{-\Delta\varphi(s_1)} K_{j_p}(s_1) \int_0^{s_1} e^{\Delta\varphi(s_2)} K_{j_{p-1}}(s_2)\ldots$$
$$\int_0^{s_{p-1}} e^{(-1)^p \Delta\varphi(s_p)} K_{j_1}(s_p) ds_p \ldots ds_1.$$

where p is the number of integrals, $j_s = \overline{0,1}, \Delta\varphi(t) = \int_0^t (\lambda_2(s) - \lambda_1(s))ds$.

The solution is sought in the form

$$u(t,\varepsilon) = \sum_{k=0}^{\infty} \varepsilon^k \Big[\sum_{s=1}^{2} \sum_{p=0}^{\infty} \sum_{j_1,\ldots,j_p=0}^{1} z_{s,p,k}^{(j_1,\ldots,j_p)}(t) \sigma_{s,p}^{(j_1,\ldots,j_p)}(t,\varepsilon) + w_k(t) \Big]; \tag{49}$$

Substituting (49) into (41), we obtain a series of iterative problems.
$k = 0$

$$\begin{cases} (A(t) - \lambda_s(t)) z_{s,p,0}^{(j_1,\ldots,j_p)}(t) = 0, \\ A(t)w_0(t) = -h(t). \\ z_{1,0,0}(0) + z_{2,0,0}(0) = A^{-1}(0)h(0) + u^0 \\ z_{s,p,0}^{(j_1,\ldots,j_p)}(t_{j_p}), p \geq 1, s = 1,2, \ j_p = 0,1 \text{ determined in the decision process} \\ \text{iterative tasks in 1 step.} \end{cases} \tag{50}$$

Solution (50) can be written as

$$\hat{z}_0 = \sum_{s=1}^{2} \sum_{p=0}^{\infty} \sum_{j_1,\ldots,j_p=0}^{1} P_s(t) z_{s,p,0}^{(j_1,\ldots,j_p)}(t) \sigma_{s,p}^{(j_1,\ldots,j_p)}(t,\varepsilon) - A^{-1}(t)h(t) \tag{51}$$

Being undefined in this step, $P_s(t)z_{s,p,0}^{(j_1,\ldots,j_p)}(t)$ are found from the solvability theorem at the first iterative step: $k = 1$

$$\begin{cases} (A(t) - \lambda_s(t))z_{s,p,1}^{(j_1,\ldots,j_p)}(t) = \frac{d}{dt}(P_s(t)z_{s,p,0}^{(j_1,\ldots,j_p)}(t)) + \\ + \sum_{j_{p+1}=0}^{1} K_{j_{p+1}}(t)P_{3-s}(t)z_{3-s,p+1,0}^{(j_1,\ldots,j_{p+1})}(t), \\ A(t)w_1(t) = -\frac{d}{dt}(A^{-1}(t)h(t)) \\ z_{1,0,1}(0) + z_{2,0,1}(0) = ((A^{-1}(t)\frac{d}{dt})^2 \int_0^t h(s)ds)(0), \\ z_{s,p,1}(t_{j_p}), p \geq 1, s = 1,2, \; j_p = 0,1 \text{ (determined in the process} \\ \text{solving iterative problems in step 2).} \end{cases} \quad (52)$$

By the solvability theorem, we obtain Cauchy problems for determining the terms of the zeroth approximation of a solution.

$$\begin{cases} p = 0 \\ \frac{d}{dt}(P_s(t)z_{s,0,0}(t)) = \dot{P}_s(t)(P_s(t)z_{s,0,0}(t)) \\ P_s(0)z_{s,0,0}(0) = P_s(0)u^0 + \frac{P_s(0)h(0)}{\lambda_s(0)}, \; s = 1,2 \end{cases} \quad (53)$$

Solution (53) can be written as

$$P_s(t)z_{s,0,0}(t) = U_s(t,0)(P_s(0)u^0 + \frac{P_s(0)h(0)}{\lambda_s(0)}), s = 1,2 \quad (54)$$

To find the initial condition for the term $p = 1$, $P_s(t)z_{s,1,0}^{(j_1)}(t)$, we expand the first Equation (52) for understanding in more detail

$$(A(t) - \lambda_s(t))z_{s,0,1}(t) = \frac{d}{dt}(P_s(t)z_{s,0,0}(t)) + K_0(t)P_{3-s}z_{3-s,1,0}^{(0)}(t) + K_1(t)P_{3-s}z_{3-s,1,0}^{(1)}(t)$$

As $\lambda_2(t) - \lambda_1(t) = t(t-1)a(t)$, and then weaning on $P_{3-s}(t)$, putting $t = 0$ and redesignating $3 - s$ на s, we get

$$P_s(0)z_{s,1,0}^{(0)}(0) = \dot{P}_s(0)(P_{3-s}(0)z_{3-s,0,0})(0)), \; s = 1,2$$

Putting $t = 1$, we get $P_s(1)z_{s,1,0}^{(1)}(1) = \dot{P}_s(1)(P_{3-s}(1)z_{3-s,0,0}(1)), \; s = 1,2$
By induction, we obtain that for $p > 1, k = 0$

$$P_s(0)z_{s,p+1,0}^{(j_1,\ldots,j_p,0)}(0) = \dot{P}_s(0)(P_{3-s}(0)z_{3-s,p,0}^{(j_1,\ldots,j_p)}(0)), \; s = 1,2$$

$$P_s(1)z_{s,p+1,0}^{(j_1,\ldots,j_p,1)}(1) = \dot{P}_s(1)(P_{3-s}(1)z_{3-s,p,0}^{(j_1,\ldots,j_p)}(1)), \; s = 1,2$$

From here are the terms of the zeroth approximation of the solution to the problem (42)

$$P_s(t)z_{s,p,0}^{(j_1,\ldots,j_p)}(t) = U_s(t,t_{j_p})P_s(t_{j_p})z_{s,p,0}^{(j_1,\ldots,j_p)}(t_{j_p}), \; j_p = 0,1$$

As a result, the leading term of the asymptotics of solution (42) has the form

$$u_{gl}(t,\varepsilon) = \sum_{s=1}^{2} U_s(t,0)(P_s(0)u^0 + \frac{P_s(0)h(0)}{\lambda_s(0)})e^{\frac{1}{\varepsilon}\varphi_s(t)} +$$
$$\sum_{s=1}^{2}\sum_{p=1}^{\infty}\sum_{j_1,\ldots,j_p=0}^{1} U_s(t,t_{j_p})P_s(t_{j_p})z_{s,p,0}^{(j_1,\ldots,j_p)}(t_{j_p})\sigma_{s,p}^{(j_1,\ldots,j_p)}(t,\varepsilon) - A^{-1}(t)h(t)$$

8. Conclusions

In this paper, we considered the singularly perturbed Cauchy problem in the presence of a "weak" turning point for the limit operator $A(t)$. It turns out that the nature of the "weak" turning point strongly affects the structure of the regularizing functions describing the singular dependence of the solution on the parameter ε. In contrast to the singularly perturbed Cauchy problems with a "simple" rational point, the rotations of the limit operator, in which the singularities are described by a finite number of regularizing functions, in this case there are countably many such functions. This greatly complicates the asymptotic behavior of the solution of the Cauchy problem at $\varepsilon \to 0$. Understanding the nature of the "weak" turning point will help in studying future studies of the fractional "weak" turning point and the "strong" turning point. We expect that our approach can be adapted to other related problems, for example, in the context of constructing difference schemes when solving equations numerically.

Funding: This research received no external funding.

Conflicts of Interest: The author declares no conflict of interest.

References

1. Lomov, S.A. *Introduction to the General Theory of Singular Perturbations*; Nauka: Moscow, Russia, 1981.
2. Bobodzhanov, A.A.; Safonov, V.F. Regular asymptotic behavior of solutions of partial differential integro-differential equations with rapidly changing kernels. *Ufa Math. J.* **2018**, *10*, 3–13. [CrossRef]
3. Lioville, J. Second memoire sur le development des fonctions en series dont divers termes sont assujettis, a une meme equation. *J. Math. Pure Appl.* **1837**, *2*, 16–35.
4. Eliseev, A.G.; Lomov, S.A. Theory of singular perturbations in the case of spectral singularities of the limit operator. *Math. USSR-Sb.* **1986**, *173*, 544–557. [CrossRef]
5. Eliseev, A.G.; Ratnikova, T.A. A singularly perturbed Cauchy problem in the presence of a rational "simple" turning point at the limit operator. *Differ. Equ. Control Process.* **2019**, *3*, 63–73.
6. Eliseev, A.G. Regularized Solution of a singularly perturbed Cauchy problem in the presence of irrational simple turning point at the limit operator. *Differ. Equ. Control Process.* **2020**, *2*, 15–32.
7. Eliseev, A.G.; Kirichenko, P.V. Paper Solution of a singularly perturbed Cauchy problem in the presence of a "weak" turning point at the limit operator. *Sib. Electron. Mate. Izv.* **2020**, *17*, 51–60. [CrossRef]

© 2020 by the authors. Licensee MDPI, Basel, Switzerland. This article is an open access article distributed under the terms and conditions of the Creative Commons Attribution (CC BY) license (http://creativecommons.org/licenses/by/4.0/).

Article

The Modified Helmholtz Equation on a Regular Hexagon—The Symmetric Dirichlet Problem

Konstantinos Kalimeris [1,*] and Athanassios S. Fokas [1,2,3]

[1] Research Center of Pure and Applied Mathematics, Academy of Athens, 11527 Athens, Greece; tf227@cam.ac.uk
[2] Department of Applied Mathematics and Theoretical Physics, University of Cambridge, Cambridge CB3 0WA, UK
[3] Viterbi School of Engineering, University of Southern California, Los Angeles, CA 90089-2560, USA
* Correspondence: kk364@cam.ac.uk or kkalimeris@academyofathens.gr

Received: 9 June 2020; Accepted: 27 June 2020; Published: 28 July 2020

Abstract: Using the unified transform, also known as the Fokas method, we analyse the modified Helmholtz equation in the regular hexagon with symmetric Dirichlet boundary conditions; namely, the boundary value problem where the trace of the solution is given by the same function on each side of the hexagon. We show that if this function is odd, then this problem can be solved in closed form; numerical verification is also provided.

Keywords: unified transform; modified Helmholtz equation; global relation

1. Introduction

We analyse the modified Helmholtz equation in a regular hexagon using the unified transform, also known as the Fokas method. This method was introduced by one of the authors [1], for analysing integrable nonlinear partial differential equations (PDEs) [2]. Later, it was realized that it also yields novel results for linear evolution PDEs [3]; results in this direction are obtained by several authors [4–10]. Furthermore, it yields new integral representations for the solution of linear elliptic PDEs in polygonal domains [11], which in the case of simple domains can be used to obtain the analytical solution of several problems which apparently cannot be solved by the standard methods [12,13]. Recently, researchers utilised the integral representations provided by the Fokas method for the local and global wellposedness analysis of Korteweg-de Vries and nonlinear Schrödinger type PDEs [14–18], as well as for studying problems from control theory [19].

The Fokas method is based on two basic ingredients:

(1) a global relation, which is an algebraic equation that involves certain transforms of all (known and unknown) boundary values.
(2) an integral representation of the solution, which involves transforms of all boundary values.

For linear PDEs, the Fokas method involves the following:

- Given a PDE, define its formal adjoint and construct a one parameter family of solutions of this equation.
- By employing the given PDE and its adjoint, obtain a one parameter family of equations in conservation form. This family, together with Green's theorem, yield the global relation.
- The above family also gives rise to a certain closed differential form. The spectral analysis of this form gives rise to a scalar Riemann–Hilbert problem, which consequently yields an integral representation of the solution. This representation involves integral transforms of all the boundary values, and since some of them are not prescribed as boundary conditions, this form of solution is not yet effective.

- The explicit solution of the problem is derived by determining the contribution of the unknown boundary values to the integral representation. This can be achieved by using the global relation, as well as equations obtained from the global relation through certain invariant transformations.

The global relation has had important analytical and numerical implications: first, it has led to novel analytical formulations of a variety of important physical problems from water waves [20–26] to three-dimensional layer scattering [27]. Second, it has led to the development of new techniques for the Laplace, modified Helmholtz, Helmholtz, biharmonic equations, both analytical [28–35] and numerical [36–47].

The above analytical solutions are given in terms of infinite series; this is to be contrasted to other techniques based on the eigenvalues of the Laplace operator that yield the solution as a bi-infinite series. The eigenvalues of the Laplace operator for the Dirichlet, Neumann and Robin problems in the interior of an equilateral triangle were first obtained by Lamé in 1833 [48]; these results have also been derived using the Fokas method [49]. Completeness for the associated expansions for the Dirichlet and Neumann problems was obtained in [50–53] using group theoretic techniques. McCartin rederived these results [54,55] and studied the connection of the eigen-structure of the equilateral triangle with that of the regular hexagon [56]. The above remarks indicate that the existing literature is based on an implicit way for deriving the solution of specific BVPs of the regular hexagon in terms of bi-infinite series. This is to be contrasted with our work which presents a direct approach for deriving explicit integral representations of the solution of a special BVP on the regular hexagon; the extension of the current methodology to more general problems is under investigation.

Organisation of the Paper

In Section 2 we implement the four steps discussed above for solving the symmetric Dirichlet problem of the modified Helmholtz equation in a regular hexagon. The main achievement of this work is presented in Section 3 and concerns the fourth step: our analysis yields the solution for the case of odd symmetric Dirichlet data in the closed form (34). We study the case of even symmetric data in Section 4, where we derive the expression (37); this expression in addition to known terms also involves an unknown term. In Section 5, Figures 1 and 2 depict the numerical verification of the main result of Section 3; also, Figures 7 and 8 indicate that the unknown term in the expression (37) is exponentially small in the high frequency limit, and hence this result provides an excellent approximation for this physically significant limit.

2. The Basic Elements

The equation investigated here is the modified Helmholtz equation in the interior of the regular hexagon, D, namely,

$$q_{xx} + q_{yy} - 4\beta^2 q = 0, \quad (x,y) \in D, \tag{1}$$

where $q(x,y)$ is a real valued function and $\beta > 0$.

Using complex coordinates,

$$z = x + iy, \quad \bar{z} = x - iy,$$

Equation (1) becomes

$$q_{z\bar{z}} - \beta^2 q = 0. \tag{2}$$

2.1. The Global Relation and the Integral Representation of the Solution in the Interior of a Convex Polygon

We first derive the global relation:
The formal adjoint also satisfies the modified Helmholtz equation

$$\tilde{q}_{z\bar{z}} - \beta^2 \tilde{q} = 0. \tag{3}$$

Multiplying Equation (2) by $\tilde{q}$, Equation (3) by q and subtracting, we find

$$\tilde{q}q_{z\bar{z}} - q\tilde{q}_{z\bar{z}} = 0, \tag{4}$$

or equivalently

$$\frac{\partial}{\partial z}(\tilde{q}q_{\bar{z}} - \tilde{q}_{\bar{z}}q) + \frac{\partial}{\partial \bar{z}}(q\tilde{q}_z - q_z\tilde{q}) = 0. \tag{5}$$

Using in (5) the special solution $\tilde{q} = e^{-i\beta\left(kz - \frac{\bar{z}}{k}\right)}$ and employing Green's theorem, we obtain

$$\int_{\partial\Omega} W(z,\bar{z},k) = 0, \quad k \in \mathbb{C}, \tag{6}$$

where W is defined by

$$W(z,\bar{z},k) = e^{-i\beta\left(kz - \frac{\bar{z}}{k}\right)}\left[(q_z + ik\beta q)\,dz - \left(q_{\bar{z}} + \frac{\beta}{ik}q\right)d\bar{z}\right], \quad k \in \mathbb{C}. \tag{7}$$

Suppose that Ω is the polygon defined via the points $z_1, z_2, \ldots, z_n, z_{n+1} = z_1$. Then (6) gives the following global relation for the modified Helmholtz in this polygon:

$$\sum_{j=1}^{n} \hat{q}_j(k) = 0, \quad k \in \mathbb{C}, \tag{8}$$

where $\{\hat{q}_j(k)\}_1^n$ are defined by

$$\hat{q}_j(k) = \int_{z_j}^{z_{j+1}} e^{-i\beta\left(kz - \frac{\bar{z}}{k}\right)}\left[(q_z + ik\beta q)\,dz - \left(q_{\bar{z}} + \frac{\beta}{ik}q\right)d\bar{z}\right], \quad k \in \mathbb{C}, \tag{9}$$

or equivalently (in local coordinates) by

$$\hat{q}_j(k) = \int_{z_j}^{z_{j+1}} e^{-i\beta\left(kz - \frac{\bar{z}}{k}\right)}\left[iq_N^{(j)}(s) + i\beta\left(\frac{1}{k}\frac{d\bar{z}}{ds} + k\frac{dz}{ds}\right)q^{(j)}(s)\right]ds, \; k \in \mathbb{C},$$
$$j = 1, \ldots, n. \tag{10}$$

In Equation (10) we have used the identity

$$q_z\,dz - q_{\bar{z}}\,d\bar{z} = iq_N\,ds,$$

where s is the arclength on the boundary $z(s) = x(s) + iy(s)$ of the polygon and q_N denotes the derivative in the outward normal direction to the boundary of the polygon.

In order to derive the integral representation of the solution one has to implement the spectral analysis of the differential form

$$d\left[e^{-i\beta\left(kz - \frac{\bar{z}}{k}\right)}\mu(z,k)\right] = W(z,\bar{z},k), \quad k \in \mathbb{C}. \tag{11}$$

This procedure yields the following theorem, proven in [6]:

Theorem 1. *Let Ω be the interior of a convex closed polygon in the complex z-plane, with corners $z_1, \ldots, z_n, z_{n+1} \equiv z_1$. Assume that there exists a solution $q(z,\bar{z})$ of the modified Helmholtz equation, i.e., of Equation (2) with $\beta > 0$, valid on Ω, and suppose that this solution has sufficient smoothness on the boundary of the polygon.*

Then, q can be expressed in the form

$$q(z, \bar{z}) = \frac{1}{4\pi i} \sum_{j=1}^{n} \int_{l_j} e^{i\beta\left(kz - \frac{\bar{z}}{k}\right)} \hat{q}_j(k) \frac{dk}{k}, \quad (12)$$

where $\{\hat{q}_j(k)\}_1^n$ are defined by (10), and $\{l_j\}_1^n$ are the rays in the complex k-plane

$$l_j = \{k \in \mathbb{C} : \arg k = -\arg(z_{j+1} - z_j)\}, \quad j = 1, \ldots, n$$

oriented from zero to infinity.

Observe that the solution given in (12) is given in terms of $\{\hat{q}_j\}_1^n$ which involve integral transforms of both q and q_N on the boundary, i.e., both known and unknown functions.

2.2. The Dirichlet Problem on a Regular Hexagon

Let $D \subset \mathbb{C}$ be the interior of a regular hexagon with vertices $\{z_j\}_1^6$,

$$z_1 = \frac{l\sqrt{3}}{2} - i\frac{l}{2} = l e^{\frac{-i\pi}{6}} \quad \text{and} \quad z_j = \omega^{j-1} z_1, \quad (13)$$

where l is the length of the side and $\omega = e^{\frac{i\pi}{3}}$. The sides $\{(z_j, z_{j+1})\}_1^6$, $z_7 \equiv z_1$ will be referred to as sides $\{(j)\}_1^6$.

For the sides $\{(j)\}_1^6$ the following parametrizations will be used:

$$z_1(s) = \frac{l\sqrt{3}}{2} + is, \quad z_j(s) = \left(\frac{l\sqrt{3}}{2} + is\right)\omega^{j-1}, \quad s \in \left[-\frac{l}{2}, \frac{l}{2}\right].$$

The general Dirichlet problem can be uniquely decomposed to 6 simpler Dirichlet problems, by employing the decomposition

$$q^{(j)}(s) = \sum_{i=1}^{6} \omega^{(j-1)(i-1)} g_i(s), \quad j = 1, \ldots, 6, \quad s \in \left[-\frac{l}{2}, \frac{l}{2}\right];$$

indeed the determinant of the matrix $\left[\omega^{(j-1)(i-1)}\right]_{i,j=1,\ldots,6}$ is non-zero (Its value is $216 = 6^3$, and for the general case Det $\left[\omega^{(j-1)(i-1)}\right]_{i,j=1,\ldots,n} = i^{\frac{2-n(n+1)}{2}} n^{n/2}$).

The existence and uniqueness of the solution of the modified Helmholtz equation shows that it is sufficient to solve each one of the above Dirichlet problems. The first of them is the symmetric Dirichlet problem, where the value $g_1(s) = d(s)$ is prescribed on each side. This symmetric problem is analysed in the next section.

2.3. The Symmetric Dirichlet Problem

The problem analysed in this subsection is the symmetric Dirichlet problem for the modified Helmholtz equation in the regular hexagon ($\Omega \equiv D$). Let $d(s)$ be a real function with sufficient smoothness and compatibility at the vertices of the hexagon, i.e., $d\left(\frac{l}{2}\right) = d\left(-\frac{l}{2}\right)$. We prescribe the boundary conditions

$$q^{(j)}(s) = d(s), \quad s \in \left[-\frac{l}{2}, \frac{l}{2}\right], \quad j = 1, \ldots, 6.$$

The above 'symmetry' property also holds for the Neumann boundary values. This fact is the consequence of the following three observations:

- The modified Helmholtz operator $\left(\frac{\partial^2}{\partial z \partial \bar{z}} - \beta^2 \mathrm{Id}\right)$ is invariant under the transformation $z \to \omega z$, namely under rotation of $2\pi/3$. Since the Dirichlet data are invariant under this rotation, then the (unique) solution $q(z, \bar{z})$ of the Helmholtz equation is also invariant under this rotation.
- If q is invariant under this transformation, then the differential form $q_z dz$ is also invariant under the transformation $z \to \omega z$:

$$\frac{\partial q(z)}{\partial z} dz = \frac{\partial q(\omega z)}{\partial z} dz = \frac{\partial (\omega z)}{\partial z} \frac{\partial q(\omega z)}{\partial (\omega z)} \frac{1}{\omega} d(\omega z) = \frac{\partial q(\omega z)}{\partial (\omega z)} d(\omega z).$$

- Evaluating the above differential form on each side we obtain

$$q_z dz = \frac{1}{2}\left(\dot{q}^{(j)}(s) + i q_N^{(j)}(s)\right) ds = \frac{1}{2}\left(\dot{d}'(s) + i q_N^{(j)}(s)\right) ds,$$

where the second equality is a direct consequence of the fact that the Dirichlet data are invariant under this rotation.

Thus,

$$q_N^{(j)}(s) = u(s), \quad s \in \left[-\frac{l}{2}, \frac{l}{2}\right], \quad j = 1, \ldots, 6.$$

Applying the parametrization of the regular hexagon on Equation (10) we obtain:

$$\hat{q}_1(k) = \hat{q}(k), \qquad \hat{q}_j(k) = \hat{q}\left(\omega^{j-1} k\right), \qquad j = 1, \ldots, 6, \tag{14}$$

with

$$\hat{q}(k) = E(-ik)[iU(k) + D(k)], \tag{15}$$

where $E(k)$, $D(k)$ and $U(k)$ are defined by

$$E(k) = e^{\beta \left(k + \frac{1}{k}\right) \frac{l\sqrt{3}}{2}},$$

$$D(k) = \beta \left(\frac{1}{k} - k\right) \int_{-\frac{l}{2}}^{\frac{l}{2}} e^{\beta \left(k + \frac{1}{k}\right) s} d(s) ds, \tag{16}$$

$$U(k) = \int_{-\frac{l}{2}}^{\frac{l}{2}} e^{\beta \left(k + \frac{1}{k}\right) s} u(s) ds, \qquad k \in \mathbb{C}.$$

The function $D(k)$ is known, whereas the unknown function $U(k)$ contains the unknown Neumann boundary value $u(s) = q_N$.

Using (15), the global relation (8) takes the form

$$\begin{aligned} & E(-ik)U(k) + E(-i\omega k)U(\omega k) + E(-i\omega^2 k)U(\omega^2 k) \\ & + E(ik)U(-k) + E(i\omega k)U(-\omega k) + E(i\omega^2 k)U(-\omega^2 k) = iG(k), \quad k \in \mathbb{C}, \end{aligned} \tag{17}$$

where the known function $G(k)$ is defined by

$$G(k) = \sum_{j=1}^{6} E\left(-i\omega^{j-1} k\right) D\left(\omega^{j-1} k\right), \qquad k \in \mathbb{C}. \tag{18}$$

The integral representation (12) of the solution takes the form

$$q(z, \bar{z}) = \frac{1}{4\pi i} \sum_{j=1}^{6} \int_{l_j} e^{i\beta \left(kz - \frac{\bar{z}}{k}\right)} E(-i\omega^{j-1} k) \left[D\left(\omega^{j-1} k\right) + iU\left(\omega^{j-1} k\right)\right] \frac{dk}{k}, \tag{19}$$

where $\{l_j\}_1^6$ are the rays defined by

$$l_j = \left\{k \in \mathbb{C} : \arg k = \frac{11-2j}{6}\pi\right\}, \quad j = 1,\ldots,6, \tag{20}$$

oriented from zero to infinity. The principal arguments of $\{l_1, l_2, l_3, l_4, l_5, l_6\}$ are $\left\{\frac{3\pi}{2}, \frac{7\pi}{6}, \frac{5\pi}{6}, \frac{\pi}{2}, \frac{\pi}{6}, \frac{11\pi}{6}\right\}$, respectively.

Since the function $d(s)$ can be uniquely written as a sum of an odd and an even function, we will only consider two particular cases:

(i) the odd case, $d(-s) = -d(s)$;
(ii) the even case $d(-s) = d(s)$.

The solution and the Neumann boundary values inherit the analogous properties:

(i) in the odd case, $u(-s) = -u(s)$, which yields $U(-k) = -U(k)$;
(ii) in the even case, $u(-s) = u(s)$, which yields $U(-k) = U(k)$ for all $k \in \mathbb{C}$.

3. Derivation of the Solution for the Symmetric Odd Case

In what follows we will show that the contribution of the unknown functions $\{U(\omega^{j-1}k)\}_1^6$ to the solution representation (19) can be computed explicitly.

Applying the condition $U(-k) = -U(k)$ in (17) we obtain the equation

$$\Delta(ik)U(k) + \Delta(i\omega k)U(\omega k) + \Delta(i\omega^2 k)U(\omega^2 k) = -iG(k), \quad k \in \mathbb{C}, \tag{21}$$

where $G(k)$ is given in (18) and $\Delta(k)$ is defined by

$$\Delta(k) = E(k) - E(-k).$$

Solving (21) for $U(k)$ and substituting the resulting expression in (15) we find

$$\hat{q}(k) = E(-ik)D(k) + \frac{E(-ik)G(k)}{\Delta(ik)}$$
$$+ i[E(-ik)E(-i\omega k) - E(-ik)E(i\omega k)]\frac{U(\omega k)}{\Delta(ik)}$$
$$+ i[E(-ik)E(-i\omega^2 k) - E(-ik)E(i\omega^2 k)]\frac{U(\omega^2 k)}{\Delta(ik)}. \tag{22}$$

The functions $\hat{q}_j(k)$ can be obtained from (22) by replacing k with $\omega^{j-1}k$ for $j = 1,\ldots,6$.

Regarding the integral representation of the solution, we restrict our attention to the first integral of (19), namely the integral along l_1 (the negative imaginary axis).

Let

$$\mathcal{P} = e^{i\beta\left(kz - \frac{z}{k}\right)}.$$

Solving (21) for $U(k)$ and substituting the resulting expression in the first integral of (19) we find that the known part of this integral is given by the expression

$$F_1 = \frac{1}{4\pi i}\int_{l_1} \mathcal{P}E(-ik)\left[D(k) + \frac{G(k)}{\Delta(ik)}\right]\frac{dk}{k}. \tag{23}$$

The unknown part involves the functions $U(\omega k)$ and $U(\omega^2 k)$ and is given by

$$C_1 = \frac{1}{4\pi}\int_{l_1} \mathcal{P}\left[E(-ik)E(-i\omega k)\frac{U(\omega k)}{\Delta(ik)} + E(-ik)E(-i\omega^2 k)\frac{U(\omega^2 k)}{\Delta(ik)}\right]\frac{dk}{k}$$
$$-\frac{1}{4\pi}\int_{l_1}\mathcal{P}\left[E(-ik)E(i\omega k)\frac{U(\omega k)}{\Delta(ik)} + E(-ik)E(i\omega^2 k)\frac{U(\omega^2 k)}{\Delta(ik)}\right]\frac{dk}{k}.$$

In what follows we will show that the contribution of the unknown functions, namely of the sum $\sum_1^6 C_j$, can be computed in terms of the given boundary conditions.

The first integral in the rhs of C_1 can be deformed from l_1 to l_1', where l_1' is a ray with $\frac{7\pi}{6} \le \arg k \le \frac{3\pi}{2}$; choosing $l_1' \equiv l_2$ we obtain

$$C_1 = \hat{C}_1 + \check{C}_1, \tag{24}$$

where

$$\hat{C}_1 = \frac{1}{4\pi}\int_{l_2}\mathcal{P}\left[E(-ik)E(-i\omega k)\frac{U(\omega k)}{\Delta(ik)} + E(-ik)E(-i\omega^2 k)\frac{U(\omega^2 k)}{\Delta(ik)}\right]\frac{dk}{k}$$

and

$$\check{C}_1 = -\frac{1}{4\pi}\int_{l_1}\mathcal{P}\left[E(-ik)E(i\omega k)\frac{U(\omega k)}{\Delta(ik)} + E(-ik)E(i\omega^2 k)\frac{U(\omega^2 k)}{\Delta(ik)}\right]\frac{dk}{k}.$$

The above deformation is justified, since it can be shown that the integrand of $\hat{C}_1$ is bounded and analytic in the region where $\arg k \in [\frac{7\pi}{6}, \frac{3\pi}{2}]$: letting $a = e^{i\frac{\pi}{6}}$, we can rewrite the first term of the integrand of $\hat{C}_1$ in the form

$$\mathcal{P}E^{-\frac{2}{\sqrt{3}}}(iak)\frac{E^{\frac{2}{\sqrt{3}}}(iak)E(-ik)E(-i\omega k)E^{\frac{1}{\sqrt{3}}}(\omega k)}{\Delta(ik)}E^{-\frac{1}{\sqrt{3}}}(\omega k)U(\omega k).$$

We observe the following:

- The zeros of $\Delta(ik)$ occur when $ik + \frac{1}{ik} \in e^{-i\frac{\pi}{2}}\mathbb{R}$, thus $k \in \mathbb{R}$.
- The function $\mathcal{P}E^{-\frac{2}{\sqrt{3}}}(iak) = e^{i\beta k(z-z_2) + \frac{\beta}{ik}(\bar{z}-\bar{z}_2)}$ is bounded and analytic for $\arg k \in [\frac{7\pi}{6}, \frac{3\pi}{2}]$.

  Indeed, if $z \in D$, then $\frac{5\pi}{6} \le \arg(z-z_2) \le \frac{3\pi}{2}$. Thus, if $\frac{7\pi}{6} \le \arg k \le \frac{3\pi}{2}$, it follows that $2\pi \le \arg[k(z-z_2)] \le 3\pi$. Hence, $\text{Re}\{ik(z-z_2)\} \le 0$.

  Therefore, the exponentials $e^{i\beta k(z-z_2)}$ and $e^{\frac{\beta}{ik}(\bar{z}-\bar{z}_2)}$ are bounded.

- The function $E^{-\frac{1}{\sqrt{3}}}(\omega k)U(\omega k)$ is bounded and analytic for $\arg k \in [\frac{7\pi}{6}, \frac{13\pi}{6}]$, namely in the region where $\text{Re}(\omega k) \ge 0$.

  Indeed, this expression involves the exponentials $e^{\beta \omega k(s-\frac{1}{2})}$ and $e^{\beta \frac{1}{\omega k}(s-\frac{1}{2})}$, which are bounded in this region, since $s \le \frac{1}{2}$.

- The function

$$\frac{E^{\frac{2}{\sqrt{3}}}(iak)E(-ik)E(-i\omega k)E^{\frac{1}{\sqrt{3}}}(\omega k)}{\Delta(ik)} = \frac{E^{\frac{1}{\sqrt{3}}}(k)}{\Delta(ik)},$$

is bounded and analytic for $\arg k \in [\frac{7\pi}{6}, \frac{3\pi}{2}]$.

Indeed, since k is at the lower half plane, then

$$\frac{E^{\frac{1}{\sqrt{3}}}(k)}{\Delta(ik)} \sim \frac{E^{\frac{1}{\sqrt{3}}}(k)}{E(ik)} = E^{-\frac{2}{\sqrt{3}}}(\omega^2 k), \quad k \to \infty,$$

which is bounded if $\text{Re}(\omega^2 k) \ge 0$.

If $\arg k \in [\frac{7\pi}{6}, \frac{3\pi}{2}]$, then $\arg(\omega^2 k) \in [\frac{11\pi}{6}, \frac{13\pi}{6}]$, which yields $\text{Re}(\omega^2 k) > 0$.

Similar considerations apply to the second term of the integrand of $\hat{C}_1$; this term can be rewritten in the form

$$\mathcal{P}E^{-\frac{2}{\sqrt{3}}}(iak)\frac{E^{\frac{2}{\sqrt{3}}}(iak)E(-ik)E(-i\omega^2 k)E^{\frac{1}{\sqrt{3}}}(\omega^2 k)}{\Delta(ik)}E^{-\frac{1}{\sqrt{3}}}(\omega^2 k)U(\omega^2 k).$$

We observe the following:

- The function $\mathcal{P}E^{-\frac{2}{\sqrt{3}}}(iak) = e^{i\beta k(z-z_2) + \frac{\beta}{ik}(z-z_2)}$ is bounded and analytic for $\arg k \in [\frac{7\pi}{6}, \frac{3\pi}{2}]$.
- The function $E^{-\frac{1}{\sqrt{3}}}(\omega^2 k)U(\omega^2 k)$ is bounded and analytic for $\arg k \in [\frac{5\pi}{6}, \frac{11\pi}{6}]$, namely in the region where $\text{Re}(\omega^2 k) \geq 0$.
- In the lower half plane

$$\frac{E^{\frac{2}{\sqrt{3}}}(iak)E(-ik)E(-i\omega^2 k)E^{\frac{1}{\sqrt{3}}}(\omega^2 k)}{\Delta(ik)} \sim 1, \quad k \to \infty.$$

Thus, it is bounded and analytic for $\arg k \in [\frac{7\pi}{6}, \frac{3\pi}{2}]$.

Using the underlined symmetries, we can express the integral representation of the solution in the form

$$q = \sum_{j=1}^{6} F_j + \sum_{j=1}^{6} C_j = \sum_{j=1}^{6} F_j + \sum_{j=1}^{6} \left(\hat{C}_j + \check{C}_j\right), \tag{25}$$

where F_j and C_j are given by applying in (23) and (24) the following rotations:

$$k \to \omega^{j-1} k, \quad l_1 \to l_j, \quad l_2 \to l_{j+1}, \quad j = 2, \ldots, 6; \quad l_7 := l_1.$$

We define $\tilde{C}_j = \hat{C}_{j-1} + \check{C}_j$, $j = 1, \ldots, 6$, where we employ the notation $\hat{C}_0 = \hat{C}_6$. Then, we rewrite the expression in (25) in the form

$$q = \sum_{j=1}^{6} F_j + \sum_{j=0}^{5} \hat{C}_j + \sum_{j=1}^{6} \check{C}_j = \sum_{j=1}^{6} F_j + \sum_{j=1}^{6} \left(\hat{C}_{j-1} + \check{C}_j\right) = \sum_{j=1}^{6} F_j + \sum_{j=1}^{6} \tilde{C}_j. \tag{26}$$

Thus, it is sufficient to compute the contribution $\{\tilde{C}_j\}_1^6$. In this direction we find (via rotation) that

$$\check{C}_2 = -\frac{1}{4\pi}\int_{l_2} \mathcal{P}\left[E(-i\omega k)E(i\omega^2 k)\frac{U(\omega^2 k)}{\Delta(i\omega k)} + E(-i\omega k)E(i\omega^3 k)\frac{U(\omega^3 k)}{\Delta(i\omega k)}\right]\frac{dk}{k}.$$

Thus

$$\tilde{C}_2 = \hat{C}_1 + \check{C}_2$$

$$= \frac{1}{4\pi}\int_{l_2} \mathcal{P}\left[E(-ik)E(-i\omega k)\frac{U(\omega k)}{\Delta(ik)} + E(-ik)E(-i\omega^2 k)\frac{U(\omega^2 k)}{\Delta(ik)}\right]\frac{dk}{k}$$

$$- \frac{1}{4\pi}\int_{l_2} \mathcal{P}\left[E(-i\omega k)E(i\omega^2 k)\frac{U(\omega^2 k)}{\Delta(i\omega k)} + E(-i\omega k)E(i\omega^3 k)\frac{U(\omega^3 k)}{\Delta(i\omega k)}\right]\frac{dk}{k}.$$

Using that $\omega^3 = -1$ and $U(-k) = -U(k)$ the above expression is simplified to

$$\tilde{C}_2 = \frac{1}{4\pi}\int_{l_2} \mathcal{P}E(-ik)E(-i\omega k)\frac{\Delta(ik)U(k) + \Delta(i\omega k)U(\omega k) + \Delta(i\omega^2 k)U(\omega^2 k)}{\Delta(ik)\Delta(i\omega k)}\frac{dk}{k}. \tag{27}$$

Employing the global relation (21) we obtain

$$\tilde{C}_2 = \frac{1}{4\pi i}\int_{l_2} \mathcal{P}E(-ik)E(-i\omega k)\frac{G(k)}{\Delta(ik)\Delta(i\omega k)}\frac{dk}{k}. \tag{28}$$

In summary, the solution takes the form

$$q = \sum_{j=1}^{6} F_j + \sum_{j=1}^{6} \tilde{C}_j, \tag{29}$$

where F_j is defined by

$$F_j = \frac{1}{4\pi i} \int_{l_j} \mathcal{P} E(-i\omega^{j-1}k) \left[D(\omega^{j-1}k) + \frac{G(\omega^{j-1}k)}{\Delta(i\omega^{j-1}k)} \right] \frac{dk}{k} \tag{30}$$

and $\tilde{C}_j$ is defined by

$$\tilde{C}_j = \frac{1}{4\pi i} \int_{l_j} \mathcal{P} E(-i\omega^{j-2}k) E(-i\omega^{j-1}k) \frac{G(\omega^{j-2}k)}{\Delta(i\omega^{j-1}k)\Delta(i\omega^{j-2}k)} \frac{dk}{k}. \tag{31}$$

Note also that the integrals of $\tilde{C}_j$ can be deformed on a sector of angle $\frac{2\pi}{3}$. For example, in $\tilde{C}_2$ the ray l_2 can be deformed in a ray l'_2 in the sector $\arg k \in (\pi, \frac{5\pi}{3})$; analogous results are valid for the remaining $\{\tilde{C}_j\}_1^6$.

Observing that $G(\omega k) = G(k)$, Equation (29) can be further simplified to

$$q = \frac{1}{4\pi i} \sum_{j=1}^{6} \int_{l_j} \mathcal{P} \left[E(-i\omega^{j-1}k) D(\omega^{j-1}k) + \frac{E(-i\omega^j k) G(\omega^{j-1}k)}{\Delta(i\omega^{j-1}k)\Delta(i\omega^{j-2}k)} \right] \frac{dk}{k}. \tag{32}$$

In order to write the integral representation in a more compact form we make the change of variables $k \to \omega^{1-j}k$ in the integrals in F_j and $\tilde{C}_j$. In this procedure:

1. the fraction $\frac{dk}{k}$ remains invariant;
2. the rays l_j become l_1;
3. the exponent $\mathcal{P} = e^{i\beta(kz - \frac{z}{k})}$ becomes $e^{i\beta\left(\omega^{1-j}kz - \frac{z}{\omega^{1-j}k}\right)}$;
4. the remaining integrands are equal to the corresponding integrands in F_1 and $\tilde{C}_1$.

Thus, we obtain

$$q = \frac{1}{4\pi i} \int_{l_1} \mathcal{T} \left[E(-ik) D(k) - \frac{E(-i\omega k)}{\Delta(ik)\Delta(i\omega^2 k)} G(k) \right] \frac{dk}{k}. \tag{33}$$

where

$$\mathcal{T} = \sum_{j=1}^{6} e^{i\beta\left(\omega^{1-j}kz - \frac{z}{\omega^{1-j}k}\right)}.$$

We make the change of variables $k \to -ik$ in the integrand of (33), so that the contour of integration transforms from the negative imaginary axis l_1 to the real imaginary axis, and we summarize the above result in the form of a proposition.

Proposition 1. *Let q satisfy the modified Helmholtz Equation (2) in the interior of a regular hexagon defined in (13). Assume that on each side of this hexagon an odd symmetric Dirichlet boundary condition is prescribed, namely,*

$$q^{(j)}(s) = d(s), \qquad s \in \left[-\frac{l}{2}, \frac{l}{2}\right], \quad j = 1, \ldots, 6,$$

with $d(-s) = -d(s)$ and $d\left(-\frac{l}{2}\right) = d\left(\frac{l}{2}\right) = 0$.

The solution q can be computed in closed form:

$$q(z,\bar{z}) = \frac{1}{4\pi i}\int_0^\infty R(k,z,\bar{z})\left[E(-k)D(-ik) - \frac{E(-\omega k)}{\Delta(k)\Delta(\omega^2 k)}G(-ik)\right]\frac{dk}{k}, \tag{34}$$

where $R(k,z,\bar{z})$, $D(k)$, $E(k)$, $G(k)$, $\Delta(k)$ are defined as follows:

$$R(k,z,\bar{z}) = \sum_{j=1}^6 e^{\beta\left(\omega^{1-j}kz + \frac{\bar{z}}{\omega^{1-j}k}\right)}$$

$$E(k) = e^{\beta(k+\frac{1}{k})\frac{l\sqrt{3}}{2}}, \quad D(k) = \beta\left(\frac{1}{k} - k\right)\int_{-\frac{l}{2}}^{\frac{l}{2}} e^{\beta(k+\frac{1}{k})s}d(s)ds,$$

$$G(k) = \sum_{j=1}^6 E\left(-i\omega^{j-1}k\right)D\left(\omega^{j-1}k\right), \quad \Delta(k) = E(k) - E(-k), \quad k \in \mathbb{C}.$$

4. The Symmetric Even Case

Applying the condition $U(-k) = U(k)$ in (17) we obtain the following equation

$$\Delta^+(ik)U(k) + \Delta^+(i\omega k)U(\omega k) + \Delta^+(i\omega^2 k)U(\omega^2 k) = iG(k), \quad k \in \mathbb{C}, \tag{35}$$

where

$$\Delta^+(k) = E(k) + E(-k)$$

and $G(k)$ is known and given in (18).

Following the same stems used in Section 3 we derive the analogue of (28), which yields the following formula for $\widetilde{C}_2$:

$$\widetilde{C}_2 = \frac{1}{4i\pi}\int_{l_2} \mathcal{P}E(-ik)E(-i\omega k)\frac{G(k)}{\Delta^+(ik)\Delta^+(i\omega k)}\frac{dk}{k}$$
$$+ \frac{1}{2\pi}\int_{l_2}\mathcal{P}\frac{U(\omega^2 k)}{\Delta^+(ik)\Delta^+(i\omega k)}\frac{dk}{k}, \tag{36}$$

where in addition to the known part which involves $G(k)$, there now exists an unknown part which involves $U(\omega^2 k)$.

Thus, the analogue of (29) now takes the form

$$q = \sum_{j=1}^6 F_j + \sum_{j=1}^6 A_j + \sum_{j=1}^6 B_j, \tag{37}$$

where F_j is known function defined by

$$F_j = \frac{1}{4\pi i}\int_{l_j}\mathcal{P}E(-i\omega^{j-1}k)\left[D(\omega^{j-1}k) + \frac{G(\omega^{j-1}k)}{\Delta^+(i\omega^{j-1}k)}\right]\frac{dk}{k}, \tag{38}$$

A_j is also known and defined by

$$A_j = \frac{1}{4\pi i}\int_{l_j}\mathcal{P}E(-i\omega^{j-2}k)E(-i\omega^{j-1}k)\frac{G(\omega^{j-2}k)}{\Delta^+(i\omega^{j-1}k)\Delta^+(i\omega^{j-2}k)}\frac{dk}{k}, \tag{39}$$

whereas B_j is the unknown function defined by

$$B_j = \frac{1}{2\pi}\int_{l_j}\mathcal{P}\frac{U(\omega^j k)}{\Delta^+(i\omega^{j-1}k)\Delta^+(i\omega^{j-2}k)}\frac{dk}{k}. \tag{40}$$

It can be shown that each of B_j decays exponentially fast as $\beta \to \infty$. The rigorous proof of this statement will be presented elsewhere. In the next section, this fact will be indicated via the numerical evaluation of each of the terms appearing in Equation (37).

5. Illustration of the Results

5.1. Odd Case

Below we depict the solution obtained by (34) for various choices of the Dirichlet datum $d(s)$ and of the parameter β. At all the examples we have fixed the length of the side of the hexagon $l = 2$.

For the first example we employ the Dirichlet datum $d(s) = \sin(\pi s)$ and the parameter $\beta = 1$; see Figure 1.

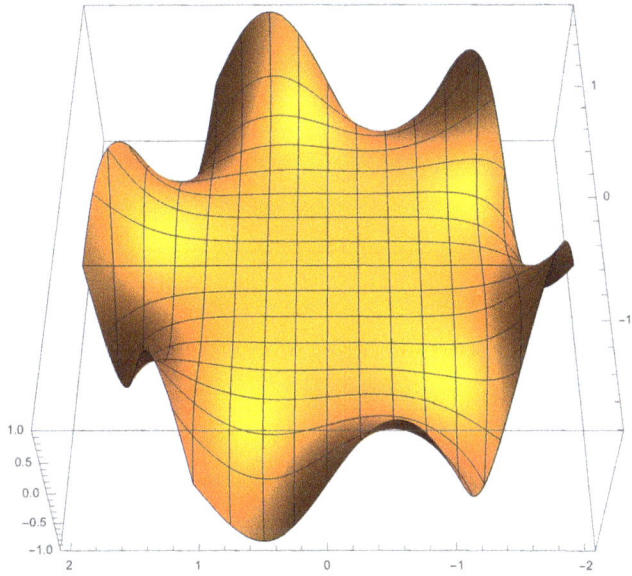

Figure 1. The solution q given by (34) for $d(s) = \sin(\pi s)$, $l = 2$ and $\beta = 1$.

We also depict the deviation of $d(s)$ from the function obtained by the integral representation (34) evaluated at the side of the hexagon, namely at $x = \frac{l\sqrt{3}}{2} = \sqrt{3}$ and $y = s \in \left[-\frac{l}{2}, \frac{l}{2}\right] \equiv [-1, 1]$; see Figure 2.

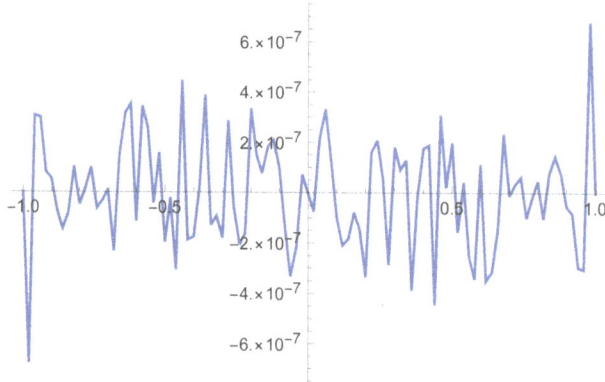

Figure 2. The deviation of q (given by (34)) from the actual Dirichlet datum $d(s)$ evaluated at the side of the hexagon; here we employ $d(s) = \sin(\pi s)$, $l = 2$ and $\beta = 1$.

For the second example we employ the Dirichlet datum $d(s) = \sin(\pi s)$ and the parameter $\beta = 1/5$; see Figure 3.

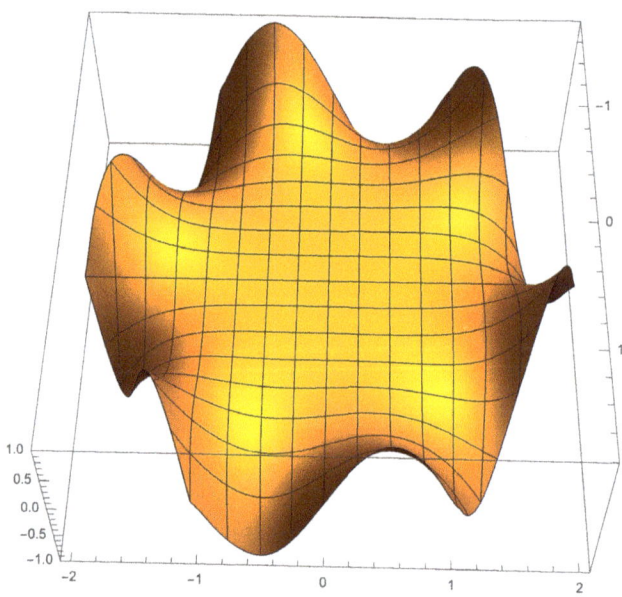

Figure 3. The solution q given by (34) for $d(s) = \sin(\pi s)$, $l = 2$ and $\beta = 1/5$.

For the third example we employ the Dirichlet datum $d(s) = \sin(2\pi s)$ and the parameter $\beta = 1$; see Figure 4.

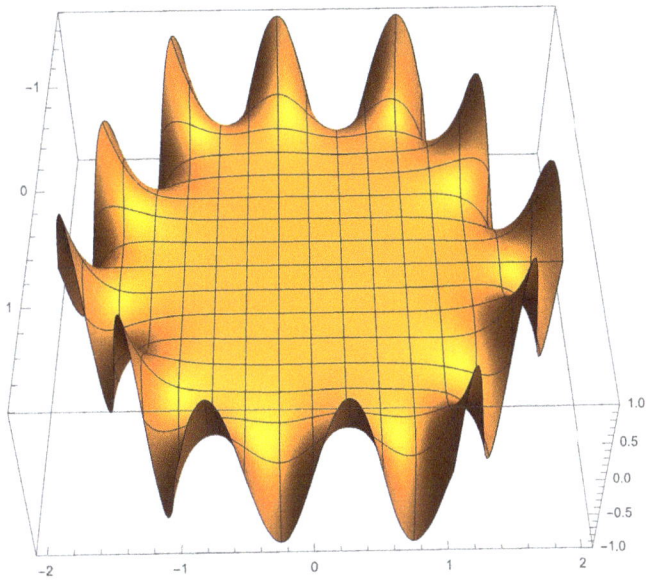

Figure 4. The solution q given by (34) for $d(s) = \sin(2\pi s)$, $l = 2$ and $\beta = 1$.

5.2. Even Case

In this case we employ the Dirichlet datum $d(s) = \cos\left(\frac{\pi}{2}s\right)$ and the parameter $\beta = 1$ at the known part of the rhs of the formula (37), namely the expression

$$\sum_{j=1}^{6} F_j + \sum_{j=1}^{6} A_j, \qquad (41)$$

where F_j and A_j are given by (38) and (39), respectively; see Figure 5.

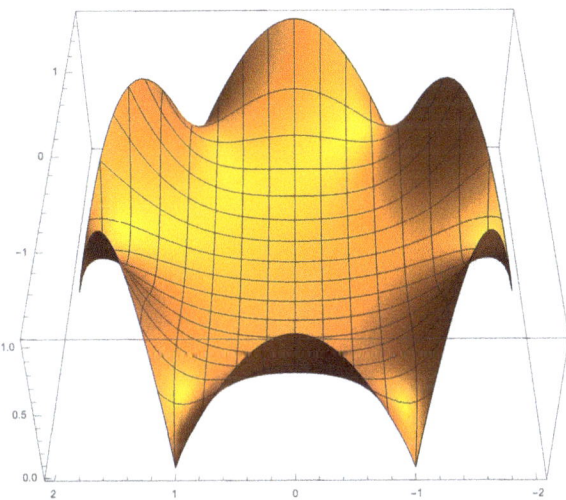

Figure 5. The known part of the solution q given by (41) for $d(s) = \cos\left(\frac{\pi}{2}s\right)$, $l = 2$ and $\beta = 1$.

We also depict the deviation of $d(s)$ from the above expression evaluated at the side of the hexagon, namely at $x = \sqrt{3}$ and $y = s \in [-1, 1]$. This is equal to the contribution $\sum_{j=1}^{6} B_j$, with B_j given by (40); see Figure 6.

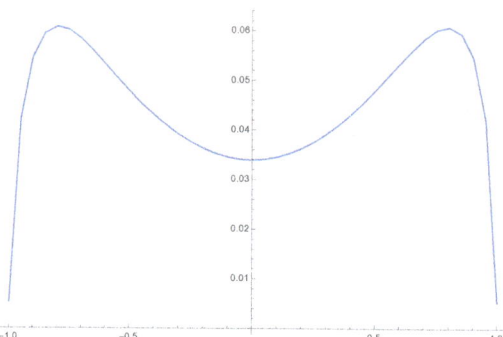

Figure 6. The deviation of the known part of the solution q given by (41) from the actual Dirichlet datum $d(s) = \cos\left(\frac{\pi}{2}s\right)$, $l = 2$ and $\beta = 1$, evaluated at the side of the hexagon.

Furthermore, we depict the latter contribution for the different values of $\beta = \frac{1}{4}, \frac{1}{2}, 1, 2, 4$, where it is clearly shown that the error decreases drastically with the increase of β; see Figure 7. We observe exponential decay for $z \neq z_j$, $j = 1, \ldots, 6$: in Figure 8 we depict the deviation from the actual Dirichlet data for three points on side (1) of the hexagon, namely $\alpha_1 = \left(\sqrt{3}, 0\right)$, $\alpha_2 = \left(\sqrt{3}, \frac{3}{10}\right)$, $\alpha_3 = \left(\sqrt{3}, \frac{9}{10}\right)$, with β in the intervals $I_1 = [1, 8]$, $I_2 = [1, 10]$, $I_3 = [1, 58]$, respectively.

Figure 7. The deviation of the known part of the solution q given by (41) from the actual Dirichlet datum $d(s) = \cos\left(\frac{\pi}{2}s\right)$ and $l = 2$, evaluated at the side of the hexagon. This deviation is depicted for the different values of $\beta = \frac{1}{4}, \frac{1}{2}, 1, 2, 4$, and it decreases drastically with the increase of β.

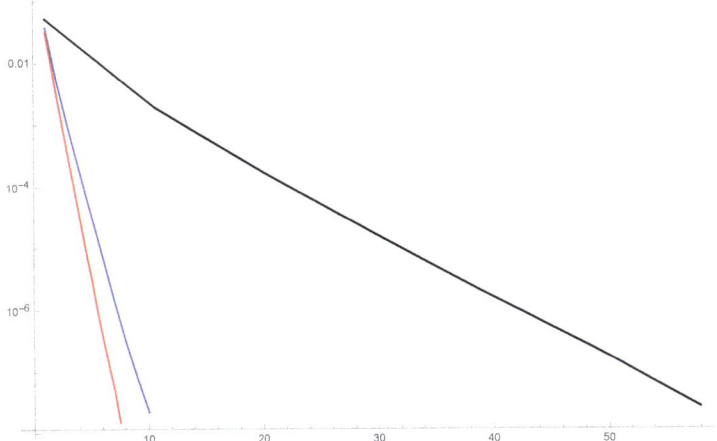

Figure 8. The deviation of the known part of the solution q given by (41) from the actual Dirichlet datum $d(s)$, evaluated at three points of side (1) of the hexagon, namely $\alpha_1 = \left(\sqrt{3}, 0\right)$ in red, $\alpha_2 = \left(\sqrt{3}, \frac{3}{10}\right)$ in blue, $\alpha_3 = \left(\sqrt{3}, \frac{9}{10}\right)$ in black. The deviation is depicted against β and it displays exponential decay.

6. Conclusions

In this work we have presented the explicit solution of a particular boundary value problem for the modified Helmholtz equation in a regular hexagon: we have solved the case where the same Dirichlet datum $d(s)$ is prescribed in all sides of the hexagon, and this function is odd. This explicit solution is described in Proposition 1. We have also obtained an approximate analytical representation for the solution for the case that $d(s)$ is even. The exact representation is given by Equation (37), where the terms F_j and A_j are given in terms of $d(s)$, but the terms B_j involve the unknown Neumann boundary value. However, these terms are exponentially small as $\beta \to \infty$. Thus, for the case of large β, Equation (37) provides the solution to this problem with an exponentially small error. The above analytical results were verified numerically in Section 5. The rigorous investigation on the analytical and numerical accuracy of the latter approximate formula will be presented in future work.

It should be noted that the arbitrary Dirichlet problem can be decomposed into 6 separate and simpler Dirichlet BVPs, which are defined in Section 2.3; the first of these BVPs is the symmetric Dirichlet problem. The analysis of the remaining problems is a work in progress.

Author Contributions: Conceptualization, K.K. and A.S.F.; methodology, K.K. and A.S.F.; validation, K.K. and A.S.F.; formal analysis, K.K. and A.S.F.; investigation, K.K. and A.S.F.; writing–original draft preparation, K.K. and A.S.F.; writing–review and editing, K.K. and A.S.F.; visualization, K.K. and A.S.F. All authors have read and agreed to the published version of the manuscript.

Funding: A.S.F. was supported by EPSRC, UK, via a senior fellowship.

Conflicts of Interest: The authors declare no conflict of interest.

References

1. Fokas, A.S. A unified transform method for solving linear and certain nonlinear PDEs. *Proc. R. Soc. Lond. A Math. Phys. Eng. Sci.* **1997**, *453*, 1411–1443. [CrossRef]
2. Fokas, A.S. On the integrability of linear and nonlinear partial differential equations. *J. Math. Phys.* **2000**, *41*, 4188–4237. [CrossRef]
3. Fokas, A.S. A new transform method for evolution partial differential equations. *IMA J. Appl. Math.* **2002**, *67*, 559–590. [CrossRef]
4. Biondini, G.; Wang, D. Initial-boundary-value problems for discrete linear evolution equations. *IMA J. Appl. Math.* **2010**, *75*, 968–997. [CrossRef]

5. Deconinck, B.; Trogdon, T.; Vasan, V. The method of Fokas for solving linear partial differential equations. *SIAM Rev.* **2014**, *56*, 159–186. [CrossRef]
6. Fokas, A.S. *A Unified Approach to Boundary Value Problems*; SIAM: Garden Grove, CA, USA, 2008.
7. Pelloni, B. The spectral representation of two-point boundary-value problems for third-order linear evolution partial differential equations. *Proc. R. Soc. Lond. A Math. Phys. Eng. Sci.* **2005**, *461*, 2965–2984. [CrossRef]
8. Pelloni, B.; Smith, D.A. Spectral theory of some non-selfadjoint linear differential operators. *Proc. R. Soc. A Math. Phys. Eng. Sci.* **2013**, *469*, 20130019. [CrossRef]
9. Pelloni, B.; Smith, D.A. Evolution PDEs and augmented eigenfunctions. Half-line. *J. Spectr. Theory* **2016**, *6*, 185–213. [CrossRef]
10. Smith, D.A. Well-posed two-point initial-boundary value problems with arbitrary boundary conditions. In *Mathematical Proceedings of the Cambridge Philosophical Society*; Cambridge University Press: Cambridge, UK, 2012; Volume 152, pp. 473–496.
11. Fokas, A.S. Two-dimensional linear partial differential equations in a convex polygon. *Proc. R. Soc. Lond. Ser. A Math. Phys. Eng. Sci.* **2001**, *457*, 371–393. [CrossRef]
12. Kalimeris, K. INITIAL and Boundary Value Problems in Two and Three Dimensions. Ph.D. Thesis, University of Cambridge, Cambridge, UK, 2010.
13. Spence, E.A. Boundary Value Problems for Linear Elliptic PDEs. Ph.D. Thesis, University of Cambridge, Cambridge, UK, 2011.
14. Batal, A.; Fokas, A.S.; Özsari, T. Uniform transform method for boundary value problems involving mixed derivatives. *arXiv* **2020**, arXiv:2002.01057.
15. Fokas, A.S.; Himonas, A.A.; Mantzavinos, D. The nonlinear Schrödinger equation on the half-line. *Trans. Am. Math. Soc.* **2017**, *369*, 681–709. [CrossRef]
16. Himonas, A.A.; Mantzavinos, D. On the Initial-Boundary Value Problem for the Linearized Boussinesq Equation. *Stud. Appl. Math.* **2015**, *134*, 62–100. [CrossRef]
17. Himonas, A.A.; Mantzavinos, D.; Yan, F. Initial-boundary value problems for a reaction-diffusion equation. *J. Math. Phys.* **2019**, *60*, 081509. [CrossRef]
18. Özsari, T.; Yolcu, N. The initial-boundary value problem for the biharmonic Schrödinger equation on the half-line. *Commun. Pure Appl. Anal.* **2019**, *18*, 3285–3316. [CrossRef]
19. Kalimeris, K.; Özsarı, T. An elementary proof of the lack of null controllability for the heat equation on the half line. *Appl. Math. Lett.* **2020**, *104*, 106241. [CrossRef]
20. Crowdy, D.G.; Davis, A.M. Stokes flow singularities in a two-dimensional channel: A novel transform approach with application to microswimming. *Proc. R. Soc. A* **2013**, *469*, 20130198.
21. Deconinck, B.; Oliveras, K. The instability of periodic surface gravity waves. *J. Fluid Mech.* **2011**, *675*, 141–167. [CrossRef]
22. Fokas, A.S.; Kalimeris, K. Water waves with moving boundaries. *J. Fluid Mech.* **2017**, *832*, 641–665. [CrossRef]
23. Oliveras, K. Stability of Periodic Surface Gravity Water Waves. Ph.D. Thesis, University of Washington, Seattle, WA, USA, 2009.
24. Plümacher, D.; Oberlack, M.; Wang, Y.; Smuda, M. On a non-linear droplet oscillation theory via the unified method. *Phys. Fluids* **2020**, *32*, 067104. [CrossRef]
25. Vasan, V.; Deconinck, B. The inverse water wave problem of bathymetry detection. *J. Fluid Mech.* **2013**, *714*, 562–590. [CrossRef]
26. Ablowitz, M.J.; Fokas, A.S.; Musslimani, Z.H. On a new non-local formulation of water waves. *J. Fluid Mech.* **2006**, *562*, 313–343. [CrossRef]
27. Nicholls, D. A high-order perturbation of surfaces (HOPS) approach to Fokas integral equations: Three-dimensional layered-media scattering. *Q. Appl. Math.* **2016**, *74*, 61–87. [CrossRef]
28. Ashton, A.C.L. Laplace's equation on convex polyhedra via the unified method. *Proc. R. Soc. A Math. Phys. Eng. Sci.* **2015**, *471*, 20140884. [CrossRef]
29. Crowdy, D. A transform method for Laplace's equation in multiply connected circular domains. *IMA J. Appl. Math.* **2015**, *80*, 1902–1931. [CrossRef]
30. Dassios, G.; Fokas, A.S. The basic elliptic equations in an equilateral triangle. *Proc. R. Soc. A Math. Phys. Eng. Sci.* **2005**, *461*, 2721–2748. [CrossRef]
31. Fokas, A.S.; Kapaev, A.A. A Riemann-Hilbert approach to the Laplace equation. *J. Math. Anal. Appl.* **2000**, *251*, 770–804. [CrossRef]

32. Fokas, A.S.; Kapaev, A.A. On a transform method for the Laplace equation in a polygon. *IMA J. Appl. Math.* **2003**, *68*, 355–408. [CrossRef]
33. Luca, E.; Crowdy, D.G. A transform method for the biharmonic equation in multiply connected circular domains. *IMA J. Appl. Math.* **2018**, *83*, 942–976. [CrossRef]
34. Antipov, Y.A.; Fokas, A.S. The modified Helmholtz equation in a semi-strip. In *Mathematical Proceedings of the Cambridge Philosophical Society*; Cambridge University Press: Cambridge, UK, 2005; Volume 138, pp. 339–365.
35. Ashton, A.C.L. On the rigorous foundations of the Fokas method for linear elliptic partial differential equations. *Proc. R. Soc. A Math. Phys. Eng. Sci.* **2012**, *468*, 1325–1331. [CrossRef]
36. Ashton, A.C.L. The Spectral Dirichlet–Neumann Map for Laplace's Equation in a Convex Polygon. *SIAM J. Math. Anal.* **2013**, *45*, 3575–3591. [CrossRef]
37. Colbrook, M.J. Extending the unified transform: Curvilinear polygons and variable coefficient PDEs. *IMA J. Numer. Anal.* **2020**, *40*, 976–1004. [CrossRef]
38. Colbrook, M.J.; Flyer, N.; Fornberg, B. On the Fokas method for the solution of elliptic problems in both convex and non-convex polygonal domains. *J. Comput. Phys.* **2018**, *374*, 996–1016. [CrossRef]
39. Colbrook, M.J.; Fokas, A.S. Computing eigenvalues and eigenfunctions of the Laplacian for convex polygons. *Appl. Numer. Math.* **2018**, *126*, 1–17. [CrossRef]
40. Colbrook, M.J.; Fokas, A.S.; Hashemzadeh, P. A Hybrid Analytical-Numerical Technique for Elliptic PDEs. *SIAM J. Sci. Comput.* **2019**, *41*, A1066–A1090. [CrossRef]
41. de Barros, F.P.J.; Colbrook, M.J.; Fokas, A.S. A hybrid analytical-numerical method for solving advection-dispersion problems on a half-line. *Int. J. Heat Mass Transf.* **2019**, *139*, 482–491. [CrossRef]
42. Davis, C.I.R.; Fornberg, B. A spectrally accurate numerical implementation of the Fokas transform method for Helmholtz-type PDEs. *Complex Var. Elliptic Equ.* **2014**, *59*, 564–577. [CrossRef]
43. Fokas, A.S.; Nachbin, A. Water waves over a variable bottom: A non-local formulation and conformal mappings. *J. Fluid Mech.* **2012**, *695*, 288–309. [CrossRef]
44. Fornberg, B.; Flyer, N. A numerical implementation of Fokas boundary integral approach: Laplace's equation on a polygonal domain. *Proc. R. Soc. A Math. Phys. Eng. Sci.* **2011**, *467*, 2983–3003. [CrossRef]
45. Grylonakis, E.N.G.; Filelis-Papadopoulos, C.K.; Gravvanis, G.A. A class of unified transform techniques for solving linear elliptic PDEs in convex polygons. *Appl. Numer. Math.* **2018**, *129*, 159–180. [CrossRef]
46. Hashemzadeh, P.; Fokas, A.S.; Smitheman, S.A. A numerical technique for linear elliptic partial differential equations in polygonal domains. *Proc. Math. Phys. Eng. Sci.* **2015**, *471*, 20140747. [CrossRef]
47. Trogdon, T.; Biondini, G. Evolution partial differential equations with discontinuous data. *Q. Appl. Math.* **2019**, *77*, 689–726. [CrossRef]
48. Lamé, G. Mémoire sur la propagation de la chaleur dans les polyèdres, et principalement dans le prisme triangulaire régulier. *J. l'Ecole Poly Tech.* **1833**, *22*, 194–251.
49. Fokas, A.S.; Kalimeris, K. Eigenvalues for the Laplace operator in the interior of an equilateral triangle. *Comput. Methods Funct. Theory* **2014**, *14*, 1–33. [CrossRef]
50. Pinsky, M.A. The eigenvalues of an equilateral triangle. *SIAM J. Math. Anal.* **1980**, *11*, 819–827. [CrossRef]
51. Pinsky, M.A. Completeness of the eigenfunctions of the equilateral triangle. *SIAM J. Math. Anal.* **1985**, *16*, 848–851. [CrossRef]
52. Práger, M. Eigenvalues and eigenfunctions of the Laplace operator on an equilateral triangle. *Appl. Math.* **1998**, *43*, 311–320. [CrossRef]
53. Terras, R.; Swanson, R. Image methods for constructing Green's functions and eigenfunctions for domains with plane boundaries. *J. Math. Phys.* **1980**, *21*, 2140–2153. [CrossRef]
54. McCartin, B.J. Eigenstructure of the equilateral triangle, Part I: The Dirichlet problem. *Siam Rev.* **2003**, *45*, 267–287. [CrossRef]
55. McCartin, B.J. Eigenstructure of the equilateral triangle, Part II: The Neumann problem. *Math. Probl. Eng.* **2002**, *8*, doi:10.1080/1024123021000053664. [CrossRef]
56. McCartin, B.J. *Laplacian Eigenstructure of the Equilateral Triangle*; Hikari Limited: Rousse, Bulgaria, 2011.

© 2020 by the authors. Licensee MDPI, Basel, Switzerland. This article is an open access article distributed under the terms and conditions of the Creative Commons Attribution (CC BY) license (http://creativecommons.org/licenses/by/4.0/).

Article

On the Triple Lauricella–Horn–Karlsson q-Hypergeometric Functions

Thomas Ernst

Department of Mathematics, Uppsala University, P.O. Box 480, SE-751 06 Uppsala, Sweden; thomas@math.uu.se

Received: 29 May 2020; Accepted: 2 July 2020; Published: 31 July 2020

Abstract: The Horn–Karlsson approach to find convergence regions is applied to find convergence regions for triple q-hypergeometric functions. It turns out that the convergence regions are significantly increased in the q-case; just as for q-Appell and q-Lauricella functions, additions are replaced by Ward q-additions. Mostly referring to Krishna Srivastava 1956, we give q-integral representations for these functions.

Keywords: triple q-hypergeometric function; convergence region; Ward q-addition; q-integral representation

MSC: 33D70; 33C65

1. Introduction

This is part of a series of papers about q-integral representations of q-hypergeometric functions. The first paper [1] was about q-hypergeometric transformations involving q-integrals. Then followed [2], where Euler q-integral representations of q-Lauricella functions in the spirit of Koschmieder were presented. Furthermore, in [3], Eulerian q-integrals for single and multiple q-hypergeometric series were found. However, this subject is by no means exhausted, and in the same proceedings, [4], concise proofs for q-analogues of Eulerian integral formulas for general q-hypergeometric functions corresponding to Erdélyi, and for two of Srivastavas triple hypergeometric functions were given. Finally, in [5], single and multiple q-Eulerian integrals in the spirit of Exton, Driver, Johnston, Pandey, Saran and Erdélyi are presented. All proofs use the q-beta integral method.

The history of the subject in this article started in 1889 when Horn [6] investigated the domain of convergence for double and triple q-hypergeometric functions. He invented an ingenious geometric construction with five sets of convergence regions in three dimensions which was successfully used by Karlsson [7] in 1974 to explicitly state the convergence regions for the known functions of three variables. We adapt this approach to the q-case, by replacing additions by q-additions and exactly stating the convergence sets for each function. Obviously combinations of the q-deformed rhombus in dimension three appear several times. It is not possible to depict the q-additions in diagrams, not even in dimension two; they depend on the parameter q. We recall Karlssons paper, which seems to have fallen into oblivion. We give proofs for all the convergence regions, and our proofs also work for Karlssons equations by putting $q = 1$.

Saran [8], followed by Exton [9] gave less correct convergence criteria. By giving q-integral representations for these functions, we also correct and give proofs for the formulas in K.J. Srivastava [10] (not Hari Srivastava). He did not give many proofs, and our proofs also work for his equations by putting $q = 1$.

2. Definitions

Definition 1. *We define 10 q-analogues of the three-variable Lauricella–Saran functions of three variables plus two G-functions. Each function is defined by*

$$F \equiv \sum_{m,n,p=0}^{+\infty} \Psi \frac{x^m y^n z^p}{\langle 1;q\rangle_m \langle 1;q\rangle_n \langle 1;q\rangle_p}. \tag{1}$$

As a result of lack of space, for every row, we first give the generic name, the function parameters, followed by the corresponding Ψ according to (1).

Function	Ψ
$\Phi_E(\alpha_1,\alpha_1,\alpha_1,\beta_1,\beta_2,\beta_2;\gamma_1,\gamma_2,\gamma_3\|q;x,y,z)$	$\dfrac{\langle\alpha_1;q\rangle_{m+n+p}\langle\beta_1;q\rangle_m\langle\beta_2;q\rangle_{n+p}}{\langle\gamma_1;q\rangle_m\langle\gamma_2;q\rangle_n\langle\gamma_3;q\rangle_p}$
$\Phi_F(\alpha_1,\alpha_1,\alpha_1,\beta_1,\beta_2,\beta_1;\gamma_1,\gamma_2,\gamma_2\|q;x,y,z)$	$\dfrac{\langle\alpha_1;q\rangle_{m+n+p}\langle\beta_1;q\rangle_{m+p}\langle\beta_2;q\rangle_n}{\langle\gamma_1;q\rangle_m\langle\gamma_2;q\rangle_{n+p}}$
$\Phi_G(\alpha_1,\alpha_1,\alpha_1,\beta_1,\beta_2,\beta_3;\gamma_1,\gamma_2,\gamma_2\|q;x,y,z)$	$\dfrac{\langle\alpha_1;q\rangle_{m+n+p}\langle\beta_1;q\rangle_m\langle\beta_2;q\rangle_n\langle\beta_3;q\rangle_p}{\langle\gamma_1;q\rangle_m\langle\gamma_2;q\rangle_{n+p}}$
$\Phi_K(\alpha_1,\alpha_2,\alpha_2,\beta_1,\beta_2,\beta_1;\gamma_1,\gamma_2,\gamma_3\|q;x,y,z)$	$\dfrac{\langle\alpha_1;q\rangle_m\langle\alpha_2;q\rangle_{n+p}\langle\beta_1;q\rangle_{m+p}\langle\beta_2;q\rangle_n}{\langle\gamma_1;q\rangle_m\langle\gamma_2;q\rangle_n\langle\gamma_3;q\rangle_p}$
$\Phi_M(\alpha_1,\alpha_2,\alpha_2,\beta_1,\beta_2,\beta_1;\gamma_1,\gamma_2,\gamma_2\|q;x,y,z)$	$\dfrac{\langle\alpha_1;q\rangle_m\langle\alpha_2;q\rangle_{n+p}\langle\beta_1;q\rangle_{m+p}\langle\beta_2;q\rangle_n}{\langle\gamma_1;q\rangle_m\langle\gamma_2;q\rangle_{n+p}}$
$\Phi_N(\alpha_1,\alpha_2,\alpha_3,\beta_1,\beta_2,\beta_1;\gamma_1,\gamma_2,\gamma_2\|q;x,y,z)$	$\dfrac{\langle\alpha_1;q\rangle_m\langle\alpha_2;q\rangle_n\langle\alpha_3;q\rangle_p\langle\beta_1;q\rangle_{m+p}\langle\beta_2;q\rangle_n}{\langle\gamma_1;q\rangle_m\langle\gamma_2;q\rangle_{n+p}}$
$\Phi_P(\alpha_1,\alpha_2,\alpha_1,\beta_1,\beta_1,\beta_2;\gamma_1,\gamma_2,\gamma_2\|q;x,y,z)$	$\dfrac{\langle\alpha_1;q\rangle_{m+p}\langle\alpha_2;q\rangle_n\langle\beta_1;q\rangle_{m+n}\langle\beta_2;q\rangle_p}{\langle\gamma_1;q\rangle_m\langle\gamma_2;q\rangle_{n+p}}$
$\Phi_R(\alpha_1,\alpha_2,\alpha_1,\beta_1,\beta_2,\beta_1;\gamma_1,\gamma_2,\gamma_2\|q;x,y,z)$	$\dfrac{\langle\alpha_1;q\rangle_{m+p}\langle\alpha_2;q\rangle_n\langle\beta_1;q\rangle_{m+p}\langle\beta_2;q\rangle_n}{\langle\gamma_1;q\rangle_m\langle\gamma_2;q\rangle_{n+p}}$
$\Phi_S(\alpha_1,\alpha_2,\alpha_2,\beta_1,\beta_2,\beta_3;\gamma_1,\gamma_1,\gamma_1\|q;x,y,z)$	$\dfrac{\langle\alpha_1;q\rangle_m\langle\alpha_2;q\rangle_{n+p}\langle\beta_1;q\rangle_m\langle\beta_2;q\rangle_n\langle\beta_3;q\rangle_p}{\langle\gamma_1;q\rangle_{m+n+p}}$
$\Phi_T(\alpha_1,\alpha_2,\alpha_2,\beta_1,\beta_2,\beta_1;\gamma_1,\gamma_1,\gamma_1\|q;x,y,z)$	$\dfrac{\langle\alpha_1;q\rangle_m\langle\alpha_2;q\rangle_{n+p}\langle\beta_1;q\rangle_{m+p}\langle\beta_2;q\rangle_n}{\langle\gamma_1;q\rangle_{m+n+p}}$
$G_A(\alpha;\beta_1,\beta_2;\gamma\|q;x,y,z)$	$\dfrac{\langle\alpha;q\rangle_{n+p-m}\langle\beta_1;q\rangle_{m+p}\langle\beta_2;q\rangle_n}{\langle\gamma;q\rangle_{n+p-m}}$
$G_B(\alpha;\beta_1,\beta_2,\beta_3;\gamma\|q;x,y,z)$	$\dfrac{\langle\alpha;q\rangle_{n+p-m}\langle\beta_1;q\rangle_m\langle\beta_2;q\rangle_n\langle\beta_3;q\rangle_p}{\langle\gamma;q\rangle_{n+p-m}}$

In the whole paper, $A_{q,m,n,p}$ denotes the coefficient of $x^m y^n z^p$ for the respective function. In the following, we follow the notation in Karlsson [7].

Discarding possible discontinuities, we introduce the following three rational functions:

$$\Psi_1(m,n,p) \equiv \lim_{\epsilon \to +\infty} \frac{A_{1,\epsilon m+1,\epsilon n,\epsilon p}}{A_{\epsilon m,\epsilon n,\epsilon p}}, \quad m > 0,\ n \geq 0,\ p \geq 0,$$

$$\Psi_2(m,n,p) \equiv \lim_{\epsilon \to +\infty} \frac{A_{1,\epsilon m,\epsilon n+1,\epsilon p}}{A_{\epsilon m,\epsilon n,\epsilon p}}, \quad m \geq 0,\ n > 0,\ p \geq 0, \tag{2}$$

$$\Psi_3(m,n,p) \equiv \lim_{\epsilon \to +\infty} \frac{A_{1,\epsilon m,\epsilon n,\epsilon p+1}}{A_{\epsilon m,\epsilon n,\epsilon p}}, \quad m \geq 0,\ n \geq 0,\ p > 0.$$

For $0 < q < 1$ fixed, exactly as in Karlsson [7], construct the following subsets of $\mathbb{R}_+^3$:

$$C_q \equiv \{(r,s,t)|\ 0 < r < |\Psi_1(1,0,0)|^{-1} \wedge 0 < s < |\Psi_2(0,1,0)|^{-1} \wedge \\ \wedge 0 < t < |\Psi_3(0,0,1)|^{-1}\}, \tag{3}$$

$$X_q \equiv \{(r,s,t)|\ \forall(n,p) \in \mathbb{R}_+^2 : 0 < s < |\Psi_2(0,n,p)|^{-1} \vee 0 < t < |\Psi_3(0,n,p)|^{-1}\}, \tag{4}$$

$$Y_q \equiv \{(r,s,t)|\ \forall(m,p) \in \mathbb{R}_+^2 : 0 < r < |\Psi_1(m,0,p)|^{-1} \vee 0 < t < |\Psi_3(m,0,p)|^{-1}\}, \tag{5}$$

$$Z_q \equiv \{(r,s,t)|\ \forall(m,n) \in \mathbb{R}_+^2 : 0 < r < |\Psi_1(m,n,0)|^{-1} \vee 0 < s < |\Psi_2(m,n,0)|^{-1}\}, \tag{6}$$

$$E_q \equiv \{(r,s,t)| \ \forall (m,n,p) \in \mathbb{R}_+^3 : 0 < r < |\Psi_1(m,n,p)|^{-1} \vee \tag{7}$$
$$\vee 0 < s < |\Psi_2(m,n,p)|^{-1} \vee 0 < t < |\Psi_3(m,n,p)|^{-1}\},$$

$$D'_q \equiv E_q \cap X_q \cap Y_q \cap Z_q \cap C_q; \tag{8}$$

Then let $D_q \subseteq (\mathbb{R}_+ \cup \{0\})^3$ denote the union of D'_q and its projections onto the coordinate planes. Horn's theorem adapted to the q-case then states that the region D_q is the representation in the absolute octant of the convergence region in C_q^3. We will describe D'_q, and D_q by that part S_q of $\partial D'_q$ which is not contained in coordinate planes.

Theorem 1. *For every row, we first give the generic name, D'_q, followed by the corresponding q-Cartesian equations of S_q.*

Function name	D'_q	qCartesian equation of S_q
Φ_E	E_q	$r \oplus_q s \oplus_q t \oplus_q 2\sqrt{s}\sqrt{t} = 1$
Φ_F	$E_q \cap Y_q$	$\frac{rs}{t} = 1$
Φ_G	$Y_q \cap Z_q$	$r \oplus_q t = 1$, $r \oplus_q s = 1$
Φ_K	E_q	$\frac{rs}{t} = 1$
Φ_M	$Y_q \cap C_q$	$r \oplus_q t = 1, s = 1$
Φ_N	$Y_q \cap C_q$	$r \oplus_q t = 1, s = 1$
Φ_P	$Y_q \cap Z_q$	$r \oplus_q t = 1$, $r \oplus_q s = 1$
Φ_R	$Y_q \cap C_q$	$\sqrt{r} \oplus_q \sqrt{t} = 1, s = 1$
Φ_S	C_q	$r = 1, s = 1, t = 1$
Φ_T	C_q	$r = 1, s = 1, t = 1$
G_A	$Y_q \cap C_q$	$r \oplus_q t = 1, s = 1$
G_B	C_q	$r = 1, s = 1, t = 1$

The idea is to follow Karlsson's proofs and then replace the additions by the respective q-additions. This gives identical convergence regions as for q-Appell and q-Lauricella functions. For each function, for didactic reasons, we first compute the quotient of corresponding coefficients.

Proof. For the notation we refer to [2]. Consider the function Φ_E. We have

$$\frac{A_{q,m+1,n,p}}{A_{q,m,n,p}} = \frac{\langle \alpha_1 + m + n + p, \beta_1 + m; q \rangle_1}{\langle \gamma_1 + m, 1 + m; q \rangle_1},$$

$$\frac{A_{q,m,n+1,p}}{A_{q,m,n,p}} = \frac{\langle \alpha_1 + m + n + p, \beta_2 + n + p; q \rangle_1}{\langle \gamma_2 + n, 1 + n; q \rangle_1}, \tag{9}$$

$$\frac{A_{q,m,n,p+1}}{A_{q,m,n,p}} = \frac{\langle \alpha_1 + m + n + p, \beta_2 + n + p; q \rangle_1}{\langle \gamma_3 + p, 1 + p; q \rangle_1}.$$

Then we have

$$C_q = \{(r,s,t)| \ 0 < r < 1 \wedge 0 < s < 1 \wedge 0 < t < 1\}$$

$$X_q = \{(r,s,t)| \ 0 < s < \left(\frac{n}{n+p}\right)^2 \wedge 0 < t < \left(\frac{p}{n+p}\right)^2\}$$

$$Y_q = \{(r,s,t)| \ 0 < r < \frac{m}{m+p} \wedge 0 < t < \frac{p}{m+p}\}$$

$$Z_q = \{(r,s,t)| \ 0 < r < \frac{m}{m+n} \wedge 0 < s < \frac{n}{m+n}\} \tag{10}$$

$$E_q = \{(r,s,t)| \ 0 < r < \frac{m}{m+n+p} \wedge 0 < s < \frac{n^2}{(m+n+p)(n+p)} \wedge$$

$$\wedge 0 < t < \frac{p^2}{(m+n+p)(n+p)}\}.$$

We have convergence domain $\left(r \oplus_q s \oplus_q t \oplus_q 2\sqrt{s}\sqrt{t} \right)^n < 1$.

In the following, we do not write regions which are obviously bounded by $0 < x < 1$. Consider the function Φ_F. We have

$$\frac{A_{q,m+1,n,p}}{A_{q,m,n,p}} = \frac{\langle \alpha_1 + m + n + p, \beta_1 + m + p; q \rangle_1}{\langle \gamma_1 + m, 1 + m; q \rangle_1},$$

$$\frac{A_{q,m,n+1,p}}{A_{q,m,n,p}} = \frac{\langle \alpha_1 + m + n + p, \beta_2 + n; q \rangle_1}{\langle \gamma_2 + n + p, 1 + n; q \rangle_1}, \qquad (11)$$

$$\frac{A_{q,m,n,p+1}}{A_{q,m,n,p}} = \frac{\langle \alpha_1 + m + n + p, \beta_1 + m + p; q \rangle_1}{\langle \gamma_2 + n + p, 1 + p; q \rangle_1}.$$

Then we have the following regions

$$Y_q = \{(r,s,t) \mid 0 < r < \left(\frac{m}{m+p}\right)^2 \wedge 0 < t < \left(\frac{p}{m+p}\right)^2\}$$

$$Z_q = \{(r,s,t) \mid 0 < r < \frac{m}{m+n} \wedge 0 < s < \frac{n}{m+n}\}$$

$$E_q = \{(r,s,t) \mid 0 < r < \frac{m^2}{(m+n+p)(m+p)} \wedge 0 < s < \frac{n+p}{m+n+p} \wedge \qquad (12)$$

$$\wedge 0 < t < \frac{(n+p)p}{(m+n+p)(m+p)}\}.$$

We have convergence domain $\frac{rs}{t} < 1$.

Consider the function Φ_G. We have

$$\frac{A_{q,m+1,n,p}}{A_{q,m,n,p}} = \frac{\langle \alpha_1 + m + n + p, \beta_1 + m; q \rangle_1}{\langle \gamma_1 + m, 1 + m; q \rangle_1},$$

$$\frac{A_{q,m,n+1,p}}{A_{q,m,n,p}} = \frac{\langle \alpha_1 + m + n + p, \beta_2 + n; q \rangle_1}{\langle \gamma_2 + n + p, 1 + n; q \rangle_1}, \qquad (13)$$

$$\frac{A_{q,m,n,p+1}}{A_{q,m,n,p}} = \frac{\langle \alpha_1 + m + n + p, \beta_3 + p; q \rangle_1}{\langle \gamma_2 + n + p, 1 + p; q \rangle_1}.$$

Then we have the following regions

$$Y_q = \{(r,s,t) \mid 0 < r < \frac{m}{m+p} \wedge 0 < t < \frac{p}{m+p}\}$$

$$Z_q = \{(r,s,t) \mid 0 < r < \frac{m}{m+n} \wedge 0 < s < \frac{n}{m+n}\}$$

$$E_q = \{(r,s,t) \mid 0 < r < \frac{m}{m+n+p} \wedge 0 < s < \frac{n+p}{m+n+p} \wedge \qquad (14)$$

$$\wedge 0 < t < \frac{n+p}{m+n+p}\}.$$

We have convergence domain $r \oplus_q t < 1$, $r \oplus_q s < 1$.

Consider the function Φ_K. We have

$$\frac{A_{q,m+1,n,p}}{A_{q,m,n,p}} = \frac{\langle \alpha_1 + m, \beta_1 + m + p; q \rangle_1}{\langle \gamma_1 + m, 1 + m; q \rangle_1},$$

$$\frac{A_{q,m,n+1,p}}{A_{q,m,n,p}} = \frac{\langle \alpha_2 + n + p, \beta_2 + n; q \rangle_1}{\langle \gamma_2 + n, 1 + n; q \rangle_1}, \qquad (15)$$

$$\frac{A_{q,m,n,p+1}}{A_{q,m,n,p}} = \frac{\langle \alpha_2 + n + p, \beta_1 + m + p; q \rangle_1}{\langle \gamma_3 + p, 1 + p; q \rangle_1}.$$

Then we have the following regions

$$\begin{aligned}
X_q &= \{(r,s,t) \mid 0 < s < \frac{n}{n+p} \wedge 0 < t < \frac{p}{n+p}\} \\
Y_q &= \{(r,s,t) \mid 0 < r < \frac{m}{m+p} \wedge 0 < t < \frac{p}{m+p}\} \\
E_q &= \{(r,s,t) \mid 0 < r < \frac{m}{m+p} \wedge 0 < s < \frac{n}{n+p} \wedge \\
&\quad \wedge 0 < t < \frac{p^2}{(m+p)(n+p)}\}.
\end{aligned} \tag{16}$$

We have convergence domain $\frac{rs}{t} < 1$.

Consider the function Φ_M. We have

$$\begin{aligned}
\frac{A_{q,m+1,n,p}}{A_{q,m,n,p}} &= \frac{\langle \alpha_1 + m, \beta_1 + m + p; q\rangle_1}{\langle \gamma_1 + m, 1 + m; q\rangle_1}, \\
\frac{A_{q,m,n+1,p}}{A_{q,m,n,p}} &= \frac{\langle \alpha_2 + n + p, \beta_2 + n; q\rangle_1}{\langle \gamma_2 + n + p, 1 + n; q\rangle_1}, \\
\frac{A_{q,m,n,p+1}}{A_{q,m,n,p}} &= \frac{\langle \alpha_2 + n + p, \beta_1 + m + p; q\rangle_1}{\langle \gamma_2 + n + p, 1 + p; q\rangle_1}.
\end{aligned} \tag{17}$$

We have the following regions

$$\begin{aligned}
Y_q &= \{(r,s,t) \mid 0 < r < \frac{m}{m+p} \wedge 0 < t < \frac{p}{m+p}\} \\
E_q &= \{(r,s,t) \mid 0 < r < \frac{m}{m+p} \wedge 0 < s < 1 \wedge 0 < t < \frac{p}{m+p}\}.
\end{aligned} \tag{18}$$

We have convergence domain $r \oplus_q t < 1$, $s < 1$.

Consider the function Φ_N. We have

$$\begin{aligned}
\frac{A_{q,m+1,n,p}}{A_{q,m,n,p}} &= \frac{\langle \alpha_1 + m, \beta_1 + m + p; q\rangle_1}{\langle \gamma_1 + m, 1 + m; q\rangle_1}, \\
\frac{A_{q,m,n+1,p}}{A_{q,m,n,p}} &= \frac{\langle \alpha_2 + n, \beta_2 + n; q\rangle_1}{\langle \gamma_2 + n + p, 1 + n; q\rangle_1}, \\
\frac{A_{q,m,n,p+1}}{A_{q,m,n,p}} &= \frac{\langle \alpha_3 + p, \beta_1 + m + p; q\rangle_1}{\langle \gamma_2 + n + p, 1 + p; q\rangle_1}.
\end{aligned} \tag{19}$$

We have the following regions

$$\begin{aligned}
X_q &= \{(r,s,t) \mid 0 < s < \frac{n+p}{n} \wedge 0 < t < \frac{n+p}{p}\} \\
Y_q &= \{(r,s,t) \mid 0 < r < \frac{m}{m+p} \wedge 0 < t < \frac{p}{m+p}\}, \\
E_q &= \{(r,s,t) \mid 0 < r < \frac{m}{m+p} \wedge 0 < s < \frac{n+p}{n} \wedge 0 < t < \frac{n+p}{m+p}\}.
\end{aligned} \tag{20}$$

We have convergence domain $r \oplus_q t < 1$, $s < 1$.

Consider the function Φ_P. We have

$$\frac{A_{q,m+1,n,p}}{A_{q,m,n,p}} = \frac{\langle \alpha_1+m+p, \beta_1+m+n; q\rangle_1}{\langle \gamma_1+m, 1+m; q\rangle_1},$$

$$\frac{A_{q,m,n+1,p}}{A_{q,m,n,p}} = \frac{\langle \alpha_2+n, \beta_1+m+n; q\rangle_1}{\langle \gamma_2+n+p, 1+n; q\rangle_1}, \qquad (21)$$

$$\frac{A_{q,m,n,p+1}}{A_{q,m,n,p}} = \frac{\langle \alpha_1+m+p, \beta_2+p; q\rangle_1}{\langle \gamma_2+n+p, 1+p; q\rangle_1}.$$

We have the following regions

$$X_q = \{(r,s,t)\mid 0<s<\frac{n+p}{n} \wedge 0<t<\frac{n+p}{p}\}$$

$$Y_q = \{(r,s,t)\mid 0<r<\frac{m}{m+p} \wedge 0<t<\frac{p}{m+p}\}$$

$$Z_q = \{(r,s,t)\mid 0<r<\frac{m}{m+n} \wedge 0<s<\frac{n}{m+n}\} \qquad (22)$$

$$E_q = \{(r,s,t)\mid 0<r<\frac{m^2}{(m+p)(m+n)} \wedge 0<s<\frac{n+p}{m+n} \wedge$$

$$\wedge 0<t<\frac{n+p}{m+p}\}.$$

We have convergence domain $r \oplus_q t < 1$, $r \oplus_q s < 1$.

Consider the function Φ_R. We have

$$\frac{A_{q,m+1,n,p}}{A_{q,m,n,p}} = \frac{\langle \alpha_1+m+p, \beta_1+m+p; q\rangle_1}{\langle \gamma_1+m, 1+m; q\rangle_1},$$

$$\frac{A_{q,m,n+1,p}}{A_{q,m,n,p}} = \frac{\langle \alpha_2+n, \beta_2+n; q\rangle_1}{\langle \gamma_2+n+p, 1+n; q\rangle_1}, \qquad (23)$$

$$\frac{A_{q,m,n,p+1}}{A_{q,m,n,p}} = \frac{\langle \alpha_1+m+p, \beta_1+m+p; q\rangle_1}{\langle \gamma_2+n+p, 1+p; q\rangle_1}.$$

We have the following regions

$$X_q = \{(r,s,t)\mid 0<s<\frac{n+p}{n} \wedge 0<t<\frac{n+p}{p}\}$$

$$Y_q = \{(r,s,t)\mid 0<r<\left(\frac{m}{m+p}\right)^2 \wedge 0<t<\left(\frac{p}{m+p}\right)^2\}$$

$$E_q = \{(r,s,t)\mid 0<r<\left(\frac{m}{m+p}\right)^2 \wedge 0<s<\frac{n+p}{n} \wedge \qquad (24)$$

$$\wedge 0<t<\frac{p(n+p)}{(m+p)^2}\}.$$

We have convergence domain $\sqrt{r} \oplus_q \sqrt{t} < 1$, $s < 1$.

The convergence regions for the following two functions are obvious.

Consider the function Φ_S. We have

$$\frac{A_{q,m+1,n,p}}{A_{q,m,n,p}} = \frac{\langle \alpha_1+m, \beta_1+m; q\rangle_1}{\langle \gamma_1+m+n+p, 1+m; q\rangle_1},$$
$$\frac{A_{q,m,n+1,p}}{A_{q,m,n,p}} = \frac{\langle \alpha_2+n+p, \beta_2+n; q\rangle_1}{\langle \gamma_1+m+n+p, 1+n; q\rangle_1}, \qquad (25)$$
$$\frac{A_{q,m,n,p+1}}{A_{q,m,n,p}} = \frac{\langle \alpha_2+n+p, \beta_3+p; q\rangle_1}{\langle \gamma_1+m+n+p, 1+p; q\rangle_1}.$$

Consider the function Φ_T. We have

$$\frac{A_{q,m+1,n,p}}{A_{q,m,n,p}} = \frac{\langle \alpha_1+m, \beta_1+m+p; q\rangle_1}{\langle \gamma_1+m+n+p, 1+n; q\rangle_1},$$
$$\frac{A_{q,m,n+1,p}}{A_{q,m,n,p}} = \frac{\langle \alpha_2+n+p, \beta_2+n; q\rangle_1}{\langle \gamma_1+m+n+p, 1+n; q\rangle_1}, \qquad (26)$$
$$\frac{A_{q,m,n,p+1}}{A_{q,m,n,p}} = \frac{\langle \alpha_2+n+p, \beta_1+m+p; q\rangle_1}{\langle \gamma_1+m+n+p, 1+p; q\rangle_1}.$$

Consider the function Φ_{G_A}. We have

$$\frac{A_{q,m+1,n,p}}{A_{q,m,n,p}} = \frac{\langle \gamma+n+p-m-1, \beta_1+m+p; q\rangle_1}{\langle \alpha+n+p-m-1, 1+m; q\rangle_1},$$
$$\frac{A_{q,m,n+1,p}}{A_{q,m,n,p}} = \frac{\langle \alpha+n+p-m, \beta_2+n; q\rangle_1}{\langle \gamma+n+p-m, 1+n; q\rangle_1}, \qquad (27)$$
$$\frac{A_{q,m,n,p+1}}{A_{q,m,n,p}} = \frac{\langle \alpha+n+p-m, \beta_1+m+p; q\rangle_1}{\langle \gamma+n+p-m, 1+p; q\rangle_1}.$$

We have the following regions

$$Y_q = \{(r,s,t)|\ 0 < r < \frac{m}{m+p} \wedge 0 < t < \frac{p}{m+p}\}$$
$$E_q = \{(r,s,t)|\ 0 < r < \frac{m}{m+p} \wedge 0 < s < 1 \wedge 0 < t < \frac{p}{m+p}\}. \qquad (28)$$

We have convergence domain $r \oplus_q t < 1$, $s < 1$.

Consider the function Φ_{G_B}. We have

$$\frac{A_{q,m+1,n,p}}{A_{q,m,n,p}} = \frac{\langle \gamma+n+p-m-1, \beta_1+m; q\rangle_1}{\langle \alpha+n+p-m-1, 1+m; q\rangle_1},$$
$$\frac{A_{q,m,n+1,p}}{A_{q,m,n,p}} = \frac{\langle \alpha+n+p-m, \beta_2+n; q\rangle_1}{\langle \gamma+n+p-m, 1+n; q\rangle_1}, \qquad (29)$$
$$\frac{A_{q,m,n,p+1}}{A_{q,m,n,p}} = \frac{\langle \alpha+n+p-m, \beta_3+p; q\rangle_1}{\langle \gamma+n+p-m, 1+p; q\rangle_1}.$$

The convergence region is obvious. □

The convergence region $xy < z$ for functions Φ_F and Φ_K is shown in Figure 1.

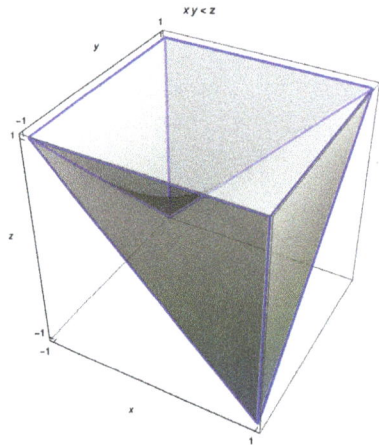

Figure 1. Convergence region $xy < z$ for functions Φ_F and Φ_K.

3. q-Integral Representations

We now turn to q-integral expressions of the respective functions. Sometimes we abbreviate the integral ranges by vectors with numbers of elements equal to the numbers of q-integrals.

Theorem 2. *A triple q-integral representation of Φ_K. A q-analogue of Dwivedi, Sahai ([11] 4.33). Put*

$$C \equiv \Gamma_q \begin{bmatrix} c_1, c_2, c_3 \\ a_1, b_1, b_2, c_1 - a_1, c_2 - b_2, c_3 - b_1 \end{bmatrix}. \tag{30}$$

Then

$$\Phi_K = C \sum_{m,n,p=0}^{+\infty} \frac{\langle b_1 + p; q \rangle_m \langle a_2; q \rangle_{n+p} x^m y^n z^p}{\langle 1; q \rangle_m \langle 1; q \rangle_n \langle 1; q \rangle_p} \int_{\vec{0}}^{\vec{1}} u^{a_1 + m - 1} (qu; q)_{c_1 - a_1 - 1}$$
$$v^{b_2 + n - 1} (qv; q)_{c_2 - b_2 - 1} \omega^{b_1 + p - 1} (q\omega; q)_{c_3 - b_1 - 1} \, d_q(u) \, d_q(v) \, d_q(\omega). \tag{31}$$

Proof. The equation numbers in the proof refer to the authors book [12]

$$\text{LHS} \stackrel{\text{by (1.46)}}{=} \sum_{m,n,p=0}^{+\infty} \frac{\langle a_2; q \rangle_{n+p} \overrightarrow{\langle b_1 + p; q \rangle}_m x^m y^n z^p}{\langle 1; q \rangle_m \langle 1; q \rangle_n \langle 1; q \rangle_p}$$
$$\Gamma_q \begin{bmatrix} c_1, c_2, c_3, a_1 + m, b_1 + p, b_2 + n \\ a_1, b_1, b_2, c_1 + m, c_2 + n, c_3 + p \end{bmatrix} \stackrel{\text{by 3}\times (7.55)}{=} \text{RHS.} \tag{32}$$

□

Definition 2. *Assume that $\vec{m} \equiv (m_1, \ldots, m_n)$, $m \equiv m_1 + \ldots + m_n$ and $a \in \mathbb{R}^\star$. The vector q-multinomial-coefficient $\binom{a}{\vec{m}}_q^\star$ [3] is defined by the symmetric expression*

$$\binom{a}{\vec{m}}_q^\star \equiv \frac{\langle -a; q \rangle_m (-1)^m q^{-\binom{\vec{m}}{2} + am}}{\langle 1; q \rangle_{m_1} \langle 1; q \rangle_{m_2} \ldots \langle 1; q \rangle_{m_n}}. \tag{33}$$

The following formula applies for a q–deformed hypercube of length 1 in $\mathbb{R}^n$. Note that formulas (34) and (35) are symmetric in the x_i.

Definition 3 ([3]). *Assuming that the right hand side converges, and $a \in \mathbb{R}^\star$:*

$$(1 \boxminus_q q^a x_1 \boxminus_q \ldots \boxminus_q q^a x_n)^{-a} \equiv \sum_{m_1,\ldots,m_n=0}^{\infty} \prod_{j=1}^{n}(-x_j)^{m_j} \binom{-a}{\vec{m}}_q^{\star} q^{\binom{\vec{m}}{2}+am}. \tag{34}$$

The following corollary prepares for the next formula.

Corollary 1. *A generalization of the q-binomial theorem [3]:*

$$(1 \boxminus_q q^a x_1 \boxminus_q \ldots \boxminus_q q^a x_n)^{-a} = \sum_{\vec{m}=\vec{0}}^{\vec{\infty}} \frac{\langle a;q \rangle_m \vec{x}^{\vec{m}}}{\langle \vec{1};q \rangle_{\vec{m}}}, \ a \in \mathbb{R}^\star. \tag{35}$$

Proof. Use formulas (33) and (34), the terms with factors $q^{-\binom{\vec{m}}{2}+am}$ cancel each other. □

Theorem 3. *A double q-integral representation of Φ_M with q-additions. A q-analogue of Saran ([8] 2.13).*

$$\Phi_M = \Gamma_q \begin{bmatrix} \gamma_1, \gamma_2 \\ \alpha_1, \alpha_2, \gamma_1 - \alpha_1, \gamma_2 - \alpha_2 \end{bmatrix} \int_0^1 \int_0^1 u^{\alpha_1-1}(qu;q)_{\gamma_1-\alpha_1-1} v^{\alpha_2-1} \tag{36}$$

$$(qv;q)_{\gamma_2-\alpha_2-1} \frac{1}{(vy;q)_{\beta_2}} (1 \boxminus_q q^{\beta_1} ux \boxminus_q q^{\beta_1} vz)^{-\beta_1} d_q(u) \, d_q(v).$$

Proof. The equation numbers in the proof refer to the authors book [12]

$$\text{LHS} = \sum_{\vec{m}=\vec{0}}^{+\infty} \frac{\langle \beta_2;q \rangle_n \langle \beta_1;q \rangle_{m+p} \langle \alpha_1;q \rangle_m \langle \alpha_2;q \rangle_{n+p}}{\langle 1,\gamma_1;q \rangle_m \langle 1;q \rangle_n \langle 1;q \rangle_p \langle \gamma_2;q \rangle_{n+p}} x^m y^n z^p$$

$$\stackrel{\text{by (1.46)}}{=} \sum_{\vec{m}=\vec{0}}^{+\infty} \frac{\langle \beta_2;q \rangle_n \langle \beta_1;q \rangle_{m+p}}{\langle 1;q \rangle_m \langle 1;q \rangle_n \langle 1;q \rangle_p} x^m y^n z^p \, \Gamma_q \begin{bmatrix} \gamma_1, \gamma_2, \alpha_1+m, \alpha_2+n+p \\ \alpha_1, \alpha_2, \gamma_1+m, \gamma_2+n+p \end{bmatrix}$$

$$\stackrel{\text{by (7.55)}}{=} \Gamma_q \begin{bmatrix} \gamma_1, \gamma_2 \\ \alpha_1, \alpha_2, \gamma_1-\alpha_1, \gamma_2-\alpha_2 \end{bmatrix} \tag{37}$$

$$\int_0^1 \int_0^1 u^{\alpha_1-1} (qu;q)_{\gamma_1-\alpha_1-1} v^{\alpha_2-1} (qv;q)_{\gamma_2-\alpha_2-1}$$

$$\sum_{\vec{m}=\vec{0}}^{+\infty} \frac{\langle \beta_2;q \rangle_n \langle \beta_1;q \rangle_{m+p}}{\langle 1;q \rangle_m \langle 1;q \rangle_n \langle 1;q \rangle_p} (ux)^m (vy)^n (vz)^p \stackrel{\text{by (7.27),(35)}}{=} \text{RHS}.$$

□

Remark 1. *Saran ([8] 2.12) gives a similar formula for Φ_K without proof. It is, however, not clear how it is proved.*

All the following vector q-integrals have dimension three. We denote $\vec{s} \equiv (s,t,u)$. The short expression to the left always means the definition.

Theorem 4. *A q-integral representation of Φ_E. A q-analogue of ([9] (3.11) p. 22).*

$$\Phi_E(\alpha_1, \alpha_1, \alpha_1, \beta_1, \beta_2, \beta_2; \gamma_1, \gamma_2, \gamma_3 | q; x, y, z)$$

$$\Gamma_q \begin{bmatrix} \gamma_1, \gamma_2, \gamma_3 \\ \nu_1, \nu_2, \nu_3, \gamma_1-\nu_1, \gamma_2-\nu_2, \gamma_3-\nu_3 \end{bmatrix} \int_{\vec{0}}^{\vec{1}} \vec{s}^{\vec{\nu}-\vec{1}} (q\vec{s};q)_{\vec{\gamma}-\vec{\nu}-\vec{1}} \tag{38}$$

$$\Phi_E(\alpha_1, \alpha_1, \alpha_1, \beta_1, \beta_2, \beta_2; \nu_1, \nu_2, \nu_3 | q; sx, ty, uz) \, d_q(\vec{s}).$$

Proof. Put

$$D \equiv \Gamma_q \begin{bmatrix} \gamma_1, \gamma_2, \gamma_3 \\ \nu_1, \nu_2, \nu_3, \gamma_1 - \nu_1, \gamma_2 - \nu_2, \gamma_3 - \nu_3 \end{bmatrix}$$
$$\sum_{m,n,p=0}^{+\infty} \frac{\langle \alpha_1; q \rangle_{m+n+p} \langle \beta_1; q \rangle_m \langle \beta_2; q \rangle_{n+p}}{\langle 1, \nu_1; q \rangle_m \langle 1, \nu_2; q \rangle_n \langle 1, \nu_3; q \rangle_p} x^m y^n z^p. \qquad (39)$$

Then we have (The equation numbers in the proof refer to the authors book [12])

$$\text{RHS} \stackrel{\text{by (6.54)}}{=} D(1-q)^3 \sum_{k,i,j=0}^{+\infty} q^{k(\nu_1+m)+i(\nu_2+n)+j(\nu_3+p)}$$
$$\langle 1+k; q \rangle_{\gamma_1-\nu_1-1} \langle 1+i; q \rangle_{\gamma_2-\nu_2-1} \langle 1+j; q \rangle_{\gamma_3-\nu_3-1}$$
$$\stackrel{\text{by (6.8,6.10)}}{=} D(1-q)^3 \sum_{k,i,j=0}^{+\infty} q^{k(\nu_1+m)+i(\nu_2+n)+j(\nu_3+p)} \qquad (40)$$
$$\frac{\langle \gamma_1-\nu_1; q \rangle_k \langle \gamma_2-\nu_2; q \rangle_i \langle \gamma_3-\nu_3; q \rangle_j \langle 1,1,1; q \rangle_\infty}{\langle 1; q \rangle_k \langle 1; q \rangle_i \langle 1; q \rangle_j \langle \gamma_1-\nu_1, \gamma_2-\nu_2, \gamma_3-\nu_3; q \rangle_\infty}$$
$$\stackrel{\text{by (7.27)}}{=} D(1-q)^3 \frac{\langle m+\gamma_1, n+\gamma_2, p+\gamma_3, 1, 1, 1; q \rangle_\infty}{\langle \nu_1+m, \nu_2+n, \nu_3+p, \gamma_1-\nu_1, \gamma_2-\nu_2, \gamma_3-\nu_3; q \rangle_\infty}$$
$$\stackrel{\text{by (1.45,1.46)}}{=} \text{LHS.}$$

□

Theorem 5. *A q-integral representation of Φ_K. A q-analogue of ([9] (3.13) p. 23).*

$$\Phi_K = \Gamma_q \begin{bmatrix} \gamma_1, \gamma_2, \gamma_3 \\ \nu_1, \nu_2, \nu_3, \gamma_1 - \nu_1, \gamma_2 - \nu_2, \gamma_3 - \nu_3 \end{bmatrix} \int_{\vec{0}}^{\vec{1}} \vec{s}^{\vec{\nu}-\vec{1}} (q\vec{s}; q)_{\vec{\gamma}-\vec{\nu}-\vec{1}} \qquad (41)$$
$$\Phi_K(\alpha_1, \alpha_2, \alpha_2, \beta_1, \beta_2, \beta_1; \nu_1, \nu_2, \nu_3 | q; sx, ty, uz) \, d_q(\vec{s}).$$

Proof. See the proof (40). □

Theorem 6. *A q-integral representation of Φ_G. A q-analogue of ([9] (3.12) p. 22).*

$$\Phi_G(\alpha_1, \alpha_1, \alpha_1, \beta_1, \beta_2, \beta_3; \gamma_1, \gamma_2, \gamma_2 | q; x, y, z)$$
$$= \Gamma_q \begin{bmatrix} \lambda_1, \lambda_2, \lambda_3 \\ \beta_1, \beta_2, \beta_3, \lambda_1 - \beta_1, \lambda_2 - \beta_2, \lambda_3 - \beta_3 \end{bmatrix} \int_{\vec{0}}^{\vec{1}} \vec{s}^{\vec{\beta}-\vec{1}} (q\vec{s}; q)_{\vec{\lambda}-\vec{\beta}-\vec{1}} \qquad (42)$$
$$\Phi_G(\alpha_1, \alpha_1, \alpha_1, \lambda_1, \lambda_2, \lambda_3; \gamma_1, \gamma_2, \gamma_2 | q; sx, ty, uz) \, d_q(\vec{s}).$$

Proof. Put

$$D \equiv \Gamma_q \begin{bmatrix} \lambda_1, \lambda_2, \lambda_3 \\ \beta_1, \beta_2, \beta_3, \lambda_1 - \beta_1, \lambda_2 - \beta_2, \lambda_3 - \beta_3 \end{bmatrix}$$
$$\sum_{m,n,p=0}^{+\infty} \frac{\langle \alpha_1; q \rangle_{m+n+p} \langle \lambda_1; q \rangle_m \langle \lambda_2; q \rangle_n \langle \lambda_3; q \rangle_p}{\langle 1, \gamma_1; q \rangle_m \langle 1; q \rangle_n \langle 1; q \rangle_p \langle \gamma_2; q \rangle_{n+p}} x^m y^n z^p. \qquad (43)$$

Then we have (The equation numbers in the proof refer to the authors book [12])

$$\text{RHS} \stackrel{\text{by (6.54)}}{=} D(1-q)^3 \sum_{k,i,j=0}^{+\infty} q^{k(\beta_1+m)+i(\beta_2+n)+j(\beta_3+p)}$$

$$\langle 1+k;q\rangle_{\lambda_1-\beta_1-1}\langle 1+i;q\rangle_{\lambda_2-\beta_2-1}\langle 1+j;q\rangle_{\lambda_3-\beta_3-1}$$

$$\stackrel{\text{by (6.8,6.10)}}{=} D(1-q)^3 \sum_{k,i,j=0}^{+\infty} q^{k(\beta_1+m)+i(\beta_2+n)+j(\beta_3+p)}$$

$$\frac{\langle \lambda_1-\beta_1;q\rangle_k \langle \lambda_2-\beta_2;q\rangle_i \langle \lambda_3-\beta_3;q\rangle_j \langle 1,1,1;q\rangle_\infty}{\langle 1;q\rangle_k \langle 1;q\rangle_i \langle 1;q\rangle_j \langle \lambda_1-\beta_1,\lambda_2-\beta_2,\lambda_3-\beta_3\rangle_\infty}$$

$$\stackrel{\text{by (7.27)}}{=} D(1-q)^3 \frac{\langle m+\lambda_1, n+\lambda_2, p+\lambda_3, 1,1,1;q\rangle_\infty}{\langle \beta_1+m, \beta_2+n, \beta_3+p, \lambda_1-\beta_1, \lambda_2-\beta_2, \lambda_3-\beta_3;q\rangle_\infty}$$

$$\stackrel{\text{by (1.45,1.46)}}{=} \text{LHS.}$$

(44)

□

Theorem 7. *A q-integral representation of* Φ_N. *A q-analogue of ([9] (3.14) p. 23).*

$$\Phi_N(\alpha_1,\alpha_2,\alpha_3,\beta_1,\beta_2,\beta_1;\gamma_1,\gamma_2,\gamma_2|q;x,y,z)$$
$$=\Gamma_q\begin{bmatrix} \lambda_1,\lambda_2,\lambda_3 \\ \alpha_1,\alpha_2,\alpha_3,\lambda_1-\alpha_1,\lambda_2-\alpha_2,\lambda_3-\alpha_3 \end{bmatrix} \int_{\vec{0}}^{\vec{1}} \vec{s}^{\vec{\alpha}-\vec{1}}(q\vec{s};q)_{\vec{\lambda}-\vec{\alpha}-\vec{1}}$$
$$\Phi_N(\lambda_1,\lambda_2,\lambda_3,\beta_1,\beta_2,\beta_1;\gamma_1,\gamma_2,\gamma_2|q;sx,ty,uz)\,d_q(\vec{s}).$$

(45)

Proof. See the proof (44). □

Theorem 8. *A q-integral representation of* Φ_S. *A q-analogue of ([9] (3.15) p. 23).*

$$\Phi_S(\alpha_1,\alpha_2,\alpha_2,\beta_1,\beta_2,\beta_3;\gamma_1,\gamma_1,\gamma_1|q;x,y,z)$$
$$=\Gamma_q\begin{bmatrix} \lambda_1,\lambda_2,\lambda_3 \\ \beta_1,\beta_2,\beta_3,\lambda_1-\beta_1,\lambda_2-\beta_2,\lambda_3-\beta_3 \end{bmatrix} \int_{\vec{0}}^{\vec{1}} \vec{s}^{\vec{\beta}-\vec{1}}(q\vec{s};q)_{\vec{\lambda}-\vec{\beta}-\vec{1}}$$
$$\Phi_S(\alpha_1,\alpha_2,\alpha_2,\lambda_1,\lambda_2,\lambda_3;\gamma_1,\gamma_1,\gamma_1|q;sx,ty,uz)\,d_q(\vec{s}).$$

(46)

Proof. See the proof (44). □

Theorem 9. *A q-integral representation of* Φ_F. *A q-analogue of ([9] (3.16) p. 24).*

$$\Phi_F(\alpha_1,\alpha_1,\alpha_1,\beta_1,\beta_2,\beta_1;\gamma_1,\gamma_2,\gamma_2|q;x,yz,z)$$
$$=\Gamma_q\begin{bmatrix} \gamma_1,\gamma_2,\gamma_2 \\ \nu_1,\nu_2,\beta_2,\gamma_1-\nu_1,\gamma_2-\nu_2,\gamma_2-\beta_2 \end{bmatrix}$$
$$\int_{\vec{0}}^{\vec{1}} s^{\nu_1-1}t^{\beta_2-1}u^{\nu_2-1}(qs;q)_{\gamma_1-\nu_1-1}(qt;q)_{\gamma_2-\beta_2-1}(qu;q)_{\gamma_2-\nu_2-1}$$
$$\Phi_F(\alpha_1,\alpha_1,\alpha_1,\beta_1,\gamma_2,\beta_1;\nu_1,\nu_2,\nu_2|q;sx,tuyz,uz)\,d_q(\vec{s}).$$

(47)

Proof. Put

$$D \equiv \Gamma_q \left[\begin{array}{c} \gamma_1, \gamma_2, \gamma_2 \\ \nu_1, \nu_2, \beta_2, \gamma_1 - \nu_1, \gamma_2 - \nu_2, \gamma_2 - \beta_2 \end{array} \right]$$
$$\sum_{m,n,p=0}^{+\infty} \frac{\langle \alpha_1; q \rangle_{m+n+p} \langle \beta_1; q \rangle_{m+p} \langle \gamma_2; q \rangle_n}{\langle 1, \nu_1; q \rangle_m \langle 1; q \rangle_n \langle 1; q \rangle_p \langle \nu_2; q \rangle_{n+p}} x^m y^n z^{n+p}.$$
(48)

Then we have (The equation numbers in the proof refer to the authors book [12])

$$\text{RHS} \stackrel{\text{by (6.54)}}{=} D(1-q)^3 \sum_{k,i,j=0}^{+\infty} q^{k(\nu_1+m)+i(\beta_2+n)+j(\nu_2+n+p)}$$

$$\langle 1+k; q \rangle_{\gamma_1-\nu_1-1} \langle 1+i; q \rangle_{\gamma_2-\beta_2-1} \langle 1+j; q \rangle_{\gamma_2-\nu_2-1}$$

$$\stackrel{\text{by (6.8,6.10)}}{=} D(1-q)^3 \sum_{k,i,j=0}^{+\infty} q^{k(\nu_1+m)+i(\beta_2+n)+j(\nu_2+n+p)}$$

$$\frac{\langle \gamma_1 - \nu_1; q \rangle_k \langle \gamma_2 - \beta_2; q \rangle_i \langle \gamma_2 - \nu_2; q \rangle_j \langle 1,1,1; q \rangle_\infty}{\langle 1; q \rangle_k \langle 1; q \rangle_i \langle 1; q \rangle_j \langle \gamma_1 - \nu_1, \gamma_2 - \beta_2, \gamma_2 - \nu_2; q \rangle_\infty}$$

$$\stackrel{\text{by (7.27)}}{=} D(1-q)^3 \frac{\langle m+\gamma_1, n+\gamma_2, n+p+\gamma_2, 1,1,1; q \rangle_\infty}{\langle \nu_1+m, \beta_2+n, \nu_2+n+p, \gamma_1-\nu_1, \gamma_2-\nu_2, \gamma_2-\beta_2; q \rangle_\infty}$$

$$\stackrel{\text{by (1.45,1.46)}}{=} \text{LHS.}$$
(49)

□

Theorem 10. *A q-integral representation of Φ_M. A q-analogue of ([9] (3.17) p. 25).*

$$\Phi_M(\alpha_1, \alpha_2, \alpha_2, \beta_1, \beta_2, \beta_1; \gamma_1, \gamma_2, \gamma_2 | q; x, yz, z)$$
$$= \Gamma_q \left[\begin{array}{c} \gamma_1, \gamma_2, \gamma_2 \\ \nu_1, \nu_2, \beta_2, \gamma_1 - \nu_1, \gamma_2 - \nu_2, \gamma_2 - \beta_2 \end{array} \right]$$
$$\int_{\vec{0}}^{\vec{1}} s^{\nu_1-1} t^{\beta_2-1} u^{\nu_2-1} (qs; q)_{\gamma_1-\nu_1-1} (qt; q)_{\gamma_2-\beta_2-1} (qu; q)_{\gamma_2-\nu_2-1}$$
$$\Phi_M(\alpha_1, \alpha_2, \alpha_2, \beta_1, \gamma_2, \beta_1; \nu_1, \nu_2, \nu_2 | q; sx, tuyz, uz) \, \vec{d_q}(s).$$
(50)

Proof. Put

$$D \equiv \Gamma_q \left[\begin{array}{c} \gamma_1, \gamma_2, \gamma_2 \\ \nu_1, \nu_2, \beta_2, \gamma_1 - \nu_1, \gamma_2 - \nu_2, \gamma_2 - \beta_2 \end{array} \right]$$
$$\sum_{m,n,p=0}^{+\infty} \frac{\langle \alpha_1; q \rangle_m \langle \alpha_2; q \rangle_{n+p} \langle \beta_1; q \rangle_{m+p} \langle \gamma_2; q \rangle_n}{\langle 1, \nu_1; q \rangle_m \langle 1; q \rangle_n \langle 1; q \rangle_p \langle \nu_2; q \rangle_{n+p}} x^m y^n z^{n+p}.$$
(51)

Then we have [12]

$$\text{RHS} \overset{\text{by (6.54)}}{=} D(1-q)^3 \sum_{k,i,j=0}^{+\infty} q^{k(v_1+m)+i(\beta_2+n)+j(v_2+n+p)}$$

$$\langle 1+k;q\rangle_{\gamma_1-v_1-1}\langle 1+i;q\rangle_{\gamma_2-\beta_2-1}\langle 1+j;q\rangle_{\gamma_2-v_2-1}$$

$$\overset{\text{by (6.8,6.10)}}{=} D(1-q)^3 \sum_{k,i,j=0}^{+\infty} q^{k(v_1+m)+i(\beta_2+n)+j(v_2+n+p)}$$

$$\frac{\langle \gamma_1-v_1;q\rangle_k\langle \gamma_2-\beta_2;q\rangle_i\langle \gamma_2-v_2;q\rangle_j\langle 1,1,1;q\rangle_\infty}{\langle 1;q\rangle_k\langle 1;q\rangle_i\langle 1;q\rangle_j\langle \gamma_1-v_1,\gamma_2-\beta_2,\gamma_2-v_2\rangle_\infty}$$

$$\overset{\text{by (7.27)}}{=} D(1-q)^3 \frac{\langle m+\gamma_1,n+\gamma_2,n+p+\gamma_2,1,1,1;q\rangle_\infty}{\langle v_1+m,\beta_2+n,v_2+n+p,\gamma_1-v_1,\gamma_2-v_2,\gamma_2-\beta_2;q\rangle_\infty}$$

$$\overset{\text{by (1.45,1.46)}}{=} \text{LHS}.$$

(52)

□

Theorem 11. *A q-integral representation of* Φ_P. *Almost a q-analogue of ([9] (3.18) p. 25).*

$$\Phi_P(\alpha_1,\alpha_2,\alpha_1,\beta_1,\beta_1,\beta_2;\gamma_1,\gamma_2,\gamma_2|q;x,zy,z)$$

$$= \Gamma_q \begin{bmatrix} \gamma_1,\gamma_2,\gamma_2 \\ \alpha_2,v_1,v_2,\gamma_1-v_1,\gamma_2-\alpha_2,\gamma_2-v_2 \end{bmatrix}$$

$$\int_{\vec{0}}^{\vec{1}} s^{v_1-1}t^{\alpha_2-1}u^{v_2-1}(qs;q)_{\gamma_1-v_1-1}(qt;q)_{\gamma_2-\alpha_2-1}(qu;q)_{\gamma_2-v_2-1}$$

$$\Phi_P(\alpha_1,\gamma_2,\alpha_1,\beta_1,\beta_1,\beta_2;v_1,v_2,v_2|q;sx,tuyz,uz)\,d_q\vec{(s)}.$$

(53)

Proof. Put

$$D \equiv \Gamma_q \begin{bmatrix} \gamma_1,\gamma_2,\gamma_2 \\ \alpha_2,v_1,v_2,\gamma_1-v_1,\gamma_2-\alpha_2,\gamma_2-v_2 \end{bmatrix}$$

$$\sum_{m,n,p=0}^{+\infty} \frac{\langle \alpha_1;q\rangle_{m+p}\langle \gamma_2;q\rangle_n\langle \beta_1;q\rangle_{m+n}\langle \beta_2;q\rangle_p}{\langle 1,v_1;q\rangle_m\langle 1;q\rangle_n\langle 1;q\rangle_p\langle v_2;q\rangle_{n+p}} x^m y^n z^{n+p}.$$

(54)

Then we have [12]

$$\text{RHS} \overset{\text{by (6.54)}}{=} D(1-q)^3 \sum_{k,i,j=0}^{+\infty} q^{k(v_1+m)+i(\alpha_2+n)+j(v_2+n+p)}$$

$$\langle 1+k;q\rangle_{\gamma_1-v_1-1}\langle 1+i;q\rangle_{\gamma_2-\alpha_2-1}\langle 1+j;q\rangle_{\gamma_2-v_2-1}$$

$$\overset{\text{by (6.8,6.10)}}{=} D(1-q)^3 \sum_{k,i,j=0}^{+\infty} q^{k(v_1+m)+i(\alpha_2+n)+j(v_2+n+p)}$$

$$\frac{\langle \gamma_1-v_1;q\rangle_k\langle \gamma_2-\alpha_2;q\rangle_i\langle \gamma_2-v_2;q\rangle_j\langle 1,1,1;q\rangle_\infty}{\langle 1;q\rangle_k\langle 1;q\rangle_i\langle 1;q\rangle_j\langle \gamma_1-v_1,\gamma_2-\alpha_2,\gamma_2-v_2\rangle_\infty}$$

$$\overset{\text{by (7.27)}}{=} D(1-q)^3 \frac{\langle m+\gamma_1,n+\gamma_2,n+p+\gamma_2,1,1,1;q\rangle_\infty}{\langle v_1+m,\alpha_2+n,v_2+n+p,\gamma_1-v_1,\gamma_2-\alpha_2,\gamma_2-v_2;q\rangle_\infty}$$

$$\overset{\text{by (1.45,1.46)}}{=} \text{LHS}.$$

(55)

□

Theorem 12. *A q-integral representation of* Φ_R. *A q-analogue of ([9] (3.19) p. 26).*

$$\Phi_R(\alpha_1, \alpha_2, \alpha_1, \beta_1, \beta_2, \beta_1; \gamma_1, \gamma_2, \gamma_2 | q; x, zy, z)$$

$$= \Gamma_q \begin{bmatrix} \gamma_1, \gamma_2, \gamma_2 \\ \beta_2, \nu_1, \nu_2, \gamma_1 - \nu_1, \gamma_2 - \beta_2, \gamma_2 - \nu_2 \end{bmatrix}$$

$$\int_{\vec{0}}^{\vec{1}} s^{\nu_1-1} t^{\beta_2-1} u^{\nu_2-1} (qs;q)_{\gamma_1-\nu_1-1} (qt;q)_{\gamma_2-\beta_2-1} (qu;q)_{\gamma_2-\nu_2-1}$$

$$\Phi_R(\alpha_1, \alpha_2, \alpha_1, \beta_1, \gamma_2, \beta_1; \nu_1, \nu_2, \nu_2 | q; sx, tuyz, uz) \, d_q\vec{(s)}.$$

(56)

Proof. See formula (49). □

Theorem 13. *A q-integral representation of* Φ_T. *A q-analogue of ([9] (3.20) p. 27).*

$$\Phi_T(\alpha_1, \alpha_2, \alpha_2, \beta_1, \beta_2, \beta_1; \gamma_1, \gamma_1, \gamma_1 | q; xz, yz, z)$$

$$= \Gamma_q \begin{bmatrix} \xi, \eta, \gamma_1 \\ \nu_1, \alpha_1, \beta_2, \xi - \alpha_1, \eta - \beta_2, \gamma_1 - \nu_1 \end{bmatrix}$$

$$\int_{\vec{0}}^{\vec{1}} s^{\alpha_1-1} t^{\beta_2-1} u^{\nu_1-1} (qs;q)_{\xi-\alpha_1-1} (qt;q)_{\eta-\beta_2-1} (qu;q)_{\gamma_1-\nu_1-1}$$

$$\Phi_T(\xi, \alpha_2, \alpha_2, \beta_1, \eta, \beta_1; \nu_1, \nu_1, \nu_1 | q; suxz, tuyz, uz) \, d_q\vec{(s)}.$$

(57)

Proof. Put

$$D \equiv \Gamma_q \begin{bmatrix} \xi, \eta, \gamma_1 \\ \nu_1, \alpha_1, \beta_2, \xi - \alpha_1, \eta - \beta_2, \gamma_1 - \nu_1 \end{bmatrix}$$

$$\sum_{m,n,p=0}^{+\infty} \frac{\langle \xi; q \rangle_m \langle \alpha_2; q \rangle_{n+p} \langle \beta_1; q \rangle_{m+p} \langle \eta; q \rangle_n}{\langle 1; q \rangle_m \langle 1; q \rangle_n \langle 1; q \rangle_p \langle \nu_1; q \rangle_{m+n+p}} x^m y^n z^{m+n+p}.$$

(58)

Then we have [12]

$$\text{RHS} \stackrel{\text{by (6.54)}}{=} D(1-q)^3 \sum_{k,i,j=0}^{+\infty} q^{k(\alpha_1+m)+i(\beta_2+n)+j(\nu_1+m+n+p)}$$

$$\langle 1+k; q \rangle_{\xi-\alpha_1-1} \langle 1+i; q \rangle_{\nu_1-\beta_2-1} \langle 1+j; q \rangle_{\gamma_1-\nu_1-1}$$

$$\stackrel{\text{by (6.8,6.10)}}{=} D(1-q)^3 \sum_{k,i,j=0}^{+\infty} q^{k(\alpha_1+m)+i(\beta_2+n)+j(\nu_1+m+n+p)}$$

$$\frac{\langle \xi-\alpha_1; q \rangle_k \langle \eta-\beta_2; q \rangle_i \langle \gamma_1-\nu_1; q \rangle_j \langle 1,1,1; q \rangle_\infty}{\langle 1; q \rangle_k \langle 1; q \rangle_i \langle 1; q \rangle_j \langle \xi-\alpha_1, \eta-\beta_2, \gamma_1-\nu_1 \rangle_\infty}$$

$$\stackrel{\text{by (7.27)}}{=} D(1-q)^3 \frac{\langle m+\xi, n+\eta, m+n+p+\gamma_1, 1, 1, 1; q \rangle_\infty}{\langle \alpha_1+m, \beta_2+n, \nu_1+m+n+p, \xi-\alpha_1, \gamma_1-\nu_1, \eta-\beta_2; q \rangle_\infty}$$

$$\stackrel{\text{by (1.45,1.46)}}{=} \text{LHS}.$$

(59)

□

4. Discussion

We have successfully combined the convergence condition [13] $(r \oplus_q t)^n < 1$ with the Horn–Karlsson convergence rules for most of the known triple q-hypergeometric functions. The Cartesian equation $r + s + t = 1$ is thereby replaced by its q-analogue $r \oplus_q s \oplus_q t$ in the spirit of Rota. The graph for the convergence region $xy/z < 1$ could also be of interest for the case $q = 1$.

Similarly, the proofs for q-Beta integrals also work for the case $q = 1$. These proofs have the same form as in previous and future papers of the author.

5. Conclusions

In the book [14] more triple hypergeometric functions are discussed. It would be interesting to compute convergence regions for their q-analogues. From our convergence theorems it is obvious that the following theorem from ([14], p. 108) can be extended to the q-case. The region of convergence for a hypergeometric series is independent of the parameters, exceptional parameter values being excluded. In this way, we plan to write a book on multiple q-hypergeometric series.

Funding: This research received no external funding

Conflicts of Interest: The author declares no conflict of interest.

References

1. Ernst, T. Multiple q-hypergeometric transformations involving q-integrals. In Proceedings of the 9th Annual Conference of the Society for Special Functions and their Applications (SSFA), Gwalior, India, 21–23 June 2010; Volume 9, pp. 91–99.
2. Ernst, T. On the symmetric q-Lauricella functions. *Proc. Jangjeon Math. Soc.* **2016**, *19*, 319–344.
3. Ernst, T. On Eulerian q-integrals for single and multiple q-hypergeometric series. *Commun. Korean Math. Soc.* **2018**, *33*, 179–196.
4. Ernst, T. On various formulas with q-integrals and their applications to q-hypergeometric functions. *Eur. J. Pure Appl. Math.* **2020**.
5. Ernst, T. *Further Results on Multiple q–Eulerian Integrals for Various q–Hypergeometric Functions*; Publications de l'Institut Mathématique: Beograd, Serbia, 2020.
6. Horn, J. Über die Convergenz der hypergeometrischen Reihen zweier und dreier Veränderlichen. *Math. Ann.* **1889**, *XXXIV*, 577–600.
7. Karlsson, P.W. Regions of convergence for hypergeometric series in three variables. *Math. Scand.* **1974**, *34*, 241–248. [CrossRef]
8. Saran, S. Transformations of certain hypergeometric functions of three variables. *Acta Math.* **1955**, *93*, 293–312. [CrossRef]
9. Srivastava, K.J. On certain hypergeometric Integrals. *Ganita* **1956**, *7*, 13–28.
10. Horn, J. Hypergeometrische Funktionen zweier Veränderlichen. *Math. Ann.* **1931**, *105*, 381–407. [CrossRef]
11. Dwivedi, R.; Sahai, V. On the hypergeometric matrix functions of several variables. *J. Math. Phys.* **2018**, *59*, 023505. [CrossRef]
12. Ernst, T. *A Comprehensive Treatment of q-Calculus*; Birkhäuser: Basel, Switzerland, 2012.
13. Ernst, T. Convergence aspects for q-Appell functions I. *J. Indian Math. Soc. New Ser.* **2014**, *81*, 67–77.
14. Srivastava, H.M.; Karlsson, P.W. *Multiple Gaussian Hypergeometric Series*; Ellis Horwood: New York, NY, USA, 1985.

© 2020 by the authors. Licensee MDPI, Basel, Switzerland. This article is an open access article distributed under the terms and conditions of the Creative Commons Attribution (CC BY) license (http://creativecommons.org/licenses/by/4.0/).